Y0-AAF-757

MULTIPLICATIVE IDEAL THEORY IN COMMUTATIVE ALGEBRA

A Tribute to the Work of Robert Gilmer

MULTIPLICATIVE IDEAL THEORY IN COMMUTATIVE ALGEBRA

A Tribute to the Work of Robert Gilmer

Edited by

JAMES W. BREWER
Florida Atlantic University, Boca Raton, Florida

SARAH GLAZ
University of Connecticut, Storrs, Connecticut

WILLIAM J. HEINZER
Purdue University, West Lafayette, Indiana

BRUCE M. OLBERDING
New Mexico State University, Las Cruces, New Mexico

Library of Congress Control Number: 2006929313

ISBN-10: 0-387-24600-2 e-ISBN: 0-387-36717-9

ISBN-13: 978-0-387-24600-0

Printed on acid-free paper.

AMS Subject Classifications: 13

Printed in the United States of America.

9 8 7 6 5 4 3 2 1

springer.com

Robert Gilmer

Photograph by Olan Mills
4325 Amnicola Highway, Chattanooga, TN 37422-3456

Preface

This volume consists of a collection of articles centered around topics in commutative ring theory influenced or inspired by Robert Gilmer. The articles were solicited by the editors from experts in the field and each represents the state of the topic at the time of publication. Some of the articles are original research papers and some are expository. Several of the expository articles also contain original research. It is the editors' hope and intention that the volume will be useful to current researchers in the field and an inspiration to others to study this beautiful area of mathematics of which we are so fond.

The volume is also intended to be a tribute to our friend and colleague, Robert Gilmer, whose work and presence has inspired us as well as others.

We would like to thank the authors for their contributions, the referees for their work and Dr. John Martindale, Senior Editor, and Mr. Robert Saley, Assistant Editor, of Springer for their encouragement and help with this project. We thank also Jim Coykendall, Mitch Keller and The Mathematics Genealogy Project for providing us with Robert Gilmer's mathematics genealogy.

Finally, we would like to express our appreciation to Robert himself for his contributions to the volume, both inspirational and actual.

Boca Raton, Florida	*Jim Brewer*
Storrs, Connecticut	*Sarah Glaz*
West Lafayette, Indiana	*William Heinzer*
Las Cruces, New Mexico	*Bruce Olberding*

April 2006

Contents

List of Contributors

D. D. Anderson
Department of Mathematics
The University of Iowa
Iowa City, IA 52242
ddanders@math.uiowa.edu

David F. Anderson
Department of Mathematics
The University of Tennessee
Knoxville, TN 37996–1300
anderson@math.utk.edu

Valentina Barucci
Dipartimento di Matematica
Università di Roma La Sapienza
Piazzale A. Moro 2
00185 Roma, Italy
barucci@mat.uniroma1.it

Silvana Bazzoni
Dipartimento di Matematica Pura e Applicata
Università di Padova
Via Belzoni 7
35131 Padova, Italy
bazzoni@math.unipd.it

Jim Brewer
Department of Mathematical Sciences
Florida Atlantic University
Boca Raton, FL 33431–6498
brewer@fau.edu

Paul-Jean Cahen
Université Paul Cézanne Aix-Marseille III
LATP CNRS-UMR 6632
Faculté des Sciences et Techniques
13397 Marseille Cedex 20, France
paul-jean.cahen@univ.u-3mrs.fr

Jean-Luc Chabert
Université de Picardie
LAMFA CNRS-UMR 6140
Faculté de Mathématiques
33 rue Saint Leu
80039 Amiens Cedex 01, France
jean-luc.chabert@u-picardie.fr

Scott T. Chapman
Trinity University
Department of Mathematics
One Trinity Place
San Antonio, TX 78212-7200
schapman@trinity.edu

Jim Coykendall
Department of Mathematics
North Dakota State University
Fargo, ND 58105-5075
jim.coykendall@ndsu.edu

D. E. Dobbs
Department of Mathematics
University of Tennessee
Knoxville, TN 37996-1300
dobbs@math.utk.edu

Alberto Facchini
Dipartimento di Matematica Pura e Applicata
Università di Padova
Via Belzoni 7
35131 Padova, Italy
facchini@math.unipd.it

Marco Fontana
Dipartimento di Matematica
Università degli Studi Roma Tre
Largo San Leonardo Murialdo, 1
00146 Roma, Italy
fontana@mat.uniroma3.it

Stefania Gabelli
Dipartimento di Matematica
Università degli Studi Roma Tre
Largo San Leonardo Murialdo, 1
00146 Roma, Italy
gabelli@mat.uniroma3.it

Alfred Geroldinger
Institut für Mathematik
Karl-Franzens-Universität Graz
Heinrichstraße 36
8010 Graz, Austria
alfred.geroldinger@uni-graz.at

Robert Gilmer
Department of Mathematics
Florida State University
Tallahassee, FL 32306-4510
gilmer@math.fsu.edu

Sarah Glaz
Department of Mathematics
University of Connecticut
Storrs, CT 06269
glaz@math.uconn.edu

Franz Halter-Koch
Institut für Mathematik
Karl-Franzens-Universität Graz
Heinrichstraße 36
8010 Graz, Austria
franz.halterkoch@uni-graz.at

Wolfgang Hassler
Institut für Mathematik und Wissenschaftliches Rechnen
Karl-Franzens-Universität Graz
Heinrichstraße 36
8010 Graz, Austria
wolfgang.hassler@uni-graz.at

William Heinzer
Department of Mathematics
Purdue University
West Lafayette IN 47907-1395
heinzer@math.purdue.edu

Evan Houston
Department of Mathematics
University of North Carolina at Charlotte
Charlotte, NC 28223
eghousto@email.uncc.edu

S. Kabbaj
Department of Mathematics
King Fahd University of Petroleum & Minerals
Dhahran 31261, Saudi Arabia
P.O. Box 5046
kabbaj@kfupm.edu.sa

Lee Klingler
Department of Mathematical Sciences
Florida Atlantic University
Boca Raton, FL 33431-6498
klingler@fau.edu

K. Alan Loper
Department of Mathematics
Ohio State University-Newark
Newark, Ohio 43055
lopera@math.ohio-state.edu

Thomas G. Lucas
Department of Mathematics
University of North Carolina Charlotte
Charlotte, NC 28223
tglucas@email.uncc.edu

Jack Ohm
900 Fort Pickens Rd, 215
Pensacola Bch, FL 32561
veebc@earthlink.net

Bruce Olberding
Department of Mathematical Sciences
New Mexico State University
Las Cruces, New Mexico 88003-8001
olberdin@nmsu.edu

Gabriel Picavet
Laboratoire de Mathématiques
Université Blaise Pascal
63177 Aubière Cedex, France
Gabriel.Picavet@math.univ-bpclermont.fr

Martine Picavet-L'Hermitte
Laboratoire de Mathématiques
Université Blaise Pascal
63177 Aubière Cedex, France
Martine.Picavet@math.univ-bpclermont.fr

Vadim Ponomarenko
Department of Mathematics
Trinity University
One Trinity Place
San Antonio, TX 78212-7200,
vadim123@gmail.com

Fred Richman
Department of Mathematical Sciences
Florida Atlantic University
Boca Raton, FL 33431-6498
richman@fau.edu

Moshe Roitman
Department of Mathematics
University of Haifa
Haifa, 31905 Israel
mroitman@math.haifa.ac.il

Christel Rotthaus
Department of Mathematics
Michigan State University
East Lansing, MI 488824-1024
rotthaus@math.msu.edu

William W. Smith
Department of Mathematics
The University of North Carolina at Chapel Hill
Chapel Hill, North Carolina 27599-3250
wwsmith@email.unc.edu

Roger Wiegand
Department of Mathematics
University of Nebraska
Lincoln, NE 68588-0323
rwiegand@math.unl.edu

Sylvia Wiegand
Department of Mathematics
University of Nebraska
Lincoln, NE 68588-0130
swiegand@math.unl.edu

Muhammad Zafrullah
57 Colgate Street
Pocatello, ID 83201
zafrullah@lohar.com

Commutative rngs

D. D. Anderson

Department of Mathematics, The University of Iowa, Iowa City, IA 52242
ddanders@math.uiowa.edu

1 Introduction

The purpose of this article is to discuss commutative rngs (that is, commutative rings that do not have an identity) and especially Robert Gilmer's work in this area. To avoid confusion we adopt the following terminology. The word "ring" will be used in the neutral sense that there may or may not be an identity. The term "rng" or "ring without identity" will mean a ring that does not have an identity and we will always explicitly say "ring with identity" when that is the case. The term "rng" appears in Jacobson [37, Section 2.17] where he suggests the pronunciation rŭng and says that the term was suggested to him by Louis Rowen. Bourbaki [6, Chapter 1] uses the term "pseudo-ring"for rings without identity. While we will be mostly concerned with commutative rings, in several places we consider noncommutative rings. This should be clear from context.

Today the word "ring" usually means ring with identity. This was not the case so long ago. I remember reading Lambek's book *Lectures on Rings and Modules* [38] in 1968 and being taken back by the fact that he assumed the existence of an identity element. When I taught material on the Jacobson radical thirty years ago I did it via quasi-regular elements. Today I usually assume the existence of an identity. I suspect the insistence on an identity today is in part due to the larger role played by homological methods in ring theory. Also, the existence of an identity entails the quasi-compactness of the spectrum of a ring, a useful property for algebraic geometry.

Now there are plenty of good rings that don't have an identity such as the even integers $2\mathbb{Z}$ and an infinite direct sum $\oplus R_\alpha$ of rings. On the other hand, since any ring can naturally be embedded in a ring with identity (see Section 2), many mathematicians no doubt take the point of view that there is no loss in generality in assuming the existence of an identity. The extent to which this is true depends on the context.

When asked to write an article for this volume, I gave considerable thought to the choice of topics. Certainly a number of suitable topics came to mind:

dimension theory, Prüfer domains and valuation domains, polynomial rings and power series rings, and semigroup rings. But I finally chose commutative rings without identity. About thirty of Robert's almost two hundred papers involve rngs to some extent or another – twenty-six which involve rngs in a significant way are listed in the references. One can of course debate which papers to include since many results on rings do not depend on the existence of an identity. Perhaps I picked commutative rngs for a topic because I remembered the material on rngs in Chapter 1 of *Multiplicative Ideal Theory* [25] and consulting his paper "Eleven nonequivalent conditions on a commutative ring" [15] (maybe even chuckling at the title).

I e-mailed Robert informing him of my article and asked for any thoughts. He sent the following. "The first thought that comes to mind is that there was always a bit of tension, to me, in whether to look at a given condition/problem/whatever under consideration in rngs, or only in rings. In most cases this wasn't an issue. Sometimes it was. I remember discussing with Jack Ohm on numerous occasions whether or not rngs were worthy of study. If you know Jack very well, you may know that he will take either side of a question, just for the sake of a good 'discussion.' A summary of the two positions would generally come down to something like (1) The fact that rngs were good enough for Emmy Noether (and other illustrious ones, though Noether was the usual citation) to consider validates their worth or (2) Life is too short for me to consider such ignoble algebraic structures; besides most questions about rngs can be reduced to questions about rings."

Finally, let me quote from the Preface of *Multiplicative Ideal Theory* [25]. "Some readers may find the failure to restrict to rings with identity noisome, so perhaps a word of explanation on this subject is in order. Examples of rings without identity abound, and on many occasions the author has found them to be useful, especially in the construction of examples (even examples of commutative rings with identity). Moreover, in many results, perhaps even a majority, the assumption that the ring contains an identity is not pertinent; it plays no role in the proof. Like most commutative algebraists, the author prefers to work in rings with identity, with all modules and homomorphisms unitary as well, but certainly is not averse to dropping these conditions, either."

In Section 2 we discuss adjoining an identity to a ring. Let R be a ring that is a subring of the ring S with identity 1. We say that S is a *unital extension of* R if $S = R[1] = \{r + n1 \mid r \in R, n \in \mathbb{Z}\}$. Besides giving Dorroh's [9] construction of a unital ring extension we give an algebra version. We also survey the work of Brown and McCoy [7] and Arnold and Gilmer [5] on unital extensions of commutative rings. In Section 3 we consider conditions on a commutative ring that are weaker than the existence of an identity. This section closely follows Gilmer's paper "Eleven nonequivalent conditions on a commutative ring" [15]. A new result is that a ring R has the property that for each $a \in R$, there exists an $r_a \in R$ with $r_a a = a$ if and only if for each ring S, every left ideal of $R \times S$ has the form $I \times J$ for left ideals I and J of R and S, respectively.

In Section 4 we consider some classical rngs of commutative ideal theory. We give a complete characterization of rings (general Z.P.I.-rings) in which every ideal is a product of prime ideals, rings (π-rings) in which every principal ideal is a product of prime ideals, and rings in which every ideal has a unique normal (primary) decomposition. We also discuss Gilmer's work on multiplication rings and related rings and on commutative rings in which every prime ideal is principal. In the final Section 5 a number of miscellaneous results (mostly due to Gilmer) on commutative rngs are given.

2 Adjunction of an Identity

In this section we discuss different ways of adjoining an identity to a ring A. This amounts to embedding A into a ring B with identity 1. For then $A^* = A[1] = \{a + m1 \mid a \in A,\ m \in \mathbb{Z}\}$ is a ring with identity. We call B a *unital ring extension of* A if $B = A^*$. Most of the results of this section do not require the rings to be commutative. We first handle the case where A is a commutative ring with a regular element and then give Dorroh's unital extension. We also give an "algebra version" of Dorroh's construction. We end this section with a survey of the work of Brown and McCoy [7] and Arnold and Gilmer [5] that determines all the unital extensions of a commutative ring.

Let A be a commutative ring containing a regular element r_0. Let N be the set of regular elements (that is, nonzero divisors) of A. Then $A_N = T(A)$ the total quotient ring of A has an identity element, namely $1 = r_0/r_0$. Now $A^* = A[1] = \{a + m1 \mid a \in A,\ m \in \mathbb{Z}\}$ is a commutative ring with identity. So $A \subseteq A^*$ is a unital extension. Note that $\operatorname{char} A = \operatorname{char} A^*$. Moreover, if A had an identity to start with, then $A^* = A$. Note that $T(A) = T(A^*)$ and $F(A) = F(A^*)$ ($F^*(A) = F^*(A^*)$) where $F(A)$ ($F^*(A)$) is the set of (finitely generated) fractional ideals of A or A^*. Thus A and A^* are simultaneously Noetherian. Here A is an ideal of A^* and $A^*/A \approx \mathbb{Z}/I$ where I is the ideal $\{m \in \mathbb{Z} \mid m1 \in A\}$.

But what do we do if A doesn't have a regular element or isn't even commutative? Dorroh [9] gave a simple way to embed any ring into a ring with identity. Let A be a ring and let $A^* = A \times \mathbb{Z}$. On A^* define $(a, n) + (b, m) = (a + b, n + m)$ and $(a, n)(b, m) = (ab + nb + ma, nm)$. It is easily checked the A^* is a ring with identity $(0, 1)$ with a natural embedding $A \longrightarrow A^*$ given by $a \longrightarrow (a, 0)$ with image $\bar{A} = \{(a, 0) \mid a \in A\}$ which is an ideal of A^*. Note that $\bar{A}[(0, 1)] = \bar{A} + \mathbb{Z}(0, 1) = A^*$ and $A^*/\bar{A} \approx \mathbb{Z}$. So A^* is a unital extension of A. Note that unlike the construction in the previous paragraph, if A has an identity, the Dorroh construction does not preserve the identity of A and $A \neq A^*$. Here A and A^* are $\mathbb{Z}$-algebras. Dorroh remarks that if A is a $\mathbb{Q}$-algebra, then we can replace $\mathbb{Z}$ by $\mathbb{Q}$ and $A \times \mathbb{Q}$ is then a

$\mathbb{Q}$-algebra with identity. In [10] Dorroh mentions that if $\operatorname{char} A = n$, then A^*/nA^* is a ring with identity where $\operatorname{char} A = \operatorname{char}(A^*/nA^*)$ and that A can be naturally embedded into A^*/nA^*. Observe that A^*/nA^* is naturally isomorphic to $A \times (\mathbb{Z}/n\mathbb{Z})$ with the product similar to that of the Dorroh extension. Of course $A \subseteq A^*/nA^*$ is a unital extension. Put another way, if A is a $\mathbb{Z}/n\mathbb{Z}$-algebra, so is A^*/nA^*. Stone [47] showed that a Boolean ring can be embedded in a Boolean ring with identity. Brown and McCoy [7] gave a detailed study of unital ring extensions. They observed that if $\operatorname{char} A = n > 0$ and $k \geq 1$, then we can replace $\mathbb{Z}$ by $\mathbb{Z}/kn\mathbb{Z}$ so that $A^* = A \times \mathbb{Z}/kn\mathbb{Z}$ is a ring with identity where $\operatorname{char} A^* = kn$. Observe that if $\operatorname{char} A = n$, then A is a $\mathbb{Z}/kn\mathbb{Z}$-algebra as is A^*.

Of course in hindsight it is clear where the addition and multiplication in the Dorroh extension are coming from. Formally adjoin 1 to A to get $A^* = A[1] = \{a + n1 \mid n \in \mathbb{Z}\}$. Then addition and multiplication are forced on us by $(a + n1) + (b + m1) = a + b + (m + n)1$ and $(a + n1)(b + m1) = ab + (n1)b + (m1)a + (n1)(m1) = ab + nb + ma + (nm)1$. So 1 becomes $(0, 1)$ and a becomes $(a, 0)$ as reminiscent in the construction of the complexes from the reals.

In each of the cases in the previous paragraph we are starting with a ring A that is an R-algebra ($R = \mathbb{Z}$, $\mathbb{Q}$, or $\mathbb{Z}/n\mathbb{Z}$) and are embedding A into an R-algebra A^* with identity. This suggests the following theorem. The details of the proof are left to the reader.

Theorem 2.1. *Let R be a ring with identity and let A be a ring that is a unitary R-R-bimodule. On $A^* = A \times R$ define $(a_1, r_1) + (a_2, r_2) = (a_1 + a_2, r_1 + r_2)$ and $(a_1, r_1)(a_2, r_2) = (a_1a_2 + r_1a_2 + a_1r_2, r_1r_2)$. Then A^* is a ring with identity $(0, 1)$. Here A^* is a unitary R-R-bimodule with $r(a, s) = (ra, rs)$ and $(a, s)r = (ar, sr)$. Hence if R is a commutative ring with 1 and A is a unitary R-algebra, A^* is a unitary R-algebra with identity. The map $A \longrightarrow A^*$ given by $a \longrightarrow (a, 0)$ is an R-algebra embedding, $A \times 0$ is a two-sided ideal of A^*, $R(0, 1) + (A \times 0) = A^*$, and $A^*/(A \times 0) \approx R$.*

Let R be a ring with identity and let A be an ideal of R. Then A is a ring. Observe that A has an identity if and only if $A = Re$ is generated by a central idempotent e of R. Suppose we apply the R-algebra version of the Dorroh extension to A. Here $A^* = A \times R$. Considering $R \times R$ with the usual sum and product, we get an R-algebra monomorphism $A^* \longrightarrow R \times R$ given by $(a, r) \longrightarrow (a, 0) + r(1, 1) = (a + r, r)$ with image $\{(r, s) \in R \times R \mid r - s \in A\}$. Thus in the case of $R = \mathbb{Z}$ and $A = 2\mathbb{Z}$, the Dorroh extension of $2\mathbb{Z}$ is naturally isomorphic to $\{(n, m) \in \mathbb{Z} \times \mathbb{Z} \mid n \equiv m \bmod 2\}$. This is quite different from the first method of adjoining an identity in the case where the ring contains a regular element; for here $(2\mathbb{Z})[1] = \mathbb{Z}$.

For the next example let $A = \bigoplus_{n=1}^{\infty} \mathbb{Z}$. Here A contains no regular elements, but $A \subseteq \prod_{n=1}^{\infty} \mathbb{Z}$, a ring with identity $(1, 1, \ldots)$. We have $(\bigoplus_{n=1}^{\infty} \mathbb{Z})[(1, 1, \ldots,)] = \{(m_i) \in \prod_{n=1}^{\infty} \mathbb{Z} \mid m_i \text{ is eventually constant}\}$. We leave it to the interested reader to check that this last ring is isomorphic to the ring obtained by the Dorroh extension.

We next observe that adjoining an identity does not commute with adjoining an indeterminate. Consider $2\mathbb{Z}$, so $(2\mathbb{Z})[1] = \mathbb{Z}$. Now $((2\mathbb{Z})[X])[1] = \mathbb{Z} + 2X\mathbb{Z}[X] \subsetneq \mathbb{Z}[X] = ((2\mathbb{Z})[1])[X]$. A similar result holds for power series adjunction. For the Dorroh extension we have $(2\mathbb{Z})^* \approx \{(n,m) \in \mathbb{Z} \times \mathbb{Z} \mid n \equiv m \bmod 2\}$ and $(2\mathbb{Z})^*[X] \approx \{(f(X), g(X)) \in \mathbb{Z}[X] \times \mathbb{Z}[X] \mid f(X) \equiv g(X) \bmod 2\mathbb{Z}[X]\}$. It is easily checked that the map $((2\mathbb{Z})[X])^* = (2\mathbb{Z})[X] \times \mathbb{Z} \longrightarrow (2\mathbb{Z})^*[X]$ given by $(f(X), n) \longrightarrow (f(X), 0) + n(1,1) = (f(X)+n, n)$ is a ring monomorphism with image $\{(f(X), g(X)) \in \mathbb{Z}[X] \times \mathbb{Z}[X] \mid f(X) \equiv g(X) \bmod 2\mathbb{Z}[X], \deg g(X) \leq 0\}$. For any unital extension $R \subseteq S = R[1]$, $R[X][1] = R[X] + \mathbb{Z}1 = S + XR[X] \subseteq S[X]$ with equality if and only if $R = S$.

These above examples seem to indicate that all questions concerning rngs cannot be resolved by just adjoining an identity. Note that for any unital extension $R \subseteq S$, R is Noetherian if and only if S is Noetherian. Indeed, a subset E of R is an (left, right, two-sided) ideal of R if and only if E is an ideal of S. So if S is Noetherian, so is R. Next suppose that R is Noetherian. Let A be an ideal of S. Then $A \cap R$ is finitely generated. But $A/A \cap R \approx (A+R)/R \subseteq S/R$ and hence is a cyclic abelian group; so A is finitely generated. Hence S is Noetherian. But Gilmer [16] has shown that $R[X]$ is Noetherian if and only if R is Noetherian with an identity.

Recall that a ring extension $R \subseteq S$ is *unital* if S is a ring with identity 1 and $S = R + \mathbb{Z}1 = R[1]$. Note that if S is a unital extension of R, then R is an ideal of S. If R_1 is a subring of R_2, we say that an ideal A_2 of R_2 *lies over* an ideal A_1 of R_1 if $A_2 \cap R_1 = A_1$. We next determine the unital extensions of a commutative ring R. This was done by Brown and McCoy [7], but we follow Arnold and Gilmer's treatment [5].

Proposition 2.2. *Let R be a unital extension of the commutative ring A and let B be an ideal of A. Then B is an ideal of R. If C is an ideal of R lying over B, then $C \subseteq (B{:}_R A)$ and C is principal modulo B. Now $(B{:}_R A)$ lies over B if and only if $(B{:}_A A) = B$. If $(B{:}_A A) = B$ and hence $(B{:}_R A) = B + R\alpha$ for some $\alpha \in R$, then $\{B + Rn\alpha\}_{n=0}^{\infty}$ is the set of ideals of R lying over B in A.*

Proof. This is essentially [5, Proposition 2.2]. Clearly B is an ideal of R, if C lies over B, then $C \subseteq (B{:}_R A)$, and $(B{:}_R A)$ lies over B if and only if $(B{:}_A A) = B$ since $(B{:}_A A) = (B{:}_R A) \cap A$. Observe that if C lies over B, then $C/B = C/C \cap A \approx (C+A)/A \subseteq R/A \approx \mathbb{Z}/m\mathbb{Z}$ for some $m \geq 0$. Hence C/B is a cyclic R/A-module and hence a cyclic abelian group. So if $(B{:}_A A) = B$, $(B{:}_R A) = B + R\alpha$ for some $\alpha \in R$. Observe that the set of R/A-submodules of $(B{:}_R A)/B$ is $\{Rn\bar{\alpha}\}_{n=0}^{\infty}$.

Let $R^* = R \times \mathbb{Z}$ be the Dorroh extension of R where R is identified with the ideal $R \times 0$ of R^*. So $R \subset R^*$ is a unital extension. Let S be a unital extension of R. Then the map $f\colon R^* \longrightarrow S$ given by $(r,n) \longrightarrow r + n1$ is a ring epimorphism. Note that $\ker f$ is an ideal of R^* lying over 0 in R and $S \approx R^*/\ker f$. Conversely, if C is an ideal of R^* lying over 0 in R, R^*/C is a unital extension of R. Hence if $\{C_\beta\}$ is the set of ideals of R^* lying over 0 in R, $\{R^*/C_\beta\}$ is the set of unital extensions of R. We can now apply Proposition 2.2 for the case $B = 0$. Now each C_β is a principal ideal of R^* and is a cyclic abelian group. So $C_\beta = R^*(r_\beta, k_\beta) = \{(nr_\beta, nk_\beta) \mid n \in \mathbb{Z}\}$ for some $r_\beta \in R$ and $k_\beta \geq 0$. If some $k_\beta > 0$, then the smallest such positive integer k_{β_0} is a divisor of each k_β (use the division algorithm) and is called the *mode of* R. If $k_\beta = 0$, then $C_\beta \subseteq R$, so that $C_\beta = C_\beta \cap R = 0$ and there is only one such C_β. If each $k_\beta = 0$, we say that R has *mode* 0. Observe that for any $(r_\beta, k_\beta) \in R^*$ with $k_\beta > 0$, $R^*(r_\beta, k_\beta) \cap R = 0$ if and only if $(-r_\beta)x = k_\beta x$ for each $x \in R$. Thus R has mode 0 if no nonzero element of R acts as a nonzero integer under multiplication and R has mode $k > 0$ if k is the smallest positive integer such that for some $r \in R$, $rx = kx$ for all $x \in R$. Note that the mode of R divides char R and R has an identity if and only if R has mode 1. If R has mode 0, R^* is the unique unital extension of R.

Further suppose that $(0{:}_R R) = 0$. By Proposition 2.2 an ideal C of R^* lies over 0 in R if and only if $C \subseteq (0{:}_{R^*} R)$; and $(0{:}_{R^*} R) = (\alpha)$ where $\alpha = (r,k) \in R^*$ and k is the mode of R. Now $\{(n\alpha)\}_{n=0}^{\infty}$ is the set of ideals of R^* lying over 0 in R and $\{R^*/(n\alpha)\}_{n=0}^{\infty}$ is the set of unital extensions of R. Put $R_n = R^*/(n\alpha)$. If $k = 0$, then $\alpha = 0$; and so each $R_n = R^*$. If $k > 0$, then $R_n/R = R^*/(R + (n\alpha)) = R^*/R \times nk\mathbb{Z} \approx \mathbb{Z}/nk\mathbb{Z}$. So up to R-isomorphism, the R_n's are distinct. We sum up these results in the next theorem.

Theorem 2.3. *Let R be a commutative ring and let R^* be the Dorroh extension of R.*
(1) If $\{C_\beta\}$ is the set of ideals of R^* lying over 0 in R, then $\{R^*/C_\beta\}$ is the set of unital extensions of R. Each C_β is principal as an ideal of R^* and is cyclic as an abelian group. If some C_β is nonzero, then there is a positive integer k, the mode of R, such that each $C_\beta = ((r_\beta, kn_\beta))$.
(2) If $(0{:}_R R) = 0$, then there is an $\alpha = (r,k) \in R^*$ such that $((n\alpha))_{n=0}^{\infty}$ is the set of ideals of R^* lying over 0 in R. If $k = 0$, then R^* is the unique unital extension of R. If $k > 0$, then $\{R_n{:}= R^*/(n\alpha)\}_{n=0}^{\infty}$ is the set of unital extensions of R and the rings R_n are distinct up to R-isomorphism.

Suppose that R is a commutative ring with regular element r. So $(0{:}_R R) = 0$. Consider the unital extension $R[1] = R + \mathbb{Z}1 \subseteq T(R)$. We claim that $R[1] \approx R_1$. Define $\varphi\colon R^* \longrightarrow R[1]$ by $\varphi((s,m)) = s + m1$. Then $\ker\varphi \subseteq (\alpha) = (0{:}_{R^*} R)$ where $\alpha = (x,k) \in R^*$. Now $(r,0)(x,k) \in (R \times 0) \cap (\alpha) = 0$, so $0 = rx + kr$. Then $0 = (rx + kr)r^{-1} = x + k1 = \varphi(\alpha)$. Hence $\ker\varphi = (\alpha)$, so $R[1] \approx R^*/(\alpha) = R_1$.

We end this section with the following example from [5].

Example 2.4. [5, Example 2.6] A ring T of mode $k > 0$ with $(0{:}_T T) = 0$ so that $T_n \approx T_m$ for all $n, m \geq 0$ even though T_n and T_m are not T-isomorphic for $n \neq m$. Let S be a ring with mode k (for example, $S = k\mathbb{Z}$). Set $S^{(n)} = \prod_{m=1}^{\infty} S_n$ and $R = \prod_{n=1}^{\infty} S^{(n)}$, so R is a ring with identity. Put $T = R \times S$, so $(0:_T T) = 0$ and T has mode k. Now for each $n \geq 0$. $T_n \approx R \times S_n$ [5, Proposition 2.5]. But certainly $R \times S_n \approx R$. So $T_n \approx T_m$ for each $n, m \geq 0$.

3 Conditions Weaker than Having an Identity

Commutative rings with identity have many familiar properties that we use everyday without much thought such as maximal ideals are prime, every proper ideal is contained in a maximal ideal, if $\sqrt{A}$ is maximal then A is primary, and if A and B are comaximal ideals, then $A \cap B = AB$. An examination of the proof of each of these above properties will show that some consequence of the existence of an identity is used. For example, if A and B are comaximal ideals, then the fact that $R(A \cap B) = A \cap B$ is used: $A \cap B = R(A \cap B) = (A + B)(A \cap B) = A(A \cap B) + B(A \cap B) \subseteq AB$.

In [15], Gilmer considered eleven conditions on a commutative ring R, the first of which that R has an identity implies the other ten:

A. R has an identity.
B. R is generated by idempotents, that is, for $r \in R$, $r = r_1 e_1 + \cdots + r_n e_n$ where $r_i \in R$ and e_i is idempotent. But since $Re_1 + \cdots + Re_n = Re$ for some idempotent e this is equivalent to $r = se$ where e is idempotent or to $r = re$ for some idempotent e.
C. If A is a nonzero (principal) ideal of R such that $\sqrt{A} \neq R$, then R/A has an identity.
D. If $x \in R$, there exists $y \in R$ with $x = xy$. Equivalently, given $x_1, \cdots, x_n \in R$ there exists $y \in R$ with $x_i = x_i y$; $RA = A$ for each ideal A of R; or $AB = A \cap B$ for comaximal ideals A and B of R.
E. R is a *u-ring*, that is, if A is a proper ideal of R then $\sqrt{A} \neq R$, or equivalently, if A is a proper ideal of R, then A is contained in a proper prime ideal of R.
F. $R = R^2$.
G. Maximal ideals are prime.
H. If P is a nonzero prime ideal of R, R/P has an identity. Equivalently, if Q is P-primary where $P \neq R$, then R/Q has an identity.
J. If A is an ideal of R with $\sqrt{A}$ maximal, then A is primary, or equivalently, R/A has an identity.
K. Each ideal is contained in a maximal ideal.
L. If A and B are comaximal proper ideals, then $AB = A \cap B$.

In the above conditions with equivalences such as D, it is the first property listed that Gilmer labeled with the letter. He then showed that the other properties were equivalent. Note that all eleven conditions are preserved by homomorphism. By choosing an appropriate ring with zero products, it is easy to construct commutative rings that fail to satisfy A, B, D, E, F, G, K, or

L. (Note that C, H and J hold vacuously in such a ring.) Gilmer gave the following diagram which completely describes the implications between these properties.

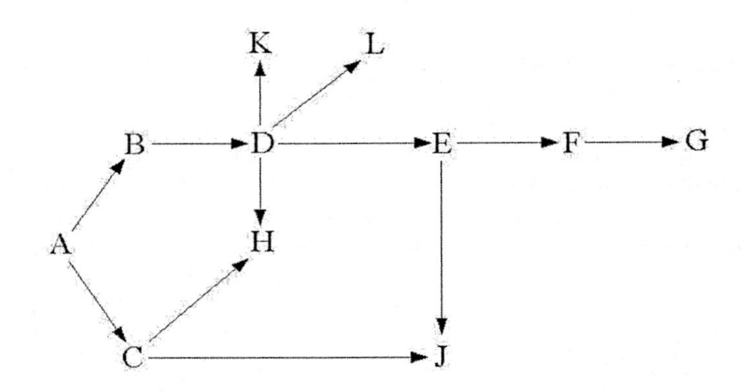

One of the useful properties of rings with identity is that ideals of $R \times S$ have the form $I \times J$ where I is an ideal of R and J is an ideal of S. This is of course not true in general since $\{(\bar{0},\bar{0}),(\bar{1},\bar{1})\}$ is an ideal of the ring $\mathbb{Z}/2\mathbb{Z}\times\mathbb{Z}/2\mathbb{Z}$ with the zero product. The next result (which the author believes is new) characterizes the rings whose ideals have the desired form.

Proposition 3.1. *For a commutative ring R the following conditions are equivalent.*

(1) *R satisfies condition D.*

(2) *For each commutative ring S, every ideal of $R \times S$ has the form $I \times J$ where I is an ideal of R and J is an ideal of S.*

(3) *Every ideal of $R \times R$ has the form $I \times J$ where I and J are ideals of R.*

Proof. (1)$\Rightarrow$(2) Let L be an ideal of $R \times S$. Let $(a,b) \in L$. Choose $r \in R$ with $ra = a$. Then $(a,0) = (ra,0) = (r,0)(a,b) \in L$ and $(0,b) = (a,b) - (a,0) \in L$. It easily follows that L has the desired form. (2)$\Rightarrow$(3) Clear. (3)$\Rightarrow$(1) Let $a \in R$. Since $\langle(a,a)\rangle = I \times J$ for ideals I and J of R, $(a,0) \in \langle(a,a)\rangle$. So $(a,0) = (r,s)(a,a) + n(a,a)$ for some $r,s \in R$ and $n \in \mathbb{Z}$. Hence $a = ra + na$ and $0 = sa + na$. So $na = -sa$ gives $a = ra + na = ra - sa = (r-s)a$.

Remark 3.2. There is a noncommutative version of Proposition 3.1 for left and right ideals which we leave to the interested reader.

While there are no simple implications other than those given in the diagram, combinations of properties give other implications. For example, Gilmer shows that $K + G \Rightarrow E$, $C + F \Leftrightarrow A$, and $E + H \Leftrightarrow D$. Also, we get other implications by putting finiteness conditions on R. For example, Gilmer shows that $E + ACC$ on prime ideals $\Rightarrow D$ and if R is finitely generated ($R = Rr_1 + \cdots + Rr_n$), then $G \Rightarrow A$. Also, a Noetherian ring (or more generally, finitely generated ring) or a ring containing a regular element satisfying D has an identity. Thus a commutative Noetherian ring R has the property that each ideal of $R \times S$ has the form $I \times J$ if and only if R has an identity.

4 Classical Rngs of Ideal Theory

In this section we consider some of the classical rings of ideal theory without the hypothesis of them containing an identity. We begin with a discussion of π-rings and general Z.P.I.-rings.

Recall that a commutative ring R is called a (*π-ring*) *general Z.P.I.-ring* if every (principal) ideal of R is a product of prime ideals. In a series of papers, S. Mori [39]-[43] studied π-rings and general Z.P.I.-rings. In [42] he showed that a commutative ring with identity is a π-ring if and only if it is a finite direct product of π-domains and SPIR's (principal ideal rings with a single prime ideal and that prime ideal is nilpotent). A very reasonable account can be found in [25, Section 46]. Examples of π-domains with identity include UFD's and Dedekind domains. Many characterizations of π-domains with identity are collected in [1, Theorem 3.1]. For example, for an integral domain D with identity the following are equivalent: (1) D is a π-domain, (2) every invertible ideal of D is a product of invertible prime ideals, (3) every nonzero prime ideal of D contains an invertible prime ideal, (4) D is a Krull domain with each height-one prime ideal invertible, (5) D is a locally factorial Krull domain, and (6) $D(X) = \{f/g \mid f, g \in D[X]$, the coefficients of g generate $D\}$ is a UFD.

Less well known is Mori's characterization of π-rings without identity. In [43], Mori showed that a π-ring without identity is either (1) an integral domain, (2) a ring $R = (p)$ where every ideal of R including 0 is a power of R, (3) $K \times R$, where K is a field and R is a ring as in (2), or (4) $K \times D$, where K is a field and $D = (p)$ is a π-domain where every nonzero ideal of D is a power of D. He showed that in a π-domain without identity every principal ideal is actually a product of principal prime ideals (here the whole ring itself is considered to be a prime ideal) and that an integral domain D without identity is a π-domain if and only if (1) every nonzero element of D is a product of irreducible elements and (2) each irreducible element of D generates a principal prime ideal. Mori also remarked that $2\mathbb{Z}_{(2)}$ is an example of a π-domain without identity in which the factorization of an element into irreducible elements is not unique but for each irreducible element p of $2\mathbb{Z}_{(2)}, (p) = 2\mathbb{Z}_{(2)}$. In his review of Mori's paper O. F. G. Schilling [46] stated that in a π-domain D without identity, $D = (p)$ for each irreducible element p of D. Mori did not make this assertion which we shall later see is not correct.

Certainly a general Z.P.I.-ring is a π-ring. It is well-known that an integral domain with identity is a general Z.P.I.-ring if and only if it is a Dedekind domain. Thus a commutative ring with identity is a general Z.P.I.-ring if and only if it is a finite direct product of Dedekind domains and SPIR's. Note that the π-rings without identity given in (2), (3), and (4) of the previous paragraph are actually general Z.P.I.-rings. Thus a general Z.P.I.-ring without identity is either (1) an integral domain, (2) a ring $R = (p)$ where every ideal of R including 0 is a power of R, (3) $K \times R$, where K is a field and R is a ring as in (2), or (4) $K \times D$, where K is a field and $D = (p)$ is a general Z.P.I.-ring which is an integral domain where every nonzero ideal of D is a power of D.

Now there are some difficulties with several of Mori's results in [41]; see Wood [48] for details. Wood gave two treatments of general Z.P.I.-rings; one based on Mori's results and one independent of them. Some of the results from [48] appear in [49]. In [12], Gilmer characterized the integral domains D without identity in which every ideal is a product of prime ideals (here D is considered to be a prime ideal). He showed that an integral domain without identity has the property that each ideal is a product of prime ideals if and only if each nonzero ideal of D is a power of D; moreover D is a PID. He showed that $D^* = D[1]$ is a rank-one discrete valuation domain with D as its maximal ideal. Conversely, if (S, M) is a rank-one discrete valuation domain with $S = M[1]$, then M is an integral domain without identity. Here a subset $A \subseteq M$ is an ideal of M if and only if it is an ideal of S; so $\{M^n\}_{n=1}^{\infty} \cup \{0\}$ is the set of ideals of M. Hence M is a general Z.P.I.-ring. Gilmer also classified these rank-one discrete valuation domains (V, M) with $V = M[1]$. See [12, page 582].

It is easily seen that if (D, M) is a quasilocal UFD with $D = M[1]$, then M is a π-domain without an identity. For let $0 \neq d \in M$. Then $d = p_1 \cdots p_n$ where p_i is a principal prime of D. Since $M[1] = D$, for $a \in M$, $aD = (a)$, the principal ideal of M generated by a. Hence $(d) = (p_1) \cdots (p_n)$ where each (p_i) is a prime ideal of D and hence of M. The converse is also true [2, Theorem 8]: let R be π-domain without identity, then $R[1]$ is a quasilocal UFD with R as its maximal ideal. Note that Gilmer's result characterizing domains that are general Z.P.I.-rings is a special case. For each n let $D = \mathbb{Z}_p[[X_1, \cdots, X_n]]$, $p > 0$ prime, and $M = (X_1, \cdots, X_n)$. Then M is a π-domain without identity, but is a general Z.P.I.-ring if and only if $n = 1$.

Recently, the author and John Kintzinger [3] have given a similar characterization of π-rings R of type (2):$R = (p)$ where every ideal of R including 0 is a power of R. Here there is an SPIR S containing R with $R[1] = S$ and R is the maximal ideal of S. Moreover, there is a complete DVR (D, M) with $M[1] = D$ and an epimorphism $\pi{:}D \to S$ with $\pi(M) = R$.

Let D be an integral domain with identity. It is well known that D is a π-domain (UFD) if and only if every nonzero prime ideal of D contains an invertible (nonzero principal) prime ideal. Thus it seems reasonable to conjecture that if D is an integral domain without identity, then D is a π-domain if and only if each nonzero prime ideal contains a nonzero principal prime ideal. However, this need not be the case as shown by an example of Gilmer [19, Example 5.3]. Let $D = X\mathbb{Z}[[X]]$. Then $D^* = \mathbb{Z}+X\mathbb{Z}[[X]] = \mathbb{Z}[[X]]$ is not quasilocal, so D is not a π-domain, but every prime ideal of D is principal. This paper [19] contains a detailed study of rings in which every prime ideal is principal. Gilmer calls such rings F-ring. Of course an F-ring with identity is a principal ideal ring. He shows that an F-ring R without identity which contains a regular element has $R^* = R[1]$ (1 the identity in $T(R)$) a PIR if and only if R is a PIR and that if every primary ideal of R is principal, then R is a PIR.

We previously mentioned that $2\mathbb{Z}_{(2)}$ is a π-domain without unique factorization into irreducibles. Let D be an integral domain without identity. Since D has no units, two elements are associates if and only if they are equal. Thus to say that D has unique factorization into irreducible elements would mean that each nonzero element of D is a product of irreducible elements and that this factoriztion is unique up to order of factors. Anderson [2, Proposition 10] showed that an integral domain without identity cannot have unique factorization.

While an integral domain without identity cannot have unique factorization into irreducible elements, we have the following result from Gilmer [21]. Suppose that R is a commutative ring without identity and that B is a finitely generated regular ideal of R that is representable as a finite product of prime ideals of R (here R is considered to be a prime ideal). Then this representation is unique.

Let D be an integral domain that may not have an identity. Then D satisfies the *cancellation law* (*CL*) *for ideals* if for ideals A, B, and C of D with $A \neq 0$, $AB = AC$ implies $B = C$. In [14] Gilmer considered CL. He showed [14, Theorem 3] that if D has an identity, then D satisfies CL if and only if D is almost Dedekind (that is, D_M is a rank-one discrete valuation domain for each maximal ideal M of D). Suppose that D doesn't have an identity and let $D^* = D[1]$ where 1 is the identity of the quotient field of D. It is easily seen that CL holds for D if and only if CL holds for D^*. Gilmer showed that if CL holds in D, then D^*/D is finite and conversely, if J is an integral domain with identity and A is an ideal of J with J/A finite and CL holds in A, then $J = A[1]$ [14, Theorem 5]. If D is an integral domain without identity which satisfies CL, then D_M is a rank-one discrete valuation domain for each maximal ideal M of D [14, Theorem 6]. However, the converse is false. For example, let K be a finite field, $J = K[X]$, and $A = (X^2)$. Then A_M is a rank-one discrete valuation domain for each maximal ideal M of A and J/A is finite. But $A^* = K + X^2K[X]$ is not almost Dedekind and hence doesn't satisfy CL; hence neither does A.

An ideal in a commutative ring R has finite norm if R/A is finite and we then set $N(A) = |R/A|$. Also, let $L(A)$ be the length of R/A. In [17], Gilmer showed that if a commutative ring R satisfied either (1) every nonzero ideal of R has finite norm and $N(AB) = N(A)N(B)$ for nonzero ideals A and B of R, or (2) every nonzero ideal of R has finite length and $L(AB) = L(A)+L(B)$

for nonzero ideals A and B of R, then R is Noetherian and has an identity and has the property that there are no ideals properly between M and M^2 for each maximal ideal M of R. Thus if R is a domain, R is a Dedekind domain.

In [8] Butts and Gilmer considered commutative rings R that satisfy Property (α) which states that every primary ideal of R is a power of its radical or Property (δ) that states that every ideal of R is a finite intersection of powers of prime ideals. [8, Corollary 4] gave that an integral domain D with ACC on prime ideals satisfying Property (α) has D_P a valuation domain for each proper prime ideal P of D. The main result [8, Theorems 11, 13, and 14] is that a ring R satisfies Property (δ) if and only if either (1) R is a general Z.P.I.-ring with identity or (2) $R = F_1 \times \cdots \times F_k \times S$ where $F_1, \cdots, F_k$ are fields and S is either a nonzero domain in which every nonzero ideal is a power of S or S is a nonzero ring in which every ideal is a power of S. Note that in case (2), R is a general Z.P.I.-ring if and only if $k \leq 1$.

In [11] Gilmer considered primary rings. A commutative ring R is a *primary ring* if R has at most two prime ideals where R is counted as a prime ideal. So either R is the only prime ideal of R which is equivalent to every element of R being nilpotent or R has two prime ideals P and R. So in this case, if A is an ideal of R, $\sqrt{A} = P$ or $\sqrt{A} = R$, that is, $b \in R$ is either nilpotent or $\sqrt{(b)} = R$. Of course, if R has an identity, R is primary if and only if R is zero-dimensional quasi-local. However, $2\mathbb{Z}_{(2)}$ is a primary integral domain. The following realization of primary domains is given. If J is an integral domain and if M is the intersection of all nonzero prime ideals of J and D is an ideal of J contained in M, then D is a primary domain. Conversely, if D_0 is a primary domain, then there exists an integral domain J_0 with identity such that D_0 is an ideal of J_0 contained in all the nonzero prime ideals of J_0 [11, Theorem 3]. In the case where the ring is Noetherian, more can be said; see the paper for details.

In [13] Gilmer considered commutative rings in which semi-primary ideals (ideals with prime radical) are primary. Recall that R is a *u-ring* if each proper ideal has proper radical, or equivalently each proper ideal is contained in a proper prime ideal (see Section 3). The main result [13, Theorem 7] is that a commutative ring R has every semi-primary ideal primary if and only if R is of one of the following types: (A) a ring in which every element is nilpotent; (B) a primary domain, (C) a zero-dimensional u-ring, or (D) a one-dimensional u-ring with the following property: if $P \subset M$ are proper prime ideals and if $p \in P$, then $p = pm$ for some $m \in M$.

Let R be a commutative ring. If an ideal A of R has a primary decomposition, then A has a normal decomposition $A = Q_1 \cap \cdots \cap Q_n$ where Q_i is P_i-primary, the P_i's are distinct, and $Q_i \not\supseteq \cap Q_j$ for each i. Here the length n and primes $P_1, \cdots, P_n$ are uniquely determined. If P_i is minimal over A, then Q_i is uniquely determined, but this need not be true in general. In [20] Gilmer determined the rings, which he called W-rings, in which every ideal has a unique normal decomposition. The main result is that an indecomposable W-ring is of one of the following types: (1) a primary ring with identity,

(2) a ring in which every element is nilpotent, (3) a primary domain, or (4) a one-dimensional integral domain in which the residue class ring of each maximal ideal is a field and in which every nonzero element belongs to only finitely many maximal prime ideals. Further, an arbitrary ring R is a W-ring if and only if R is a finite direct product of indecomposable W-rings, at most one of which has no identity.

Recall that a commutative ring R is an *AM-ring* if for each pair of ideals $A \subsetneq B$ of R, $A = BC$ for some ideal C of R and that an AM-ring R is a *multiplication ring* if $RA = A$ for each ideal of A of R. For an ideal A of R, the *kernel of* A is $\bigcap\{Q \mid Q \supseteq A$ where Q is P-primary, P a minimal prime of $A\}$. The following results come from Gilmer and Mott [31]. A ring R has the property that every semiprimary ideal is primary if and only if each ideal of R is equal to its kernel [31, Theorem 4]. For a commutative ring R, the following conditions are equivalent: (1) R is an AM-ring, (2) if $A \subsetneq P$ are ideals of R with P prime, then $A = PB$ for some ideal B of R, (3) (a) every semiprimary ideal of R is primary, (b) every primary ideal of R is a prime power, and (c) if $P \neq R$ is a prime ideal and if A is an ideal of R with $A \subseteq P^n$, but $A \not\subseteq P^{n+1}$, then $P^n = (A{:}(y))$ for some $y \in R - P$ [31, Theorems 12 and 13]. Along the way the following interesting result is proved [31, Theorem 3]. Let R be a u-ring in which the set of prime ideals is inductive. If the zero ideal of R is a finite product of prime ideals, then R has an identity. Wood [50] and Wood and Bertholf [51] considered rings whose proper homomorphic images satisfy (a) or (b) or is a multiplication ring.

Griffin [35, 36] studied valuation rings, Prüfer rings, and multiplication rings in the context of commutative rings R that have *fixing elements* (that is, for each $a \in R$, there exists $r \in R$ with $ra = a$, Condition D in Section 3) or in which R is generated by idempotents (Condition B of Section 3). Suppose that R is generated by idempotents. For an idempotent $e \in R$, call $a \in R$, *e-regular* if $ae = a$ and $ax = 0$ implies $ex = 0$. Griffin defined a commutative ring K containing R to be the *total quotient ring* of R if (i) for $x \in K$, there exist $e, a, b \in R$ with $ex = x$, $e^2 = e$, and b is e-regular with $bx = a$, and (ii) if $a \in R$ is e-regular there exists $x \in K$ with $ax = e$. If R is generated by idempotents, then R has a total quotient ring unique up to isomorphism which can be constructed as follows. The set of idempotents of R is directed by the partial order $e \leq f \Leftrightarrow ef = e$. The ring Re has an identity and hence a total quotient ring K_e. If $e \leq f$, we have a natural inclusion $K_e \longrightarrow K_f$. The total quotient ring of R is then the direct limit $\varinjlim K_e$ over the directed set of idempotents of R. If R has an identity, this is just the usual total quotient ring of R. Griffin extended the notions of (Manis) valuation and Prüfer rings (finitely generated regular ideals are invertible) to rings with fixing elements. The second paper gave a detailed study of multiplication rings.

5 Miscellaneous Papers

In this final section we give a brief overview of Gilmer's other papers concerning rings that may not have an identity.

For a positive integer n, let $R(n)$ be a complete set of representatives of isomorphism classes of associative rings of order n and let $\rho(n) = |R(n)|$. If $n = p_1^{e_1} \cdots p_k^{e_k}$ is the prime factorization of n, then a ring R of order n is uniquely decomposable as the direct sum of ideals $I_1, \cdots, I_k$ of orders $p_1^{e_1}, \cdots, p_k^{e_k}$. Thus to determine $R(n)$ or $\rho(n)$ it suffices to determine $R(p_i^{e_i})$ or $\rho(p_i^{e_i})$ for $1 \leq i \leq k$. For a prime p, $R(p)$ and $R(p^2)$ are known. It is not hard to show that $|R(p)| = 2$ and $\left|R(p^2)\right| = 11$. In [32], Gilmer and Mott determined $R(p^3)$ and $\rho(p^3)$. (The paper contains some errors that are corrected in an unpublished Addendum.) The determination is straightforward but rather long and tedious with many cases. Unlike $\rho(p)$ and $\rho(p^2)$, $\rho(p^3)$ depends on the prime p.

In [5], Arnold and Gilmer investigated the dimension theory of rngs. Let R be a commutative ring. Then $\dim R$ is defined as usual except that R may have no proper ideals in which case we define $\dim R = -1$. Let A be an ideal of R. Then $\dim A \leq \dim R \leq \dim A + \dim R/A + 1$ [5, Corollary 3.4]. Hence if S is a unital extension of R, $\dim R \leq \dim S \leq \dim R + 2$ with $\dim S \leq \dim R + 1$ if S/R is not isomorphic to $\mathbb{Z}$ [5, Proposition 4.1].

Let $R^{(m)} = R[X_1, \cdots, X_m]$ and $n_m = \dim R^{(m)}$. Then $\{n_i\}_{i=0}^{\infty}$ is called the *dimension sequence for* R. If $\dim R = -1$, then each element of R is nilpotent, so each element of $R^{(m)}$ is also nilpotent, and hence $\dim R^{(m)} = -1$. Let S be a unital ring extension of R. In general the sequence $\{\dim S^{(m)} - \dim R^{(m)}\}_{m=0}^{\infty}$ is not well-behaved. But we do have $0 \leq \dim S^{(m)} - \dim R^{(m)} \leq m + 2$ [5, Proposition 5.13]. However, the main result of [5] is that a sequence of non-negative integers is the dimension sequence for a ring without identity if and only if it is the dimension sequence for a ring with identity [5, Theorem 5.10]. Since Arnold and Gilmer [4] had previously determined the set of dimension sequences for commutative rings with identity, the problem of finding the possible dimension sequences for rngs is solved.

Let R be a commutative ring with set of zero divisors $Z(R)$. Let $\alpha = |R|$ and $\beta = |Z(R)|$. In [23] Gilmer considered the following questions. (1) What conditions are necessary on α and β? (2) If a pair (α, β) of cardinals satisfies these conditions, can we find a corresponding pair $(R, Z(R))$ so that $|R| = \alpha$ and $|Z(R)| = \beta$? He obtained an answer to Question 1 so that the answer to Question 2 is "yes". If $\beta = 1$, R is an integral domain and so α is either infinite (note that the ring of polynomials over $\mathbb{Z}$ in α indeterminates has cardinality α) or is a power of a prime. So assume $\beta > 1$. Then Gilmer showed that $\alpha \leq \beta^2$. So if β is infinite, $\alpha = \beta$ (and we can take our ring to be any abelian group of cardinality α with zero products), and if β is finite, then α is finite. In the case where α is finite we can reduce to the case where R is local so $\alpha = p^t$ (p a prime) and $Z(R)$ is an ideal of R so $\beta = p^s$ where $0 \leq s < t$.

Then Gilmer [23, Theorem 3] showed that such a pair exists if and only if $t-s$ divides s.

In [22] Gilmer showed that if R is a ring (not necessarily commutative) with only finitely many subrings, then R is finite. This had previously been proved by A. Rosenfeld [45] for the case of rings with identity, but Gilmer's proof is independent of [45].

In [29] Gilmer, Lea, and O'Malley considered rings (not necessarily commutative) whose proper subrings or ideals satisfy certain properties. Let $(P1)$ be the property "has finite characteristic" and $(P2)$ be the property "has no proper zero divisor". They proved the following.

Theorem 5.1. *Let R be a ring.*

(1) [29, Corollary 2.3] *Every proper subring of R satisfies property $(P1)$ if and only if*
  (i) *R has finite characteristic, or*
  (ii) *R is the zero ring on the p-quasicyclic group $\mathbb{Z}(p^\infty)$, p a prime.*

(1) [29, Proposition 2.5 (Corollary 2.7)] *Every proper (left) ideal of R satisfies property $(P1)$ if and only if*
  (i) *R has finite characteristic,*
  (ii) *R is the zero ring on $\mathbb{Z}(p^\infty)$, p a prime, or*
  (iii) *R is a simple ring having no nonzero elements of finite order (R is a division ring of characteristic zero).*

(3) [29, Corollary 2.11] *Every proper subring of R satisfies $(P2)$ if and only if*
  (i) *R satisfies $(P2)$,*
  (ii) *$R \approx \mathbb{Z}/(p) \oplus \mathbb{Z}/(q)$ where p and q are primes, or*
  (iii) *R is the zero ring on $\mathbb{Z}/(p)$ where p is prime.*

(4) [29, Corollary 2.10 (Corollary 2.12)] *Every proper (left) ideal of R satisfies property $(P2)$ if and only if*
  (i) *R satisfies property $(P2)$,*
  (ii) *R is the zero ring on $\mathbb{Z}/(p)$, p a prime,*
  (iii) *R is the direct product of two simple rings, each of which satisfies property $(P2)$ (R is a direct product of two division rings), or*
  (iv) *R does not satisfy property $(P2)$ and R is a simple ring for which $R^2 = R$.* (This case cannot occur for the case of left ideals.)

In [30] Gilmer and O'Malley considered rings (not necessarily commutative) which satisfy property $(C1)$: R does not satisfy ACC on ideals, but each proper subring of R does or property $(C2)$: R does not satisfy ACC on left ideals, but each proper left ideal of R satisfies ACC on left ideals. They proved the following result.

Theorem 5.2. [30, Theorem 3.2] *For a ring R, the following conditions are equivalent.*

(a) *R has property $(C1)$.*

(b) *R has property* $(C2)$.
(c) *R is the zero ring on* $\mathbb{Z}(p^\infty)$, *p a prime.*

Let R be a commutative ring. Then R is *hereditarily Noetherian* if each subring of R is Noetherian. In [27] Gilmer and Heinzer determined the hereditarily Noetherian commutative rings. For this question they remarked that it is no restriction to consider only commutative rings with identity and subrings that contain that identity. For let R^* be the Dorroh extension of R and let e be the identity element of R^*. Let S be a subring of R. Now S is Noetherian if and only if $S[e]$ is Noetherian. Thus R is hereditarily Noetherian if and only if each subring S of R^* containing e is Noetherian. They showed [27, Theorem 2.4] that a commutative ring T is hereditarily Noetherian if and only if T is isomorphic to a subring of a ring of the form $D_1 \times \cdots \times D_n \times R$ where $D_1, \cdots, D_n$ are hereditarily Noetherian integral domains with identity and R is a unitary ring with finitely generated additive group.

In [28] Gilmer and Heinzer proved that if R is an uncountable commutative ring (with identity e) of cardinality ω, then there exists a proper subring S of R (containing e) with $|S| = \omega$.

In [18] Gilmer investigated R-automorphisms on $R[X]$. Let R be a commutative ring with 1. Then an R-endomorphism φ on $R[X]$ is completely determined by $\varphi(X)$; for if $\varphi(X) = t$, then $\varphi(g(X)) = g(t)$. Denote this map by f_t. Gilmer showed that if $t = t_0 + t_1X + \cdots + t_nX^n \in R[X]$, then f_t is an R-automorphism on $R[X]$ if and only if t_1 is a unit and t_i is nilpotent for $i \geq 2$. So each R-automorphism of $R[X]$ has this form. He then extended this result to commutative rings containing a regular element. Let R be such a ring with total quotient ring T. Gilmer showed that each R-endomorphism of $R[X]$ is induced by a T-endomorphism of $T[X]$. For $t = t_0 + t_1X + \cdots + t_nX^n \in T[X]$, $f_t|_{R[X]}$ is an R-automorphism on $R[X]$ if and only if $Rt_i \subseteq R$ for each i, $Rt_1 = R$, and t_i is nilpotent for each $i \geq 2$. The paper also discussed the difficulties in extending these results to rngs not containing a regular element.

By now it should be evident that sometimes the existence or lack of an identity element plays a major role and sometimes it doesn't. Gilmer has several papers concerning commutative rings that need not have an identity for the simple reason that the results don't depend on the existence of an identity. Two good examples are [26] and [44]. In fact, we quote from [44, page 97]: "On the other hand, the assumption that the rings under consideration have an identity plays no essential role, and therefore it will not be made."

We end by discussing some of Gilmer's work related to semigroup rings: [24], [33], [34], and [44]. Let R be a commutative ring and S a commutative semigroup written additively. The semigroup ring $R[X; S] = \{\Sigma_{s \in S} r_s X^s | r_s \in R\}$ has an identity if and only if both R and S do. The semigroup S is *idempotent* if $S + S = S$. It is easy to adjoin an identity to a semigroup S. Just let 0 be an element not in S and let $S^1 = S \cup \{0\}$ where $0 + 0 = 0$ and $0 + s = s + 0$ for all $s \in S$. Now if R has an identity and S is a monoid, it is not hard to prove that $R[X; S]$ is Noetherian (Artinian) if and only if

R is Noetherian (Artinian) and S is finitely generated (finite). The problem of determining when $R[S]$ is Noetherian or Artinian in the case where R or S doesn't have an identity is more complicated and cannot be solved by simply adjoining an identity. There are several reasons for this. First, while a (left) Artinian ring with identity is (left) Noetherian, this need not be the case in general; to wit, $\mathbb{Z}(p^\infty)$ with the zero product. Second, we have already remarked that in the case of $S = \mathbb{N}_0$, $R[X;\mathbb{N}_0] = R[X]$ Noetherian implies that R has an identity [16]. In [24], Gilmer determined when $R[X;S]$ is Noetherian or Artinian.

Theorem 5.3. *Let R be a commutative ring and S a commutative semigroup.*

(1) *Suppose that S is idempotent.*
  (a) *If R has (doesn't have) an identity, then $R[X;S]$ is Noetherian if and only if R is Noetherian and S is finitely generated (S is finite).*
  (b) *$R[X;S]$ is Artinian if and only if R is Artinian and S is finite.*
(2) *Suppose that S is not idempotent.*
  (a) *If R has (doesn't have) an identity, then $R[X;S]$ is Noetherian if and only if S is finitely generated (finite) and $(R,+)$ is finitely generated.*
  (b) *$R[X;S]$ is Artinian if and only if S is finite and $(R,+)$ is Artinian.*

In [44], Parker and Gilmer determined the nilradical of $R[X;S]$ for R any commutative ring and S any commutative semigroup. Here the existence of identity plays no role. Let p be a prime number. For $a,b \in S$ define $a{\sim_p}b$ if $p^k a = p^k b$ for all $k \geq 1$ and define $a \sim b$ if $na = nb$ for large n. Then $\sim_p$ and $\sim$ are congruences on S. The nilradical of $R[X;S]$ is $\bigcap_{\lambda\in\Lambda}\{P_\lambda[X;S] + I_{P_\lambda}\}$ where $\{P_\lambda\}_{\lambda\in\Lambda}$ is the set of prime ideals of R, p_λ is the characteristic of R/P_λ, and $I_{P_\lambda} = (\{rX^a - rX^b | r \in R \text{ and } a \sim_{p_\lambda} b\})$ if $p_\lambda > 0$ and $I_{P_\lambda} = (\{rX^a - rX^b | r \in R, a \sim b\})$ if $p_\lambda = 0$ [44, Theorem 3.14].

In [34], Gilmer and Teply characterized when $R[X;S]$ is von Neumann regular for R a commutative ring and S a commutative monoid (written additively). Let $\sim_p$ and $\sim$ be as in the preceding paragraph. They showed [34, Theorem 8] that if $\{M_\lambda\}_{\lambda\in\Lambda}$ is the set of maximal ideals of R, and $p_\lambda = \operatorname{char} R/M_\lambda$, then R is von Neumann regular if and only if (1) R is von Neumann regular, (2) S is free of asymptotic torsion (if $x \sim y$, then $x = y$), (3) S is p-torsion-free for each prime p in $\{p_\lambda\}_{\lambda\in\Lambda}$ ($x \sim_p y \Rightarrow x = y$), and (4) S is a torsion semigroup (for each $s \in S$, there exist distinct positive integers m and n with $ms = ns$).

Let R_1 and R_2 be commutative rings and S a not necessarily commutative semigroup. In [33] Gilmer and Spiegel investigated the question of when $R_1[X;S] \approx R_2[X;S]$ implies $R_1 \approx R_2$. Of course, well-known examples show that R_1 need not be isomorphic to R_2 even if $S = \mathbb{N}_0$ or $\mathbb{Z}$. The main result [33, Theorem 17] of this paper is that if F and K are fields and S and T are not necessarily commutative semigroups with S containing a periodic element (an element s with $s^n = s^m$ for some $n > m > 0$) with $F[X;S] \approx K[X;T]$, then $F \approx K$.

ACKNOWLEDGMENT. I would like to thank the anonymous referee. His careful reading greatly improved the article and saved me from having several embarassing mistakes appear in print.

References

1. D. D. Anderson, Globalization of some local properties in Krull domains, Proc. Amer. Math. Soc. 85 (1982), 141–145.
2. D. D. Anderson, π-Domains without identity, *Advances in Commutative Ring Theory* (eds. D. E. Dobbs, M. Fontana, and S-E. Kabbaj) Marcel Dekker, 1999, 25–30.
3. D. D. Anderson and J. S. Kintzinger, General ZPI-rings without identity, preprint.
4. J. T. Arnold and R. Gilmer, Dimension sequence of a commutative ring, Amer. J. Math. 96 (1974), 385–408.
5. J. T. Arnold and R. Gilmer, Dimension theory of commutative rings without identity, J. Pure Appl. Algebra 5 (1974), 209–231.
6. N. Bourbaki, *Algebra. I. Chapters 1-3*, Springer-Verlag, Berlin, 1989.
7. B. Brown and N. H. McCoy, Rings with unit element which contain a given ring, Duke Math. J. 13 (1946), 9–20.
8. H. S. Butts and R. Gilmer, Primary ideals and prime power ideals, Canad. J. Math. 18 (1966), 1183–1195.
9. J. L. Dorroh, Concerning adjuctions to algebras, Bull. Amer. Math. Soc. 38 (1932), 85–88.
10. J. L. Dorroh, Concerning the direct product of algebras, Ann. Math. 36 (1935), 882–885.
11. R. Gilmer, Commutative rings containing at most two prime ideals, Michigan Math. J. 10 (1963), 263–268.
12. R. Gilmer, On a classical theorem of Noether in ideal theory, Pacific J. Math. 13 (1963), 579–583.
13. R. Gilmer, Extension of results concerning rings in which semi-primary ideals are primary, Duke Math. J. 31 (1964), 73–78.
14. R. Gilmer, The cancellation law for ideals in a commutative ring, Canad. J. Math. 17 (1965), 281–287.
15. R. Gilmer, Eleven nonequivalent conditions on a commutative ring, Nagoya Math J. 26 (1966), 183–194.
16. R. Gilmer, If $R[X]$ is Noetherian, R contains an identity, Amer. Math. Monthly 74 (1967), 700.
17. R. Gilmer, A note on two criteria for Dedekind domains, L'Enseignement Mathématique 13 (1967), 253–256.
18. R. Gilmer, R-automorphisms of $R[X]$, Proc. London Math. Soc. 18 (1968), 328–336.
19. R. Gilmer, Commutative rings in which each prime ideal is principal, Math. Ann. 183 (1969), 151–158.
20. R. Gilmer, The unique primary decomposition theorem in commutative rings without identity, Duke Math. J. 36 (1969), 737–747.

21. R. Gilmer, On factorization into prime ideals, Comment. Math. Helv. 47 (1972), 70–74.
22. R. Gilmer, A note on rings with only finitely many subrings, Scripta Math. 29 (1973), 37–38.
23. R. Gilmer, Zero divisors in commutative rings, Amer. Math. Monthly 93 (1986), 382–387.
24. R. Gilmer, Chain conditions in commutative semigroup rings, J. Algebra 103 (1986), 592–599.
25. R. Gilmer, *Multiplicative Ideal Theory*, Queen's Papers in Pure and Applied Mathematics, vol. 90, Queen's University, Kingston, Ontario, 1992.
26. R. Gilmer, A. Grams, and T. Parker, Zero divisors in power series rings, J. Reine Angew. Math. 278/279 (1975), 145–164.
27. R. Gilmer and W. Heinzer, Noetherian pairs and hereditarily Noetherian rings, Arch. Math. 41 (1983), 131–138.
28. R. Gilmer and W. Heinzer, On the cardinality of subrings of a commutative ring, Canad. Math. Bull. 29 (1986), 102–108.
29. R. Gilmer, R. Lea, and M. O'Malley, Rings whose proper subrings have property P, Acta Math. Sci. (Szeged) 33 (1972), 69–75.
30. R. Gilmer and M. O'Malley, Non-Noetherian rings for which each proper subring is Noetherian, Math. Scand. 31 (1972), 118–122.
31. R. Gilmer and J. L. Mott, Multiplication rings as rings in which ideals with prime radical are primary, Trans. Amer. Math. Soc. 114 (1965), 40–52.
32. R. Gilmer and J. L. Mott, Associative rings of order p^3, Proc. Japan Acad. 49 (1973), 795–799. With mimeographed Addendum.
33. R. Gilmer and E. Spiegel, Coefficient rings in isomorphic semigroup rings, Comm. Algebra 13 (1985), 1789–1809.
34. R. Gilmer and M. L. Teply, Idempotents of commutative semigroup rings, Houston J. Math. 3 (1977), 369–385.
35. M. Griffin, Valuation rings and Prüfer rings, Canad. J. Math. 26 (1974), 412–429.
36. M. Griffin, Multiplication rings via their total quotient rings, Canad. J. Math. 26 (1974), 430–449.
37. N. Jacobson, *Basic Algebra I*, Second Edition, W. H. Freeman and Company, New York, 1985.
38. J. Lambek, *Lectures on Rings and Modules*, Blaisdell Publishing Company, Waltham, MA, Toronto, London, 1966.
39. S. Mori, Über die Produktzerlegung der Hauptideale, J. Sci. Hiroshima Univ. Ser. A. 8 (1938), 7–13.
40. S. Mori, Über die Produktzerlegung der Hauptideale, II, J. Sci. Hiroshima Univ. Ser. A. 9 (1939), 145–155.
41. S. Mori, Allgemeine Z.P.I.-ringe, J. Sci. Hiroshima Univ. Ser. A. 10 (1940), 117–136.
42. S. Mori, Über die Produktzerlegung der Hauptideale, III, J. Sci. Hiroshima Univ. Ser. A. 10 (1940), 85–94.
43. S. Mori, Über die Produktzerlegung der Hauptideale, IV, J. Sci. Hiroshima Univ. Ser. A. 11 (1941), 7–14.
44. T. Parker and R. Gilmer, Nilpotent elements of commutative semigroup rings, Michigan Math. J. 22 (1975), 97–108.
45. A. Rosenfeld, A note on two special types of rings, Scripta Math. 28 (1967), 51–54.

46. O. F. G. Schilling, Review in *Mathematical Reviews* of [43], Math Review. 2 (1941), 121.
47. M. H. Stone, The theory of representations for Boolean algebras, Trans. Amer. Math. Soc. 40 (1936), 37–111.
48. C. A. Wood, *On General Z.P.I.-Rings*, Dissertation, Florida State University, Tallahassee, Florida, December 1967.
49. C. A. Wood, On general Z.P.I.-rings, Pacific J. Math. 30 (1969), 837–846.
50. C. A. Wood, Commutative rings for which each proper homomorphic image is a multiplication ring, J. Sci. Hiroshima Univ. Ser. A.-I 33 (1969), 85–94.
51. C. A. Wood and D. E. Bertholf, Commutative rings for which each proper homomorphic image is a multiplication ring. II, Hiroshima Math. J. 1 (1971), 1–4.

Robert Gilmer's work on semigroup rings

David F. Anderson

Department of Mathematics, The University of Tennessee, Knoxville, TN 37996–1300 anderson@math.utk.edu

1 Introduction

Group rings, and more generally semigroup rings, have played an important role in modern algebra and topology. In this article, we are interested in Robert Gilmer's pioneering work on semigroup rings. This includes his two papers with T. Parker [30, 31] on divisibility properties in semigroup rings, submitted in March and May of 1973, respectively; his semigroup ring example of a two-dimensional non-Noetherian UFD [24], submitted in May of 1973; his work with J. T. Arnold on the (Krull) dimension of semigroup rings [12], submitted in September of 1975; and his book *Commutative Semigroup Rings* [25], finished in the summer of 1983 and published in 1984. Arnold and Parker (see [45]) were both PhD students of Gilmer.

In the introduction, we give a leisurely motivation for semigroup rings and establish notation. In the second section, we cover Gilmer's work with T. Parker on divisibility properties in semigroup rings. In the third section, we discuss Gilmer's construction of a two-dimensional non-Noetherian UFD, his work with J. T. Arnold on the dimension of a semigroup ring, and his book on semigroup rings. In the final section, we consider generalizations to Krull semigroup rings, graded rings, and divisibility properties in semigroups. We also discuss the (t-)class group and Picard group of monoid domains.

The polynomial ring $\mathbb{Q}[X]$ over the field $\mathbb{Q}$ of rational numbers is a PID. Varying the coefficients produces different ring-theoretic properties. For example, the polynomial ring $D[X]$ over an integral domain D is a UFD (resp., GCD-domain, Krull domain, PVMD) if and only if D is a UFD (resp., GCD-domain, Krull domain, PVMD). One often tries to prove that $D[X]$ satisfies a certain property $\mathcal{P}$ if and only if D satisfies property $\mathcal{P}$. Sometimes this holds; other times it does not. For example, because of dimension constraints, $D[X]$ is a PID or a Dedekind domain only in the trivial case when D is a field.

In $\mathbb{Q}[X]$, the exponents are nonnegative integers. Instead of just varying the coefficients, why not also vary the set S of exponents? For example, if we let $S = \mathbb{Z}$, then we get the Laurent polynomial ring $\mathbb{Q}[X, X^{-1}]$. So how should we vary the exponents? We will follow the notation of Northcott [44] which emphasizes that semigroup rings are generalized polynomial rings. Let R be a commutative ring with $1 \neq 0$. Then $R[X;S]$ will be the ring of all formal polynomials $\sum r_\alpha X^\alpha$ with each $\alpha \in S$ and $r_\alpha \in R$, almost all $r_\alpha = 0$, addition defined by $\sum r_\alpha X^\alpha + \sum s_\alpha X^\alpha = \sum (r_\alpha + s_\alpha) X^\alpha$, and with multiplication defined using the distributive law and $(r_\alpha X^\alpha)(r_\beta X^\beta) = r_\alpha r_\beta X^{\alpha+\beta}$. Let $\alpha, \beta, \gamma \in S$. Since $X^\alpha X^\beta = X^{\alpha+\beta}$, the set S must be closed under addition. The commutative and associative laws in $R[X;S]$, $X^\alpha X^\beta = X^\beta X^\alpha$ and $X^\alpha(X^\beta X^\gamma) = (X^\alpha X^\beta)X^\gamma$, yield $\alpha + \beta = \beta + \alpha$ and $\alpha + (\beta + \gamma) = (\alpha + \beta) + \gamma$ in S, respectively. Also, we want 1 to be X^0; so S should be an additive commutative monoid. We call $R[X;S]$ a *semigroup* (or *monoid*) *ring*.

We usually want $R[X;S]$ to be an integral domain; so R would have to be an integral domain. If $\alpha + \beta = \alpha + \gamma$ in S, then $X^\alpha X^\beta = X^\alpha X^\gamma$ in $R[X;S]$ would yield $X^\beta = X^\gamma$, and hence $\beta = \gamma$. Thus S must be a cancellative monoid. Also, S must be *torsionfree*, in the sense that $n\alpha = n\beta$ for n a positive integer and $\alpha, \beta \in S$ implies that $\alpha = \beta$ (since $X^\alpha - X^\beta$ divides $X^{n\alpha} - X^{n\beta}$ in $R[X;S]$). Conversely, let R be an integral domain and S an additive commutative torsionfree cancellative monoid. Then S may be totally ordered, and hence it is easily seen that the product of two nonzero elements in $R[X;S]$ is nonzero. Thus we have shown the following theorem (also see [23] or [25, Theorem 8.1]).

Theorem 1.1. *The semigroup ring $R[X;S]$ is an integral domain if and only if R is an integral domain and S is a commutative torsionfree cancellative monoid.*

Given an additive commutative cancellative monoid S, let $\langle S \rangle = \{s - t \mid s, t \in S\}$ be its quotient group. The fact that S is torsionfree is equivalent to $\langle S \rangle$ being torsionfree, i.e., S is a submonoid of a torsionfree abelian group. Let $U(S)$ be the set of invertible elements of S; then $U(S)$ is the maximal subgroup of S and $U(S) = S \cap -S$.

In the integral domain case, it is easy to determine the units of $R[X;S]$. They are precisely the monomials rX^α, where $r \in U(R)$ and $\alpha \in U(S)$. If $R[X;S]$ is not an integral domain, then it is of interest to investigate special types of elements of $R[X;S]$ such as units, zero-divisors, nilpotent elements, and idempotent elements. Gilmer has investigated these in joint papers with R. Heitmann [28], T. Parker [46], and M. Teply [33, 34] (also see [25, Chapter 2]).

Note that the polynomial ring $R[\{X_\alpha\}]$ is the semigroup ring $R[X; \oplus_\alpha \mathbb{Z}_+]$ and the Laurent polynomial ring $R[\{X_\alpha, X_\alpha^{-1}\}]$ is the group ring $R[X; \oplus_\alpha \mathbb{Z}]$. More generally, let A be a subring of $R[\{X_\alpha\}]$ generated by monomials over R. Then $A = R[X;S]$, where $S = \{(n_\alpha) \in \oplus_\alpha \mathbb{Z}_+ \mid \sum X_\alpha^{n_\alpha} \in A\}$. In particular, a subring A of $R[X]$ generated by monomials over R is $R[X;S]$ for $S = \{n \in \mathbb{Z}_+ \mid X^n \in A\}$. We view semigroup rings as a generalization of polynomial rings. The reason that things work so nicely in the polynomial ring case is that $S = \mathbb{Z}_+$ is the nicest possible semigroup. Although polynomial rings are semigroup rings, we are primarily interested in the general case when S is not $\oplus_\alpha \mathbb{Z}_+$.

Let D be an integral domain with quotient field K and S a commutative torsionfree cancellative monoid. Then $K[X;S] = D[X;S]_T$ and $D[X;\langle S\rangle] = D[X;S]_{T'}$, where $T = D \setminus \{0\}$ and $T' = \{X^\alpha \,|\, \alpha \in S\}$ are multiplicative subsets of $D[X;S]$. Also, note that $D[X;S] = K[X;S] \cap D[X;\langle S\rangle]$. We can thus sometimes reduce questions about monoid domains to group rings or to monoid domains over a field, often using a "Nagata-type" theorem (i.e., if R_T satisfies a certain property $\mathcal{P}$ for a "nice" multiplicative set T, then R also satisfies property $\mathcal{P}$). Since S is totally ordered, $D[X;S]$ becomes a graded integral domain with $\deg(dX^\alpha) = \alpha$ for $0 \neq d \in D$ and $\alpha \in S$. Thus graded ring techniques often play an important role in studying semigroup rings (cf. Section 4).

In $R[X;S]$, we can vary both the coefficients and the exponents. Thus semigroup rings provide a very handy way to construct examples since ring-theoretic properties of $R[X;S]$ are determined by properties of both R and S, and hence we have much more freedom than in polynomial rings. This is very pretty mathematics which illustrates the interplay between ring-theoretic and semigroup-theoretic techniques. It has certainly played a major role in my research activity.

For notation, R will be a commutative ring with nonzero identity and $U(R)$ its group of units, D will be an integral domain with quotient field $qf(D)$, and K will be a field. The dimension of R, $\dim(R)$, will always mean Krull dimension, and $\mathrm{char}(R)$ will be the characteristic of R. We let S denote a commutative cancellative monoid, written additively, with group of invertible elements $U(S)$ and quotient group $\langle S\rangle$. Let G denote an abelian group (usually torsionfree) and $\mathrm{rank}(G) = \dim_{\mathbb{Q}}(\mathbb{Q} \otimes_{\mathbb{Z}} G)$. For a set A, let $A^* = A \setminus \{0\}$; and for a partially ordered monoid S, let S_+ be its set of nonnegative elements. As usual, $\mathbb{Z}$ and $\mathbb{Q}$ will denote the integers and rational numbers, respectively. For more on semigroups, see [25, 37]; and for abelian groups, see [20, 21]. For any undefined notions or notation, see Gilmer's "other" book [27]; see [19] for Krull domains. In most cases, we will cite both the original reference and the corresponding result in [25].

2 Divisibility in Semigroup Rings

In this section, we discuss Gilmer's two papers with T. Parker [30, 31] on divisibility properties in semigroup rings. The main goal of [30] is to determine necessary and sufficient conditions for $D[X;S]$ to be a UFD. But first we consider GCD-domains. For most of our results, we will first consider the group ring case, and then the general monoid ring result.

Theorem 2.1. *Let D be an integral domain and G a torsionfree abelian group. Then $D[X;G]$ is a GCD-domain if and only if D is a GCD-domain.*

Proof. We may reduce to the case where G is finitely generated, and hence free. In this case, the result follows easily from the well-known polynomial ring case. For more details, see either [30, Proposition 5.1 and Theorem 6.1] or [25, Theorems 14.1 and 14.2].

In particular, $K[X;G]$ is a GCD-domain for any field K and any torsionfree abelian group G. We next consider the case for monoid domains. Let $R = K[X^2, X^3] = K[X;S]$, where K is a field and $S = \{0, 2, 3, 4, \ldots\} \subset \mathbb{Z}_+$. Then R is not a GCD-domain since X^5 and X^6 have no GCD in R. In analogy for integral domains, define a torsionfree cancellative monoid S to be a *GCD-monoid* if each pair of elements of S has a GCD(equivalently, an LCM). Then any free abelian monoid or torsionfree abelian group is a GCD-monoid. However, $S = \{0, 2, 3, 4, \ldots\}$ is not a GCD-monoid since 5 and 6 have no GCD in S. Our next result is somewhat typical in that $D[X;S]$ satisfies a certain ring-theoretic property $\mathcal{P}$ if and only if D satisfies property $\mathcal{P}$ and S satisfies the additive monoid analog of property $\mathcal{P}$ (also see Section 4).

Theorem 2.2. *Let D be an integral domain and S a torsionfree cancellative monoid. Then $D[X;S]$ is a GCD-domain if and only if D is a GCD-domain and S is a GCD-monoid.*

Proof. If $D[X;S]$ is a GCD-domain, then D must be a GCD-domain and S a GCD-monoid. The converse follows from Theorem 2.1 using a "Nagata-type" theorem that $D[X;S]_T = D[X;\langle S\rangle]$, where $T = \{X^\alpha \mid \alpha \in S\}$, is a GCD-domain implies that $D[X;S]$ is a GCD-domain. See either [30, Theorems 6.1 and 6.4] or [25, Theorems 14.1 and 14.5] for more details.

An integral domain D is a UFD if and only if D is a GCD-domain and D satisfies the ascending chain condition on principal ideals (ACCP). Note that $R = K[X;\mathbb{Q}]$ is a GCD-domain for any field K by Theorem 2.1, but R is not a UFD since ACCP fails. For example, we have the strictly ascending chain of principal ideals $(1 - X) \subset (1 - X^{1/2}) \subset (1 - X^{1/4}) \subset \cdots$ in R, which corresponds to the strictly ascending chain of cyclic subgroups $\langle 1\rangle \subset \langle 1/2\rangle \subset \langle 1/4\rangle \subset \cdots$ in $\mathbb{Q}$. Similarly, $K[X;\mathbb{Q}_+]$ is a GCD-domain, but not a UFD.

Let G be a torsionfree abelian group. Recall that every nonzero element of G has *type* $(0, 0, 0, \ldots)$ means that for each $0 \neq g \in G$, there is a largest positive integer n_g such that the equation $n_g x = g$ is solvable in G (see [21, Section 85]). This is equivalent to each rank-one subgroup of G is cyclic (free), or more suggestively, G satisfies ACC on cyclic subgroups, or ACC on cyclic submonoids [25, Theorem 14.10]. This property plays an important role in properties related to chain conditions since (as above) a strictly ascending chain of cyclic subgroups $\langle g_1\rangle \subset \langle g_2\rangle \subset \cdots$ in G gives rise to a strictly ascending chain of principal ideals $(1 - X^{g_1}) \subset (1 - X^{g_2}) \subset \cdots$ in $D[X;G]$.

Note that any subgroup of a torsionfree abelian group which satisfies ACC on cyclic subgroups also satisfies ACC on cyclic subgroups. In particular, if S is a torsionfree cancellative monoid such that $\langle S\rangle$ satisfies ACC on cyclic subgroups, then so does $U(S)$. However, the converse is false since $S = \mathbb{Q}_+$ has $U(S) = 0$ which certainly satisfies ACC on cyclic subgroups, but $\langle S \rangle = \mathbb{Q}$ does not. Although the results in [30, 31] are stated using the type $(0, 0, 0, \ldots)$ terminology, we use the more suggestive ACC on cyclic subgroups or cyclic submonoids terminology as in [25].

Theorem 2.3. *Let D be an integral domain and G a torsionfree abelian group. Then $D[X; G]$ is a UFD if and only if D is a UFD and G satisfies ACC on cyclic subgroups.*

Proof. If $D[X; G]$ is a UFD, then D must be a UFD, and G satisfies ACC on cyclic subgroups by the above remarks. The converse is much more difficult and finally reduces to the case where D is an algebraically closed field. For more details, see either [30, Theorem 7.13] or [25, Theorem 14.16].

As in the GCD-domain case, we may define the analog of unique factorization for a torsionfree cancellative monoid. We call such a monoid a *factorial monoid.* Note that a factorial monoid has the form $G \oplus F_+$, where G is any torsionfree abelian group and $F = \oplus_\alpha \mathbb{Z}$ is a free abelian group with the usual product order. We are now ready for the main result of both [30] and this section. The Krull domain analog of Theorem 2.4 will be discussed in Section 4.

Theorem 2.4. *Let D be an integral domain and S a torsionfree cancellative monoid. Then $D[X; S]$ is a UFD if and only if D is a UFD, S is a factorial monoid, and $U(S)$ satisfies ACC on cyclic subgroups.*

Proof. Again, the "$\Rightarrow$" implication is fairly clear. The converse follows from Theorem 2.3 via a "Nagata-type" theorem as in the proof of Theorem 2.2. See either [30, Theorem 7.17] or [25, Theorem 14.16] for more details.

Theorem 2.4 just says that a factorial monoid domain looks like the ring $D[X; G][\{Y_\alpha\}]$, where D is a UFD, G is a torsionfree abelian group which satisfies ACC on cyclic subgroups, and $\{Y_\alpha\}$ is a family of indeterminates. Note that if S is a factorial monoid, then $U(S)$ satisfies ACC on cyclic subgroups if and only if $\langle S \rangle$ satisfies ACC on cyclic subgroups, if and only if S satisfies ACC on cyclic submonoids.

As a corollary of Theorem 2.3, the group ring $D[X; G]$ satisfies ACCP if and only if D satisfies ACCP and G satisfies ACC on cyclic subgroups ([30, Corollary 7.14] or [25, Theorem 14.17]). What about $D[X; S]$? Several partial results are given in [30, pages 77 and 82]. For example, $D[X; S]$ satisfies ACCP if D satisfies ACCP, S satisfies ACC on cyclic submonoids, and $\langle S \rangle$ satisfies ACC on cyclic subgroups. However, the converse fails. Let $S = \{q \in \mathbb{Q} \mid q \geq 1\} \cup \{0\}$, and let K be a field. Then one easily checks that $K[X; S]$ satisfies ACCP, S satisfies ACC on cyclic submonoids, but $\langle S \rangle = \mathbb{Q}$ does not satisfy ACC on cyclic subgroups. This gives an easy example to show that ACCP is not preserved by localization since $K[X; \mathbb{Q}] = K[X; S]_T$, where

$T = \{ X^\alpha \mid \alpha \in S \}$. Other chain conditions in semigroup rings are investigated in [25, 26].

Article [30] concludes with a characterization of when $D[X;S]$ is a PID (or Dedekind domain or Euclidean domain). This happens only in the trivial case when D is a field and S is isomorphic to either $\mathbb{Z}_+$ or $\mathbb{Z}$, i.e., $D[X;S]$ is either $K[X]$ or $K[X, X^{-1}]$ for some field K.

Theorem 2.5. *Let D be an integral domain and S a nonzero torsionfree cancellative monoid. Then the following statements are equivalent.*

(1) *$D[X;S]$ is a Euclidean domain.*
(2) *$D[X;S]$ is a PID.*
(3) *$D[X;S]$ is a Dedekind domain.*
(4) *D is a field and S is isomorphic to either $\mathbb{Z}_+$ or $\mathbb{Z}$.*

Proof. This follows since $D[X;S]$ must be integrally closed and one-dimensional. For more details, see either [30, Theorem 8.4] or [25, Theorem 13.8].

There were two immediate sequels to [30]. First, in [31], Gilmer and Parker considered several additional divisibility properties and also allowed the coefficient rings to have zero-divisors. Secondly, in [24], Gilmer used results from [30] to construct a two-dimensional non-Noetherian UFD; this example will be discussed in the next section.

We first state the main results from [31] in the integral domain setting. These results are similar to that for PIDs in Theorem 2.5, but there is a little more freedom on the monoid S since these rings need not be Noetherian. Recall that we have observed that $K[X;\mathbb{Q}]$ and $K[X;\mathbb{Q}_+]$ are both GCD-domains for any field K, but are not UFDs. They are also Bezout domains since they are ascending unions of PIDs (for example, $K[X;\mathbb{Q}] = \bigcup_{n=1}^{\infty} K[X;(1/n!)\mathbb{Z}]$). It will be convenient to call a monoid S a *Prüfer submonoid of* $\mathbb{Q}$ if $S = G \cap \mathbb{Q}_+$, where G is a subgroup of $\mathbb{Q}$ containing $\mathbb{Z}$. This just means that S is the union of an ascending sequence of cyclic submonoids [25, Theorem 13.5].

Theorem 2.6. *Let D be an integral domain and S a nonzero torsionfree cancellative monoid. Then the following statements are equivalent.*

(1) *$D[X;S]$ is a Bezout domain.*
(2) *$D[X;S]$ is a Prüfer domain.*
(3) *D is a field and S is isomorphic to either a subgroup of $\mathbb{Q}$ containing $\mathbb{Z}$ or a Prüfer submonoid of $\mathbb{Q}$.*

Proof. See either [31, Theorem] or [25, Theorem 13.6] for details.

We next allow R to have zero-divisors, but S will still be a torsionfree cancellative monoid. In this case, the only change is that "field" gets replaced by "von Neumann regular ring". We say that a commutative ring R is a *Prüfer ring* if each finitely generated regular ideal of R is invertible and that R is a *Bezout ring* if each finitely generated ideal of R is principal.

Theorem 2.7. *Let R be a commutative ring and S a nonzero torsionfree cancellative monoid. Then the following statements are equivalent.*

(1) *$R[X;S]$ is a Bezout ring.*
(2) *$R[X;S]$ is a Prüfer ring.*
(3) *R is von Neumann regular and S is isomorphic to either a subroup of $\mathbb{Q}$ containing $\mathbb{Z}$ or a Prüfer submonoid of $\mathbb{Q}$.*

Proof. For details, see either [31, Corollary 3.1] or [25, Theorem 18.9].

In [31], Gilmer and Parker also determined when a monoid ring is either an almost Dedekind domain or a general ZPI-ring. Recall that an integral domain D is an *almost Dedekind domain* if D_M is a Noetherian valuation domain for each maximal ideal M of D, and that a commutative ring R is a *general ZPI-ring* if each ideal of R is a finite product of prime ideals, equivalently, if R is a finite direct sum of Dedekind domains and special principal ideal rings.

Theorem 2.8. *Let D be an integral domain and S a nonzero torsionfree cancellative monoid. Then $D[X;S]$ is an almost Dedekind domain if and only if D is a field and S is isomorphic to either $\mathbb{Z}_+$ or a subgroup of $\mathbb{Q}$ containing $\mathbb{Z}$ such that if char$(D) = q$ is nonzero, then $1/q^k \notin S$ for some positive integer k.*

Proof. See either [31, Theorem], or [25, Corollary 20.15] for the group ring case.

For example, for any field K, the monoid domain $K[X;\mathbb{Q}_+]$ is an almost Dedekind domain which is not a Dedekind domain. They also gave more technical conditions for almost Dedekind semigroup rings; the interested reader may consult [31, Theorem 4.2].

Theorem 2.9. *Let R be a commutative ring and S a nonzero torsionfree cancellative monoid. Then $R[X;S]$ is a general ZPI-ring if and only if R is a finite direct sum of fields and S is isomorphic to either $\mathbb{Z}$ or $\mathbb{Z}_+$. In particular, $R[X;S]$ is a general ZPI-ring if and only if $R[X;S]$ is a principal ideal ring.*

Proof. If $R[X;S]$ is a general ZPI-ring, then it is a Noetherian Prüfer ring. The result then follows from Theorems 2.5 and 2.7 since R is a finite direct sum of fields. The converse is clear. See either [31, Theorem 5.1 and Corollary 5.1] or [25, Theorem 18.10] for more details.

Several related conditions are also investigated in [25, Sections 18 and 19]. For example, in Theorem 2.7, we may add the equivalence that $R[X;S]$ is arithmetical (recall that a ring T is *arithmetical* if $A \cap (B + C) = (A \cap B) + (A \cap C)$ for all ideals A, B, C of T) [25, Theorem 18.9]. In [25, Section 19], arithmetical monoid rings are studied in the case where the cancellative monoid S is not torsionfree. The treatment of these topics is considerably reorganized in [25] from that in [31] (see the comments in [25, page 251]).

3 Non-Noetherian UFDs, Dimension Theory, and the Book

In this section, we first discuss Gilmer's example of a two-dimensional non-Noetherian UFD. It is a direct application of results in [30] and is one of my favorite examples in ring theory. Later, we will discuss Gilmer's work with J. T. Arnold on the dimension theory of semigroup rings and his book on semigroup rings.

The standard example of a non-Noetherian UFD is $R = K[\{X_n\}_{n=1}^{\infty}]$, the polynomial ring over a field K in infinitely many indeterminates. Unfortunately, R has infinite Krull dimension. So what's an example of a finite-dimensional non-Noetherian UFD? Examples of 3-dimensional non-Noetherian quasilocal UFDs in characteristic 0 and 2 were given by J. David [17, 18] in 1972–1973. Note that a one-dimensional UFD is a PID, and hence Noetherian. So what about the two-dimensional case?

The idea is to construct a group ring $D[X;G]$ which is a two-dimensional UFD, but not Noetherian. Recall that $D[X;G]$ is Noetherian if and only if D is Noetherian and G is finitely generated (i.e., free of finite rank) [25, Theorem 7.7]. So first we need to know a little about the Krull dimension of a group ring. Let G be a torsionfree abelian group with $\text{rank}(G) = \gamma$. Then there is free abelian subgroup F of G with $\text{rank}(F) = \gamma$ and G/F is a torsion group. Hence $R[X;F] \subseteq R[X;G]$ is an integral extension, and thus $\dim(R[X;G]) = \dim(R[X;F])$. Let $\{X_\alpha\}_{\alpha \in A}$ be a family of indeterminates with $|A| = \gamma$. Then $\dim(R[\{X_\alpha, X_\alpha^{-1}\}_{\alpha \in A}]) = \dim(R[\{X_\alpha\}_{\alpha \in A}])$, and hence $\dim(R[X;G]) = \dim(R[X;F]) = \dim(R[\{X_\alpha\}_{\alpha \in A}])$. As a special case, if D is either a Prüfer domain or a Noetherian integral domain and G is a torsionfree abelian group of finite rank n, then $\dim(D[X;G]) = \dim(D)+n$. In particular, if K is a field, then $\dim(K[X;G]) = \text{rank}(G)$.

By the above paragraph, we need to find a torsionfree abelian group G of rank two which is not finitely generated, but satisfies ACC on cyclic subgroups. Which, if any, abelian groups G satisfy these conditions? Fortunately, there is such a rank-two abelian group G. So in this case, $R = K[X;G]$ is a two-dimensional non-Noetherian UFD for any field K. Let L be a rank-two torsionfree abelian group which is not free (and hence not finitely generated), but every rank-one subgroup of L is free (cyclic), and hence L satisfies ACC on cycic subgroups. Such an abelian group exists (see [47] or [21, Section 88]). Let $L_n = L \oplus \mathbb{Z}^n$ for each integer $n \geq 0$. Then L_n is not finitely generated, $\text{rank}(L_n) = n + 2$, and L_n satisfies ACC on cyclic subgroups.

Theorem 3.1. *Let K be a field. Then $K[X;L_n]$ is a non-Noetherian UFD of Krull dimension $n+2$.*

Proof. By Theorem 2.3, $R = K[X;L_n]$ is a UFD. By our earlier remarks, R is not Noetherian since L_n is not finitely generated and $\dim(R) = \text{rank}(L_n) = n+2$. See [31, Theorem 4] for more details.

If $\text{char}(K) = p > 0$, then the group ring $R = K[X; L_n]$ in Theorem 3.1 may be localized at a suitable maximal ideal M so that the quasilocal domain R_M is an n-dimensional UFD, but not Noetherian.

Theorem 3.2. *Let p be prime and $n \geq 2$ an integer. Then there is a non-Noetherian quasilocal UFD of Krull dimension n and characteristic p.*

Proof. See [31, Theorem 4] for more details.

Theorem 3.2 leaves open the characteristic 0 case for quasilocal domains. In [14, Theorem D and Example], J. W. Brewer, D. L. Costa, and E. L. Lady showed that for each integer $n \geq 2$, there is a non-Noetherian quasilocal UFD with characteristic 0 and Krull dimension n. (Brewer was also a PhD student of Gilmer.) Their example is based on a localization of the group ring $\mathbb{Z}[G]$, where $G = L$ when $n = 2$ and $G = \mathbb{Z}[1/p] \oplus \mathbb{Z}[1/p] \oplus \mathbb{Z}[1/p]$ for p a prime when $n \geq 3$. In fact, they showed that the technique used in Theorem 3.2 of localizing a group ring over a field will not work in characteristic 0 [14, Theorem A]. Several other examples of 3-dimensional non-Noetherian quasilocal UFDs have been given in the literature (see [11]).

We next briefly discuss Gilmer's work with J. T. Arnold [12] on computing the Krull dimension of $R[X; S]$. This generalizes earlier work on Krull dimension mentioned in this section and does not assume that the cancellative monoid S is torsionfree. Theorem 3.3 reduces the calculation of the Krull dimension of a semigroup ring to that of a group ring, which in turn reduces it to the calculation of the Krull dimension of a polynomial ring since $\dim(R[X; G]) = \dim(R[\{X_\alpha\}_{\alpha \in A}])$, where $|A| = \text{rank}(G)$. In [12], they also extended several results about chains of prime ideals in polynomial rings to semigroup rings $R[X; S]$, where S is a finitely generated torsionfree cancellative monoid.

Theorem 3.3. *Let R be a commutative ring and S a cancellative monoid with quotient group G. Then* $\dim(R[X; S]) = \dim(R[X; G])$.

Proof. This is proved in several reductions; first to the case where R is a finite-dimensional integral domain and S is finitely generated and torsionfree, and then to showing that $\dim(R[X_1, X_1^{-1}, \ldots, X_n, X_n^{-1}]) = \dim(R[X_1, \ldots, X_n, h_1, \ldots, h_j])$, where $h_1, \ldots, h_j$ are pure monomials in $X_1, \ldots, X_n$. For more details, see either [12] or [25, Theorem 21.4].

We conclude this section with a short discussion of Gilmer's book *Commutative Semigroup Rings* [25]. It was written when most of the topics were available only in their original research articles, and it is still the only other reference for many of these topics. Like *Multiplicative Ideal Theory* [27], it is still *the* reference in the field. It gives a unified, self-contained treatment of semigroups and semigroup rings. Many proofs are modified or simplified, sometimes to correct gaps of previous proofs in the literature.

The book consists of 25 sections grouped in 5 chapters. The chapters are (I) Commutative semigroups, (II) Semigroup rings and their distinguished elements, (III) Ring-theoretic properties of monoid domains, (IV) Ring-theoretic properties of monoid rings, and (V) Dimension theory and the isomorphism problems. Except for the second part of Chapter (V), the chapter titles are fairly self-explanatory and much of their content is discussed in this article. The "isomorphism problems" concerns the question of when does $R_1[X;S] \cong R_2[X;S]$ (resp., $R[X;S] \cong R[X;T]$) imply that $R_1 \cong R_2$ (resp., $S \cong T$) (also see [32]).

4 Generalizations

In this final section, we discuss three types of extensions or generalizations of Gilmer's work on semigroup rings. The first is to other divisibility properties for monoid domains, with emphasis on when a monoid domain is a Krull domain. The second is to graded integral domains and their divisibility properties, and the third is to divisibility in monoids.

After determining when a monoid domain is a UFD or a Dedekind domain, the next natural question is: when is $D[X;S]$ a Krull domain, and if so, how do we calculate its divisor class group $Cl(D[X;S])$? Again, we first state the group ring case, which is due to R. Matsuda [40].

Theorem 4.1. *Let D be an integral domain and G a torsionfree abelian group. Then $D[X;G]$ is a Krull domain if and only if D is a Krull domain and G satisfies ACC on cyclic subgroups. Moreover, if $D[X;G]$ is a Krull domain, then $Cl(D[X;G]) = Cl(D)$.*

Proof. Suppose that $R = D[X;G]$ is a Krull domain with $qf(D) = K$. Then D is a Krull domain since $D = R \cap K$, and G satisfies ACC on cyclic subgroups since R satisfies ACCP. Conversely, if G satisfies ACC on cyclic subgroups, then $K[X;G]$ is a UFD by Theorem 2.3. The f-adic discrete valuations on $qf(R)$ induced by the irreducible elements f of the UFD $K[X;G]$ together with the discrete valuations on $qf(R)$ induced by the height-one prime ideals of the Krull domain D show that R is a Krull domain. The divisor class group result follows from Nagata's Theorem [19, Corollary 7.2]. For more details, see either [40, Propositions 3.3 and 5.3] or [25, Theorems 15.1, 15.4, and 16.2].

Theorem 4.1, together with ideas from the previous section, can be used to construct a 3-dimensional non-Noetherian Krull domain with any given divisor class group. Let G be an abelian group and D a Dedekind domain with class group G (such a D exists by Claborn's Theorem [19, Theorem 14.10]). Let $L = L_0$ be as in Theorem 3.1. Then $R = D[X;L]$ is a Krull domain with $Cl(R) = Cl(D) = G$ by Theorem 4.1, and R is non-Noetherian with $\dim(R) = 3$ for reasons discussed in the previous section.

An integral domain is a Krull domain if and only if it is completely integrally closed and satisfies ACC on integral v-ideals. We thus define a torsionfree cancellative monoid to be a *Krull monoid* if it is completely integrally closed and satisfies ACC on integral v-ideals (for other equivalent conditions, see [15, 25, 37]). Our next theorem is due to L. G. Chouinard [15].

Theorem 4.2. *Let D be an integral domain with quotient field K and S a torsionfree cancellative monoid. Then $D[X;S]$ is a Krull domain if and only if D is a Krull domain, S is a Krull monoid, and $U(S)$ satisfies ACC on cyclic subgroups. Moreover, $Cl(D[X;S]) = Cl(D) \oplus Cl(K[X;S])$ and $Cl(K[X;S])$ is independent of the field K.*

Proof. See either [15, Theorem 1] or [25, Theorem 15.6] for details. The "moreover" statement is from [3, Proposition 7.3].

Krull monoids have the form $G \oplus T$, where G is any torsionfree abelian group and T is a submonoid of a free abelian group $F = \oplus_\alpha \mathbb{Z}$ with the usual product order such that $T = \langle T \rangle \cap F_+$. Thus a Krull monoid domain is just a subring of a polynomial ring over a Krull group ring generated by monomials. For a Krull monoid S, it is easy to see that $U(S)$ satisfies ACC on cyclic subgroups if and only if $\langle S \rangle$ satisfies ACC on cyclic subgroups.

Let K be a field and $S \subseteq F_+$ a Krull monoid with $S = \langle S \rangle \cap F_+$ (here $F = \oplus_\alpha \mathbb{Z}$ is a free abelian group with the usual product order and each pr_α is the natural projection map) such that the $pr_\alpha|_{\langle S \rangle}$'s are distinct essential valuations of S. Then $Cl(K[X;S]) \cong F/\langle S \rangle$ (see [15, Theorem 2], [25, Section 16], and [4]). This fact may be used to show that any abelian group G is the divisor class group of a quasilocal Krull domain. Let K be a field and G an abelian group. Then one can construct a Krull domain $R = K[X;S]$ with $Cl(R) = G$ [15, Corollary 2]. In fact, one may then localize R to obtain a quasilocal Krull domain A with $Cl(A) = G$. See [4] for some specific calculations.

An integral domain D is a *Prüfer v-multiplication domain* (PVMD) if the monoid of finite-type v-ideals of D forms a group under v-multiplication. Thus a Krull domain or a Prüfer domain is a PVMD. Analogously, we define a torsionfree cancellative monoid S to be a *PVMD monoid* if the monoid of finite type v-ideals of S forms a group under v-multiplication. Theorem 4.2 then generalizes to PVMDs.

Theorem 4.3. *Let D be an integral domain, S a torsionfree cancellative monoid, and G a torsionfree abelian group. Then $D[X;S]$ is a PVMD if and only if D is a PVMD and S is a PVMD monoid. In particular, $D[X;G]$ is a PVMD if and only if D is a PVMD.*

Proof. The first "$\Rightarrow$" implication, the "in particular" statement, and several other results about PVMD semigroup rings were proved by S. Malik (see [38, Chapter 14] and [39]). The converse of the first implication was also conjectured by Malik, and it was proved in [2, Proposition 6.5].

By Theorems 2.6, 4.1, and 4.3, the group rings $\mathbb{Z}[X;\mathbb{Q}]$, $\mathbb{Q}[X;\mathbb{Z}\oplus\mathbb{Q}] = \mathbb{Q}[X;\mathbb{Q}][Y]$, and $\mathbb{Q}[X;\mathbb{Q}\oplus\mathbb{Q}]$ are all two-dimensional PVMDs which are neither Krull domains nor Prüfer domains.

For any integral domain D, let $T(D)$ be the group of t-invertible fractional t-ideals of D under t-multiplication, and let $Prin(D)$ be its subgroup of nonzero principal fractional ideals. Then the *(t-)class group* of D is the abelian group $Cl(D) = T(D)/Prin(D)$. Let $Inv(D) \subseteq T(D)$ be the subgroup of invertible ideals of D. Then $Pic(D) = Inv(D)/Prin(D)$, the *Picard group* or *ideal class group* of D, is a subgroup of $Cl(D)$. If D is either a Prüfer domain or a one-dimensional integral domain, then $Cl(D) = Pic(D)$, and $Cl(D)$ is the usual divisor class group if D is a Krull domain. The class group is important because ring-theoretic properties of D are often reflected in group-theoretic properties of $Cl(D)$. For example, if D is a PVMD, then $Cl(D) = 0$ (resp., is torsion) if and only if D is a GCD-domain (resp., AGCD-domain). Recall that D is an *almost GCD-domain* (AGCD-domain) if for any $0 \neq a, b \in D$, there is a positive integer $n = n(a,b)$ such that $a^nD \cap b^nD$ is principal. For more on the class group, see the survey article [7].

We next discuss the class group of a monoid domain. As a first step, S. Gabelli [22] showed that $Cl(D) = Cl(D[X])$ if and only if D is integrally closed. In analogy with Theorem 4.2 for Krull domains, our next result, due to S. El Baghdadi, L. Izelgue, and S. Kabbaj [13], gives a very satisfactory answer for the class group of an integrally closed monoid domain.

Theorem 4.4. *Let D be an integral domain with quotient field K and S a torsionfree cancellative monoid. If $D[X;S]$ is integrally closed, then $Cl(D[X;S]) = Cl(D) \oplus Cl(K[X;S])$ and $Cl(K[X;S])$ is independent of the field K.*

Proof. For details, see [13, Corollaries 2.8 and 2.10].

For the non-integrally closed monoid domain case, we include a result from [9]. Recall that an additive submonoid S of $\mathbb{Z}_+$ is called a *numerical semigroup* if $\mathbb{Z}_+ \setminus S$ is finite.

Theorem 4.5. *Let D be an integral domain with quotient field K and S a numerical semigroup. Then $Cl(D[X;S]) = Cl(D[X]) \oplus Pic(K[X;S])$. In particular, if D is integrally closed, then $Cl(D[X;S]) = Cl(D) \oplus Pic(K[X;S])$.*

Proof. Let $N = \{\, X^\alpha \,|\, \alpha \in S \,\}$ and $T = D^*$. Then the natural homomorphism $Cl(D[X;S]) \longrightarrow Cl(D[X;S]_N) \oplus Cl(D[X;S]_T) = Cl(D[X,X^{-1}]) \oplus Cl(K[X;S])$, given by $[I] \mapsto ([I_N],[I_T])$, is an isomorphism. Also, $Cl(D[X]) = Cl(D[X,X^{-1}])$ for any integral domain D, and $Cl(K[X;S]) = Pic(K[X;S])$ since $K[X;S]$ is one-dimensional. For more details, see [9, Theorem 5]. The "in particular" statement follows from the result of Gabelli [22] mentioned above.

Note that $Pic(K[X;S])$ in Theorem 4.5 may be computed (for example, by using the Mayer-Vietoris exact sequence for (U, Pic)). As a special case, we have that $Cl(D[X^2,X^3]) = Cl(D[X]) \oplus K$.

We next consider the Picard group of a monoid domain. In [29, Theorem 1.6], Gilmer and R. Heitmann showed that $Pic(D) = Pic(D[X])$ if and only if D is seminormal (recall that an integral domain D is *seminormal* if $x^2, x^3 \in D$ for $x \in qf(D)$ implies $x \in D$). Analogously, define a torsionfree cancellative monoid S to be *seminormal* if $2x, 3x \in S$ for $x \in \langle S \rangle$ implies $x \in S$. Then $D[X; S]$ is seminormal if and only if D and S are seminormal [2, Corollary 6.2].

Theorem 4.6. *Let D be an integral domain and S a nonzero torsionfree cancellative monoid. Then $Pic(D) = Pic(D[X; S])$ if and only if $D[X; S]$ is seminormal and $Pic(D) = Pic(D[X; U(S)])$. Moreover, if $U(S) \neq 0$, then $Pic(D) = Pic(D[X; S])$ if and only if $Pic(D) = Pic(D[X; \mathbb{Z}])$.*

Proof. This result is from [6, Corollary]. Also see [5].

Chouinard also determined the projective modules over certain Krull monoid domains [16]. This was later generalized by J. Gubeladze [36] to monoid domains of the form $D[X; S]$, where D is a PID and S is seminormal. R. G. Swan [48] has given a detailed exposition of Gubeladze's work; our next result is [48, Theorem 1.1].

Theorem 4.7. *Let D be a Dedekind domain and S a torsionfree cancellative monoid. Then all finitely generated projective $D[X; S]$-modules are extended from D if and only if S is seminormal.*

Closedness properties usually behave fairly well in that $D[X; S]$ often satisfies a property $\mathcal{P}$ if and only D satisfies $\mathcal{P}$ and S satisfies the additive semigroup analog of $\mathcal{P}$. For example, this holds for integrally closed, completely integrally closed, root closed, and seminormal. This is because these properties are "homogeneous" in the sense that a graded integral domain R satisfies them if and only if R satisfies them for homogeneous elements (see below). However, as we have seen, things do not behave as well for chain conditions. Many other ring-theoretic properties have been studied for semigroup rings (see [8, 25]). Much of this work has been done by R. Matsuda; we cite only [40, 41]. The interested reader should check Math Reviews (or MathSciNet) for more of his work.

We have already given several instances in this article where semigroup rings have been used to construct examples (also see the examples in [8, Section 6]). Another of my favorite examples is (the localization of) a monoid domain used by A. Grams [35] to construct an atomic integral domain which does not satisfy ACCP (recall that an integral domain is *atomic* if each nonzero nonunit is a product of irreducible elements). (Grams was also a PhD student of Gilmer.)

Much of the work on monoid domains generalizes to the context of graded integral domains. By a *(S-)graded integral domain*, we mean an integral domain R graded by a torsionfree cancellative monoid S. That is, $R = \oplus_{\alpha \in S} R_\alpha$, where each R_α is an additive abelian subgroup of R and $R_\alpha R_\beta \subseteq R_{\alpha+\beta}$ for all $\alpha, \beta \in S$. Each nonzero $x \in R_\alpha$ is homogeneous of $\deg(x) = \alpha$. Let H be the set of nonzero homogeneous elements of R. Then H is a submonoid of R^* under multiplication. We call R_H the *homogeneous quotient field* of R. It is $\langle S \rangle$-graded in the natural way by $\deg(r/s) = \deg(r) - \deg(s)$ for $r, s \in H$, and every nonzero homogeneous element of R_H is a unit. Also, R_H is a completely integrally closed GCD-domain [3, Propositions 3.2 and 3.3] and $R_H \cong (R_H)_0^\gamma[X; \langle S \rangle]$, a twisted group ring over the field $(R_H)_0$. Moreover, R_H is a Laurent polynomial ring over the field $(R_H)_0$ (and hence is a PID) when R is $\mathbb{Z}_+$- or $\mathbb{Z}$-graded.

The monoid domain $R = D[X; S]$ is S-graded with $\deg(dX^\alpha) = \alpha$ for each $d \in D^*$ and $\alpha \in S$. In this case, $H = \{\, dX^\alpha \mid d \in D^* \text{ and } \alpha \in S\}$, and thus R has homogeneous quotient field $R_H = K[X; G]$, where $K = qf(D)$ and $G = \langle S \rangle$. However, $R = D[X; S]$ is a very special graded ring in that it is an inert extension of $R_0 = D$ (an extension $A \subseteq B$ of integral domains is *inert* if whenever $xy \in A$ for $x, y \in B$, then $x = ru$ and $y = su^{-1}$ for some $r, s \in A$ and $u \in U(B)$) and $R_\alpha = DX^\alpha \cong D$ for each $\alpha \in S$. Other graded domain constructions that have received considerable attention include the $A + XB[X]$ construction (see [8]).

Given a divisibility property, we can define the corresponding homogeneous divisibility property in the obvious manner. For example, we say that R is a *graded GCD-domain* if any two nonzero homogeneous elements of R have a (necessarily homogeneous) GCD, R is a *graded UFD* if every nonzero nonunit homogeneous element of R is a product of (necessarily homogeneous) prime elements, R is a *graded Krull domain* if R is completely integrally closed with respect to homogeneous elements of R_H and R satisfies ACC on homogeneous integral v-ideals, and R is a *graded PVMD* if the monoid of homogeneous finite-type v-ideals of R forms a group under v-multiplication. We can ask if R satisfies a given divisibilty property if and only if either R^* or H satisfies the corresponding homogeneous divisibility property. Our next result gives some examples, for others, see [1, 2, 8].

Theorem 4.8. *Let R be a graded integral domain and H its monoid of nonzero homogeneous elements. Then*

(a) *R is a GCD-domain if and only if R is a graded GCD-domain (i.e., H is a GCD-monoid).*
(b) *R is a UFD if and only if R is a graded UFD (i.e., H is a factorial monoid) and R_H is a UFD.*
(c) *R is a Krull domain if and only if R is a graded Krull domain and R_H is a Krull domain.*
(d) *R is a PVMD if and only if R is a graded PVMD.*

Proof. See [1, Theorems 3.4, 4.4, and 5.8] for parts (a), (b), and (c), respectively. Part (d) is proved in [2, Theorem 6.4].

Since R_H is always a GCD-domain, parts (a) and (d) have the same form as (b) and (c) in Theorem 4.8. Also, if R is $\mathbb{Z}_+$- or $\mathbb{Z}$-graded, then R is a UFD (resp., Krull domain) if and only if R is graded UFD (resp., Krull domain) since R_H is a PID. However, $K[X;\mathbb{Q}]$ is a graded UFD for any field K, but not a UFD. Theorems 2.2, 2.4, 4.2, and 4.3 follow easily from Theorem 4.8, see [1, Propositions 3.5, 4.7, and 5.11] and [2, Proposition 6.5] for details.

One can also consider conditions on the nonzero homogeneous ideals of a graded integral domain R. This leads to the study of the homogeneous class group $HCl(R)$ and the homogeneous Picard group $HPic(R)$ (see [2, 3, 7, 8, 10, 13]).

Recently, there has been considerable activity on generalizing ring-theoretic properties to the context of semigroups or monoids. This comes about for (at least) two reasons. First, we have seen that $R[X;S]$ satisfies a certain ring-theoretic property often implies that S satisfies the corresponding additive monoid property. For example, we have seen that if $D[X;S]$ is a GCD-domain (resp., UFD, Krull domain, PVMD, seminormal domain), then S is a GCD (resp., factorial, Krull, PVMD, seminormal) monoid. This holds for many more properties. In fact, R. Matsuda [42, 43] has recast much of *Multiplicative Ideal Theory* [27] in the context of torsionfree cancellative monoids. Again, the interested reader should consult Math Reviews (or MathSciNet) for other work of Matsuda.

Secondly, divisibility properties of an integral domain D are often equivalent to the corresponding divisibility properties in the multiplicative monoid D^*. For example, D is a UFD (resp., GCD-domain, Krull domain) if and only if D^* is a factorial (resp., GCD, Krull) monoid. Thus, nowadays much of the research on non-unique factorization in integral domains is done in the more general setting of commutative cancellative monoids. For recent such work and additional references, see [8] and F. Halter-Koch's book *Ideal Systems* [37].

References

1. D. D. Anderson and D. F. Anderson, Divisibility properties of graded domains, Canad. Math. J. 34 (1982), 196–215.
2. D. D. Anderson and D. F. Anderson, Divisiorial and invertible ideals in a graded integral domain, J. Algebra 76 (1982), 549–569.
3. D. F. Anderson, Graded Krull domains, Comm. Algebra 7 (1979), 1–19.
4. D. F. Anderson, The divisor class group of a semigroup ring, Comm. Algebra 17 (1980), 1179–1185.
5. D. F. Anderson, The Picard group of a monoid domain, J. Algebra 115 (1988), 342–351.
6. D. F. Anderson, The Picard group of a monoid domain, II, Arch. Math. 55 (1990), 143–145.

7. D. F. Anderson, The class group and local class group of an integral domain, *Non-Noetherian Ring Theory* (eds. S.T. Chapman and S. Glaz) Kluwer Acad. Publ., 2000, 33–55.
8. D. F. Anderson, Divisibility properties in graded integral domains, *Arithmetical Properties of Commutative Rings* (ed. S. T. Chapman) Chapman and Hall, 2005, 22–45.
9. D. F. Anderson and G. W. Chang, The class group of $D[\Gamma]$ for D an integral domain and Γ a numerical semigroup, Comm. Algebra 32 (2004), 787–792.
10. D. F. Anderson and G. W. Chang, Homogeneous splitting sets of a graded integral domain, J. Algebra 288 (2005), 527–544.
11. D. F. Anderson and S. B. Mulay, Non-catenary factorial domains, Comm. Algebra 8 (1980), 467–476.
12. J. T. Arnold and R. Gilmer, Dimension theory of commutative semigroup rings, Houston J. Math. 2 (1976), 299–313.
13. S. El Baghdadi, L. Izelgue, and S. Kabbaj, On the class group of a graded domain, J. Pure Appl. Algebra 171 (2002), 171–184.
14. J. W. Brewer, D. L. Costa, and E. L. Lady, Prime ideals and localizations in commutative group rings, J. Algebra. 34 (1975), 300–308.
15. L. G. Chouinard II, Krull semigroups and divisor class groups, Canad. J. Math. 33 (1981), 1459–1468.
16. L. G. Chouinard II, Projective modules over Krull semigroup rings, Michigan Math. J. 29 (1982), 495–502.
17. J. E. David, A non-Noetherian factorial ring, Trans. Amer. Math. Soc. 169(1972), 495–502.
18. J. E. David, A characteristic zero non-Noetherian factorial ring of dimension 3, Trans. Amer. Math. Soc. 180(1973), 315–325.
19. R. Fossum, *The Divisor Class Group of a Krull Domain*, Springer-Verlag, New York, 1973.
20. L. Fuchs, *Infinite Abelian Groups, Vol. I*, Academic Press, New York, 1970.
21. L. Fuchs, *Infinite Abelian Groups, Vol. II*, Academic Press, New York, 1973.
22. S. Gabelli, On divisorial ideals in polynomial rings over Mori domains, Comm. Algebra 15 (1987), 2349–2370.
23. R. Gilmer, A note on semigroup rings, Amer. Math. Monthly 75 (1969), 36–37.
24. R. Gilmer, A two-dimensional non-Noetherian factorial ring, Proc. Amer. Math. Soc. 44 (1974), 25–30.
25. R. Gilmer, *Commutative Semigroup Rings*, Chicago Lectures in Mathematics, Univ. Chicago Press, Chicago, 1984.
26. R. Gilmer, Chain conditions in commutative semigroup rings, J. Algebra 103 (1986), 592–599.
27. R. Gilmer, *Multiplicative Ideal Theory*, Queen's Papers in Pure and Applied Mathematics, vol. 90, Queen's University, Kingston, Ontario, 1992.
28. R. Gilmer and R. C. Heitmann, The group of units of a commutative semigroup ring, Pacific J. Math. 85 (1979), 65–86.
29. R. Gilmer and R. C. Heitmann, On $Pic(R[X])$ for R seminormal, J. Pure Appl. Algebra 16 (1980), 251–257.
30. R. Gilmer and T. Parker, Divisibility properties in semigroup rings, Michigan Math. J. 21 (1974), 49–64.
31. R. Gilmer and T. Parker, Semigroup rings as Prüfer rings, Duke Math. J. 41 (1974), 219–230.

32. R. Gilmer and E. Spiegel, Coefficient rings in isomorphic semigroup rings, Comm. Algebra 13 (1985), 1789–1809.
33. R. Gilmer and M. L. Tepley, Units of semigroup rings, Comm. Algebra 5 (1977), 1275–1303.
34. R. Gilmer and M. L. Tepley, Idempotents of commutative semigroup rings, Houston J. Math. 3 (1977), 369–385.
35. A. Grams, Atomic domains and the ascending chain condition, Proc. Cambridge Philos. Soc. 75 (1974), 321–329.
36. J. Gubeladze, Anderson's conjecture and the maximal class of monoids over which projective modules are free, Mat. Sbornik. 35 (177) (1988), 169–185. (Russian)
37. F. Halter-Koch, *Ideal Systems*, Marcel Dekker, New York, 1998.
38. S. Malik, *A Study of Strong S-Rings and Prüfer v-Multiplication Domains*, Dissertation, Florida State University, Tallahassee, Florida, August, 1979.
39. S. Malik, Properties of commutative group rings and semigroup rings, *Algebra and Its Applications* (eds. H. L. Manocha and J. B. Srivastava) Marcel Dekker, 1984, 133–146.
40. R. Matsuda, On algebraic properties of infinite group rings, Bull. Fac. Sci. Ibaraki Univ. Ser. A 7 (1975), 29–37.
41. R. Matsuda, Torsion-free abelian semigroup rings, VI, Bull. Fac. Sci. Ibaraki Univ. Ser. A 18 (1986), 23–43.
42. R. Matsuda, Note on Gilmer's Multiplicative Ideal Theory, I, Arab. J. Sci. Eng. Sect. C Theme Issues 26 (2001), 127–140.
43. R. Matsuda, Note on Gilmer's Multiplicative Ideal Theory, II, Arab. J. Sci. Eng. Sect. C Theme Issues 26 (2001), 141–154.
44. D. G. Northcott, *Lessons on Rings, Modules, and Multiplicities*, Cambridge Univ. Press, London, 1968.
45. T. Parker, *The Semigroup Ring*, Dissertation, Florida State University, Tallahassee, Florida, December, 1973.
46. T. Parker and R. Gilmer, Nilpotent elements of commutative semigroup rings, Michigan Math. J. 22 (1975), 97–108.
47. L. Pontryagin, The theory of topological commutative groups, Ann. Math. 35 (1934), 361–388.
48. R. G. Swan, Gubeladze's proof of Anderson's conjecture, Contemp. Math. 124 (1992), 215–250.

Numerical semigroup algebras

Valentina Barucci

Dipartimento di Matematica, Università di Roma La Sapienza, Piazzale A. Moro 2, 00185 Roma, Italy, barucci@mat.uniroma1.it

1 Introduction

Given a ring R and a semigroup S the semigroup ring $R[S]$ inherits the properties of S and R. If no restrictions are posed on the semigroup S and on the ring R the class of semigroup rings is very large. There is a wide literature on this subject. Gilmer's book [15] is the classical reference in the commutative case, i.e. when both the ring R and the semigroup S are commutative. The book contains a deep study of the conditions under which the semigroup ring $R[S]$ has given ring theoretic properties. The following short survey considers a very particular class of semigroup rings: the ring of coefficient R is supposed to be a field and, for the main part of the results, the semigroup S is supposed to be a numerical semigroup, i.e. an additive submonoid of $\mathbb{N}$, with finite complement in $\mathbb{N}$. Also within this particular class of semigroup algebras over a field, the paper deals only with some themes and it is far from being complete. Results and proofs are ranged through two different sources. On one side there is the classical ring theory, a rich, historically settled and well known theory. On the other side there are more elementary but simetimes quicker arguments which come from studying a less rich structure as that of semigroups. In doing so I hope to be not too far from R. Gilmer's open research attitude and from his tolerant view. Here, as in many other fields of human life, there is not a "Unique Thought", but several different points of view and techniques may coexist and be reciprocally useful.

2 Semigroup algebras

If k is a field and S an (additive) semigroup, the *semigroup algebra* $k[S]$ is defined as the set of functions from S into k that are finitely nonzero, with addition and multiplication defined as follows

$$(f+g)(s) = f(s) + g(s)$$

$$(fg)(s) = \sum_{t+u=s} f(t)g(u)$$

Since f and g are finitely nonzero, only a finite number of summands of $\sum_{t+u=s} f(t)g(u)$ are nonzero. The elements of the semigroup algebra $k[S]$ are usually written in the form $\sum_i^n f(s_i)x^{s_i}$ and the operations defined above coincide with addition and multiplication defined as for polynomials. In all this paper S is an (additive) commutative semigroup with zero and k is a field. As usual an equivalence relation on S compatible with the operation is called a *congruence* on S.

If S is a finitely generated semigroup, $S = \langle s_1, \ldots, s_\nu \rangle$, then the semigroup homomorphism

$$\phi: \mathbb{N}^\nu \longrightarrow S$$

defined by $\phi(1, 0, \ldots, 0) = s_1, \ldots, \phi(0, \ldots, 0, 1) = s_\nu$ is surjective and

$$S \cong \mathbb{N}^\nu / \rho \tag{1}$$

where ρ is the congruence on $\mathbb{N}^\nu$ defined by $a\rho b \Leftrightarrow \phi(a) = \phi(b)$.

Recall that a polynomial g in n indeterminates over the field k, $g \in k[x_1, \ldots, x_\nu]$ is called a *pure difference binomial* if it is of the form

$$x_1^{e_1} \ldots x_\nu^{e_\nu} - x_1^{f_1} \ldots x_\nu^{f_\nu}$$

for some nonnegative integers e_i and f_i.

Theorem 2.1. [15, Theorems 7.7 and 7.11] *The following conditions are equivalent:*

(1) *S is a finitely generated semigroup.*
(2) *$k[S]$ is a Noetherian ring.*
(3) *$k[S] \cong k[x_1, \ldots, x_\nu]/I$, where I is an ideal generated by pure difference binomials.*

The proof of Theorem 2.1 is based on the mentioned characterization of a finitely generated semigroup. More precisely it turns out that, if $S \cong \mathbb{N}^\nu / \rho$, then the ideal I of condition 3) in Theorem 2.1, called the *kernel ideal* of the congruence ρ or the *toric ideal*, is the ideal of $k[\mathbf{x}] = k[x_1, \ldots, x_\nu]$ generated by $\mathbf{x}^a - \mathbf{x}^b$, where $a, b \in \mathbb{N}^\nu$ and $a\rho b$. Since by the Hilbert's Basis Theorem the ideal I is finitely generated, also the congruence ρ ($\rho \subseteq \mathbb{N}^\nu \times \mathbb{N}^\nu$) is finitely generated, i.e. there exists a finite set $\sigma \subseteq \mathbb{N}^\nu \times \mathbb{N}^\nu$ such that the smallest congruence on $\mathbb{N}^\nu$ which contains σ is ρ. Such σ is called a *presentation* for the semigroup S. A direct, not easy proof that ρ is finitely generated is in [25]. The natural problem of computing the cardinality of a minimal set of generators for I or equivalently the cardinality of a minimal presentation σ has been largely studied. We will say something for particular cases in Section 3.

Example 1. The semigroup $S = \mathbb{Z} = \langle 1, -1 \rangle$ is isomorphic to $\mathbb{N}^2/\rho$, where $(a_1, a_2)\rho(b_1, b_2) \Leftrightarrow a_1 - a_2 = b_1 - b_2$. Thus $k[\mathbb{Z}] \cong k[x_1, x_2]/I$, where I is generated by the binomials

$$\mathbf{x}^{(a_1,a_2)} - \mathbf{x}^{(b_1,b_2)} = x_1^{a_1} x_2^{a_2} - x_1^{b_1} x_2^{b_2}$$

with $a_1 - a_2 = b_1 - b_2$, that is I is generated by $x_1 x_2 - 1$.

The finitely generated semigroup S of a Noetherian semigroup algebra $k[S]$ can be or not to be cancellative, torsion free, reduced (it is *reduced* when the only invertible element is zero). In case S is cancellative, S can be embedded in its quotient group $G(S)$, the smallest group (up to isomorphism) which contains S and for the Krull dimension, it is $\dim k[S] = \dim k[G(S)]$ (cf. [15, Theorem 21.4]). On the other hand, for a not necessarily finitely generated semigroup S:

Theorem 2.2. [15, Theorem 81] *$k[S]$ is an integral domain if and only if S is torsion free and cancellative.*

An important class of commutative semigroups is the class of *affine* semigroups, which are defined as finitely generated, cancellative, torsion free and reduced semigroups. In some literature (e.g. [10]) such semigroups are called positive affine semigroups. They have a nice characterization:

Theorem 2.3. (cf. e.g.[29, Theorem 3.11]) *S is an affine semigroup if and only if S is isomorphic to a finitely generated submonoid of $\mathbb{N}^n$, for some positive integer n.*

If S is an affine semigroup, then the quotient group $G(S)$ is a finitely generated torsion free group, thus $G(S) \cong \mathbb{Z}^d$, for some $d \in \mathbb{N}$. In this case the semigroup algebra $k[S]$ is a Noetherian domain and $\dim k[S] = \dim k[G(S)] = d$. We say in this case that S is an affine semigroup of *rank d*.

Recall that if S is a cancellative semigroup of quotient group $G = G(S)$, then the *integral closure* S' of S in G is the set of elements of G integral over S, where $t \in G$ is said to be *integral* over S if $nt \in S$, for some positive integer n. It is easy to verify that S' is a semigroup and S is *integrally closed* or *normal* if $S = S'$ (cf. [15], p. 151). Moreover:

Theorem 2.4. [15, Theorem 12.4 (1) and Corollary 12.11] *If S is a torsion free and cancellative semigroup then*

(1) *The integral closure of $k[S]$ is $k[S']$.*
(2) *$k[S]$ is integrally closed in its field of quotients if and only if S is integrally closed.*

Recall also that (if S is a cancellative semigroup of quotient group $G = G(S)$) the *complete integral closure* S^* of S in G is the set of elements of G almost integral over S, where $t \in G$ is said to be *almost integral* over S if there exists $s \in S$ such that $s + nt \in S$, for each positive integer n. It is immediate to verify that S^* is a semigroup and S is *completely integrally closed* if $S = S^*$ (cf. [15], p. 151). If $t \in G$ is integral over S, then t is almost integral, but the converse is not true in general and S^* may be even not completely integrally closed, as Example 2 below shows. Moreover:

Theorem 2.5. [15, Theorem 12.4 (3) and Corollary 12.7] *If S is a torsion free and cancellative semigroup then*

(1) *The complete integral closure of $k[S]$ is $k[S^*]$.*
(2) *$k[S]$ is completely integrally closed in its field of quotients if and only if S is completely integrally closed.*

Example 2. [16] The ring $k[\{x^{2n+1}y^{n(2n+1)}; n \geq 0\}]$ is the semigroup ring $k[S]$ where $S \subset \mathbb{Z}\times\mathbb{Z}$ is the semigroup generated by $\{(2n+1, n(2n+1)); n \geq 0\}$. The complete integral closure of this ring is $k[\{xy^n; n \geq 0\}] = k[S^*]$, where $S^* = \{(0,0)\} \cup \{(m,n); m \geq 1, n \geq 0\}$ is the complete integral closure of S in $G(S) = \mathbb{Z}\times\mathbb{Z}$. The ring $k[S^*]$ (and the semigroup S^* as well) is not completely integrally closed, its complete integral closure is $k[S^{**}] = k[x, y] = k[\mathbb{N} \times \mathbb{N}]$.

Now we go back to affine semigroups. If S is an affine semigroup, we can suppose without loss of generality that the quotient group of S is $\mathbb{Z}^d$. For example if $S = \langle 4, 6\rangle \subseteq \mathbb{N}$, then $G(S) = 2\mathbb{Z} \cong \mathbb{Z}$. But $S = \langle 4, 6\rangle \cong \langle 2, 3\rangle$ and the quotient group of this is $\mathbb{Z}$.

Let $S = \langle s_1, \dots, s_\nu\rangle \subseteq \mathbb{N}^d$ be an affine semigroup of quotient group $\mathbb{Z}^d$ and let $C(S) = \{q_1 s_1 + \cdots + q_\nu s_\nu;\ q_i \in \mathbb{Q}, q_i \geq 0\}$ be the *cone* generated by S. Then it is not difficult to prove that $S' = C(S) \cap \mathbb{Z}^d$, cf. [29, Chaper 7.4] or [10, Proposition 2.1.1]. Moreover S' can be proved to be finitely generated and so it is an affine (normal) semigroup [10, Proposition 2.1.1].

If S is an affine normal semigroup, then $k[S]$ is an integrally closed Noetherian domain, thus it is a Krull domain. Hochster [20] proved that in this case $k[S]$ is a Cohen Macaulay ring.

The divisor class group of such Krull domain is the subject of some recent work of W. Bruns, who studies certain *conic* elements of the divisor class group which are finite in number and "close" to the trivial class, cf. [11].

If S is an affine semigroup, for each element $s \in S$, $s = t + u$ for a finite number of pairs (t, u) of elements of S, so in this case $k[[S]]$ can be defined, as the ring of all (not necessarily finitely nonzero) functions from S to k with the operations defined as for the semigroup ring $k[S]$ (cf. [14, Chapter I, 1, Exercise 6]). If $S = \mathbb{N}^d$, then $k[[S]] = k[[x_1, \dots, x_d]]$, the ring of formal power series in d indeterminates over k. Notice that in our terminology for example $\mathbb{Z}$ is not an affine semigroup (is not reduced) and $k[[\mathbb{Z}]]$ can't be defined.

Thus, if S is an affine semigroup, $k[[S]]$ is a local Noetherian domain.

Although for several problems considering $k[S]$ or $k[[S]]$ is equivalent, the local structure of $k[[S]]$ makes it closer than $k[S]$ to the semigroup S. A semigroup is in fact always local in the sense that the non invertible elements (the nonzero elements in case of a reduced semigroup) form an ideal, the unique maximal ideal of S. Some classical results for local rings have a corresponding formulation, in general less profound and easier to prove, for semigroups. In the next section we will give an example of this energy-saving philosophy in case of the simplest affine semigroups, those of rank 1, i.e. the numerical semigroups. Despite the apparent simplicity of such semigroups, several important examples were found among these and several unsolved problems still exist about them.

3 Numerical semigroups

3.1 Main notions

The results of this section are expressed for a numerical semigroup S, but all of them can be given in terms of the semigroup algebra $k[[S]]$, which by the results recalled in Section 2 is a one-dimensional Noetherian domain of integral closure $k[[S']] = k[[\mathbb{N}]] = k[[t]]$, the ring of formal power series in one indeterminate over k.

We fix for all the section the following notation. S is a numerical semigroup, i.e. a subsemigroup of $\mathbb{N}$, with zero and with finite complement in $\mathbb{N}$. $M = S \setminus \{0\}$ is the *maximal ideal* of S, e is the *multiplicity* of S, that is the smallest positive integer of S, g is the *Frobenius number* of S, that is the greatest integer which does not belong to S. Moreover we set $n = \mathrm{Card}(\{s \in S \mid s < g\})$ and denote by $\delta = g + 1 - n$ the number of gaps of S, that is in ring terms the length of the $k[[S]]$-module $k[[t]]/k[[S]]$.

A *relative ideal* of S is a nonempty subset I of $\mathbb{Z}$ (which is the quotient group of S) such that $I + S \subseteq I$ and $I + s \subseteq S$, for some $s \in S$. A relative ideal which is contained in S is an *integral ideal* of S. If I is a relative ideal of S, then $\mathrm{Ap}_e(I) = \mathrm{Ap}(I) = I \setminus (e + I)$ is the set of the e smallest elements in I in the e congruence classes mod e and is called the *Apery set* of I (with respect to e). In particular $\mathrm{Ap}(S)$ is the Apery set of S. Since g is the greatest gap of S, $g + e$ is the largest element in $\mathrm{Ap}(S)$.

The following Lemma, corresponding to Nakayama's Lemma for local rings, is very easy to prove for numerical semigroups:

Lemma 3.1. *If I is a relative ideal of S, then the unique minimal set of generators of I is $I \setminus (M + I)$.*

Since $e + I \subseteq M + I$, then $I \setminus (M + I) \subseteq I \setminus (e + I) = \mathrm{Ap}_e(I)$ and by Lemma 3.1 each relative ideal I of S is generated by at most e elements. I is *maximally generated*, that is I needs e generators, if and only if $e + I = M + I$.

In particular $M \setminus 2M$ is the minimal set of generators of M and its cardinality is the *embedding dimension* ν of S. We have $\nu \leq e$ and if equality holds the semigroup S is called of *maximal embedding dimension*.

Problem 1. The following problem was asked by Wilf in [36] and is still unsolved, althought the answer is positive for several classes of numerical semigroups: is $g + 1 \leq n\nu$?

A particular relative ideal of S plays a special role. It is the *canonical ideal* $\Omega = \{z \in \mathbb{Z} \mid g - z \notin S\}$. Of course it is $S \subseteq \Omega \subseteq \mathbb{N}$.

If I, J are relative ideals of S, set $I - J = \{z \in \mathbb{Z} \mid z + J \in I\}$, which is still a relative ideal of S.

Proposition 3.2. *The following conditions are equivalent for a relative ideal J of S:*

(1) $J - (J - I) = I$, *for each relative ideal I of S.*
(2) *J is an irreducible relative ideal, i.e. if $J = I_1 \cap I_2$, with I_i relative ideals of S, then $J = I_1$ or $J = I_2$.*
(3) $J = \Omega + h$, *for some $h \in \mathbb{Z}$.*
(4) $\mathrm{Card}((J - M) \setminus J) = 1$.

Proof. The equivalence of 1) and 3) is proved in [21] (cf. also [3, Lemma 1]). A ring theoretic proof of the equivalence of 1), 2) and 4) is in [19, Satz 3.3].

The *type* t of a numerical semigroup S is the minimal number of generators of the canonical ideal, that is $t = \mathrm{Card}(\Omega \setminus (\Omega + M))$.

Proposition 3.3. *For the type t of S the following holds:*

(1) $t = \mathrm{Card}((S - M) \setminus S)$
(2) $S = (h_1 + \Omega) \cap \cdots \cap (h_t + \Omega)$, *for some $h_i \in \mathbb{Z}$, where the intersection is irredundant.*
(3) *Each relative principal ideal of S is an irredundant intersection of t irreducible relative ideals.*
(4) $1 \leq t \leq e - 1$.

Proof. For 1), cf. e.g. [3, Lemma 1]. A ring theoretic proof of 2) and 3) is in [5, Proposition 2.6]. For 4). cf. e.g. [1, Remark I.2.7].

Problem 2. It was asked (in ring terms) in [24] whether

$$t \leq \mathrm{Card}(2\Omega \setminus (2\Omega + M))$$

i.e. if the number of generators of Ω is always smaller than the number of generators of 2Ω. Also in this case the answer is proved to be positive only for some classes of semigroups.

If $t = 1$ or equivalently $\Omega = S$, then the numerical semigroup S is classically called *symmetric*. By Proposition 3.3 the type of a numerical semigroup is the Cohen Macaulay type of the semigroup algebra $k[[S]]$, so symmetric semigroups correspond to Gorenstein rings.

Proposition 3.4. (cf. e. g. [1, Lemma I.1.8]) *The following conditions are equivalent for S:*

(1) *S is symmetric.*
(2) *For each $z \in \mathbb{Z}$, $z \in S$ if and only if $g - z \notin S$.*
(3) *g is odd and S is maximal in the set of numerical semigroups with Frobenius number g.*

J. C. Rosales and M. B. Branco noticed that a semigroup S is maximal in the set of numerical semigroups with Frobenius number g if and only if S is an irreducible semigroup, i.e. if $S = S_1 \cap S_2$, with S_i numerical semigroups, then $S = S_1$ or $S = S_2$. In [28] the semigroups that can be expressed as an intersection of symmetric semigroups are studied.

Proposition 3.5. [35] *If $\nu(S) = 2$ then S is symmetric.*

The symmetric semigroups of embedding dimension 3 were characterized by J. Herzog in [17]. An important class of symmetric semigroups are those which arise as value semigroups of irreducible plane algebroid curves. They were characterized by Zariski (cf. e.g [2, Proposition 4.8]) and may have arbitrary high multiplicity or embedding dimension. For example $S = \langle 30, 42, 280, 855 \rangle$ is the value semigroup of $\mathbb{C}[[t^{30}, t^{42} + t^{112} + t^{127}]]$ and is symmetric. Symmetric semigroups of maximal embedding dimension are of the form $S = \langle 2, 2k+1 \rangle$ (cf. e.g.[1, Theorem I.4.2]). On the other hand it is proved in [27] that, if e and ν are integers such that $2 \le \nu \le e-1$, then there exists a symmetric semigroup with $e(S) = e$ and $\nu(S) = \nu$.

In case the Frobenius number g is even, particular semigroups of type 2, the *pseudosymmetric semigroups* take the place of the symmetric ones. They are defined by one of the conditions of next Proposition.

Proposition 3.6. [1, Lemma I.1.9] *The following conditions are equivalent for a numerical semigroup S:*

(1) $(S - M) \setminus S) = \{g, g/2\}$.
(2) *g is even and for each $z \in \mathbb{Z}$, $z \neq g/2$, $z \in S$ if and only if $g - z \notin S$.*
(3) *g is even and S is maximal in the set of numerical semigroups with Frobenius number g.*

The *almost symmetric* semigroups generalize the symmetric and the pseudosymmetric ones. They are defined by one of the conditions of the next Proposition.

Proposition 3.7. [3, Proposition 4 and Definition-Proposition 20] *The following conditions are equivalent for a numerical semigroup S:*

(1) $\delta = n + t - 1$.
(2) $t = \mathrm{Card}(\Omega \setminus S) + 1$.
(3) $M + \Omega = M$.

(4) $\Omega \subseteq S - M$.

For almost symmetric semigroups the mentioned Problems 1 and 2 have positive answers, cf. [24] and [22] respectively.

3.2 Minimal presentations

If S is a numerical semigroup, in the isomorphism (1) of Section 1, $\mathbb{N}^\nu/\rho \cong S$, we denote by μ the minimal number of generators of the congruence ρ or, equivalently, the minimal number of generators of the toric ideal I (where $K[[x_1, \ldots, x_\nu]]/I \cong k[[S]]$). It is well known that $\mu \geq \nu - 1$ (cf.[17, Lemma 1.9]). If $\mu = \nu - 1$, the semigroup S and the corresponding semigroup algebra $k[[S]]$ is called a *complete intersection.* If S is complete intersection, then S is symmetric (cf. e.g. [17]), but not every symmetric semigroup is a complete intersection, although the semigroups of values of irreducible plane algebroid curves are complete intersections (cf.[17] or [2]).

It is proved in [29, Theorem 10.15] that

$$\mu \leq \frac{(2e - \nu)(\nu - 1)}{2} + 1$$

but an upper bound for μ which depends only on ν does not exist if $\nu \geq 4$ (cf. [7]). On the other hand, for $\nu = 2$, then I is a principal ideal and so $\mu = 1$ and for $\nu = 3$, Herzog [17] proved that $\mu \leq 3$. A similar problem can be posed in the symmetric case.

Problem 3. (cf. [8, pag.218]). If S is symmetric, does there exist an upper bound for μ which depends only on ν?

Also this problem has answers only for particular classes of semigroups. For three generated semigroups, i.e. $\nu = 3$, it is proved in [17] that symmetric is equivalent to complete intersection, so for such semigroups, $\mu = \nu - 1 = 2$. For $\nu = 4$, Bresinsky [8] proved that $\mu \leq 5$. Rosales [27] constructs a family of symmetric semigroups with $\mu \leq \frac{\nu(\nu-1)}{2} - 1$ and conjectures this is the bound, but Bresinsky [9] showed that, for $\nu = 5$, μ can be equal to 13 as e.g. in $S = \langle 19, 23, 29, 31, 37 \rangle$ which has $\mu = 13$, cf. [9]. He proved indeed that for a certain class of symmetric semigroups with $\nu = 5$, the best bound is $\mu \leq 13$.

3.3 Blowup and Hilbert function

Given a Noetherian local ring (R, m) with residue field k, a classical topic in commutative algebra is to study the associated graded ring $gr(R) = \bigoplus_{n\geq 0} m^n/m^{n+1}$ and its Hilbert function $H^0(n) = \dim_k(m^n/m^{n+1})$. In particular it is interesting to know when $gr(R)$ is Cohen Macaulay. In case S is a numerical semigroup and $R = k[[S]]$, the information for a similar study is all contained in S. Here is a list of results in terms of the semigroup. Some proofs are given, when the proof at semigroup level is simpler than the corresponding proof in the more general context of rings.

Let I be a relative ideal of S. Then $I-I$ is the smallest semigroup T such that I is a relative ideal of T. There is a chain of semigroups $S \subseteq (I-I) \subseteq (2I-2I) \subseteq \cdots \subseteq \mathbb{N}$, which stabilizes on a semigroup $\mathcal{B}(I) = (nI - nI)$, for $n >> 0$, called the semigroup obtained by S by blowing up I or the *blowup* of I. Notice that, if we replace I with a "translation" J of I, $J = I + z$, for some $z \in \mathbb{Z}$, then $(hI - hI) = (hJ - hJ)$, for each $h \geq 1$. The least integer $h \geq 1$ such that $hI - hI = \mathcal{B}(I)$ is called the *reduction number* $r(I)$ of I and it is characterized by the following equivalent conditions.

Proposition 3.8. [1, Corollary I.2.2] *Let I be a relative ideal of S and $s \in S$ such that $s + I \subseteq S$. Denoting by i the smallest element in I, for an integer $h \geq 1$ the following conditions are equivalent:*

(1) $(h+1)I = hI + i$.
(2) $\mathrm{Card}(h(s+I) \setminus (h+1)(s+I)) = s + i$.

It is immediate that, if $h \geq r(I)$, then the equivalent conditions of Proposition 3.8 hold for h.

Notice that, if $i = 0$, then condition 1) of Proposition 3.8 becomes $(h+1)I = hI$, i.e. the reduction number $r(I)$ is in this case the smallest integer $h \geq 1$ such that hI is a semigroup and the blowup $\mathcal{B}(I)$ is in this case the smallest semigroup containing I.

If I is an integral ideal of a numerical semigroup S, the *Hilbert function* of I is defined for each $h \geq 0$ by

$$H_I^0(h) = \mathrm{Card}(hI \setminus (h+1)I)$$

where for $h = 0$ we set $hI = S$.

Notice that, if I is an integral ideal of S, then condition 2) of Proposition 3.8 becomes $\mathrm{Card}(hI \setminus (h+1)I) = i$, where i ($i \geq 0$) is the smallest integer in I. Thus $H_I^0(h) = i$, for $h >> 0$. This is actually well-known in the more general ring context.

Consider now the case $I = M$, the maximal ideal of S. By the above observations, $\mathcal{B}(M)$ is the smallest semigroup containing $M - e$, that is the translation of M with smallest element equal to zero. It follows that, if $S = \langle n_1 = e, n_2, \ldots, n_\nu \rangle$, then $\mathcal{B}(M) = \langle e, n_2 - e, \ldots, n_\nu - e \rangle$, where the generators are not always all necessary.

We can also recall that:

Proposition 3.9. [6, Theorem 1] (1) *A relative ideal I of S is maximally generated if and only if I is an ideal of $\mathcal{B}(M)$.*

(2) *$S - \mathcal{B}(M)$ is the biggest maximally generated integral ideal of S.*

Proof. I is maximally generated if and only if $e + I = M + I$, i.e. if and only if $I + (M-e) = I$. This last condition is equivalent to $I + h(M-e) = I$, for each integer $h \geq 1$. Thus it is enough to recall that, for $n >> 0$, $n(M-e) = \mathcal{B}(M)$.

3.4 Good semigroups

This sections deals with numerical semigroups S such that the associated graded ring $gr(k[[S]])$ is Cohen Macaulay. For simplicity, we call *good* such semigroups in this paper. This is the same as requiring that the maximal homogeneous ideal of $gr(k[[S]])$ contains a nonzero divisor. This condition can be expressed for S in the following terms. A numerical semigroup S is good if there exists $n \in \mathbb{N}$ and $x \in nM \setminus (n+1)M$ such that, for each $s \in S$,

$$s \in hM \setminus (h+1)M \Longrightarrow s + x \in (h+n)M \setminus (h+n+1)M \qquad (2)$$

As it is well-known in ring context (cf. e.g. [13]), if such element x exists, then the multiplicity e of S has to be one of such elements. We include the elementary proof of that for semigroups.

Lemma 3.10. *A numerical semigroup S is good if and only if for each $s \in S$*

$$s \in hM \setminus (h+1)M \Longrightarrow s + e \in (h+1)M \setminus (h+2)M$$

Proof. We claim first that if $x \in S$, $x \in nM \setminus (n+1)M$, satisfies condition (2), then $x = ne$. Consider $nex \in S$. We have

$$nex = \underbrace{x + \cdots + x}_{ne \text{ times}} \in ne(nM) = n^2 eM$$

and applying ne times property (2) satisfied by x, we have that nex is a necessary generator for n^2eM, that is $nex \notin hM$, if $h > n^2e$. On the other hand

$$nex = \underbrace{ne + \cdots + ne}_{x \text{ times}} \in xnM$$

Thus $xn \leq n^2e$ and $x \leq ne$. Since $x \in nM$, it follows that $x = ne$.

If $x = e$ does not satisfy (2), then $s + e \in (h+2)M$, for some $s \in hM \setminus (h+1)M$, and $2e + s \in e + (h+2)M \subseteq (h+3)M$. By induction $ne + s \in (h+n+1)M$, so ne does not satisfy (2).

Example 3. a) In the semigroup $S = \langle 7, 10, 12 \rangle$ of maximal ideal M, we have:

$M = \{\underline{7}, \underline{10}, \underline{12}, 14, 17, 19, 20, 21, 22, 24, 26, \rightarrow\}$
$2M = \{\underline{14}, \underline{17}, \underline{19}, \underline{20}, 21, \underline{22}, 24, 26, \rightarrow\}$
$3M = \{\underline{21}, \underline{24}, \underline{26}, \underline{27}, 28, \underline{29}, \underline{30}, 31, \underline{32}, \rightarrow\}$
$4M = \{\underline{28}, \underline{31}, \underline{33}, \underline{34}, 35, \underline{36}, \underline{37}, 38, \underline{39}, \rightarrow\}$

where the minimal set of generators of hM, i.e. $hM \setminus (h+1)M$ is given by the underlined elements and where "$n, \rightarrow\}$" means that all the integers m, $m > n$ are in the set. Here $e = 7$ and if s is a necessary generator for hM then $s + 7$ is a necessary generator for $(h+1)M$. So S is a good semigroup.

b) In the semigroup $S = \langle 6, 7, 15 \rangle$ we have:

$M = \{\underline{6}, \underline{7}, 12, 13, 14, \underline{15}, 18, 19, 20, 21, 22, 24, \rightarrow\}$

$2M = \{\underline{12}, \underline{13}, \underline{14}, 18, 19, 20, 21, \underline{22}, 24, \rightarrow\}$
$3M = \{\underline{18}, \underline{19}, \underline{20}, \underline{21}, 24, 25, 26, 27, 28, \underline{29}, 30, 31, \rightarrow\}$
$4M = \{\underline{24}, \underline{25}, \underline{26}, \underline{27}, \underline{28}, 30, \rightarrow\}$
$5M = \{\underline{30}, \underline{31}, \underline{32}, \underline{33}, \underline{34}, \underline{35}, \rightarrow\}$

Here $e = 6$. If we consider $15 \in M \setminus 2M$, necessary generator of M, we get that $15 + 6 = 21 \notin 2M \setminus 3M$, so this is not a good semigroup. Notice however that this semigroup is symmetric.

If $\nu(S) = 3$, i.e. $S = \langle n_1, n_2, n_3 \rangle$, there are precise numerical conditions on n_1, n_2, n_3, found independently in [26] and [18] which characterize the goodness of S.

A good semigroup has the following well-known important property:

Proposition 3.11. *If S is a good semigroup, then the Hilbert function $H_M^0(n)$ is non decreasing.*

Proof. For each $h \geq 0$, the map

$$j{:}hM \setminus (h+1)M \longrightarrow (h+1)M \setminus (h+2)M$$

given by $j(s) = s+e$ is injective by Lemma 3.10, thus $\mathrm{Card}(hM \setminus (h+1)M) \leq \mathrm{Card}(h+1)M \setminus (h+2)M$.

Example 4. If S is not good, Proposition 3.11 does not necessarily hold. The following is an example of J. Sally, cf. [31], p. 40. If $S = \langle 15, 21, 23, 47, 48, 49, 50, 52, 54, 55, 56, 58 \rangle$, then $H_M^0(n) = \{1, 12, 11, 13, 15, 15, \ldots\}$.

As we saw in Example 3 b) a symmetric semigroup is not always good. Here is a list of sufficient conditions in order to have a good semigroup.

Theorem 3.12. *If S satisfies one of the following conditions, then S is good:*

(1) *S is a complete intersection (in particular if $\nu(S) = 2$).*
(2) *S is of maximal embedding dimension.*
(3) *S is of almost maximal embedding dimension (i.e. $\nu(S) = e(S) - 1$) and $t(S) < e(S) - 2$.*
(4) *S is symmetric and $\nu = e - 2$.*
(5) *S is generated by an arithmetic sequence.*
(6) *$e(S) \leq 4$, except the case $S = \langle 4, n_2, 3n_2 - 4 \rangle$, with n_2 odd, $n_2 > 4$.*

Proof. For 1) and 2) cf. [32]. 3) is proved in [34] and 4) in [33]. This last proof is simplified in [30]. For 5) cf. [23, Proposition 1.1] and 6) can be deduced from the mentioned characterization in [26] of a good semigroup generated by three elements, cf. also [4, Proposition 3.5].

We finish this section by giving two characterizations of good semigroups, both based on comparing S and $\mathcal{B}(M)$.

Let as before $S = \langle n_1, \ldots, n_\nu \rangle$ and $\mathcal{B}(M) = \langle e, n_2 - e, \ldots n_\nu - e \rangle$ the blowup of the maximal ideal. Consider the affine semigroup $T \subseteq \mathbb{N}^2$ generated by

$$(e, 0), (n_2, n_2 - e), \ldots, (n_\nu, n_\nu - e), (0, e)$$

Using a result of Hochster and Ratliff, it is noticed in [12] that, with our terminology, S is a good semigroup if and only if the semigroup ring $k[T]$ is Cohen Macaulay and the following nice criterion is given:

Theorem 3.13. [12, Lemma 3.1 and Theorem 3.6] *The following conditions are equivalent:*

(1) *S is a good semigroup.*
(2) *$T = (S \times \mathcal{B}(M)) \cap G(T)$, where $G(T)$ is the quotient group of T.*

Let $\mathrm{Ap}_e(S) = \{\omega_0, \ldots, \omega_{e-1}\}$ the Apery set of S and $\mathrm{Ap}_e(\mathcal{B}(M)) = \{\omega'_0, \ldots, \omega'_{e-1}\}$ the Apery set of $\mathcal{B}(M)$. Suppose that $\omega_i \equiv \omega'_i \equiv i \pmod{e}$. For $0 \le i \le e-1$, consider the integers a_i such that $\omega'_i = \omega_i - a_i e$. In other words a_i is the largest integer h such that $\omega_i - he \in \mathcal{B}(M)$. Set $b_0 = 0$ and, for $1 \le i \le e-1$ denote by b_i the maximum number h of nonzero summands s_j such that $\omega_i = \sum s_j$, with $s_j \in S$, in other words b_i is the largest integer h such that $\omega_i \in hM$.

Since $\omega_i - b_i e = s_1 + \cdots + s_{b_i} - b_i e = (s_1 - e) + \cdots + (s_{b_i} - e) \in \mathcal{B}(M)$, we have always $a_i \ge b_i$.

We have the following criterion. A more general version of it for (not necessarily semigroup) rings appears in [4].

Theorem 3.14. *S is good if and only if $a_i = b_i$, for each i, $0 \le i \le e-1$.*

Proof. Let S be good. Since $\omega_i - a_i e \in \mathcal{B}(M)$, then $\omega_i - a_i e + nM \subseteq nM$, for $n >> 0$. In particular $\omega_i + (n - a_i)e \in nM$, for $n >> 0$. Since S is good we get $\omega_i \in a_i M$ and so $a_i \le b_i$, that with $a_i \ge b_i$ which always holds gives $a_i = b_i$, for each i. Conversely, if S is not good, by Lemma 3.10 there is $s \in S$, $s = \omega_i + de \in (b_i + d)M$ for some i, such that $s + e \in (b_i + d + 2)M$. In this case $s + e = s_1 + \cdots + s_{b_i+d+2}$, for some $s_j \in S$. It follows that $z = s + e - (b_i + d + 2)e = (s_1 - e) \ldots (s_{b_i+d+2} - e) \in \mathcal{B}(M)$. On the other hand $z = \omega_i + (d+1)e - (b_i + d + 2)e = \omega_i - (b_i + 1)e$ and so $a_i > b_i$.

Corollary 3.15. *S is good if and only if* $\mathrm{Card}(\mathcal{B}(M) \setminus S) = \sum_{i=0}^{e-1} b_i$.

Problem 4. Characterize the symmetric semigroups which are good.

Problem 5. This problem was asked by Maria Evelina Rossi: Is the Hilbert function $H^0_M(n)$ non decreasing, if S is symmetric?

For a good semigroup the Hilbert function $H^0_M(n)$ is completely well-known. With the notation introduced above, the coefficients b_i's are strictly related to the Hilbert function $H^0_M(n)$. In fact for the factor semigroup

$\bar{S} = S/(e+S) = \{\overline{\omega_0} = \bar{0}, \dots, \overline{\omega_{e-1}}, \bar{e}\}$ of maximal ideal $N = \{\overline{\omega_1}, \dots, \overline{\omega_{e-1}}, \bar{e}\}$ the Hilbert function is $H_N^0(n) = \mathrm{Card}(nN \setminus (n+1)N) = \mathrm{Card}\{i \mid b_i = n\}$, as can be easily deduced from the definition of the b_i's. Since S is good, by classical methods of ring theory (cf. [33]), a discrete integration of such function gives $H_M^0(n)$, that is

$$H_M^0(n) = H_N^1(n) := \sum_{j=0}^{n} H_N^0(j)$$

Thus by Theorem 3.14 we get:

Theorem 3.16. *Let S be a good semigroup and let a_i be the coefficients defined above. Then*

$$H_M^0(n) = \mathrm{Card}\{i \mid a_i \leq n\}$$

Example 5. Consider the good semigroup $S = \langle 7, 10, 12 \rangle$ of Example 3 a). It is $\mathcal{B}(M) = \langle 3, 5, 7 \rangle$, so that the Apery sets ordered according to the class of congruence mod 7 are respectively $\mathrm{Ap}_7(S) = \{0, 22, 30, 10, 32, 12, 20\}$ and $\mathrm{Ap}_7(\mathcal{B}(M)) = \{0, 8, 9, 3, 11, 5, 6\}$. It follows that the coefficients a_i's are $a_0 = 0, a_1 = 2, a_2 = 3, a_3 = 1, a_4 = 3, a_5 = 1, a_6 = 2$ and so $H_M^0(n) = \{1, 3, 5, 7, 7, \dots\}$.

References

1. Barucci, V., Dobbs D., Fontana, M.: Maximality properties in numerical semigroups and applications to one-dimensional analytically irreducible local domains. Mem. Amer. Math. Soc. **125** 598 (1997)
2. Barucci, V., D'Anna, M., Fröberg, R.: On plane algebroid curves. In: Commutative Ring Theory and Applications. Lecture Notes Pure Appl. Math. **231** Dekker, New York, 37–50 (2002)
3. Barucci, V., Fröberg, R.:One-dimensional almost Gorenstein rings. J. Algebra **188**, 418–442 (1997)
4. Barucci, V., Fröberg, R.: Associated graded rings of one-dimensional analytically irreducible rings. Preprint
5. Barucci, V., Houston, E., Lucas T., Papick I.: m-canonical ideals in integral domains II. In Ideal theoretic methods in Commutative Algebra. Lecture Notes Pure Appl. Math. **217** Dekker, New York, 11–20 (2001)
6. Barucci, V., Pettersson, K.: On the biggest maximally generated ideal as the conductor in the blowing up ring. Manuscr. Math. **88**, 457–466 (1995)
7. Bresinsky, H.: On prime ideals with generic zero $x_i = t^{n_i}$. Proc. Amer. Math. Soc. **47**, 329–332 (1975)
8. Bresinsky, H.: Symmetric semigroups of integers generated by 4 elements. Manuscr. Math. **17**, 205–219 (1975)
9. Bresinsky, H.: Monomial Gorenstein ideals. Manuscr. Math. **29**, 159–181 (1979)
10. Bruns, W., Gubeladze, J., Trung, N.V.: Problems and algorithms for affine semigroups. Sem. Forum. **64**, 180–212 (2002)

11. Bruns, W.: Conic divisor classes of normal affine monoids. In Abstracts of Second Joint Meeting of AMS, DMV, ÖMG at Mainz, p. 211 (2005)
12. Cavaliere, M. P., Niesi G.: On form ring of a one-dimensional semigroup ring. In: Commutative Algebra, Trento 1981. Lecture Notes Pure Appl. Math. **84** Dekker, New York, 39–48 (1983)
13. Garcia, A.: Cohen-Macaulayness of the associated graded of a semigroup ring. Comm. Algebra **10**, 393–415 (1982)
14. Gilmer, R.: Multiplicative Ideal theory. Marcel Dekker, New York (1972)
15. Gilmer, R.: Commutative Semigroup Rings. U. Chicago Press (1984).
16. Gilmer, R., Heinzer, B.: On the complete integral closure of an integral domain. J. Aust. Math. Soc. **6**, 351–361 (1966)
17. Herzog, J.: Generators and relations of abelian semigroups and semigroup rings. Manuscr. Math. **3**, 175–193 (1970)
18. Herzog, J.: When is a regular sequence super regular? Nagoya Math. J. **93**, 649–685 (1971)
19. Herzog, J., Kunz, E.: Der kanonische Modul eines Cohen-Macaulay Rings. Lecture Notes in Math. **238**, Springer-Verlag, Berlin (1971)
20. Hochster, M.: Rings of invariants of tori, Cohen-Macaulay rings generated by monomials, and polytopes. Ann. Math. **96**, 318–337 (1972)
21. Jäger, J.: Längeberechnungen und kanonische Ideale in eindimensionalen Ringen. Arch. Math. **29**, 504–512 (1977)
22. La Valle, M.: Semigruppi numerici. Tesi di Laurea in Matematica, Universitá La Sapienza, Roma (1995)
23. Molinelli S., Tamone G.: On the Hilbert function of certain rings of monomial curves. J. Pure Appl. Algebra **101**, 191–206 (1995)
24. Odetti F., Oneto A., Zatini E.: Dedekind different and type sequence. Le Matematiche **LV**, 499–516 (2000)
25. Redei, L.: The theory of finitely generated commutative semigroups, Pergamon, Oxford-Edinburgh-New York (1965)
26. Robbiano, L., Valla, G.: On the equations defining tangent cones. Math. Proc. Camb. Phil. Soc. **88**, 281–297 (1980)
27. Rosales, J. C.: Symmetric numerical semigroups with arbitrary multiplicity and embedding dimension. Proc. Amer. Math. Soc. **129**, 2197–2203 (2001)
28. Rosales J. C., Branco M. B.: Numerical semigroups that can be expressed as an intersection of symmetric numerical semigroups. J. Pure Appl. Algebra **171**, 303–314 (2002)
29. Rosales, J. C., Garc´(i)a-Sánchez, P, A.: Finitely generated commutative monoids. Nova Science Publishers, New York (1999)
30. Rossi M. E., Valla, G.: Cohen-Macaulay local rings of embedding dimension $e+d-3$. Proc. London Math. Soc. **80**, 107–126 (2000)
31. Sally, J.: Numbers of generators of ideals in local rings. Lecture Notes Pure Appl. Math. **35** Dekker, New York (1978)
32. Sally, J.: On the associated graded ring of a local Cohen-Macaulay ring. J. Math. Kyoto Univ. **17**, 19–21 (1977)
33. Sally, J.: Good embedding dimension for Gorenstein singularities. Math. Ann. **249**, 95–106 (1980)
34. Sally, J.: Cohen-Macaulay local rings of embedding dimension $e+d-2$. J. Algebra **83**, 325–333 (1983)
35. Sylvester, J. J.: Mathematical questions with their solutions. Uducational Times **41**, 21 (1884)

36. Wilf, H. S.: A circle-of-lights algorithm for the money-changing problem. Amer. Math. Monthly **85:7**, 562–565 (1978)

Prüfer rings

Silvana Bazzoni[1] and Sarah Glaz[2]

[1] Università di Padova, Dipartimento di Matematica Pura e Applicata, Via Belzoni, 35131 Padova, Italy. bazzoni@math.unipd.it
[2] University of Connecticut, Department of Mathematics, Storrs, Connecticut 06269, USA. glaz@math.uconn.edu

Dedicated to Robert Gilmer

1 Introduction

In the introduction to his book, *Multiplicative Ideal Theory* [26], Robert Gilmer states: "It is possible to enumerate a few concepts which are central in our development of multiplicative ideal theory. Quotient rings and rings of quotients fall into this category, they are basic to all subsequent considerations; invertible ideals also constitute a basic tool in the presentation of the theory. *A third concept which plays a central role in the development of the classical ideal theory is that of a Prüfer domain*".

Prüfer domains were defined in 1932 by H. Prüfer [56], as domains in which every finitely generated ideal is invertible. In 1936, Krull [49] named these rings in Prüfer's honor and proved the first of the many equivalent conditions that make an integral domain Prüfer (see Theorem 1.1). Although, the new concept started to slowly appear in the literature, it reached its central role which it enjoys today in the sixties and seventies, due, in no small part, to Robert Gilmer's publications and their impact on research in commutative algebra. On one hand, Prüfer rings and related ring conditions feature in many of Robert's about 200 articles, investigating a large variety of ring properties, including the connection of the Prüfer condition to other properties of interest, see for example [5, 6, 18-36]. Many commutative algebraists followed in Robert's footsteps, using some of his methods and examples, including his emphasis on the Prüfer domain notion. On the other hand, in 1968, the first version of Robert's book: *Multiplicative Ideal Theory* [26], was published by Queen's University Press, collecting all results and references known to date on Prüfer rings and emphasizing their central role in ring theory research. This was followed by the publication of revised versions of this book in 1972 [29], and in 1992 [36]. This book became, and continues to be, one of the most influential books for research in commutative algebra, and with it the centrality of the Prüfer domain notion becomes consolidated.

The theorem below collects the equivalent definitions to the Prüfer domain notion, that appear in *Multiplicative Ideal Theory* [26, 29, 36].

Theorem 1.1. *Let R be a domain. The following conditions are equivalent:*

(1) *R is a Prüfer domain.*
(2) *Every two-generated ideal of R is invertible.*
(3) *R_P is a valuation domain for every prime ideal P of R.*
(4) *R_P is a valuation domain for every maximal ideal P of R.*
(5) *Each finitely generated non-zero ideal I of R is a cancellation ideal, that is $IJ = IK$ for ideals J and K implies $J = K$.*
(6) *If I, J and K are finitely generated ideals of R such that $IJ = IK$, and $I \neq 0$, then $J = K$.*
(7) *R is integrally closed and there is an integer $n > 1$ such that for every two elements a, $b \in R$, $(a,b)^n = (a^n, b^n)$.*
(8) *R is integrally closed and there exists an integer $n > 1$ such that for every two elements a, $b \in R$, $a^{n-1}b \in (a^n, b^n)$.*
(9) *Each ideal I of R is complete, that is $I = \cap IV_\alpha$ as V_α run over all the valuation overrings of R.*
(10) *Each finitely generated ideal of R is an intersection of valuation ideals.*
(11) *If I, J, and K are non-zero ideals of R, then $I \cap (J+K) = (I \cap J) + (I \cap K)$.*
(12) *If I, J, and K are non-zero ideals of R, then $I(J \cap K) = IJ \cap IK$.*
(13) *If I and J are non-zero ideals of R, then $(I+J)(I \cap J) = IJ$.*
(14) *If I and J are non-zero ideals of R, and K is a finitely generated non-zero ideal of R, then $(I+J):K = (I:K) + (J:K)$.*
(15) *For any two elements a, $b \in R$, $(a:b) + (b:a) = R$.*
(16) *If I and J are two finitely generated non-zero ideals of R, and K is a non-zero ideal of R, then $K:(I \cap J) = (K:I) + (K:J)$.*
(17) *R is integrally closed and each overring of R is the intersection of localizations of R.*
(18) *R is integrally closed and each overring of R is the intersection of quotient rings of R.*
(19) *Each overring of R is integrally closed.*
(20) *R is integrally closed and the prime ideals of any overring of R are extensions of the prime ideals of R.*
(21) *R is integrally closed and for each prime ideal P of R, and each overring S of R, there is at most one prime ideal of S lying over P.*
(22) *For any two polynomials f and $g \in R[x]$, $c(fg) = c(f)c(g)$; where for a polynomial $h \in R[x]$, $c(h)$ denotes the ideal of R generated by the coefficients of h.*

Recall that for a ring R, an *overring of* R is a subring of the total ring of quotients of R containing R. The equivalence of condition 1 and 2 is due to

Prüfer 1932 [56]; 3 and 4 are due to Krull 1936 [49]; 5, 6, 7 and 8 are due to Jensen 1963 [46], and Gilmer 1965 [19]; 9 and 10 are due to Gilmer and Ohm 1965 [18]; 11 - 16 are due to Jensen 1963 [46]; 17 - 21 are due to Butts and Phillips 1965 [10], and Gilmer 1966 [20, 21]; 22 is due to Tsang 1965 [58], and Arnold and Gilmer 1967 [5, 22].

Research in commutative algebra involving Prüfer or related conditions branched in the many directions glimpsed from the above characterizations. The present article is concerned with one of these directions: the extensions of the Prüfer domain notion to rings with zero divisors, and the interesting and surprising ways in which the above characterizations changed in response to the various possible extensions. The centrality of the Prüfer domain notion resulted in an abundance of extensions of this notion to both domains and general rings. Each one of the twenty two characterizations above yielded several such extensions, and a number of others arose from different sources. For that, and for reasons of mathematical preference, we restrict our article to surveying the type of Prüfer domain notion extensions that have a relation to homological algebra. By this we mean that either the methods employed in obtaining the results come from homological algebra, or if ring theoretic methods were applied, the results involve homological algebra notions. In particular, we consider the following extensions of the Prüfer domain notion:

1. R is a semihereditary ring.
2. $w.gl.dimR \leq 1$.
3. R is an arithmetical ring.
4. R is a Gaussian ring.
5. R is a Prüfer ring.

In Section 2 we exhibit the early extensions of the Prüfer domain notion to rings with zero divisors, involving the definitions of arithmetical and Prüfer rings. This approach employed the definition of an extension of the notion of a valuation domain to rings with zero divisors, that is Manis valuations. It is the work of a number of commutative algebraists, and culminates in Griffin's 1970 article [43]. Section 3 describes the extensions of the Prüfer domain notion that arise in homological algebra, namely semihereditary rings and rings of weak global dimension at most one. Classical characterizations of these rings allow for comparisons between them and the two extensions described in Section 2. Section 4 focuses on the extension of the Prüfer domain notion to Gaussian rings. This section covers a number of classical results, as well as recent work by Glaz 2005 [42]. [42] involves homological algebra methods for relating the extensions of the notion of a Prüfer domain to semihereditary rings, rings of $w.gl.\dim R \leqslant 1$, and Gaussian rings; as well as provides a homological algebra characterization of coherent Gaussian rings. In Section 5 all five extensions mentioned above appear, and implication relations between them are clarified. This section surveys briefly recent work by Glaz 2005 [41], and current work of the authors of the present article, Bazzoni and Glaz [7]. In [7] Bazzoni and Glaz introduce a new element into the investigation. This element entails of a closer scrutiny of the five Prüfer conditions mentioned above when imposed on the total ring of quotient of a ring. It is through this devise that exact conditions for reversing the implications between the five extensions were found.

2 Multiplicative Ring Extensions: Arithmetical and Prüfer Rings

The earliest extension we know of the notion of a Prüfer domain to rings with zero divisor, is the notion of an arithmetical ring, defined by Fuchs in 1949 [16]. Specifically:

Definition 2.1. *A ring R is called arithmetical if the lattice formed by the ideals of R is distributive, that is, any three ideals of R, I, J and K satisfy $I \cap (J + K) = (I \cap J) + (I \cap K)$.*

These rings provide an extension to the Prüfer domain notion via property 11 of Theorem 1.1. Early investigations into some of the properties of arithmetical rings brought out more similarities to Prüfer domains. A wealth of results, some of which we will cite in this article, and further references can be found in Jensen [47, 48], Butts and Smith [11], and Griffin [43], Glaz [41]. In Butts and Smith's article [11] arithmetical rings are called Prüfer rings, and what we call nowadays Prüfer rings are called α-rings. Theorem 2.2 below is a combination of results from the articles mentioned in this paragraph.

Theorem 2.2. *Let R be a ring. The following conditions are equivalent.*

(1) *R is an arithmetical ring.*
(2) *The ideals of R_P are totally ordered by inclusion for every prime ideal P of R.*
(3) *The ideals of R_P are totally ordered by inclusion for every maximal ideal P of R.*
(4) *Every finitely generated ideal of R is locally principal.*
(5) *If I and J are ideals of R, and K is a finitely generated ideal of R, then $(I + J){:}K = (I{:}K) + (J{:}K)$.*
(6) *If I and J are two finitely generated ideals of R, and K is an ideal of R, then $K{:}(I \cap J) = (K{:}I) + (K{:}J)$.*
(7) *Let L denote the set of all prime ideals P of R satisfying that the ideals of R_P are totally ordered by inclusion. For each ideal I of R, $I = \cap (IR_P \cap R)$, as P runs over all prime ideals in L.*
(8) *If I and J are two ideals of R with J finitely generated and $I \subset J$, then there exists an ideal K such that $I = JK$.*

The following well know result shows the connection between conditions 2 and 3 of Theorem 2.2 and Krull's characterization of a Prüfer domain given in Theorem 1.1, 3 and 4.

Proposition 2.3. *Let R be a domain. Then R is a valuation domain if and only if the ideals of R_P are totally ordered by inclusion for every prime (respectively maximal) ideal P of R.*

Definition 2.4. *Let R be a commutative ring, and denote by $Q(R)$, the total ring of quotients of R. A (fractionary) ideal I of R is invertible if $II^{-1} = R$, where $I^{-1} = \{r \in Q(R):rI \subset R\}$.*

An invertible ideal is finitely generated and contains a regular element. For an ideal I of R there is a strong relation between invertibility, projectivity, and the property of being locally principal, namely:

Theorem 2.5. *Let R be a ring, and let I be an ideal of R. Then:*

(1) *If I is invertible, then I is projective.*
(2) *If I is projective, then I is locally principal.*
(3) *If I is finitely generated and regular then:*
I is invertible if and only if I is projective if and only if I is locally principal.
In particular, the three conditions are equivalent for a finitely generated ideal of a domain R.

The second early extension of the Prüfer domain notion is the definition of a Prüfer ring. We believe its first appearance in the literature was under the name α-ring in Butts and Smith's article [11]. α-rings were named Prüfer rings by Griffin [43].

Definition 2.6. *A ring R is called a Prüfer ring if every finitely generated regular ideal of R is invertible.*

Prüfer rings appear in the literature in many sources, too numerous to cite fully. We will follow Griffin's article [43] for some of the early results, and mention other as needed in the exposition. Glaz [41], Butts and Smith [11], Huckaba [45], and Larsen and McCarthy [50] provide a more extensive treatment and references on the subject. Griffin [43] extends many of the characterizations of a Prüfer domain stated in Theorem 1.1 to Prüfer rings. He does this by using an extension of the valuation domain notion to rings with zero divisors, the notion of a Manis valuation [54], and by replacing localizations by prime ideals with other quotient rings associated with prime ideals. In order to exhibit Griffin's results on Prüfer rings we digress briefly to outline the definition and few properties of a Manis valuation ring.

Definition 2.7. *A valuation is a map v from a ring Q onto a totally ordered group G and a symbol ∞ , such that for all x and y in Q:*

(1) $v(xy) = v(x) + v(y)$.

(2) $v(x+y) \geqslant min\{v(x), v(y)\}$.
(3) $v(1) = 0$ *and* $v(0) = \infty$.
The ring $V = V_v = \{x \in Q{:}v(x) \geqslant 0\}$, *together with the ideal* $P = P_v = \{x \in Q{:}v(x) > 0\}$, *denoted* (V, P), *is called a Manis valuation pair (of* Q*).* V *is called a Manis valuation ring (of* Q*).* G *is called the value group of* V.

(V, P) is a Manis valuation pair of Q, where V is a subring of a ring Q and P is a prime ideal of V if and only if for every $x \in Q - V$, there exists a $y \in P$ such that $xy \in V - P$. In spite of the similarity to the notion of a valuation domain, there are many important ways in which Manis valuation pairs fail to satisfy the basic analogous properties of valuation domains. For example, there exist Manis valuation rings that are not Prüfer rings. And, if (V, P) is a Manis valuation pair, P may not be the unique regular maximal ideal of V, and it may happen that P is not even a maximal ideal of V. These anomalies seem to feed on each other, as for example, we see in the following result from Boisen and Larsen [8]:

Theorem 2.8. *Let* R *be a ring and let* P *be a prime ideal of* R*. The following conditions are equivalent:*

(1) (R, P) *is a Manis valuation pair and* R *is a Prüfer ring.*
(2) R *is a Prüfer ring and* P *is the unique regular maximal ideal of* R.
(3) R *is a Manis valuation ring and* P *is the unique regular maximal ideal of* R.

In [30] Gilmer provides an interesting example of an integrally closed Manis valuation ring, which is not a Prüfer ring.

Example 2.9. *An example of a ring* R *for which each ideal generated by a finite set of regular elements is invertible,* R *is a Manis valuation ring, but* R *is not a Prüfer ring.*

Let $D = k[x, y]$ be a polynomial ring in two indeterminates over a field k, and let $\{m_\lambda\}$ be the set of maximal ideals of D not containing y. Let $N = \oplus(D/m_\lambda)$, and let $R = D + N$, the idealization of D by N. In [30] it is shown that each ideal of R generated by regular elements is principal, and hence invertible. R is not a Prüfer ring since the regular ideal of R generated by x and y is not R-invertible. By the criteria described in the previous paragraph one can check that R is a Manis valuation ring.

For other early examples of this phenomenon the reader is referred to an example by Griffin appearing in Huckaba's book [45, Chapter VI, Example 7], and an example by Boisen and Larsen [8].

We conclude our short digression into the Manis valuation notion by pointing out a case where the notion of a Manis valuation behaves in a desirable domain-like fashion.

Theorem 2.10. *Let V be a ring in which every regular ideal is generated by its set of regular elements (such a ring is called a Marot ring). The following conditions are equivalent:*

(1) *V is a Manis valuation ring, and a Prüfer ring.*
(2) *V is a Manis valuation ring.*
(3) *For each regular element x in Q, the total ring of quotients of V, either x or x^{-1} lie in V.*

It is worth noting that the notion of a Manis valuation is just one of the possible extensions of the notion of a valuation domain to rings with zero divisors. Among the several candidates that were considered in the early seventies, only one other is still occasionally called in the literature a valuation ring. This is the notion called in [45] a *chained ring*, namely a ring R whose set of ideals is totally ordered by inclusion. Note that a chained ring is a local arithmetical ring. A chained ring is a Marot ring, and therefore it is a Manis valuation ring and a Prüfer ring. The converse does not hold in general as the following example from [45] shows:

Example 2.11. *An example of a ring R which is not a chained ring, but it is a Manis valuation ring and a Prüfer ring.*

Let R be a total ring of quotient that is not chained and let P be a prime ideal of R. Then R becomes a Manis valuation ring and a Prüfer ring by defining a valuation map v satisfying $v(x) = \infty$ if $x \epsilon P$ and $v(x) = 0$ if $x \epsilon R - P$.

Next we consider two quotient rings of a ring R associated to a prime ideal P of R.

Definition 2.12. *Let R be a ring and denote by S a multiplicatively closed subset of R.*

The regular quotient ring of R with respect to S, denoted by $R_{(S)}$, is the localization of R by the set of regular elements in S.

The large quotient ring of R with respect to S, denoted by $R_{[S]}$, is the set $R_{[S]} = \{z \in Q(R) : zs$ is in R for some s in $S\}$

We have the following containments $R \subset R_{(S)} \subset R_{[S]} \subset Q(R)$.

If P is a prime ideal of R and $S = R - P$, we denote the above quotients by the usual $R_{(P)}$ and respectively $R_{[P]}$.

A ring R in which all the regular elements which are not units are contained in a prime ideal P, with $R_{[P]}$ a Manis valuation ring, is called a Manis prevaluation ring.

We are now ready to state Griffin's [43] fifteen conditions equivalent to the definition of a Prüfer ring:

Theorem 2.13. *Let R be a ring. The following conditions are equivalent:*

(1) *R is a Prüfer ring.*
(2) *Every two-generated regular ideal is invertible.*

(3) *$R_{[P]}$ is a Manis valuation ring for every maximal ideal P of R.*
(4) *$R_{(P)}$ is a Manis prevaluation ring for every maximal ideal P of R.*
(5) *For every maximal ideal P of R, the regular ideals of R which are contained in P are totally ordered by inclusion.*
(6) *If I is a finitely generated regular ideal of R, and J and K are ideals of R with $J = IK$, then $J = K$.*
(7) *R is integrally closed and for any two elements a and b in R with a regular, there exists and integer $n > 1$ for which $(a,b)^n = (a^n, b^n)$.*
(8) *If I is a finitely generated regular ideal of R, and J is an ideal of R contained in I, then there exists an ideal K such that $J = KI$.*
(9) *If $J$and K are ideals of R, one of which is regular, and I is an ideal of R, then $I(J \cap K) = IJ \cap IK$.*
(10) *If I and J are ideals of R, one of which is regular, then $(I+J)(I \cap J) = IJ$.*
(11) *If I, J, and K are ideals of R with I regular and K finitely generated, then $(I + J){:}K = (I{:}K) + (J{:}K)$*
(12) *If I, J, and K are ideals of R with K regular and I and J finitely generated, then $I{:}(J \cap K) = (I{:}J) + (I{:}K)$.*
(13) *If I, J, and K are ideals of R one of which is regular, then $I \cap (J+K) = (I \cap J) + (I \cap K)$.*
(14) *Every overring of R is flat.*
(15) *Every overring of R is integrally closed.*

A comparison between Theorem 2.13 and Theorem 1.1 shows the changes which occurred in conditions 1 - 21 of Theorem 1.1 as a result of this extension of the Prüfer domain notion. In spite of the similarities, Prüfer rings are much less "tame" then Prüfer domains, since the invertibility condition affects the regular ideals, while other ideals seem to have almost random behavior. We also note that Theorem 2.13 lacks a condition analogous to condition 22 of Theorem 1.1. This lack is the driving force behind the recent and current investigations described in Sections 4 and 5.

We conclude this section by exhibiting the relation found in Griffin's article [43] between arithmetical, and Prüfer rings.

Definition 2.14. *Let R be a ring. A maximal ideal of zero is an ideal (necessarily prime) maximal with respect to not containing regular elements.*

A ring is said to have arithmetical zero divisors if for every maximal ideal of zero P, the ideals of R_P are totally ordered by inclusion.

Theorem 2.15. *Let R be a ring with total ring of quotients $Q(R)$. R is arithmetical if and only if R is a Prüfer ring and $Q(R)$ has arithmetical zero divisors.*

3 Homological Algebra Extensions: Rings With Weak Global Dimension Less Or Equal To One

Another early generalization of the Prüfer domain notion is that of a semihereditary ring. This notion comes from homological algebra, and the earliest we saw it appear in the literature is in Cartan Eilenberg's book 1956 [12].

Definition 3.1. *A ring R is called semihereditary if all finitely generated ideals of R are projective.*

Various properties of semihereditary rings were and are considered by many authors. We will restrict ourselves to a number of properties that will place this type of rings in the family of Prüfer domain extensions to rings with zero divisors. For the results in this section and additional results and references on semihereditary rings one may consult, for example, Endo [14], Marot [55], Griffin [43], and Glaz [37].

From the homological algebra point of view semihereditary rings belong to the class of rings of finite weak global dimension.

Rings of weak global dimension zero are the so called Von Neumann regular rings. These rings will play an important role in our article and we pause here to record some of their very interesting characterizations.

Theorem 3.2. *Let R be a ring. The following conditions are equivalent:*

(1) *R is Von Neumann regular.*
(2) *Every R-module is flat.*
(3) *For every element x in R, there exists an element y in R, such that $x^2y = x$.*
(4) *Every element of R can be expressed as a product of a unit and an idempotent.*
(5) *Every finitely generated ideal of R is principal generated by an idempotent.*
(6) *R_P is a field for every maximal ideal P of R.*
(7) *R is a reduced self injective ring.*

We note that Von Neumann regular rings are coherent rings.

Semihereditary rings have weak global dimension less or equal to 1. They are precisely those rings of weak global dimension less or equal to one that are coherent (see Theorem 3.3). This means that Von Neumann regular rings are semihereditary rings. From Theorem 2.5 we deduce that the notion of a semihereditary ring is an extension of the Prüfer domain notion. Theorem 3.3 below collects several characterizations of semihereditary rings. Note that like in Theorem 3.2, and contrary to the characterizations of Prüfer and arithmetical rings, the emphasis in Theorem 3.3 is on homological properties rather then on multiplicative ring properties.

Theorem 3.3. *Let R be a ring. The following conditions are equivalent:*

(1) *R is a semihereditary ring.*
(2) *$w.gl.dim R \leqslant 1$ and R is a coherent ring.*
(3) *$Q(R)$, the total ring of quotients of R, is Von Neumann regular, and R_P is a valuation domain for every maximal ideal P of R.*
(4) *$w.gl.dim R \leqslant 1$ and $w.gl.dim Q(R) = 0$.*
(5) *Every finitely generated ideal of R is a summand of an invertible ideal of R.*
(6) *Every finitely generated submodule of a projective R-module is projective.*
(7) *Every torsion-free R-module is flat.*

In general there are non-coherent rings of weak global dimension one (see for example, Glaz [41]). Nevertheless, a general ring of weak global dimension one can also be considered an extension of the Prüfer domain notion to rings with zero divisors. The easiest way to see this is to consider the following characterizations of rings with $w.gl.\dim R \leqslant 1$ [37]:

Theorem 3.4. *Let R be a ring. The following conditions are equivalent:*

(1) *$w.gl.dim R \leqslant 1$.*
(2) *Every ideal of R is flat.*
(3) *R_P is a valuation domain for all prime ideals P of R.*

We conclude this section by pointing out the implications between the four extensions discussed in sections 2 and 3, and some instances where these implications may be reversed. These, and additional results in the same direction, may be found in Glaz [41]. We obtain the following inclusions:

Semihereditary rings $\subset$ Rings with $w.gl.dim R \leqslant 1 \subset$ Arithmetical rings $\subset$ Prüfer rings

The second implication follows from Theorem 3.4 (3) and Theorem 2.2(2). Glaz [41] provides a number of examples that show that these implications cannot be in general reversed. The discussion carried out in sections 2 and 3 provides instances in which we can reverse these implications. For example, we know that if a ring is coherent the first implication can be reversed; while if the total ring of quotients has arithmetical zero divisors, the last implication can be reversed. An instance where the middle implication may be reversed was provided by Jensen [47]:

Theorem 3.5. *Let R be a ring. The following conditions are equivalent:*

(1) *$w.gl.dim R \leqslant 1$.*
(2) *R is an arithmetical reduced ring.*

4 Recent Focus On An Old Extension: Gaussian Rings

Tsang [58], in her 1965 Ph.D. thesis, defined another extension of the Prüfer domain notion to rings with zero divisors. This extension, which she called Gaussian rings, became a focus of intensive recent investigation, primarily because of its connection with the 40 years-old content conjecture of Kaplansky.

Definition 4.1. *Let R be a ring, and let x be a variable over R. Let f be a polynomial in $R[x]$.*

$c(f)$, the content of f, is the ideal of R generated by the coefficients of f.

In general $c(fg) \subset c(f)c(g)$ for every polynomial g in $R[x]$.

A polynomial f satisfying $c(fg) = c(f)c(g)$ for every polynomial g in $R[x]$ is called a Gaussian polynomial.

A ring R is called Gaussian if every polynomial with coefficients in R is a Gaussian polynomial.

Tsang [58] proceeded to prove many important and interesting results on Gaussian polynomials and Gaussian rings, some of which we will reproduce in this section. In particular she showed that a polynomial whose content ideal is invertible, or more generally, locally principal, is a Gaussian polynomial. Kaplansky's conjecture, referred to in the previous paragraph, states that the converse also holds; that is, the content ideal of a Gaussian polynomial must be an invertible, or locally principal, ideal. A number of authors contributed towards a solution of this conjecture: D.D. Anderson and Kang in 1995 [4], Glaz and Vasconcelos in 1997 - 98 [38, 39], Heinzer and Huneke in 1998 [44]. Loper and Roitman in 2005 [51] solved the question affirmatively for all domains. T.G. Lucas in 2005 [52] extended this solution to a partial positive answer for non-domains. One notes, that it is known that in general the answer is No (see Glaz and Vasconcelos [38, 39]). What is still unclear is to what extent is this conjecture and variations on this conjecture true. The following theorem summarizes the results known to date:

Theorem 4.2. *Let R be a ring, and let f be a Gaussian polynomial with coefficients in R, then:*

(1) *If $(0{:}c(f)) = 0$, then $c(f)$ is locally principal.*
(2) *If $c(f)$ is a regular ideal, then $c(f)$ is invertible.*
(3) *In particular if R is a domain, then $c(f)$ is invertible.*

A corollary of this theorem is that the notion of a Gaussian ring may be seen as an extension of the notion of a Prüfer domain. This was indeed already proved independently by both Tsang [58], and Gilmer [22], and appears as condition 22 in Theorem 1.1. Moreover, the theorem above makes some questions, but not all, on the nature of Gaussian rings easier to answer. The following result summarizes some of the characterizations of Gaussian rings found in Tsang [58] , D.D Anderson and Camillo [3], Glaz [42], and Lucas [53]. D.D Anderson articles [1,2] deal with other aspects of the Gaussian ring property.

Theorem 4.3. *Let (R, m) be a local ring with maximal ideal m. The following conditions are equivalent:*

(1) *R is a Gaussian ring.*
(2) *For any finitely generated ideal I of R, $I/I \cap (0{:}I)$ is a cyclic R-module.*
(3) *For any two-generated ideal I of R, $I/I \cap (0{:}I)$ is a cyclic R-module.*
(4) *For any two elements a and b in R, the following two properties hold:*
  (i) *$(a, b)^2 = (a^2)$ or (b^2)*
  (ii) *If $(a, b)^2 = (a^2)$ and $ab = 0$, then $b^2 = 0$.*
(5) *Every homomorphic image of R is an Armendariz ring, where a ring A is Armendariz if for any two polynomials with coefficients in A, f and g, $c(fg) = 0$ implies that $c(f)c(g) = 0$.*

It is interesting to note that the Gaussian property is a local property, and therefore shedding light on the local case provides much of the information needed to clarify the global case. For example, Tsang [58] showed that the prime ideals of a local Gaussian ring (R, m), are totally ordered by inclusion, therefore if (R, m) is not a domain, its nilradical is its unique minimal prime ideal. It follows that a local Gaussian ring modulo its nilradical is a valuation domain. In particular a reduced local Gaussian ring is a valuation domain. We conclude that a semi-local reduced Gaussian ring is a finite direct sum of Prüfer domains.

In 2005 Glaz [42] investigated to what extent the semihereditary condition and the property $w.gl.dimR \leqslant 1$ are close to the Gaussian property. By Theorem 3.4 (3) a ring of $w.gl.dimR \leqslant 1$ is locally and therefore globally Gaussian.

Before we provide the answer given in Glaz [42], we recall two zero divisor controlling conditions on a ring.

Definition 4.4. *A ring R is called a PF ring if principal ideals of R are flat.*

A ring R is called a PP ring (or a weak Bear ring) if principal ideal of R are projective.

The following two results clarify some of the consequences of the PF and PP conditions for a ring (see Glaz [37]).

Theorem 4.5. *Let R be a ring. The following conditions are equivalent:*

(1) *R is a PF ring.*
(2) *R_P is a domain for every prime ideal P of R.*
(3) *R_P is a domain for every maximal ideal P of R.*
(4) *R is reduced and every prime ideal of R contains a unique minimal prime ideal.*
(5) *R is reduced and every maximal ideal P of R contains a unique minimal prime ideal p. In this case $p = \{r \in R$: there is a $u \in R - p$ such that $ur = 0\}$ and $R_p = Q(R_P)$, the quotient field of R_P.*

Theorem 4.6. *Let R be a ring. The following conditions are equivalent:*

(1) *R is a PP ring.*
(2) *For every element a in R, the ideal $(0{:}a)$ is generated by an idempotent.*
(3) *Every element of R can be expressed as a product of a non zero divisor and an idempotent.*

The PP condition on a ring is related to the PF condition in the following theorem from Glaz [37].

Theorem 4.7. *Let R be a ring. The following conditions are equivalent:*

(1) *R is a PP ring.*
(2) *R is a PF ring and $Min\ R$, the set of minimal prime ideals of R with the induced Zarisky topology, is compact.*
(3) *R is a PF ring and $Q(R)$, the total ring of quotients of R, is Von Neumann regular.*

The reader is referred to [37] for a more extensive exposition on the relations between the conditions mentioned in Theorem 4.7, including a number of examples that show that the two conditions of Theorem 4.7 (2), as well as the two conditions of Theorem 4.7 (3), are independent of each other. We are now ready to bring the results found in Glaz [42] on the relation between the Gaussian property of a ring R and the property $w.gl.dimR \leqslant 1$.

Theorem 4.8. *Let R be a ring. The following conditions are equivalent:*

(1) *$w.gl.dimR \leqslant 1$.*
(2) *R is a Gaussian PF ring.*
(3) *R is a Gaussian reduced ring.*

Theorem 4.9. *Let R be a ring. The following conditions are equivalent:*

(1) *R is a semihereditary ring.*
(2) *R is a Gaussian PP ring.*
(3) *R is a Gaussian ring and $Q(R)$, the total ring of quotients of R, is a Von Neumann regular ring.*

The results of Theorems 4.8 and 4.9 were further used by Glaz [42] to find a classification for coherent Gaussian rings. Recall that a *regular ring* is a ring whose finitely generated ideals have finite projective dimensions.

Theorem 4.10. *Let R be a coherent Gaussian ring. Then either $w.gl.dimR \leqslant 1$, or $w.gl.dimR = \infty$. In particular, if R is a regular ring, then R is a semihereditary ring.*

Note that Theorem 4.10 states that a coherent regular Gaussian ring R has $w.gl.dimR \leqslant 1$, a statement that includes the possibility that such a ring is Von Neumann regular. For additional homological properties of coherent Gaussian rings see Glaz [42].

It is worth noting that if R is a Gaussian ring then, by Theorem 4.2 (2), every regular ideal of R is invertible. In addition, by Theorem 2.2 (4) and by Tsang's [58] result, any polynomial over an arithmetical ring is a Gaussian polynomial. We conclude that:

$$\text{Arithmetical rings} \subset \text{Gaussian rings} \subset \text{Prüfer rings}$$

Glaz [41] provides examples that show that the above implications are, in general, not reversible.

5 Current Investigation: Prüfer Conditions On Total Rings of Quotients

In this section we will describe briefly the current work of the authors Bazzoni and Glaz [7]. This work investigates the relations between all five Prüfer conditions mentioned in the introduction. Recall that we have the following implications:

$$\text{Semihereditary rings} \subset \text{Rings of } w.gl.dim R \leqslant 1 \subset \text{Arithmetical rings} \subset \text{Gaussian rings} \subset \text{Prüfer rings}$$

In [7], we shifted emphasis from considering the consequences of imposing any of these conditions on the ring itself, to considering the consequences of imposing these conditions on the total ring of quotients of the ring. In other words, we are concerned with the question: What role does any of the five Prüfer conditions on R have on a Prüfer condition on $Q(R)$, and vice-versa?

We first note [7] that it is easy to answer one direction of this question:

Theorem 5.1. *If R is a ring satisfying any of the five Prüfer conditions mentioned above, then $Q(R)$, the total ring of quotients of R, satisfies the same Prüfer condition.*

Considering the reverse question, we first note that since a total ring of quotients has no proper regular ideals, any ring which is a total ring of quotients is a Prüfer ring. Since there are rings which are not Prüfer rings, we conclude that being a Prüfer ring is not a property that descends from the total ring of quotients to the ring itself.

Although there are total rings of quotients that are not Gaussian rings (see Example 5.3), the Gaussian property still does not descend from the total ring of quotients to the ring (see Example 5.2).

Example 5.2. *A ring R with $Q(R)$ Gaussian, but R not Gaussian.*

Let R be a local, Noetherian, reduced ring which is not a domain. Such a ring cannot be Gaussian. Since R is Noetherian, $Min\ R$ is a finite set. Let $Min\ R = \{P_1, ..., P_n\}$. Then $Q(R) = R_{P_1} \times ... \times R_{P_n}$, and each R_{P_i} is a field. Thus $Q(R)$, as a direct product of fields, is Von Neumann regular and therefore Gaussian.

Example 5.3. *A non-Gaussian total ring of quotients.*

Let k be a countable, algebraically closed field. Let I be an infinite set and denote by K^I the set of all set maps from I to k. Let N denote the set of natural numbers. Quentel [57] (see Glaz [37], or Huckaba [45] for a corrected version of this ring construction), constructed an algebra $R \subset K^{I \times N^N}$, which is a reduced total ring of quotients that is not Von Neumann regular, but its $Min\ R$ is compact in the induced Zarisky topology. This ring can be shown to be non-Gaussian.

Example 5.4. *A Gaussian ring R, such that its total ring of quotients is not Von Neumann regular.*

Let R be the subring of the countable direct product $\prod Q[x]$, where $Q[x]$ is the polynomial ring in one variable x over the rational numbers Q, consisting of $(x, 0, x^2, 0, x^3, 0, \ldots\ldots)$, and all sequences that eventually consist of constants. It can be calculated that $w.gl.dimR \leqslant 1$, therefore R and its total ring of quotient $Q(R)$ are Gaussian. But $Q(R)$ is not Von Neumann regular by Theorem 3.3.

Additional examples, highlighting the Gaussian behavior of total rings of quotients appear in Bazzoni and Glaz [7].

Before answering the question posed in the first paragraph of this section we proved two results, Theorems 5.5 and 5.6, which are of interest in their own right. The proofs of the following results can be found in Bazzoni and Glaz [7].

Theorem 5.5. *Let R be a ring. Then R is an arithmetical ring if and only if R is a Gaussian ring and $Q(R)$ is an arithmetical ring.*

Theorem 5.6. *Let R be a ring. Then R is a Gaussian ring if and only if R is a Prüfer ring and $Q(R)$ is a Gaussian ring.*

Theorems 5.5 and 5.6 provide very powerful tools for the understanding of the exact relations between a ring and its total ring of quotients under any of the five Prüfer conditions considered in this article. We summarize these relations in Theorem 5.7. For a proof of Theorem 5.7 and other results on this topic see Bazzoni and Glaz [7]:

Theorem 5.7. *Let R be a ring, and let $Q(R)$ be its total ring of quotients. The following conditions hold:*

(1) *R is a Gaussian ring if and only if R is a Prüfer ring and $Q(R)$ is a Gaussian ring.*
(2) *R is an arithmetical ring if and only if R is a Prüfer ring and $Q(R)$ is an arithmetical ring.*
(3) *R has $w.gl.dimR \leqslant 1$ if and only if R is a Prüfer ring and $w.gl.dimQ(R) \leqslant 1$.*

(4) *R is a semihereditary ring if and only if R is a Prüfer ring and $Q(R)$ is a semihereditary ring.*
(5) *In the implications between Prüfer conditions displayed at the beginning of this section, we enumerate the Prüfer conditions from left to right, starting with 1 for semihereditary rings, and ending with 5 for Prüfer rings. Then, R has Prüfer condition n if and only if R has Prüfer condition $n+1$ and $Q(R)$ has Prüfer condition n, for all $1 \leqslant n \leqslant 4$.*
(6) *If $Q(R)$ is Von Neumann regular then all five Prüfer conditions on R are equivalent.*

We note that Theorem 5.7 (2) can also be deduced by showing that if a total ring of quotient is arithmetical then it must have arithmetical zero divisors, and then employing Theorem 2.15. We also note that Theorem 5.7 (6) can easily be deduced from a result of Griffins' [43].

References

1. D.D. Anderson, *Another generalization of principal ideal rings,* J. Algebra **48** (1997), 409-416
2. D.D. Anderson, *GCD domains, Gauss' Lemma, and contents of polynomials*, Non-Noetherian commutative ring theory, Kluwer Academic Publishers (2000),1-31
3. D.D. Anderson and V. Camillo, *Armendariz rings and Gaussian rings*, Comm. Algebra **26** (1998), 2265-2272
4. D.D. Anderson and B.G. Kang, *Content formulas for polynomials and power series and complete integral closure,* J. Algebra **181** (1996), 82-94
5. J.T. Arnold and R. Gilmer, *Idempotent ideals and unions of nets of Prüfer domains*, J. Sci. Hiroshima Univ. Ser. A-I Math. **31** (1967), 131-145
6. J.T. Arnold and R. Gilmer, *On the contents of polynomials*, Proc. Amer. Math. Soc. **24** (1970) 556 -562.
7. S. Bazzoni and S. Glaz, *Gaussian properties of total rings of quotients*, preprint
8. M. Boisen and M. Larsen, *Prüfer and valuation rings with zero divisors*, Pac. J. Math. **40** (1972), 7-12
9. J. Brewer and R. Gilmer, *Integral domains whose overrings are ideal transforms,* Math. Nachr. **51** (1971), 255-267
10. H.S. Butts and R.C. Phillips, *Almost multiplication rings*, Canad. J. Math. **17** (1965), 267-277
11. H.S. Butts and W. Smith, *Prüfer rings*, Math. Z **95** (1967), 196-211
12. H. Cartan and S. Eilenberg, *Homological algebra*, Princeton University Press, 1956
13. A. Corso and S. Glaz, *Gaussian ideals and the Dedekind-Mertens Lemma*, Marcel Dekker Lecture Notes Pure Appl. Math **217** (2001), 131 - 143
14. S. Endo, *On semihereditary rings*, J. Math. Soc. Japan **13** (1961), 109-119
15. M. Fontana, J.A. Huckaba and I.J. Papick, *Prüfer domains,* Marcel Dekker, 1997
16. L. Fuchs, *Uber die Ideale arithmetischer ringe*, Math. Helv. **23** (1949), 334-341

17. L. Fuchs and L. Salce, *Modules over non-Noetherian domains*, Amer. Math. Soc. Mathematical Surveys and Monographs **84**, 2001.
18. R. Gilmer and J. Ohm, *Primary ideals and valuation ideals*, Trans. Amer. Math. Soc. **117** (1965), 237-250
19. R. Gilmer, *The cancellation law for ideals in a commutative ring*, Canad. J. Math **17** (1965), 281-287
20. R. Gilmer, *Overrings of Prüfer domains*, J. Algebra **4** (1966), 331-340
21. R. Gilmer, *Domains in which valuation ideals are prime powers*, Arch. Math. (Basil) **17** (1966), 210-215
22. R. Gilmer, *Some applications of the Hilfssatz von Dedekind-Mertens*, Math. Scand. **20** (1967), 240-244
23. R. Gilmer and W. Heinzer, *Overrings of Prüfer domains II,* J. Algebra **7** (1967), 281-302
24. R. Gilmer, *On a condition of J. Ohm for integral domains,* Canad. J. Math. **20** (1968), 970-983
25. R. Gilmer and J.L. Mott, *On proper overrings of integral domains*, Monatsh. Math. **72** (1968), 61-71
26. R. Gilmer, *Multiplicative ideal theory*, Queen's Papers in Pure and Applied Mathematics, Queen's University Press, 1968
27. R. Gilmer, *Two constructions of Prüfer domains*, J. Reine Angew. Math. **239/240** (1969), 153-162
28. R. Gilmer and W. Heinzer, *On the number of generators of an invertible ideal*, J. Algebra **14** (1970), 139-151
29. R. Gilmer, *Multiplicative ideal theory*, Marcel Dekker, 1972
30. R. Gilmer, *On Prüfer rings*, Bull. Amer. Math. Soc. **78** (1972), 223-224
31. R. Gilmer, *Prüfer-like Conditions on the set of overrings of an integral domain*, Lecture Notes in Math. **311**, Springer Berlin (1973), 90-102
32. R. Gilmer and T. Parker, *Semigroup rings as Prüfer rings*, Duke Math. J. **21** (1974), 219-230
33. R. Gilmer and J.F. Hoffman, *The integral closure need not be a Prüfer domain*, Mathematika **21** (1974), 233-238
34. R. Gilmer and J.F. Hoffman, *A characterization of Prüfer domains in terms of polynomials*, Pacific J. Math. **60** (1975), 81-85
35. R. Gilmer, *Prüfer domains and rings of integer valued polynomials*, J. Algebra **129** (1990), 502-517
36. R. Gilmer, *Multiplicative ideal theory*, Queen's University Press, 1992
37. S. Glaz, *Commutative coherent rings*, Springer-Verlag Lecture Notes **1371**, 1989
38. S. Glaz and W.V. Vasconcelos, *Gaussian polynomials*, Marcel Dekker Lecture Notes **186** (1997), 325-337
39. S. Glaz and W.V. Vasconcelos, *The content of Gaussian polynomials*, J. Algebra **202** (1998), 1-9
40. S. Glaz, *Controlling the zero-divisors of a commutative ring*, Marcel Dekker Lecture Notes **231** (2003), 191-212
41. S. Glaz, *Prüfer conditions in rings with zero-divisors*, CRC Press Series of Lectures in Pure Appl. Math. **241**, (2005), 272-282
42. S. Glaz, *The weak dimension of Gaussian rings*, Proc. Amer. Math. Soc. **133** (2005), 2507-2513 (electronic)
43. M. Griffin, *Prüfer rings with zero-divisors*, J. Reine Angew Math. **239/240** (1970), 55-67

44. W. Heinzer and C. Huneke, *Gaussian polynomials and content ideals*, Proc. Amer. Math. Soc. **125** (1997), 739-745
45. J. A. Huckaba, *Commutative rings with zero-divisors*, Marcel Dekker, 1988
46. C.U. Jensen, *On characterizations of Prüfer rings*, Math. Scand. **13** (1963), 90-98
47. C.U. Jensen, *A remark on arithmetical rings*, Proc. Amer. Math. Soc. **15** (1964), 951-954
48. C.U. Jensen, *Arithmetical rings*, Acta Math. Hungr. **17** (1966), 115-123
49. W. Krull, *Beitrage zur arithmetik kommutativer integritatsbereiche,* Math. Z. **41** (1936), 545 - 577
50. M.D. Larsen and P.J. McCarthy, *Multiplicative theory of ideals*, Pure and Appl. Math. 43, Academic Press 1971
51. K. A. Loper and M. Roitman, *The content of a Gaussian polynomial is invertible*, Proc. Amer. Math. Soc. **133** (2005), 1267-1271 (electronic)
52. T.G. Lucas, *Gaussian polynomials and invertibility,* Proc. Amer. Math. Soc. **133** (2005), 1881-1886 (electronic)
53. T.G. Lucas, *The Gaussian property for rings and polynomials*, preprint
54. M.E. Manis, *Extension of valuation theory*, Bull. Amer. Math. Soc. **73** (1967), 735-736
55. J. Marot, *Sur les anneaux universellement Japonais*, Ph.D. Thesis, Universite de Paris-Sud, 1977
56. H. Prüfer, *Untersuchungen uber teilbarkeitseigenschaften in korpern*, J. Reine Angew. Math. **168** (1932), 1-36
57. Y. Quentel, *Sur la compacite du spectre minimal d'un anneau*, Bull. Soc. Math. France **99** (1971), 265 - 271
58. H. Tsang, *Gauss's Lemma*, Ph.D. Thesis, University of Chicago,1965

Subrings of zero-dimensional rings

Jim Brewer[1] and Fred Richman[2]

[1] Florida Atlantic University brewer@fau.edu
[2] Florida Atlantic University richman@fau.edu

It is an honor to write an article in a volume dedicated to the work of Robert Gilmer.

1 Introduction

When Sarah Glaz, Bill Heinzer and the junior author of this article approached Robert with the idea of editing a book dedicated to his work, we asked him to give us a list of his work and to comment on it to the extent he felt comfortable. As usual, he was extremely thorough in his response. When the authors of this article began to consider what topic we wanted to write about, we were impressed by Robert's comment that he was particularly pleased with his series of papers with Bill on the embeddability of a ring in a zero-dimensional ring. So we decided to write about that. Our task was complicated by the fact that Robert had already written several excellent expository papers on the subject [8, 9, 10].

The problem is easy to state.

Problem 1.1. Find necessary and sufficient conditions on a ring R for it to be embeddable in a ring of (Krull) dimension zero.

Note that it is not required that the total quotient ring of R be zero dimensional, only that R can be embedded in some zero-dimensional ring. Although interesting on its face, the problem appears quite innocuous. Indeed, when Robert and Bill began to look at it seriously, it had already been solved by Arapović [3, Theorem 7]. More specifically, Arapović proved the following result.

Theorem 1.2. (Arapović) *A ring R is embeddable in a zero-dimensional ring if and only if R has a family of primary ideals $\{Q_\lambda\}_{\lambda\in\Lambda}$, such that:*

A1. $\bigcap_{\lambda\in\Lambda} Q_\lambda = 0$, *and*

A2. *For each* $a \in R$, *there is* $n \in \mathbf{N}$ *such that for all* $\lambda \in \Lambda$, *if* $a \in \sqrt{Q_\lambda}$, *then* $a^n \in Q_\lambda$.

In [16] Bill says, "This result of Arapovic is definitive, but there [remain] a number of questions concerning the existence of zero-dimensional extensions."

Condition A1, as a stand-alone condition on R, simply says that the intersection of *all* primary ideals in R is equal to 0, while A2 is a uniformity condition on the primary ideals of the given family $\{Q_\lambda\}_{\lambda\in\Lambda}$. Moreover, as conditions on the family of ideals Q_λ, condition A1 is inherited by superfamilies and A2 by subfamilies, so there is some sort of balance going on. Notice that each of these conditions makes sense for any family of ideals, not just for a family of primary ideals.

Condition A2 on a family of ideals can be characterized as saying that intersection commutes with radical for any (countable) subfamily.

Theorem 1.3. *A family* $\{Q_\lambda\}_{\lambda\in\Lambda}$ *of ideals in a ring* R *satisfies A2 if and only if for each (countable) subset* $\Gamma \subset \Lambda$

$$\sqrt{\bigcap_{\lambda\in\Gamma} Q_\lambda} = \bigcap_{\lambda\in\Gamma} \sqrt{Q_\lambda}.$$

Proof. If $\{Q_\lambda\}_{\lambda\in\Lambda}$ satisfies A2 and $x \in \bigcap_{\lambda\in\Gamma}\sqrt{Q_\lambda}$, then there exists n such that $x^n \in Q_\lambda$ for each $\lambda \in \Gamma$. Hence $x \in \sqrt{\bigcap_{\lambda\in\Gamma} Q_\lambda}$. The other inclusion always holds. Conversely, given $x \in R$, let $\Gamma = \{\lambda \in \Lambda : x \in \sqrt{Q_\lambda}\}$. Then the displayed equation says that there is n such that $x^n \in Q_\lambda$ for all $\lambda \in \Gamma$.

Finally, we note that if the displayed equation fails for some subset $\Gamma \subset \Lambda$, then it fails for a countable subset of Γ. Indeed, if $x \notin \sqrt{\bigcap_{\lambda\in\Gamma} Q_\lambda}$, then for each n there exists $\lambda_n \in \Gamma$ such that $x^n \notin Q_{\lambda_n}$, so $x \notin \sqrt{\bigcap_{n=1}^{\infty} Q_{\lambda_n}}$.

2 Zero-dimensional rings

We will need an arithmetic characterization of zero-dimensional rings. Such characterizations take the form of properties satisfied by every element of the ring. The following lemma gives three of the more interesting properties.

Lemma 2.1. *Let* x *be an element of a ring* R. *The following three conditions are equivalent:*

(1) *There is an idempotent* $e \in R$ *such that* xe *is a unit in* Re *and* $x(1-e)$ *is nilpotent.*
(2) *There exists* n *such that* $Rx^n = Rx^{n+1}$.
(3) $Rx + \bigcup_{n=1}^{\infty}(0:x^n) = R$.

The idempotent in (1) is unique.

Proof. If x satisfies (1), then $x = xe + x(1-e)$. So $x^n = x^n e$, because $x(1-e)$ is nilpotent, and $rxe = e$ for some $r \in R$, because xe is a unit in Re. Thus $rx^{n+1} = rxx^n e = x^n e = x^n$. If x satisfies (2), then $x^n = rx^{n+1}$, so $x^n(1-rx) = 0$, which says that $1-rx \in (0{:}x^n)$ whence $1 \in Rx + \bigcup_{n=1}^{\infty}(0{:}x^n)$. If x satisfies (3), then $rx + a = 1$ where $ax^n = 0$. So $rx^{n+1} = x^n$. Let $e = r^n x^n$. Then $e^2 = r^n(r^n x^{2n}) = r^n x^n = e$. Moreover, $rxe = r^n rx^{n+1} = r^n x^n = e$ and $(x(1-e))^n = x^n(1-e) = x^n(1-r^n x^n) = x^n - r^n x^{2n} = 0$.

To see that the idempotent in (1) is unique, suppose $e, f \in R$ are idempotents such that xe is a unit in Re and $x(1-f)$ is nilpotent. Then $e = rxe = rxfe + rx(1-f)e$. As $x(1-f)$ is nilpotent, $e = r^n x^n fe$ whence $e \leq f$. By symmetry, the idempotent in (1) is unique.

Notice that each of the three conditions of Lemma 2.1 asserts that a certain set of natural numbers n is nonempty, and the proof of Lemma 2.1 shows that these three sets are the same.

Theorem 2.2. *A ring R is zero dimensional if and only if the conditions of Lemma 2.1 are satisfied for every $x \in R$.*

Proof. We first prove the contrapositive of the statement that if R is zero dimensional, then condition (3) of the lemma holds for all $x \in R$. If $xR + \bigcup_{n=1}^{\infty}(0{:}x^n)$ is a proper ideal, then it is contained in a prime ideal Q of R. Look at the multiplicative system $S = \{x^n y{:}n = 0, 1, 2, \ldots \text{ and } y \in R \setminus Q\}$. Now $0 \notin S$ because $y \notin Q$ and $Q \supseteq \bigcup_{n=1}^{\infty}(0{:}x^n)$. Thus there is a prime ideal P of R that misses S. Since $S \supseteq R \setminus Q$, we have $P \subseteq Q$. Moreover, $x \in Q \setminus P$ so $P \neq Q$ whence P is not maximal.

To finish, we prove the contrapositive of the statement that condition (3) for all x implies that R is zero dimensional. Suppose there exist distinct prime ideals $P \supset Q$ in R and let $x \in P \setminus Q$. Then $\bigcup_{n=1}^{\infty}(0{:}x^n) \subset Q$ and $Rx \subset P$, so $Rx + \bigcup_{n=1}^{\infty}(0{:}x^n) \subset P$.

Condition (1) seems to be the most perspicuous. The outstanding arithmetic property of zero-dimensional rings is that they have lots of idempotents while the difference between zero-dimensional rings and von Neumann regular rings is that the former have nilpotent elements. Condition (1) is reminiscent of the decomposition of a linear transformation into its semisimple and nilpotent parts. Condition (2) is the descending chain condition on the powers of Rx. A variant of (1) and (2), which Gilmer and Heinzer use, is that some power of Rx is an idempotent ideal (see Lemma 4.1).

Arapović [1, Theorem 6] showed that R is zero dimensional if and only if R is a total quotient ring and for every $x \in R$ there exists an idempotent e_x such that $x + (1 - e_x)$ is invertible and $x(1-e_x)$ is nilpotent. Clearly this is the same e_x as in (1). It's not too difficult to show that $e_{xy} = e_x e_y$.

Corollary 2.3. *The intersection of an arbitrary nonempty set of zero-dimensional subrings of a ring R is zero dimensional. More specifically, if $x \in R$ and*

$\mathcal{S}_x$ is the set of subrings containing x and satisfying condition (1) of Lemma 2.1, then $\mathcal{S}_x$ is closed under nonempty intersection.

Proof. Each $S \in \mathcal{S}_x$ contains an idempotent e such that xe is invertible in the ring Se, and $x(1-e)$ is nilpotent. Clearly xe is also invertible in the ring Re, so if $\mathcal{S}_x$ is nonempty, then $R \in \mathcal{S}_x$. Now the uniqueness of e in condition (1) implies that $e \in \bigcap \mathcal{S}_x$. Moreover, if $S \in \mathcal{S}_x$, then there is an element $s \in S$ such that $sxe = e$ so $se \in S$ is the inverse of xe in the ring Re, that is, the inverse t of xe in the ring Re lies in Se. So $t \in \bigcap \mathcal{S}_x$. But any subring containing x, e, and t is in $\mathcal{S}_x$, so $\mathcal{S}_x$ is closed under nonempty intersection.

The first part of Corollary 2.3 was proved by Arapović [2, Theorem 7] for R zero-dimensional and by Gilmer and Heinzer [13] in general. Arapović didn't really need the zero-dimensional hypothesis for his proof, as you might expect given that the general theorem is true. The element t in the proof of Corollary 2.3 is the "pointwise inverse" of x, see [20, Lemma 2] and [15, Lemma 4.3.9], used by Gilmer and Heinzer. It is characterized by the two equations $txt = x$ and $xtx = t$. These equations make sense in any semigroup, not necessarily commutative, see Clifford and Preston [4, §1.9].

Condition (2) of Lemma 2.1 with $n = 1$ is the defining condition for a von Neumann regular ring. What kinds of rings do you get if you impose condition (2) for, say, $n = 2$?

The ideal $I = \bigcup_{n=1}^{\infty}(0{:}x^n)$ in condition (3) is the kernel of the localization map $R \to R_S$ where S consists of the powers of x. So condition (3) says that the localization map $R \to R_S$ is onto for any multiplicatively closed set S. Note also that x is regular in R/I, and that if $\varphi{:}R \to R'$ takes x to a regular element, then $\varphi(I) = 0$. So condition (3) is equivalent to the condition that every regular homomorphic image of x is invertible. It follows that a ring has dimension zero if and only if every homomorphic image is a total quotient ring.

In [14, Proposition 2.4], Robert and Bill show that for products of local rings, the localization map is onto at any maximal ideal. Of course that's also true for zero-dimensional rings by condition (3). What other rings have that property?

It's not hard to see that condition (3) is equivalent to the condition $xR + \left(\sqrt{0}{:}x\right) = R$. Indeed

$$xR + \bigcup_{n=1}^{\infty}(0{:}x^n) \subset xR + \left(\sqrt{0}{:}x\right) \subset \sqrt{xR + \bigcup_{n=1}^{\infty}(0{:}x^n)}.$$

Both containments above can be strict. The quotient ring of R modulo $xR + \left(\sqrt{0}{:}x\right)$ is called the **upper boundary** $R^{\{x\}}$ of x in R by Coquand, Lombardi, and Roy in [5]. This allows an elegant inductive definition of (Krull) dimension: $\dim R \le n$ if $\dim R^{\{x\}} \le n-1$ for all $x \in R$, the dimension of the trivial ring being set equal to -1.

Arapović's theorem implies, of course, that condition A2 must hold for some family of primary ideals in any zero-dimensional ring. It is an interesting fact that A2 holds for the family of *all* ideals in a zero-dimensional ring, hence also for any family of ideals. Moreover this property characterizes zero-dimensional rings.

Theorem 2.4. *The following conditions on a ring R are equivalent.*

(1) *The ring R is zero dimensional,*
(2) *Condition A2 holds for the family of all ideals of R,*
(3) *Condition A2 holds for the family of all primary ideals of R.*

Proof. Suppose that R is zero dimensional. For $x \in R$, let n be such that $x^n R = x^{n+1} R$, as in Theorem 2.2, condition (3). If I is any ideal, and $x \in \sqrt{I}$, then $x^m \in I$ for some m, hence $x^n \in I$ also. So A2 holds for the family of all ideals of R.

Conversely, suppose A2 holds for the family of all primary ideals of R. To show that R is zero dimensional, it suffices, by Theorem 1.3, to show that if $\dim R > 0$, then there is a sequence of primary ideals Q_n such that $\sqrt{\bigcap_{n=1}^{\infty} Q_n} \neq \bigcap_{n=1}^{\infty} \sqrt{Q_n}$.

Let P' be a prime ideal of R that is not maximal. The primary ideals of R/P' are in one-to-one correspondence with the primary ideals of R that contain P', and this correspondence respects intersections and radicals, so we may assume that R is an integral domain.

Let x be a nonzero nonunit of R and let P be a minimal prime of the principal ideal Rx. For each positive integer n, let

$$Q_n = R_P x^n \cap R = \{r \in R : sr \in Rx^n \text{ for some } s \in R \backslash P\}.$$

Each Q_n is P-primary, so $\bigcap_{n=1}^{\infty} \sqrt{Q_n} = P$. To show that $x \notin \sqrt{\bigcap_{n=1}^{\infty} Q_n}$, it suffices to show that $x^{n-1} \notin Q_n$. Suppose, by way of contradiction, that $x^{n-1} \in Q_n$. Then $sx^{n-1} \in Rx^n$ for some $s \in R \backslash P$, so $s \in Rx \subset P$ because R is an integral domain, a contradiction.

The authors are indebted to the referee for the proof that the last condition in Theorem 2.4 is equivalent to the other two.

Why is the intersection of all primary ideals in a zero-dimensional ring equal to zero? At the end of [7], Robert says that it would be interesting to have a characterization of the intersection of all primary ideals in a ring R. For the nonce we will call that intersection the *Gilmer radical* of R and denote it by $G(R)$. So A1 for primary ideals is the condition that the Gilmer radical be zero.

The following theorem shows that $G(R)$ is indeed a radical, that is, $G(R/G(R)) = 0$. It also characterizes the condition $G(R) = 0$ in terms of zero-dimensional rings.

Theorem 2.5. *The following ideals of R are equal.*

(1) *The intersection of all primary ideals of* R,
(2) *The intersection of the kernels of all maps of* R *into zero-dimensional local rings,*
(3) *The intersection of the kernels of all maps of* R *into zero-dimensional rings.*

Proof. To show that (2) is contained in (1), we note that if Q is primary, and $P = \sqrt{Q}$, then R_P/Q_P is a zero-dimensional local ring and the kernel of the natural map from R to R_P/Q_P is Q because if $r/1 = q/s$, then $s'sr \in Q$ so $r \in Q$ because $s's \notin P$. To see that (1) is contained in (2), note that the kernel of a map from R into a zero-dimensional local ring is primary because zero is a primary ideal in a zero-dimensional local ring. It remains to show that (2) is contained in (3). The key observation for that is that any ring is a subring of the product of its localizations at each of its maximal ideals, that is, the natural map $R \to \prod R_M$ is one-to-one.

So $G(R) = 0$ if and only if R is a subring of a product of zero-dimensional (local) rings, as Arapović proves in [3, Theorem 13]. In particular, A1 and A2 hold for the family of all primary ideals of a zero-dimensional ring, hence for the corresponding family of primary ideals that are contractions of these in any subring. This establishes the necessity of Arapović's two conditions.

If I is an ideal of a zero-dimensional ring R, then R/I is also zero dimensional, so I is an intersection of primary ideals. Note that the nilradical of a ring is the intersection of the kernels of maps into fields, and the Jacobson radical is the intersection of the kernels of maps onto fields.

3 Products of zero-dimensional rings

If direct products of zero-dimensional rings were zero dimensional, then $G(R) = 0$ would be a necessary and sufficient condition for the embeddability of R in a zero-dimensional ring. However, there are direct products of zero-dimensional rings that are not zero dimensional. In particular, the following ring was described by Robert [9] as "a good test case for several questions Heinzer and I have considered". First note that if a ring R is zero dimensional, then its Jacobson radical $J(R)$ is equal to its nilradical $N(R)$, and that the Jacobson radical of a direct product of rings is equal to the direct product of their Jacobson radicals.

Example 3.1. Let $R = \prod_{n=1}^{\infty}(\mathbf{Z}/p^n\mathbf{Z})$. Then

$$J(R) = J\left(\prod_{n=1}^{\infty} \frac{\mathbf{Z}}{p^n\mathbf{Z}}\right) = \prod_{n=1}^{\infty} J\left(\frac{\mathbf{Z}}{p^n\mathbf{Z}}\right) = \prod_{n=1}^{\infty} \frac{p\mathbf{Z}}{p^n\mathbf{Z}},$$

so the element $p \cdot 1$ of R belongs to $J(R)$. But $p \cdot 1$ does not belong to $N(R)$, so R is not zero dimensional.

We will see below that the dimension of this ring is actually infinite. Maroscia [19] gave necessary and sufficient conditions for a direct product of zero-dimensional rings to be zero dimensional. As noted above, if a ring has dimension zero, then its Jacobson radical is equal to its nilradical. This condition on radicals is not sufficient for a ring to be zero dimensional as the ring of integers shows, but for a direct product of rings of dimension zero, it is exactly what is needed.

Theorem 3.2. (Maroscia) *Let* $\{R_\lambda\}_{\lambda\in\Lambda}$ *be a family of zero-dimensional rings. The following conditions are equivalent:*

(1) *The ring* $S = \prod R_\lambda$ *is zero dimensional.*
(2) $J(S) = N(S)$.
(3) $N(S) = \prod N(R_\lambda)$.

Proof. It is clear that (1) implies (2) because the $J = N$ in any zero-dimensional ring. The implication from (2) to (3) is true because the J commutes with products and the R_λ have dimension zero. Now suppose (3) holds. Then

$$\frac{S}{N(S)} = \frac{\prod R_\lambda}{\prod N(R_\lambda)} \simeq \prod \frac{R_\lambda}{N(R_\lambda)}.$$

For each $\lambda \in \Lambda$, the ring $R_\lambda/N(R_\lambda)$ is zero dimensional and reduced, that is, it is a von Neumann regular ring. As products of von Neumann regular rings are von Neumann regular, $\prod(R_\lambda/N(R_\lambda))$ is von Neumann regular. Since the dimension of $S/N(S)$ is zero, the dimension of S is zero.

In [12, Theorem 3.4], Gilmer and Heinzer added a fourth condition to these three: $\dim(S) < \infty$. Thus, the dimension of a direct product of zero-dimensional rings is either zero or infinite; there is no in between. This significant contribution shows that the ring in the example above is infinite dimensional. The proof is much more involved than that of the equivalence of conditions (1)–(3) and we shall omit it here in favor of an alternative proof in Section 7 using an arithmetic characterization of Krull dimension due to Lombardi [18].

Gilmer and Heinzer introduced two other equivalent conditions in [11]. One is of particular interest to us because, for our purposes, it would be enough to embed an arbitrary direct product of zero-dimensional rings in a zero-dimensional ring. But, the following theorem of Gilmer and Heinzer shows that approach won't work.

Theorem 3.3. *If* $S = \prod_{\lambda\in\Lambda} R_\lambda$ *is a product of zero-dimensional rings, then the following conditions are equivalent.*

(1) *The ring* S *can be embedded in a zero-dimensional ring;*
(2) *The set* $\Delta_m = \{\lambda \in \Lambda : x^m \neq 0$ *for some* $x \in N(R_\lambda)\}$ *is finite for some positive integer* m*;*

(3) *The ring S is zero dimensional.*

Proof. Clearly (3) implies (1). To see that (2) implies (3), choose m such that Δ_m is finite and let $\Gamma = \Lambda \setminus \Delta_m$. Then

$$S = T \times \prod_{\lambda \in \Delta_m} R_\lambda,$$

where $T = \prod_{\lambda \in \Gamma} R_\lambda$. If $x \in \prod_{\lambda \in \Gamma} N(R_\lambda)$, then $x^m = 0$ by definition of Γ so $x \in N(T)$. Since the inclusion $N(T) \subset \prod_{\lambda \in \Gamma} N(R_\lambda)$ always holds, we conclude that $N(T) = \prod_{\lambda \in \Gamma} N(R_\lambda)$, so by the result of Maroscia, $\dim(T) = 0$. As Δ_n is finite, and R_λ is zero dimensional, $\dim(S) = 0$ as well.

The implication from (1) to (2), or rather its contrapositive, is handled easily by a simple criterion that Robert introduced in [7], namely that if S can be embedded in a zero-dimensional ring, and $x \in S$, then the ascending chain of ideals $0{:}x^k$ stabilizes (see Theorem 5.1). If (2) fails, then we can find distinct $\lambda_1, \lambda_2, \lambda_3, \ldots$ in Λ, elements $x_i \in N(R_{\lambda_i})$, and positive integers $m_1 < m_2 < m_3 < \cdots$ so that $x_i^{m_i} \neq 0$ and $x_i^{m_{i+1}} = 0$. Let the λ_i-th coordinate of $x \in S$ be x_i, for $i = 1, 2, \ldots$, and the rest of the components be zero (or whatever). Then the element of S whose coordinate is 1 in R_{λ_i} and 0 elsewhere is in $0{:}x^{m_{i+1}}$ but not in $0{:}x^{m_i}$.

All this taken together gives the following lovely result about when a direct product of zero-dimensional rings has dimension zero.

Theorem 3.4. (Maroscia, Gilmer, and Heinzer) *Let $\{R_\lambda\}_{\lambda \in \Lambda}$ be a family of zero-dimensional rings and let $S = \prod_{\lambda \in \Lambda} R_\lambda$. The following conditions are equivalent.*

(1) *The ring S is zero dimensional;*
(2) *The ring S is finite dimensional;*
(3) $J(S) = N(S)$;
(4) $N(S) = \prod_{\lambda \in \Lambda} N(R_\lambda)$;
(5) *The ring S is a subring of a zero-dimensional ring;*
(6) *The set $\{\lambda \in \Lambda{:} x^m \neq 0$ for some $x \in N(R_\lambda)\}$ is finite for some positive integer m.*

4 Sufficiency

We return to the embedding problem. We have seen that if $G(R) = 0$, then R can be embedded in a direct product S of zero-dimensional rings, but that S need not be zero dimensional. Nor can that deficiency be remedied merely by enlarging S. What if we look at rings between S and R?

Gilmer and Heinzer [13, Theorem 3.1] proved a very pretty generalization of the arithmetic characterization of a zero-dimensional ring, which distills and clarifies two related constructions of Arapović in [2, Theorem 7] and [3, Theorem 7]:

> If R is a subring of S, then R is contained in a zero-dimensional subring of S if and only if for each $x \in R$, some power of xS is an idempotent ideal of S.

The condition here is the relative version of condition (1) of Lemma 2.1 and says that S, rather than R itself, has enough idempotents to decompose each element of R into a unit and a nilpotent coordinate. That's all you need for Arapović's two constructions. Gilmer and Heinzer rely on the unique "pointwise inverse" of an element x that generates an idempotent ideal: this is the inverse of x within the idempotent ideal viewed as a ring [20, Lemma 2], [15, Lemma 4.3.9]. The required zero-dimensional subring of S is generated over R by the pointwise inverses of the elements x^m where m is the smallest positive integer (or any positive integer) such that $x^m S$ is idempotent.

To see how this result relates to Arapović's constructions, we start with a simple lemma connecting the two approaches.

Lemma 4.1. *Let S be a ring and $x \in S$. Then the following three conditions on an idempotent e of S are equivalent.*

- *$x(1-e)$ is nilpotent and $x+(1-e)$ is invertible (Arapović),*
- *$x(1-e)$ is nilpotent and xe is invertible in Se,*
- *Some power of Sx is equal to Se (Gilmer-Heinzer).*

At most one such idempotent e exists.

Proof. Note that $x+(1-e)$ is invertible if and only if $xe = (x+(1-e))e$ is invertible in Se and $(1-e)+x(1-e) = (x+(1-e))(1-e)$ is invertible in $S(1-e)$. But if $x(1-e)$ is nilpotent, then $(1-e)+x(1-e)$ is invertible in $S(1-e)$. So the first two conditions are equivalent. As $x = xe + x(1-e)$, the Gilmer-Heinzer condition says that $(x(1-e))^n = 0$ and $Sx^n = Se$ for some n. The second condition says exactly the same thing because xe is invertible in Se if and only if $x^n e$ is.

The uniqueness of e follows immediately from the Gilmer-Heinzer condition and the fact that an idempotent principal ideal has a unique idempotent generator.

Arapović considers two cases $R \subset S$ in which every element $x \in R$ admits such an idempotent in the larger ring S. The first [2, Theorem 7] is where $\dim S = 0$, in which case every element of S admits such an idempotent. The second [3, Theorem 7] is where $S = \prod T(R/Q_\lambda)$ where the Q_λ are primary ideals satisfying A1 and A2. Here $T(R/Q_\lambda)$ is the total quotient ring of R/Q_λ. Because of A1, the ring R can be considered a subring of S, and because of A2, for each $x \in R$ there exists an idempotent e of S satisfying the conditions of

Lemma 4.1. This idempotent is constructed using the n in the definition of A2 and the fact that each element of $T(R/Q_\lambda)$ is either nilpotent or invertible.

In both of these cases, Arapović constructed a minimal zero-dimensional extension ring of R within S, pretty much as follows.

Theorem 4.2. *Let $R \subset S$ be rings such that for each element $x \in R$ there is an idempotent $e \in S$ satisfying the conditions of Lemma 4.1. Then there is a unique smallest zero-dimensional subring of S containing R.*

Proof. Let R' be the ring generated by R and the idempotents $e_x \in S$ for $x \in R$. Any zero-dimensional subring of S containing R must contain R' because of the uniqueness of the idempotents e_x. Let E be the boolean algebra of idempotents generated by the idempotents e_x for $x \in R$. Then each element $a \in R'$ can be written as $a = r_1 f_1 + \cdots + r_n f_n$ where the $f_i \in E$ are orthogonal. Let $e = e_{r_1} f_1 + \cdots + e_{r_n} f_n$. Then

$$a(1-e) = r_1(1-e_{r_1}) f_1 + \cdots + r_n(1-e_{r_n}) f_n$$

is nilpotent because the elements $r_i(1-e_{r_i})$ are nilpotent. Moreover, ae is invertible in Se because $r_i e_{r_i}$ is invertible in Se_{r_i}. If a is regular in R', then $e = 1$ so $f_i \leq e_{r_i}$ for each i. But $r_i e_{r_i}$ is invertible in Se_{r_1}, so $r_i f_i$ is invertible in Sf_i, whence a is invertible in S. Thus regular elements of R' are invertible in S.

The total quotient ring T of R' within S, is the desired subring. It is contained in any zero-dimensional subring of S containing R because such a subring must contain R' and be a total quotient ring. It is zero dimensional because if $x = a/b$ is in the total quotient ring of R' within S, then the e constructed above for $a \in R'$ has the property that ae is invertible in Se and $a(1-e)$ is nilpotent. so xe is invertible in Te and $x(1-e)$ is nilpotent.

In [13], Gilmer and Heinzer showed that the intersection of any nonempty family of zero-dimensional subrings of a commutative ring S is zero dimensional. This follows, rather impredicatively, from Theorem 4.2 upon taking R to be the intersection.

5 The new criterion

We turn to the construction of rings which are not embeddable in a zero-dimensional ring. For that purpose, we use a simple consequence of Theorem 2.2. This observation is due to Robert [7].

Theorem 5.1. *If the ring R is embeddable in a zero-dimensional ring S, then for each element $x \in R$, there exists a positive integer m such that x^m and x^{m+1} have the same annihilator in R.*

Proof. Let $x \in R$. By Theorem 2.2, choose a positive integer n such that $Sx^n = Sx^{n+1}$. Then $\mathrm{Ann}_S x^n = \mathrm{Ann}_S x^{n+1}$, where Ann_S denotes the annihilator in S. Consequently, $\mathrm{Ann}_R(x^n) = R \cap \mathrm{Ann}_S x^n = R \cap \mathrm{Ann}_S x^{n+1} = \mathrm{Ann}_R x^{n+1}$.

We will refer to the condition of Theorem 5.1 as "the new criterion", from the title of [7]. The general idea has a history in the study of subrings of a class of rings: you take a condition on ideals and restrict it to annihilator ideals (annulets). For example, if you want to characterize subrings of Noetherian rings, an obvious property to consider is the ascending chain condition on annulets (acc$\perp$) because annulets are contracted from any extension. This condition does not characterize subrings of Noetherian rings because of Jeanne Kerr's example [17] of a commutative Goldie ring R (acc$\perp$ and finite Goldie dimension) such that $R[X]$ does not have acc$\perp$, so acc$\perp$ is not inherited by polynomial rings but being embeddable in a Noetherian ring is (see also Moshe Roitman [21]).

Note that the new criterion for (fixed) $m = 1$ simply says that R is reduced, which is the exact condition necessary for embedding R in a von Neumann regular ring, that is, a ring S such that $Sx = Sx^2$ for all x. So it is not that far fetched that the new criterion would also be a *sufficient* condition for embeddability in a zero-dimensional ring. What can you deduce from the condition that x^2 and x^3 have the same annihilator for all x in R?

6 Valuation rings

Following Fuchs and Salce [6], we call a ring R a *valuation ring* if the (principal) ideals of R form a chain under inclusion. These are also called *uniserial rings* or *chained rings* (Robert's preference). Arapović [3, Theorem 8] gave a class of examples of rings where the intersection of all primary ideals is not zero. He then gave a generic example of a valuation ring with that property. Here is such an example:

Let E be the set of weakly positive elements of the group $\mathbf{Z} \oplus \mathbf{Z}$ under the lexicographic order, that is, the elements (a, b) such that either $a > 0$, or $a = 0$ and $b \geq 0$:

$$(0,0) < (0,1) < \cdots < (1,-2) < (1,-1) < (1,0) < (1,1) < \cdots .$$

Consider polynomials in X with coefficients in a fixed field k and exponents in E. Then allow denominators with nonzero constant terms. The result is a valuation domain R. The claim is that the ideal I generated by $X^{(1,1)}$ is not an intersection of primary ideals. Pass to the valuation ring R/I. Now $X^{(1,0)}X^{(0,1)} = 0$, and $X^{(0,1)}$ is not nilpotent, so $X^{(1,0)}$ is in every primary ideal of R/I.

It's an easy observation that in a valuation ring the prime ideals are exactly the complements of saturated submonoids that don't contain 0 (because in a valuation ring, a nonempty subset that's closed under multiplication by ring elements is an ideal—you don't need to require closure under addition). For valuation domains, that's the well-known result that prime ideals correspond to convex subgroups of the divisibility group (a totally ordered abelian group). So the total quotient ring of a valuation ring is obtained by localizing at the prime ideal which is the complement of the set of regular elements.

The following lemma was observed by Robert in his proof of [10, Result 1.8].

Lemma 6.1. *The intersection of any set (chain) of primary ideals in a valuation ring is primary.*

Proof. Let $Q = \bigcap_i Q_i$, where Q_i is primary, and let $P = \bigcap_i \sqrt{Q_i}$. Note that P is a prime ideal containing Q. If P' is any prime ideal containing Q, then, because we are in a valuation ring, either $P' = Q$ or $P' \supset Q_i$ for some i. In particular, either Q is prime, and we're done, or P is the smallest prime ideal containing Q, whence $P = \sqrt{Q}$. Moreover, either $P = Q$ or $P \supset Q_i$ for some i, so we may assume that $P = \sqrt{Q_i}$ for some i and thus we may assume that $P = \sqrt{Q_i}$ for all i. Now if $st \in Q$ and $s \notin P$, then $t \in Q_i$ for all i, hence $t \in Q$. So Q is P-primary.

In particular, there is a smallest primary ideal in any valuation ring. Another way to get this minimal primary ideal is to let P be the minimal prime ideal and $Q = \{x \in R{:}xs = 0 \text{ for some } s \notin P\}$.

It is not true that the intersection of a chain of primary ideals in a general ring need be primary. Let k be a field and R be $k[x, y]/(xy)$, the generic ring with a zero divisor. Then M^n is primary and $\bigcap M^n = 0$ but 0 is not primary. Of course we could make this example local. Note also that the family of primary ideals M^n does not satisfy Arapović's condition A2, but the pair of prime ideals (x) and (y) does, as does $(x)^n$.

The equivalence of conditions (1), (2), and (4) in the next theorem is Robert's [10, Result 1.8].

Theorem 6.2. *The following conditions are equivalent for a valuation ring R:*

(1) *The minimal primary ideal is zero,*
(2) *Zero is an intersection of primary ideals (Arapović's condition A1 holds),*
(3) *The new criterion of Theorem 5.1 holds,*
(4) *R is a subring of a zero-dimensional ring.*

Proof. Clearly (1) and (2) are equivalent and both imply (4). We know that (4) implies (3), that is, the new criterion is a necessary condition for a ring to be a subring of a zero-dimensional ring. To see that (3) implies (1), let a_1 be an element of the minimal primary ideal Q. There exists $s \notin \sqrt{Q}$ such that $sa_1 = 0$. Moreover, as R is a valuation ring, we can write $a_1 = sa_2$, so $s^2a_2 = 0$.

Continuing, we write $a_n = sa_{n+1}$. Now $0{:}s^n$ stabilizes, and $s^n a_n = 0$, so we must have $s^n a_{n+1} = 0$ for some n. But $s^n a_{n+1} = a_1$.

Here is Robert's [7, Theorem 4.3].

Theorem 6.3. *Suppose that R is a valuation ring with total quotient ring T. If $\dim T > 0$, then R does not satisfy the new criterion of Theorem 5.1, so R cannot be embedded in a zero-dimensional ring.*

Proof. We'll prove the contrapositive. If R satisfies the new criterion, then, by the preceding theorem, zero is a primary ideal of R, hence a primary ideal of T. Thus all zero divisors of T are nilpotent, so T, being a total quotient ring, has dimension zero.

Rephrased, this result says that if a valuation ring R can be embedded in a zero-dimensional ring, then each zero divisor of R must be nilpotent.

7 An arithmetic approach to Krull dimension

Lombardi [18] introduced a characterization of Krull dimension that does not refer to prime ideals. Let the polynomial $P_m(X_1, \ldots, X_m, Y_1, \ldots, Y_m)$ be defined as

$$Y_1 Y_2 \cdots Y_m + X_1 Y_1 Y_2 \cdots Y_m + X_2 Y_2 Y_3 \cdots Y_m + \cdots + X_m Y_m$$

There are $m+1$ monomials here of degrees $m, m+1, m, m-1, \ldots, 2$. We say that a sequence $x_1, \ldots, x_m$ in R is **pseudoregular** if for all elements $a_1, \ldots, a_m \in R$ and positive integers e, we have

$$P_m(a_1 x_1, \ldots, a_m x_m, x_1^e, \ldots, x_m^e) \neq 0.$$

Lombardi showed that the Krull dimension of R is at least m if and only if there exists a pseudoregular sequence of length m in R. Note that for $m = 1$ this says that the Krull dimension is greater than zero if and only if there exists an element x such that $x^n \notin Rx^{n+1}$ for all n, which is the denial of the characterization of dimension zero (see Theorem 2.2) from which Robert's new criterion was derived.

We will use this idea to show that the ring $R = \prod \mathbf{Z}_{p^n}$ is infinite dimensional. We must show how to construct pseudoregular sequences in R of arbitrary length. For $r \in (0,1) \cap \mathbf{Q}$, define $x_r \in R$ by

$$x_r(n) = \begin{cases} p^{n^r} \text{if} n^r \in \mathbf{N} \\ 0 \text{ otherwise.} \end{cases}$$

Suppose $r_1, \ldots, r_k \in (0,1) \cap \mathbf{Q}$. Then $x_{r_1}(n) \cdots x_{r_k}(n) \neq 0$ for infinitely many n. Indeed, if d is a common denominator of $r_1, \ldots, r_k$, and n is of the form

m^d, then the product is nonzero and the exponent of p in the product is equal to

$$n^{r_1} + n^{r_2} + \cdots + n^{r_k}.$$

If $s \in (0,1) \cap \mathbf{Q}$ is greater than $\max(r_1, \ldots, r_m)$, then

$$\lim_{n \to \infty} \left(n^s - n^{r_1} - n^{r_2} - \cdots - n^{r_k}\right) = \infty$$

because $n^{r_i}/n^s \to 0$. Of course this expression is only relevant for us when all the powers are integers—but that happens for infinitely many n.

Now suppose $r_1 < r_2 < \cdots < r_m$. We claim that $x_{r_1}, \ldots, x_{r_m}$ is a pseudoregular sequence in R. Let $a_1, \ldots, a_m \in R$ and $e \in \mathbf{N}$. We want to show that

$$P_m\left(a_1 x_{r_1}, \ldots, a_m x_{r_m}, x_{r_1}^e, \ldots, x_{r_m}^e\right) \neq 0.$$

Look at the terms of the polynomial

$$P_m\left(x_{r_1}, \ldots, x_{r_m}, x_{r_1}^e, \ldots, x_{r_m}^e\right).$$

The exponent of p in $x_{r_1}^e(n) \cdots x_{r_m}^e(n)$, when the latter is nonzero, is

$$n^{er_1} + n^{er_2} + \cdots + n^{er_m}.$$

The exponent of p in $x_{r_1} x_{r_1}^e(n) \cdots x_{r_m}^e(n)$ is

$$n^{(e+1)r_1} + n^{er_2} + \cdots + n^{er_m}$$

and the difference of these exponents, $n^{(e+1)r_1} - n^{er_1}$ goes to infinity. The exponent of p in $x_{r_2} x_{r_2}^e x_{r_2}^e \cdots x_{r_m}^e$ is

$$n^{(e+1)r_2} + n^{er_3} + \cdots + n^{er_m}$$

and if we subtract the first exponent from this we get $n^{(e+1)r_2} - n^{er_1} - n^{er_2}$ which goes to infinity. So, eventually, the exponent of p in $x_{r_1}^e(n) \cdots x_{r_m}^e(n)$ becomes smaller than all the exponents of the other terms. This means that the order of the n-th coordinate of $x_{r_1}^e \cdots x_{r_m}^e$ is greater than the order of the n-th coordinates of the other terms. This situation does not change if we replace the unexponentiated terms x_{r_i} by $a_i x_{r_i}$ because they do not occur in the first term. As the order of the first term is bigger than the orders of the other terms at some n, the sum of all the terms cannot be 0.

This construction easily extends to any product of rings $\prod R_i$ where $N\left(\prod R_i\right) \neq \prod N\left(R_i\right)$.

8 Inheritance by polynomial rings

In [7], Robert commented that, "In practice Theorem 3.1 [Arapović's criterion] has limited ease of application." (He and Bill did use it to prove that

$\prod \mathbf{Z}_{p^n}$ was not a subring of a zero-dimensional ring in [11]—Bill points this out in [16]—but this is now done more easily with the new criterion.) Although this seems generally to be true, it is possible to use Arapović's criterion to prove that embeddability in a zero-dimensional ring is inherited by polynomial rings. In fact, Arapović proves this [3, Theorem 12] for an arbitrary set of indeterminants. Interestingly, Arapović does not use his criterion here! Instead he uses [1, Proposition 8] which says that the total quotient ring of a polynomial ring over a zero-dimensional ring is zero dimensional.

Theorem 8.1. *If R is a subring of a zero-dimensional ring, then so is $R[X]$.*

Proof. We may think of X as standing for an arbitrary set of indeterminants. First note that if Q is a primary ideal of R, then $Q[X]$ is a primary ideal of $R[X]$. So if we have an Arapović family of primary ideals Q_λ in R, we get a family of primary ideals $Q_\lambda[X]$ in $R[X]$ whose intersection is zero. Let P_λ be the radical of Q_λ. Suppose $f[X] \in R[X]$. There exists n such that if a is a coefficient of f that is in P_λ, then $a^n \in Q_\lambda$. Let I be the ideal generated by the d coefficients of $f[X]$. Then $m = dn - 1$ has the property that if $I \subset P_\lambda$, then $I^m \subset Q_\lambda$. So if $f[X] \in P_\lambda[X]$, then $I^m \subset Q_\lambda$, so $f[X]^m \in Q_\lambda[X]$.

Can one show that the new criterion is inherited by polynomial rings? That would have to be the case if the new criterion were sufficient for embedding in a zero-dimensional ring, which it undoubtedly is not. A somewhat related question is: Does the new criterion imply that if I is a finitely generated ideal, then the ideals $\mathrm{Ann} I^n$ stabilize? In fact it does, and we will end our paper by proving this not very deep fact.

Theorem 8.2. *Let I be a finitely generated ideal in a ring R that satisfies the new criterion. Then there exists n such that $\mathrm{Ann} I^n = \mathrm{Ann} I^{n+1}$.*

Proof. Let $I = (x_1, \ldots, x_t)$ and choose n_i such that $\mathrm{Ann} x_i^{n_i} = \mathrm{Ann} x_i^{n_i+1}$. The claim is that we can take n to be $n_1 + \cdots + n_t$. Any standard generator of $I^{n_1+\cdots+n_t}$ is of the form $x_1^{e_1} \cdots x_t^{e_t}$ where $e_1 + \cdots + e_t = n_1 + \cdots + n_t$. So $e_i \geq n_i$ for some i. Therefore $\mathrm{Ann} x_1^{e_1} \cdots x_t^{e_t} = \mathrm{Ann} x_1^{e_1} \cdots x_i^{e_i+1} \cdots x_t^{e_t}$. But $x_1^{e_1} \cdots x_i^{e_i+1} \cdots x_t^{e_t} \in I^{n_1+\cdots+n_t+1}$, so anything that kills $I^{n_1+\cdots+n_t+1}$ must kill every generator of $I^{n_1+\cdots+n_t}$. (The argument also works for $n = n_1 + \cdots + n_t - t + 1$.)

References

1. Arapović, Miroslav, Characterization of the 0-dimensional rings, *Glasnik Mat.* **18** (1983), 39–46.
2. Arapović, Miroslav, The minimal 0-dimensional overrings of commutative rings, *ibid.*, 47–52.

3. Arapović, Miroslav, On the embedding of a commutative ring into a 0-dimensional ring, *ibid.*, 53–59.
4. Clifford, Alfred H. and Gordon B. Preston, *Algebraic theory of semigroups*, Volume 1, American Mathematical Society, 1961.
5. Coquand, Thierry, Henri Lombardi and Marie-Francoise Roy, An elementary characterization of Krull dimension, in *From Sets and Types to Topology and Analysis*, Oxford Logic Guides **48**, Oxford University Press, 2005.
6. Fuchs, Laszlo and Luigi Salce, *Modules over nonnoetherian domains*, AMS 2001.
7. Gilmer, Robert, A new criterion for embeddability in a zero-dimensional commutative ring, *Lecture notes in pure and applied mathematics* **220**, Marcel Decker 2001, 223–229 .
8. Gilmer, Robert, Background and preliminaries on zero-dimensional rings, in *Zero-dimensional commutative rings*, Lecture notes in pure and applied mathematics **171**, Marcel Dekker 1995, 1–13.
9. Gilmer, Robert, Zero dimensionality and products of commutative rings, *ibid.* 15–25.
10. Gilmer, Robert, Zero-dimensional extension rings and subrings, *ibid.* 27–39.
11. Gilmer, Robert and William J. Heinzer, On the imbedding of a direct product into a zero-dimensional commutative ring, *Proc. Amer. Math. Soc.* **106** (1989), 631–637.
12. Gilmer, Robert and William J. Heinzer, Products of commutative rings and zero dimensionality, *Trans. Amer. Math. Soc.* **331** (1992), 663–680.
13. Gilmer, Robert and William J. Heinzer, Zero-dimensionality in commutative rings, *Proc. Amer. Math. Soc.*, **115** (1992), 881–893.
14. Gilmer, Robert and William J. Heinzer, Imbeddability of a commutative ring in a finite-dimensional ring, *Manuscripta Math.* **84** (1994) 401–414.
15. Glaz, Sarah, *Commutative coherent rings*, Lecture notes in mathematics **1371**, Springer, 1989.
16. Heinzer, William J., Dimensions of extension rings, in *Zero-dimensional commutative rings*, Lecture notes in pure and applied mathematics **171**, Marcel Dekker 1995, 57–64.
17. Kerr, Jeanne Wald, The polynomial ring over a Goldie ring need not be Goldie, *J. Algebra* **134** (1990) 344–352.
18. Lombardi, Henri, Dimension de Krull, Nullstellensätze et évaluation dynamique, *Math. Zeit.*, **242** (2002) 23–46.
19. Maroscia, Paolo, Sur les anneaux de dimension zero, *Atti Accad. Naz. Lincei Rend. Cl. Sci. Fis. Mat. Natur.* **56** (1974), 451–459.
20. Olivier, Jean-Pierre, Anneaux absolument plats universels et épimorphismes a buts reduits, *Séminaire d'Algèbre Pierre Samuel*, Paris, 1967–68.
21. Roitman, Moshe, On polynomial extensions of Mori domains over countable fields, *J. Pure Appl. Alg.* **64** (1990) 315–328

Old problems and new questions around integer-valued polynomials and factorial sequences

Jean-Luc Chabert[1] and Paul-Jean Cahen[2]

[1] Université de Picardie, LAMFA CNRS-UMR 6140, Faculté de Mathématiques, 33 rue Saint Leu, 80039 Amiens Cedex 01, France
jean-luc.chabert@u-picardie.fr

[2] Université Paul Cézanne Aix-Marseille III, LATP CNRS-UMR 6632, Faculté des Sciences et Techniques, 13397 Marseille Cedex 20, France
paul-jean.cahen@univ.u-3mrs.fr

It is but natural, in this tribute to the work of Robert Gilmer, *to write a few words about him. Robert showed extremely helpful, in many ways, for our book on* Integer-valued Polynomials; *he made numerous useful comments and was kind enough to undertake a very scrupulous proofreading. It could also be underlined that he promoted the notation* Int(D) *which seems now to be universally adopted. It is thus our pleasure, to dedicate this paper to Robert.*

1 On Bhargava's factorials

1.1 Arithmetical viewpoint

The first generalization of the notion of factorials can probably be attributed to Carlitz [8]. It stems from the arithmetical analogy between the ring $\mathbb{Z}$ of integers and the ring $\mathbb{F}_q[T]$ of polynomials over a finite field: both rings are principal ideal domains with finite residue fields, finite group of units and an infinite number of irreducible elements. With respect to this analogy, monic polynomials correspond to natural numbers and monic irreducible polynomials to prime numbers. The construction of *Carlitz factorials* is a little mysterious. He first defines, for each positive integer j, a polynomial D_j that may be interpreted as a piece of factorial:

$$D_j = \prod_{f \text{monic}, \deg(f)=j} f \tag{1}$$

Then, for each positive integer n with q-adic expansion:

$$n = n_0 + n_1 q + \ldots + n_s q^s \qquad (0 \leq n_j < q), \tag{2}$$

Carlitz defines the n-th factorial by:

$$n!_{\mathcal{C}} = \prod_{j=0}^{s} D_j^{n_j}. \tag{3}$$

We shall clarify that this may somehow be called a factorial by relating this construction to other generalizations.

1.2 Number theoretical viewpoint

Here is another generalization replacing the ring $\mathbb{Z}$ by the ring $\mathcal{O}_K$ of integers of a number field K. We first interpret the classical factorial as a product of powers of prime numbers, the power of p being given by *Legendre's formula*:

$$n! = \prod_{p \in \mathbb{P}} p^{w_p(n)} \quad \text{where} \quad w_p(n) = \sum_{k \geq 1} \left[\frac{n}{p^k}\right]. \tag{4}$$

We then analogously define the n-th factorial with respect to K as a product of powers of maximal ideals of $\mathcal{O}_K$:

$$n!_{\mathcal{O}_K} = \prod_{\mathfrak{p} \in \mathrm{Max}(\mathcal{O}_K)} \mathfrak{p}^{w_\mathfrak{p}(n)}, \tag{5}$$

where the power of $\mathfrak{p}$ is linked to its norm $q = N(\mathfrak{p}) = Card(\mathcal{O}_K/\mathfrak{p})$ by a formula very close to Legendre's formula:

$$w_\mathfrak{p}(n) = w_q(n) = \sum_{k \geq 1} \left[\frac{n}{q^k}\right]. \tag{6}$$

Note that here factorials are not elements, like numbers in $\mathbb{Z}$ or polynomials in $\mathbb{F}_q[T]$, but ideals of the ring $\mathcal{O}_K$. These ideals may first be traced in Pólya's work on integer-valued polynomials in 1919 [33], although factorials were not mentioned, and later in papers by Gunji and McQuillan [25] in 1970 and by Zantema [40] in 1982.

1.3 Algebraic viewpoint

The previous generalization can naturally be extended with a commutative algebraic viewpoint, replacing more generally the ring $\mathcal{O}_K$ by a Dedekind domain D. The corresponding n-th factorial ideal $n!_D$ appears as a product of prime ideals as in (5), the power $w_\mathfrak{p}(n)$ of $\mathfrak{p}$ being given by formula (6), using the norm $q = N(\mathfrak{p}) = Card(D/\mathfrak{p})$. Note that, if $q = +\infty$, then $w_q(n) = 0$, so that a maximal ideal with an infinite residue field does not appear in any factorial ideal. Factorial ideals apply in particular to rings of integers of function fields, that is, finite algebraic extensions of $\mathbb{F}_q(T)$ (compare with Γ-ideals of Goss [24]).

If D is a principal ideal domain, factorials can be interpreted as elements of D. In particular, letting D be the ring $\mathbb{F}_q[T]$, we obtain Carlitz factorials thanks to *Sinott's formula* [24, Thm 9.1.1]:

$$n!_C = \prod_{f \text{ monic, irreducible}} f^{w_f(n)} \quad \text{where} \quad w_f(n) = \sum_{k \geq 1} \left[\frac{n}{q^{k \deg(f)}} \right]. \tag{7}$$

Indeed the norm of the (principal prime) ideal $f\mathbb{F}_q[t]$ is obviously given by

$$N(f\mathbb{F}_q[t]) = Card(\mathbb{F}_q[t]/f\mathbb{F}_q[t]) = q^{\deg(f)}.$$

1.4 Multiplicative viewpoint

Write the usual factorial as

$$n! = \prod_{k=0}^{n-1} (n-k).$$

If we replace the natural sequence $0, 1, 2, 3, \ldots$ by a geometrical sequence $1, q, q^2, q^3, \ldots$, where q denotes an integer, $q \geq 2$, we obtain the *Jackson factorials* [27]:

$$n!_q = \prod_{k=0}^{n-1} (q^n - q^k). \tag{8}$$

This is a different kind of generalization: now it is not the ring of integers $\mathbb{Z}$ which is replaced by a Dedekind domain D but the subset $\mathbb{N}$ of $\mathbb{Z}$ which is replaced by another subset S (as here $S = \{q^n \mid n \in \mathbb{N}\}$). In fact, all these generalizations are particular cases of the following one.

1.5 Combinatorial viewpoint

Bhargava's factorials were introduced in 1997 [4]. For a Dedekind domain D and a subset S of D, Bhargava defined factorial ideals by means of the following local notion of $\mathfrak{p}$-ordering, where $\mathfrak{p}$ is a maximal ideal of D and $v_\mathfrak{p}$ denotes the corresponding valuation.

Definition 1.1. *A* $\mathfrak{p}$-ordering *of S is a sequence $\{a_n\}_{n \in \mathbb{N}}$ in S such that, for each $n > 0$, a_n minimizes the expression*

$$v_\mathfrak{p}\left(\prod_{k=0}^{n-1} (a_n - a_k)\right).$$

Thus, a_0 being arbitrarily chosen,

$$v_\mathfrak{p}(a_1 - a_0) \quad = \quad \inf_{s \in S} v_\mathfrak{p}(s - a_0)$$

and, inductively, for each $n > 0$,

$$v_{\mathfrak{p}}\left(\prod_{k=0}^{n-1}(a_n - a_k)\right) = \inf_{s\in S} v_{\mathfrak{p}}\left(\prod_{k=0}^{n-1}(s - a_k)\right). \tag{9}$$

Obviously, such $\mathfrak{p}$-orderings always exist and are not unique. However, as we shall see in the next section, the value of (9) does not depend on the choice of the $\mathfrak{p}$-ordering of S. Thus, letting

$$w_{S,\mathfrak{p}}(n) = v_{\mathfrak{p}}\left(\prod_{k=0}^{n-1}(a_n - a_k)\right) \tag{10}$$

Bhargava defined the n-th factorial ideal of S with respect to D by the formula:

$$n!_S^D = \prod_{\mathfrak{p}\in \mathrm{Max}(D)} \mathfrak{p}^{w_{S,\mathfrak{p}}(n)}. \tag{11}$$

One may check that formula (11) generalizes all previously mentioned ones. Moreover, as shown by Bhargava [5], there are many reasons that allow us to consider it a good generalization. For instance, here are 3 of its nice properties:

It is well known that	We also have
For all $k, l \in \mathbb{N}$ $k!l!$ divides $(k+l)!$ in $\mathbb{Z}$	For all $k, l \in \mathbb{N}$ $k!_S^D l!_S^D$ divides $(k+l)!_S^D$ as ideals of D
For all $x_0, x_1, \ldots, x_n \in \mathbb{Z}$ $\prod_{0\leq i<j\leq n}(x_j - x_i)$ is divisible by $1! \times 2! \times \cdots n!$	For all $x_0, x_1, \ldots, x_n \in S$ $\prod_{0\leq i<j\leq n}(x_j - x_i)\, D$ is divisible by $1!_S^D \times 2!_S^D \times \cdots n!_S^D$
For all $f \in \mathbb{Z}[X]$ f monic, $\deg(f) = n$ the GCD of $\{f(k) \mid k \in \mathbb{Z}\}$ divides $n!$ (Pólya 1915)	For all $f \in D[X]$ f monic, $\deg(f) = n$ the ideal generated by $\{f(s) \mid s \in S\}$ divides $n!_S^D$

1.6 Last generalization: Integer-valued polynomial viewpoint

We finally allow D to be any domain with quotient field K (not restricting ourselves to Dedekind domains, D could for instance be an order of a number field). For a subset S of D we consider the ring of *integer-valued polynomials* on S with respect to D:

$$\mathrm{Int}(S, D) = \{f(X) \in K[X] \mid f(S) \subseteq D\}.$$

We then set the following.

Definition 1.2. *The n-th* factorial ideal *of S with respect to D is defined by:*

$$n!_S^D = \{a \in D \mid a\, f \in D[X],\ \forall f \in \mathrm{Int}(S,D),\ \deg(f) \leq n\}.$$

Hence, the ideals $\{n!_S^D\}_{n\in\mathbb{N}}$ form a decreasing sequence of ideals of D with $0!_S^D = D$.

Proposition 1.3. *Definition 1.2 generalizes Formula (11).*

Proof. Assume D to be a Dedekind domain. We may look at things locally since, for every maximal ideal $\mathfrak{p}$ of D, one has (see for instance [7, I.2.7]):

$$\mathrm{Int}(S,D)_{\mathfrak{p}} = \mathrm{Int}(S,D_{\mathfrak{p}}).$$

Now fix a maximal ideal $\mathfrak{p}$ of D and consider a $\mathfrak{p}$-ordering $\{a_n\}_{n\in\mathbb{N}}$ of D. It follows from the definition of $\mathfrak{p}$-orderings that the Lagrange polynomials

$$g_n(X) = \prod_{k=0}^{n-1} \frac{X - a_k}{a_n - a_k}$$

form a basis of the $D_{\mathfrak{p}}$-module $\mathrm{Int}(S,D_{\mathfrak{p}})$. Consequently, $af \in D_{\mathfrak{p}}[X]$ for every $f \in \mathrm{Int}(S,D)$ such that $\deg(f) \leq n$, if and only if, a is divisible by $\prod_{k=0}^{n-1}(a_n - a_k)$ in $D_{\mathfrak{p}}$, that is,

$$n!_S^D\, D_{\mathfrak{p}} = \mathfrak{p}^{w_{S,\mathfrak{p}}(n)}\, D_{\mathfrak{p}} \quad \text{where} \quad w_{S,\mathfrak{p}}(n) = v_{\mathfrak{p}}\left(\prod_{k=0}^{n-1}(a_n - a_k)\right). \square$$

Remark 1.4. With Definition 1.2, it is easy to see that, for a Dedekind domain:
1) The function $w_{S,\mathfrak{p}}$ defined by (10) does not depend on the choice of the $\mathfrak{p}$-ordering of S.
2) The product $k!_S^D l!_S^D$ divides $(k+l)!_S^D$ since the product of two integer-valued polynomials of respective degree k and l is an integer-valued polynomial of degree $k + l$.
3) If $S \subseteq T \subseteq D$, then $n!_T^D$ divides $n!_S^D$ since $\mathrm{Int}(T,D) \subseteq \mathrm{Int}(S,D)$.

2 Examples and questions on factorials

2.1 An example in non-commutative algebra: Hurwitz quaternions

In the previous definition of factorial ideals we just need to consider $\mathrm{Int}(S,D)$ as a D-module. This allows for instance to consider the ring $\mathbb{H}$ of *Hurwitz quaternions*, that is,

$$\mathbb{H} = \left\{ a + bi + cj + dk \mid (a,b,c,d) \in \mathbb{Z}^4 \text{ or } (\mathbb{Z} + \frac{1}{2})^4 \right\}$$

One knows that $\mathbb{H}$ is a non-commutative principal ideal domain with quotient field:

$$\mathbb{H}(\mathbb{Q}) = \{a + bi + cj + dk \mid (a,b,c,d) \in \mathbb{Q}^4\}$$

Now consider the left $\mathbb{H}$-module of integer-valued polynomials on $\mathbb{H}$:

$$\mathrm{Int}(\mathbb{H}) = \{f(X) \in \mathbb{H}(\mathbb{Q})[X] \mid f(\mathbb{H}) \subseteq \mathbb{H}\}.$$

The corresponding factorial ideals $n!_{\mathbb{H}}$ are left ideals of $\mathbb{H}$. The first ideals are:

$$0!_{\mathbb{H}} = 1!_{\mathbb{H}} = 2!_{\mathbb{H}} = 3!_{\mathbb{H}} = \mathbb{H} \ \text{ and } \ 4!_{\mathbb{H}} = \mathbb{H}\,\frac{1+i}{2}.$$

The value of $4!_{\mathbb{H}}$ follows from the fact that $\frac{1+i}{2}\,(X^4 - X)$ is integer-valued (Gerboud [19]).

Question A. Find a formula for the (principal) factorial ideals $n!_{\mathbb{H}}$ with $n \geq 5$.

2.2 Factorials of the prime numbers and Bernoulli polynomials

Factorials of the set $\mathbb{P}$ of prime numbers (with respect to $\mathbb{Z}$) are given by the formula [12]:

$$n!_{\mathbb{P}} = \prod_{p \in \mathbb{P}} p^{\omega_p(n)} \ \text{ where } \ \omega_p(n) = \sum_{k \geq 0} \left[\frac{n-1}{(p-1)p^k}\right]. \tag{12}$$

The first terms of this sequence are

1, 1, 2, 24, 48, 5 760, 11 520, 2 903 040, 5 806 080, 1 393 459 200, ...

If we look at *The On-line Encyclopedia of Integer sequences* [34], we find another sequence with the same first terms: Sequence A075265 defined by Paul D. Hanna as the sequence $(d_n)_{n \in \mathbb{N}}$ such that

$$\left(-\frac{\log(1-x)}{x}\right)^m = \left(\sum_{k=1}^{\infty} \frac{x^k}{k+1}\right)^m$$

$$= 1 + \frac{m}{2}\,x + \frac{m(3m+5)}{24}\,x^2 + \ldots = \sum_{n \geq 1} \frac{1}{d_n}\, C_n(m)\, x^n \tag{13}$$

where $d_n \in \mathbb{N}$ and the polynomial $C_n(m) \in \mathbb{Z}[m]$ is primitive of degree n. Experimental checking suggests and theoretical proof [11] shows that:

$$d_n = (n+1)!_{\mathbb{P}}$$

Moreover, superseeker@research.att.com suggests (and it may be proved) that

$$(n+1)!_{\mathbb{P}} = n! \times e_n$$

where e_n is the nth term of Sequence A0011898 formed by the denominators of Bernoulli polynomials. More precisely, the nth Bernoulli polynomial of order m, denoted by $B_n^{(m)}$, is defined by:

$$\left(\frac{t}{e^t - 1}\right)^m = \sum_{n=0}^{\infty} B_n^{(m)} \frac{t^n}{n!} \tag{14}$$

and may be written:

$$B_n^{(m)} = \frac{1}{e_n} D_n(m)$$

where $e_n \in \mathbb{N}$ and the polynomial $D_n(m) \in \mathbb{Z}[m]$ is primitive. Thus, we have:

Proposition 2.1.

$$\left(\frac{t}{e^t - 1}\right)^m = \sum_{n=0}^{\infty} D_n(m) \frac{t^n}{(n+1)!_{\mathbb{P}}} \tag{15}$$

where $D_n(m) \in \mathbb{Z}[m]$ is primitive.

Such a link with denominators of Bernoulli numbers had been previously suggested by Bhargava [5, Example 21].

Question B. Explain the relation between the sequence of factorials of $\mathbb{P}$ and the sequence of denominators of either Bernoulli numbers or Bernoulli polynomials.

2.3 Subsets with the same factorial sequences

The following question seems to be natural:

Question C. Let S and T be two subsets of an integral domain D. Under which conditions the factorial sequences of S and T are equal?

These factorial sequences are obviously equal if S and T are *polynomially equivalent*, that is, $\mathrm{Int}(S, D) = \mathrm{Int}(T, D)$ (see for instance [7, Chapter IV] or [15, section 2] with a new approach to polynomial closure). This is far from necessary! Indeed if $T = uS + a = \{us + a \mid s \in S\}$ where $a \in A$ and u is a unit of A, the factorial sequences of S and T are clearly equal.

For an infinite subset S of $\mathbb{Z}$, Gilmer and Smith conjectured [23] that, if $f \in \mathrm{Int}(S, \mathbb{Z})$ is such that $\mathrm{Int}(S, \mathbb{Z}) = \mathrm{Int}(f(S), \mathbb{Z})$, then $\deg(f) = 1$ [15, Question, p. 114]. Fares [17] proved this conjecture by establishing that, in fact, if S and $f(S)$ have the same factorial sequences, then $\deg(f) = 1$. He even recently extended this result [18] with the following:

Proposition 2.2. *Let $\mathcal{O}_K$ be the ring of integers of any number field K and let S be any infinite subset of $\mathcal{O}_K$. If $\varphi(X) \in K(X)$ is a rational function such that S and $\varphi(S)$ have the same factorial sequences, then φ is an homographic function (i.e., a rational function of the form $\frac{aX+b}{cX+d}$ with $ad - bc \neq 0$).*

This suggests two more questions:

Question C1. Does Fares' result hold for ring of integers of function fields?

Question C2. Assume that S is an infinite subset of the ring $\mathcal{O}_K$ of integers of a number field K. Let f and $g \in \mathrm{Int}(S, \mathcal{O}_K)$. Does the equality of the factorial sequences of $f(S)$ and $g(S)$ imply that $g = f \circ h$ where $\deg(h) = 1$?

Even in the case of the ring $\mathbb{Z}$, there are nevertheless examples of subsets S and T with the same factorial sequences such that T is not of the form $uS + a$ where $a \in A$ and u is a unit of A:
1) when the subsets are finite: for instance, the three subsets $S = \{0, 2, 35\}$, $T = \{0, 7, 22\}$ and $U = \{0, 10, 21\}$ have the same factorial sequences but there does not exist any polynomial f of degree 1 such that either $T = f(S)$, or $U = f(S)$, or $U = f(T)$.
2) when T is not assumed to be the image of S by a polynomial, as for instance:

$$S = 5\mathbb{Z} \cup (1 + 5\mathbb{Z}) \text{ and } T = 5\mathbb{Z} \cup (2 + 5\mathbb{Z})$$

For a subset S of $\mathbb{Z}$, A. Crabbe [16] tried to test the factorial sequence on finite subsets. For each prime number p and each $r \geq 1$, set

$$S \bmod(p^r) = \{0 \leq a < p^r \mid \exists s \in S \text{ such that } s \equiv a \pmod{p^r}\}.$$

Gilmer characterized the subsets S such that $\mathrm{Int}(S, \mathbb{Z}) = \mathrm{Int}(\mathbb{Z})$ as the subsets which are prime power complete, that is, such that $S \bmod(p^r) = \{0 \leq a < p^r\}$, for each prime p and each $r \geq 1$ [20, Theorem 2]. More generally, it follows from [7, IV.§1 and §2] that two subsets S and T of $\mathbb{Z}$ are polynomially equivalent if and only they have the same p-adic completion for each p and hence, if and only if $S \bmod(p^r) = T \bmod(p^r)$ for each $p \in \mathbb{P}$ and each $r \geq 1$. Crabbe asked the following question [16, Conjecture 3.3].

Question C3. Is the equality of the factorial sequences of two subsets S and T of $\mathbb{Z}$ characterized by the equalities of the factorial sequences of $S \bmod(p^r)$ and $T \bmod(p^r)$ for each $p \in \mathbb{P}$ and each $r \geq 1$?

He proved one way: if $S \bmod(p^r)$ and $T \bmod(p^r)$ have the same factorial sequences, for each $p \in \mathbb{P}$ and each $r \geq 1$, then S and T have the same factorial sequences. He proved the converse for $S \subseteq T = \mathbb{Z}$. In fact, whenever $S \subseteq T$, if S and T have the same factorial sequences, then $\mathrm{Int}(S, \mathbb{Z}) = \mathrm{Int}(T, \mathbb{Z})$, thus $S \bmod(p^r) = T \bmod(p^r)$ for each $p \in \mathbb{P}$ and each $r \geq 1$. A fortiori, $S \bmod(p^r)$ and $T \bmod(p^r)$ have the same factorial sequences.

2.4 Several indeterminates

Another natural question is:

Question D. What would be a good generalization of the notion of factorials to several indeterminates?

Let n be a positive integer, D be an integral domain with quotient field K, and $\underline{S}$ be a subset of D^n. Denote by $\mathrm{Int}(\underline{S}, D)$ the ring of integer-valued polynomials in several indeterminates on $\underline{S}$, that is:

$$\mathrm{Int}(\underline{S}, D) = \{f \in K[\underline{X}] \mid \forall \underline{a} \in \underline{S},\ \ f(\underline{a}) \in D\}$$

where $\underline{X} = (X_1, \ldots, X_k)$. For each $\underline{k} = (k_1, \ldots, k_n) \in \mathbb{N}^k$, a definition of the $\underline{k}$-th factorial ideal of $\underline{S}$ with respect to D could be:

$$\underline{k}!_{\underline{S}}^{D} = \{a \in D \mid \forall f \in \mathrm{Int}(\underline{S}, D), \text{ such that } \deg_{X_j}(f) \le k_j,\ \ af \in D[\underline{X}]\}.$$

As noticed by Ostrowski [35], if $D = \mathbb{Z}$ and $\underline{S} = \mathbb{Z}^n$, then one has:

$$\underline{k}! = k_1! \cdots k_n!$$

since the products $\binom{X_1}{k_1} \cdots \binom{X_n}{k_n}$ form a basis of the $\mathbb{Z}$-module $\mathrm{Int}(\mathbb{Z}^n, \mathbb{Z})$. More generally, if D is a Dedekind domain and $\underline{S}$ is of the form $S_1 \times \cdots \times S_n$, then (see [25] and [7, § XI.1]):

$$\underline{k}!_{\underline{S}}^{D} = \prod_{j=1}^{n} k_j!_{S_j}^{D}.$$

But, if $\underline{S}$ is more general, the question is much more difficult. There are some partial studies by Mulay [30] and Bhargava [5, § 12].

3 Simultaneous orderings

3.1 Newtonian orderings (or simultaneous $\mathfrak{p}$-orderings)

Recall that the polynomials

$$\binom{X}{n} = \frac{X(X-1)\ldots(X-n+1)}{n!} = \prod_{k=0}^{n-1} \frac{X-k}{n-k}.$$

form a basis of the $\mathbb{Z}$-module $\mathrm{Int}(\mathbb{Z}) = \{f \in \mathbb{Q}[X] \mid f(\mathbb{Z}) \subseteq \mathbb{Z}\}$. This is linked to Gregory-Newton interpolation formula ([7], Historical Introduction). More generally, the unique degree n polynomial that interpolates a function f at a given set of $n+1$ distinct arguments $\{a_n\}_{0 \le n \le N}$ can be written in different manners and, in particular, as the *Newton's interpolation polynomial* [32], that is, as a linear combination of the polynomials:

$$f_n(X) = \prod_{k=0}^{n-1} \frac{X - a_k}{a_n - a_k}.$$

By analogy, we introduce the following definition.

Definition 3.1. *Let D be a domain and E be a subset of D. A sequence $\{a_n\}_{n\in\mathbb{N}}$ of distinct elements of E is said to be an* infinite Newtonian ordering *for E in D, or shortly an* ordering *for E, if the polynomials*

$$f_n(X) = \prod_{k=0}^{n-1} \frac{X - a_k}{a_n - a_k}$$

form a basis of the D-module $\mathrm{Int}(E, D)$.

It is easy to establish the following.

Lemma 3.2. *A sequence $\{a_n\}$ is an ordering for E in D if and only if the polynomials f_n belong to* $\mathrm{Int}(E, D)$.

Infinite orderings do not necessarily exist and thus, for a given integer N, one may consider *orderings of length N*, that is, finite sequences $\{a_n\}_{0\leq n<N}$ such that the interpolation polynomials $\{f_n\}_{0\leq n<N}$ belong to $\mathrm{Int}(E, D)$.

Question E. [5, Quest. 30] Characterize the subsets of $\mathbb{Z}$ which admit an infinite Newtonian ordering.

Here are some examples (see [5]):
1) $0, 1, \ldots, n, \ldots$ is an infinite ordering for $\mathbb{N}$, and also for $\mathbb{Z}$.
2) $1^2, 2^2, \ldots, n^2, \ldots$ is an infinite ordering for $\{n^2 \mid n \in \mathbb{N}\}$
3) $1, q, q^2, \ldots, q^n, \ldots$ is an infinite ordering for $\{q^n \mid n \in \mathbb{N}\}$.
On the other hand, there exists an infinite ordering for $\{n^k \mid n \in \mathbb{N}\}$ or for $\{n^k \mid n \in \mathbb{Z}\}$ if and only if $k = 1$ or 2.

Of course, one can study subsets of other rings, in particular of Dedekind domains. For instance, in line with our third example above, for every non-constant polynomial $g \in \mathbb{F}_q[T]$, the sequence $1, g, g^2, \ldots, g^n, \ldots$ is an infinite ordering for $\{g^n \mid n \in \mathbb{N}\}$ in $\mathbb{F}_q[T]$.

A few facts are relevant in the study of orderings, whether infinite or of length N, for a subset of a domain D:
1) Orderings are related to factorials ideals: if $\{a_n\}$ is an infinite ordering for E in D (resp. an ordering of length N), then each factorial ideal $n!_E^D$ (resp. each factorial ideal up to N) is generated by $\prod_{k=0}^{n-1}(a_n - a_k)$ and, in particular, is principal.
2) Local behaviour: it follows from the containment [7, I.2.4]

$$\mathrm{Int}(E, D) \subseteq \mathrm{Int}(E, D_{\mathfrak{p}})$$

that if $\{a_n\}$ is an ordering for E in D (of length N or infinite) then it is an ordering for E in $D_{\mathfrak{p}}$ for each prime ideal $\mathfrak{p}$ of D. Conversely if $\mathcal{P}$ is a set of prime ideals of D such that $D = \cap_{\mathfrak{p}\in\mathcal{P}} D_{\mathfrak{p}}$, and if $\{a_n\}$ is an ordering for E in $D_{\mathfrak{p}}$ for all $\mathfrak{p} \in \mathcal{P}$ then it is an ordering for E in D.

In particular, suppose that D is a Dedekind domain with finite residue fields. Then $\{a_n\}$ is an ordering for E if and only if, for every maximal ideal $\mathfrak{p}$ of D, it is a $\mathfrak{p}$-ordering of E (Def. 1.1). Following Bhargava, this is known as a *simultaneous ordering* (note that, for a discrete valuation domain with maximal ideal $\mathfrak{p}$ and finite residue field, a Newtonian ordering is nothing else than a $\mathfrak{p}$-ordering).

3) Non-uniqueness: when they exist, orderings are not necessarily unique. For instance, if $\{a_n\}_{0\leq n\leq N}$ is an ordering for D itself in the ring D, then, for every $b \in D$ and every unit u of D, $\{ua_n + b\}$ is also an ordering for D, moreover there may also be orderings of a different type. For instance, the sequence $\{(-1)^n \left[\frac{n+1}{2}\right]\}_{n\in\mathbb{N}}$ is an ordering for $\mathbb{Z}$ [7, Exercise I.5] which is not linked by a linear transformation to the sequence $0, 1, \ldots, n, \ldots$ of natural numbers.

4) Polynomial closure: if $\{a_n\}$ is an ordering for E in D it is also an ordering for the *polynomial closure* $\overline{E}$ of E in D, that is,

$$\overline{E} = \{b \in D \mid \forall f \in \mathrm{Int}(E, D), \quad f(b) \in D\}.$$

3.2 Newtonian domains

Definition 3.3. *A domain D is said to be a* Newtonian domain *if there exists an infinite Newtonian ordering for D in D.*

Here are some examples:

1) $\mathbb{Z}$ is a Newtonian domain.

2) Every local domain D with infinite residue field is a Newtonian domain. Any sequence of elements in distinct residue classes is an ordering for D since $\mathrm{Int}(D) = D[X]$.

3) Every discrete valuation domain V is a Newtonian domain. If the residue field is infinite this follows from the previous example and if the residue field if finite, a Newtonian ordering is given by a $\mathfrak{p}$-ordering (where $\mathfrak{p}$ is the maximal ideal of V). A particular case is that of a *very well distributed and well ordered sequence* [7, Definition II.2.1]. To build such a sequence [7, Proposition II.2.3], let t be a generator of the maximal ideal $\mathfrak{p}$ and $a_0 = 0, a_1, \ldots, a_{q-1}$ be a system of representatives of V modulo $\mathfrak{p}$ then, for each $n \in \mathbb{N}$ with q-adic expansion (2), put

$$a_n = a_{n_0} + a_{n_1}t + \ldots + a_{n_k}t^k. \tag{16}$$

From the previous examples and the Chinese remainder theorem we deduce the following.

Proposition 3.4. *A semi-local principal ideal domain is a Newtonian domain.*

Let us consider now non semi-local domains. The first question is the following.

Question F. Does there exist a number field $K \neq \mathbb{Q}$ such that the ring of integers $\mathcal{O}_K$ of K is a Newtonian domain? [5, Quest. 30]

Known results are essentially negative (Wood [38]).

Proposition 3.5. *The ring of integers $\mathcal{O}_K$ of an imaginary quadratic field K is not Newtonian.*

The reason why it is easier to obtain a negative answer for imaginary quadratic fields is probably due to the fact that there are only finitely many units. Here is a positive result [14].

Proposition 3.6. *Let K be a number field and let D be a localization of the ring $\mathcal{O}_K$ of integers of K. Then, the sequence $\{n\}_{n\in\mathbb{N}}$ is a Newtonian ordering for D (and thus D is a Newtonian domain) if and only if every prime number splits completely in D.*

For instance, if S denotes the multiplicative subset generated by the prime numbers p such that $p \equiv 1 \pmod 4$, then $S^{-1}\mathbb{Z}[i]$ is a Newtonian domain. More generally, we have the following.

Proposition 3.7. *Let D be a Dedekind domain which is Newtonian, let $\{a_n\}_{n\in\mathbb{N}}$ be an ordering for D and let R be an integral domain containing D. The following assertions are equivalent:*

(i) *$\{a_n\}_{n\in\mathbb{N}}$ is an ordering for R,*
(ii) *For each $\mathfrak{p} \in \max(D)$ with finite residue field and for each $\mathfrak{m} \in \max(R)$ containing $\mathfrak{p}$, one has $R/\mathfrak{m} \simeq D/\mathfrak{p}$ and $\mathfrak{m}R_\mathfrak{m} = \mathfrak{p}R_\mathfrak{m}$.*

When R is Noetherian, this is also equivalent to:

(iii) *each $\mathfrak{p} \in \max(D)$ with finite residue field such that $\mathfrak{p}R \neq R$ splits completely in R (that is, $\mathfrak{p}R = \prod_{i=1}^r \mathfrak{m}_i$ where the $\mathfrak{m}_i$'s are distinct maximal ideals of R with norm equals to the norm of $\mathfrak{p}$).*

The proof follows from the results given in [7, §IV.3]. See also [38, Prop. 5.1] when R is a Dedekind domain.

One may also consider the question of Newtonian domains in the context of function fields with analogous results. First of all, for each finite field $\mathbb{F}_q$, the polynomial ring $\mathbb{F}_q[T]$ is a Newtonian domain. Indeed, the sequence $\{a_n\}_{n\in\mathbb{N}}$ obtained by Formula (16) for the valuation domain $\mathbb{F}_q[T]_{(T)}$, that is,

$$a_n = a_{n_0} + a_{n_1}T + \ldots + a_{n_k}T^k \tag{17}$$

is in fact a Newtonian ordering for the domain $\mathbb{F}_q[T]$ itself (see [5] or [1]). Similarly to Question F, one may then ask the following.

Question F1. Are there algebraic extensions $K \neq \mathbb{F}_q(T)$ of $\mathbb{F}_q(T)$ such that the integral closure $\mathcal{O}_K$ of $\mathbb{F}_q[T]$ in K is a Newtonian domain?

There is a result very similar to Wood's result. Recall that the extension K of $\mathbb{F}_q(T)$ is called an *imaginary extension* if the infinite place of $\mathbb{F}_q(T)$, that is the valuation associated to $\frac{1}{T}$, has only one extension to K. In this case the units of K are the elements of $\mathbb{F}_q^*$. Assume that q is odd, then a quadratic extension $K = \mathbb{F}_q(T)[Y]/(Y^2 - D(T))$ of $\mathbb{F}_q(T)$ is imaginary if and only if either $\deg(D)$ is odd or the leading coefficient of D is not a square in $\mathbb{F}_q$.

Proposition 3.8. (Adam [1]) *Assume that q is odd and consider an imaginary quadratic extension $K = \mathbb{F}_q(T)[Y]/(Y^2 - D(T))$ of $\mathbb{F}_q(T)$. Then the integral closure $\mathcal{O}_K$ of $\mathbb{F}_q[T]$ in K is not a Newtonian domain unless* $\deg(D) = 1$.

In fact, if $\deg(D) = 1$, the quadratic extension K is isomorphic to $\mathbb{F}_q(T)$.

More generally, we ask the following.

Question F2. Characterize the Dedekind domains that are Newtonian.

For a discrete valuation domain with maximal ideal $\mathfrak{p}$ and finite residue field, Newtonian orderings and $\mathfrak{p}$-orderings are the same; they were characterized by Julie Yeramian [39] and correspond to the *very well ordered sequences* defined by Y. Amice [3]. By globalization, we obtain the following partial answer:

Proposition 3.9. *Let D be a Dedekind domain with finite residue fields. A sequence $\{a_n\}_{n\in\mathbb{N}}$ in D is a Newtonian ordering for D if and only if, for each maximal ideal $\mathfrak{p}$ of D with norm q, for each $s \in \mathbb{N}$, and for each $k \in \mathbb{N}^*$, the q^k following consecutive elements form a complete system of representatives of D modulo $\mathfrak{p}^k$:*

$$a_{sq^k}, a_{sq^k+1}, \ldots, a_{(s+1)q^k-1}. \tag{18}$$

3.3 Schinzel orderings

The last proposition is obviously related to an old problem suggested by J. Browkin in 1965 for $\mathbb{Q}[i]$, which is known as Schinzel's problem [31, Problem 8].

Schinzel's problem (1969). Does there exist a number field $K \neq \mathbb{Q}$ with a sequence $\{a_n\}_{n\in\mathbb{N}}$ of elements in the ring of integers $\mathcal{O}_K$ of K such that, for each ideal I of $\mathcal{O}_K$ with norm $N = N(I) = Card(\mathcal{O}_K/I)$, the sequence $a_0, a_1, \ldots, a_{N-1}$ is a complete system of representatives of $\mathcal{O}_K$ modulo I?

Some results are known: K cannot be a quadratic field (Wantula, 1969), $\mathcal{O}_K$ must be a principal ideal domain (Wasen, 1976).

More generally, we may consider a domain D. As for Newtonian orderings, good infinite sequences may not exist and hence, we may wish to restrict ourselves to finite sequences. We thus set the following.

Definition 3.10. *Let D be a domain.*
(1) *A sequence $\{a_n\}_{n\in\mathbb{N}}$ in D is called an* infinite Schinzel ordering *for D if, for each integer k and each ideal I of D with norm $N(I) \geq k$, the elements $a_0, a_1, \ldots, a_{k-1}$ are in distinct classes modulo I.*
(2) *Given a positive integer N, a sequence $\{a_n\}_{0\leq n<N}$ in D is called a* Schinzel ordering of length N *for D if, for each $k \leq N$ and each ideal I of D with norm $N(I) \geq k$, the elements $a_0, a_1, \ldots, a_{k-1}$ are in distinct classes modulo I.*
(3) *The domain D is said to be a* Schinzel domain *if there exists an infinite Schinzel ordering for D.*

Here are some examples:
(1) $\mathbb{Z}$ is a Schinzel domain with Schinzel ordering $\{n\}_{n\in\mathbb{N}}$.
(2) A discrete valuation domain with finite residue field is a Schinzel domain. The sequence constructed by means of Formula (16) is a Schinzel ordering.
(3) 3) $\mathbb{F}_q[T]$ is a Schinzel domain: the Newtonian ordering for $\mathbb{F}_q[T]$ defined by (17) is a Schinzel ordering because, for $g \in \mathbb{F}_q[T]$ of degree d, the q^d first elements of this sequence are in distinct classes modulo g.
(4) A local domain with an infinite residue field is (trivially) a Schinzel domain (the norm of every proper ideal is infinite and a sequence of elements in distinct classes modulo the maximal ideal is a Schinzel ordering).

The following necessary condition was communicated to us by Sophie Frisch.

Proposition 3.11. *Let D be a domain with finite residue rings. If D is a Schinzel domain, then D is Euclidean for the norm.*

Proof. For $x \in D$, write $N(x)$ for the norm of the principal ideal xD. Assume there exists an infinite Schinzel ordering $\{a_n\}_{n\in\mathbb{N}}$. One may always assume that $a_0 = 0$. Then, for each k, a_k and a_0 are in the same class modulo a_kD. It follows from the definition of a Schinzel ordering that $N(a_k) \leq k$. Let $x, y \in D$, with $x \neq 0, x$ not a unit. Set $n = N(x)$. As $a_0, a_1, \ldots, a_{n-1}$ form a complete system of residues modulo xD, one may write $y = qx + r$, with $q \in D$ and $r = a_k$, for some $k \leq n-1$. Clearly, one then has $N(a_k) < N(x)$. □

In general, the ring of integers of a number field is not a Schinzel domain. Maximal lengths of Schinzel orderings are given in [28] and [37].

It may be interesting to compare Schinzel and Newtonian orderings. In particular, we pose the following.

Question G. Are the classes of Newtonian and Schinzel domains distinct?

In fact, for a Dedekind domain with finite residue fields, one can list six natural properties for a sequence $\{a_n\}_{n\in\mathbb{N}}$ in D as follows: ν denotes the norm of any nonzero ideal $\mathcal{I}$ of D, q denotes the norm of any maximal ideal $\mathfrak{p}$ of D and, 'c.s.r.' means 'is a complete system of representatives of ...'

property	for all	the sequence	c.s.r.
I	$\nu = N(\mathcal{I}), r \in \mathbb{N}$	$a_r, \ldots, a_{r+\nu-1}$	$D/\mathcal{I}$
I'	$q = N(\mathfrak{p}), s \in \mathbb{N}^*, r \in \mathbb{N}$	$a_r, \ldots, a_{r+q^s-1}$	$D/\mathfrak{p}^s$
II	$\nu = N(\mathcal{I}), k \in \mathbb{N}$	$a_{k\nu}, \ldots, a_{(k+1)\nu-1}$	$D/\mathcal{I}$
II' Newton	$q = N(\mathfrak{p}), s \in \mathbb{N}^*, k \in \mathbb{N}$	$a_{kq^s}, \ldots, a_{(k+1)q^s-1}$	$D/\mathfrak{p}^s$
III Schinzel	$\nu = N(\mathcal{I})$	$a_0, \ldots, a_{\nu-1}$	$D/\mathcal{I}$
III'	$q = N(\mathfrak{p}), s \in \mathbb{N}^*$	$a_0, \ldots, a_{q^s-1}$	$D/\mathfrak{p}^s$

One can say that a Dedekind domain D satisfies one of these properties if there exists a sequence in D which satisfies the given property. We have the following obvious implications:

$$\begin{array}{ccccc} I & \longrightarrow & II & \longrightarrow & III \\ \downarrow & & \downarrow & & \downarrow \\ I' & \longrightarrow & II' & \longrightarrow & III' \end{array}$$

Problem G1. Discuss each reverse implication.

We list below some examples for the strongest properties (I, I' and II):
Property I: a) Obviously, the sequence of natural numbers satisfies I in $\mathbb{Z}$.
b) A discrete valuation domain with a finite residue field. A sequence $\{a_n\}_{n\in\mathbb{N}}$ satisfies I if and only if it is a *very well distributed and well ordered* sequence [7, §II.2], that is if, for each n and m,

$$v(a_n - a_m) = v_q(n - m)$$

where v denotes the valuation, q denotes the cardinality of the residue field and $v_q(n-m)$ denotes the greatest k such that q^k divides $n-m$. The sequence given by Formula 16 satisfies this property.
Property I': a) A Dedekind domain D with characteristic 0 such that every prime number splits completely in D. Merely consider the sequence $\{n\}_{n\in\mathbb{N}}$.
b) A semi-local principal ideal domain D (because property I may be globalized for finitely many maximal ideals, cf. [39]).
Property II: The sequence given by (17) yields that $\mathbb{F}_q[T]$ has this property.

Here is a partial answer to question G1. The domain $\mathbb{F}_q[T]$ satisfies II, but not I' [2, Prop. 2.8]). Consequently, II$\not\to$I'. We may also notice that the negative results concerning the Newtonian property are generally obtained by using only the first terms of the sequences (that is, property III').

4 The Pólya-Ostrowski group and Pólya fields

A necessary (but not sufficient) condition for D to be a Newtonian domain is that all factorial ideals are principal. We discuss here this property.

4.1 The Pólya-Ostrowski group

Definition 4.1. *Let D be a Dedekind domain.*
(1) The factorial group *of D is the subgroup $\mathcal{F}act(D)$ of the group $\mathcal{J}(D)$ of nonzero fractional ideals of D generated by the factorial ideals.*
(2) The Pólya-Ostrowski group *of D is the image $\mathcal{P}o(D)$ of $\mathcal{F}act(D)$ in the class group $\mathcal{C}l(D) = \mathcal{J}(D)/\mathcal{P}(D)$ of D.*

It is easy to check the following (see for instance, [10, Prop. 2.2]).

Proposition 4.2. *Let D be a Dedekind domain. Then $\mathcal{F}act(D)$ is a free Abelian group with a basis formed by the non trivial ideals*

$$\Pi_q = \prod_{\mathfrak{p}\in Max(D),\, N(\mathfrak{p})=q} \mathfrak{p}.$$

Assume throughout that K is a number field and that $\mathcal{O}_K$ denotes its ring of integers. A natural question is the following.

Problem H. Describe the Pólya-Ostrowski group $\mathcal{P}o(\mathcal{O}_K)$ of the ring of integer $\mathcal{O}_K$ of a number field K.

We have some partial answers for Galois extensions $K/\mathbb{Q}$, since in this case the ideals Π_q are the *ambige Ideale* of Hilbert.
1) The Pólya-Ostrowski $\mathcal{P}o(\mathcal{O}_K)$ is generated solely by the Π_q's where q is some power of a ramified prime number [35].
2) A description of the Pólya-Ostrowski group of a quadratic number field can be found in [26, Prop. 105 and 106] or [7, II.4.4].
3) The following sequence of Abelian groups is exact:

$$1 \to H^1(G, U(\mathcal{O}_K)) \to \oplus_{p\in\mathbb{P}}\mathbb{Z}/e_p\mathbb{Z} \to \mathcal{P}o(\mathcal{O}_K) \to 1$$

where e_p denotes the ramification index of p in the extension $K/\mathbb{Q}$ (see [10] or [40]).
4) If K is the compositum of two Galois subextensions K_1 and K_2 of $\mathbb{Q}$ such that $[K_1{:}\mathbb{Q}]$ and $[K_2{:}\mathbb{Q}]$ are relatively prime, then by [10, Prop. 3.6],

$$\mathcal{P}o(\mathcal{O}_K) \simeq \mathcal{P}o(\mathcal{O}_{K_1}) \times \mathcal{P}o(\mathcal{O}_{K_2}).$$

4.2 Pólya fields

Proposition 4.3. *Let D be a Dedekind domain. The following assertions are equivalent:*

(i) $\mathrm{Int}(D)$ *admits a* regular basis, *that is, a basis* $\{f_n\}_{n\in\mathbb{N}}$ *where* $\deg(f_n) = n$.
(ii) *The ideals $n!_D$ are principal.*
(iii) *The ideals $\Pi_q = \prod_{N(\mathfrak{p})=q} \mathfrak{p}$ are principal.*
(iv) *The Pólya-Ostrowski group of D is trivial, that is, $\mathcal{P}o(D) \simeq \{1\}$.*

Returning to number fields, we use a definition of Zantema [40].

Definition 4.4. *A* Pólya field *is a number field K such that $\mathcal{P}o(\mathcal{O}_K) = \{1\}$, that is, such that* $\mathrm{Int}(\mathcal{O}_K)$ *admits a regular basis.*

Here again, we offer some answers.
1) Every cyclotomic field is a Pólya field.
2) For the characterization of the quadratic Pólya fields see [40] or [7, II.4.5].
3) It follows from the previous section that, if K_1 and K_2 are two Pólya fields such that $[K_1{:}\mathbb{Q}]$ and $[K_2{:}\mathbb{Q}]$ are relatively prime, then K_1K_2 is a Pólya field.

The notion of Pólya field may also be extended to function fields with some partial results. The first ones were given by Van der Linden [36]. More general results were obtained by Adam [2, Chapter 5]:
1) He proved that the analog of cyclotomic fields in the context of function fields are Pólya fields.

2) He characterized the analog of imaginary quadratic Kummer extensions that are Pólya fields.

Finally, let us recall the problem of *class field towers* which goes back to Kronecker and Weber. The ring of integers $\mathcal{O}_K$ of a number field K is not necessarily a principal ideal domain. However, if h_K denotes the class number of K, that is, the order of the class group $\mathcal{C}l(\mathcal{O}_K)$), there exists a Galois extension $H(K)$ of K of degree h_K, the *Hilbert class field* of K, such that, for every ideal $\mathcal{I}$ of $\mathcal{O}_K$, its extension $\mathcal{I}\mathcal{O}_{H(K)}$ is a principal ideal of $\mathcal{O}_{H(K)}$ (and moreover, the Galois group $Gal(H(K)/K)$ is isomorphic to $\mathcal{C}l(\mathcal{O}_K)$). Again, $\mathcal{O}_{H(K)}$ is not necessarily a principal ideal domain, so that, one may iterate the process and consider $H(H(K))$..., and so on. The question was as follows: does this construction of the Hilbert class field tower of K stops after a finite number of steps? The answer is no and was given by Golod and Shafarevich in 1964.

Analogously, we introduce the following definition:

Definition 4.5. *An extension of number fields L/K is called a* Pólya extension *if all the extended ideals $\Pi_q(\mathcal{O}_K)\mathcal{O}_L$ are principal.*

In other words, the extension L/K is a Pólya extension if and only if the $\mathcal{O}_L$-module $\mathrm{Int}(\mathcal{O}_K, \mathcal{O}_L)$ has a regular basis. Of course, if K is a Pólya field, then every finite extension of K is a Pólya extension and, whatever the fixed number field K, there exist a Pólya extension L of K contained in the Hilbert class field $H(K)$ of K (since $\mathcal{P}o(\mathcal{O}_K) \subseteq \mathcal{C}l(\mathcal{O}_K)$). If this extension L is not a Pólya field, we may iterate the process. This suggests two questions:

Question I1. For every number field K, is there a smallest Pólya extension of K contained in the Hilbert class field $H(K)$ of K?

Question I2. For every number field K, is there a Pólya field containing K? That is, is there a finite Pólya extension tower of K?

5 Around Prüfer domains

We end our paper with a short section which, following R. Gilmer [22], is most strongly linked to multiplicative ideal theory. In the classical case of the ring of integer-valued polynomials in a global field K (i.e., a number field or a function field), the ring $\mathrm{Int}(\mathcal{O}_K)$ is a 2-dimensional Prüfer domain. In the case of a local field (i.e., a field which is complete for a discrete valuation with a finite residue field), if V denotes the ring of the valuation, $\mathrm{Int}(V)$ provides a very natural example of a 2-dimensional Prüfer domain that is completely integrally closed but not the intersection of rank-one valuation domains. This last example answers several questions which go back to Krull.

It was then natural to ask the following: for which domain D is $\mathrm{Int}(D)$ a Prüfer domain?

When D is Noetherian, it is necessary and sufficient that D is a Dedekind domain with finite residue fields. In general, the answer is more difficult and Gilmer's paper [21] is a seminal step for this characterization (see [15, Questions, p. 113]). We state it in characteristic 0.
$\mathrm{Int}(D)$ is a Prüfer domain if and only if D is an almost Dedekind domain (each localization of D with respect to a maximal ideal $\mathfrak{p}$ is a discrete valuation domain) with finite residue fields and, for each prime number p, the following subsets are bounded: $\{|D/\mathfrak{p}| \mid p \in \mathfrak{p}\}$ and $\{v_\mathfrak{p}(p) \mid p \in \mathfrak{p}\}$ where $v_\mathfrak{p}$ denotes the normalized valuation associated to $\mathfrak{p}$ (see [9], [21] and [29]).

Recent papers deal more and more with subsets.

Problem J. Characterize the pairs (S, D), where D is a domain and S is a subset of D, such that $\mathrm{Int}(S, D)$ is a Prüfer domain.

There are several partial answers (for instance [7, V.Exercises] or [13]), but no characterization. It is however interesting to note that subsets allow one to provide examples of Prüfer domains with arbitrarily large Krull dimensions.

Finally, recall a question from Brewer and Klinger [6] which is of great interest. It concerns the question whether, as with Dedekind domains, Prüfer domains D have the *simultaneous bases property*. This property is defined as follows: for each $n \in \mathbb{N}^*$ and each sub-D-module M of D^n, one has
1) M is a projective D-module of rank $k \leq n$,
2) there exist n rank-one projective sub-D-modules $P_1, \ldots, P_n$ of D^n and a decreasing sequence $I_1 \supseteq \ldots \supseteq I_k$ of ideals of D such that

$$D^n = P_1 \oplus \cdots \oplus P_n\,,\ M = I_1 P_1 \oplus \cdots \oplus I_k P_k.$$

For a Prüfer domain, this property is equivalent to the *bcs-property* that may be formulated in the following way: for each matrix $B \in \mathcal{M}_{n \times m}(D)$ of unit content, there exists a matrix $C \in \mathcal{M}_{m \times l}(D)$ such that BC has unit content and rank one (recall that the content of a matrix is the ideal generated by its coefficients). The final question is as follows.

Question K (Brewer and Klinger). Does the very classical Prüfer domain $\mathrm{Int}(\mathbb{Z}) = \{f \in \mathbb{Q}[X] \mid f(\mathbb{Z}) \subseteq \mathbb{Z}\}$ satisfy the bcs-property?

References

1. D. Adam, Simultaneous orderings in function fields, *J. Number Theory* **112** (2005), 287–297.
2. D. Adam, Fonctions et Polynômes à Valeurs entières en caractéristique finie, Thesis, June 2004, Amiens, France.
3. Y. Amice, Interpolation p-adique, *Bull. Soc. Math. France* **92** (1964), 117–180.
4. M. Bhargava, P-orderings and polynomial functions on arbitrary subsets of Dedekind rings, *J. reine angew. Math.* **490** (1997), 101–127.

5. M. Bhargava, The factorial function and generalizations, *Amer. Math. Monthly*, **107** (2000), 783–799.
6. J. Brewer and L. Klinger, Rings of Integer-Valued Polynomials and the bcs-Property, in *Commutative ring theory and applications*, 65–75, Lecture Notes in Pure and Appl. Math. **231**, Dekker, New York, 2003.
7. P.-J. Cahen & J.-L. Chabert, *Integer-Valued Polynomials*, Amer. Math. Soc. Surveys and Monographs, **48**, Providence, 1997.
8. L. Carlitz, On certain functions connected with polynomials in a Galois field, *Duke Math. J.* **1** (1935), 137–168.
9. J.-L. Chabert, Integer-Valued Polynomials, Prüfer domains and Localization, *Proc. Amer. Math. Soc.* **118** (1993), 1061–1073.
10. J.-L. Chabert, Factorial Groups and Pólya Groups in Galoisian Extension of $\mathbb{Q}$, in *Commutative ring theory and applications, Lecture Notes in Pure and Appl. Math.* **231**, Dekker, New York, 2003.
11. J.-L. Chabert, Integer-valued polynomials on prime numbers and logarithm powers expansion, *European J. of Combinatorics*, to appear.
12. J.-L. Chabert, S. Chapman and W. Smith, A Basis for the Ring of Polynomials Integer-Valued on Prime Numbers, in *Factorization in integral domains*, 271–284, Lecture Notes in Pure and Appl. Math. **189**, Dekker, New York, 1997.
13. P.-J. Cahen, J.-L. Chabert and K.A. Loper, High dimension Prüfer domains of integer-valued polynomials, *J. Korean Math. Soc.* **38** (2001), 915–935.
14. J.-L. Chabert and G. Gerboud, Polynômes à valeurs entières et binômes de Fermat, *Canad. J. Math.* **45** (1993), 6–21.
15. S. Chapman, V. Ponomarenko and W. W. Smith, Robert Gilmer's contributions to the theory of integer-valued polynomials, this volume.
16. A. Crabbe, Generalized factorial functions and binomial coefficients, Honors Thesis, Trinity University, San Antonio, Texas, 2001.
17. Y. Fares, Factorial preservation, *Arch. Math* **83** (2004), 497–506.
18. Y. Fares, δ-rings and factorial sequences preservation, *Acta Arithmetica*, to appear.
19. G. Gerboud, Polynômes à valeurs entières sur l'anneau des quaternions de Hurwitz, preprint, Amiens, 1998.
20. R. Gilmer, Sets that determine Integer-valued Polynomials, *J. of Number Theory* **33** (1989), 95–100.
21. R. Gilmer, Prüfer domains and Rings of Integer-Valued Domains, *J. of Algebra* **129** (1990), 502–517.
22. R. Gilmer, Forty years of commutative ring theory, in *Rings, modules, algebras, and abelian groups*, 229–256, Lecture Notes in Pure and Appl. Math. **236**, Dekker, New York, 2004.
23. R. Gilmer and W. Smith, On the polynomial equivalence of subsets E and $f(E)$ of $\mathbb{Z}$, *Arch. Math.* **73** (1999), 355–365.
24. D. Goss, *Basic Structures of Function Field Arithmetic*, Springer, 1998, New York.
25. H. Gunji and D.L. McQuillan, On a Class of Ideals in an Algebraic Number Field, *J. Number Theory* **2** (1970), 207–222.
26. D. Hilbert, Die Theorie der algebraischen Zahlkörper, 1897.
27. F.H. Jackson, On q-definite integrals, *Quart. J. Pure and Appl. Math.* **41** (1910), 193–203.
28. J. Latham, On sequences of algebraic integers, *Journ. London Math. Soc.* **6** (1973), 555–560.

29. K.A. Loper, A classification of all D such that Int(D) is a Prüfer domain, *Proc. Amer. math. Soc.* **126** (1998), 657–660.
30. S.B. Mulay, Integer-Valued Polynomials in Seeral Variables, *Comm. Algebra* **27** (1999), 2409–2423.
31. W. Narkiewicz, Some unsolved problems, *Bull. Soc. Math France, Mémoire* **25** (1971), 159–164.
32. I. Newton, *Mathematical Principles of Natural Philosophy,* 1687.
33. G. Pólya, Über ganzwertige Polynome in algebraischen Zahlkörpern, *J. reine angew. Math.* **149** (1919), 97–116.
34. N.J.A. Sloane, *The On-Line Encyclopedia in Integer Sequences,* http://www.research.att.com/ñjas/sequences/index.html
35. A. Ostrowski, Über ganzwertige Polynome in algebraischen Zahlkörpern, *J. reine angew. Math.* **149** (1919), 117–124.
36. F.J. Van Der Linden, Integer valued polynomials over function fields, *Nederl. Akad. Wetensch. Indag. Math.* **50** (1988), 293–308.
37. R. Wasén, Remark on a problem of Schinzel, *Acta Arith.* **29** (1976), 425–426.
38. M. Wood, P-orderings: a metric viewpoint and the non-existence of simultaneous orderings, *J. Number Theory,* **99** (2003), 36-56.
39. J. Yeramian, Anneaux de Bhargava, *Comm. Algebra* **32** (2004), 3043–3069.
40. H. Zantema, Integer valued polynomials over a number field, *Manuscr. Math.* **40** (1982), 155–203.

Robert Gilmer's contributions to the theory of integer-valued polynomials

Scott T. Chapman[1*], Vadim Ponomarenko[2] and William W. Smith[3]

[1] Trinity University, Department of Mathematics, One Trinity Place, San Antonio, TX 78212-7200, schapman@trinity.edu
[2] Trinity University, Department of Mathematics, One Trinity Place, San Antonio, TX 78212-7200, vadim123@gmail.com
[3] The University of North Carolina at Chapel Hill, Department of Mathematics, Chapel Hill, North Carolina 27599-3250, wwsmith@email.unc.edu

1 Robert Gilmer and Int(E, D)

It is fitting in a volume dedicated to Robert Gilmer's work in commutative algebra that special mention be made of his contributions to the theory of integer-valued polynomials. To remind the reader, if D is an integral domain with quotient field K and $E \subseteq D$ a subset of D, then let

$$\text{Int}(E, D) = \{f(X) \in K[X] \mid f(a) \in D \text{ for every } a \in E\}$$

denote the ring of integer-valued polynomials on D with respect to the subset E (for ease of notation, if $E = D$, then set $\text{Int}(D, D) = \text{Int}(D)$). Gilmer's work in this area (with the assistance of various co-authors) was truly groundbreaking and led to numerous extensions and generalizations by authors such as J. L. Chabert, P. J. Cahen, D. McQuillan and A. Loper. In this paper, we will review Gilmer's papers dedicated to this subject. We close with an elementary analysis of polynomial closure in integral domains, a topic which Gilmer motivated with a characterization of which subsets S of $\mathbb{Z}$ define the ring $\text{Int}(\mathbb{Z})$ in [18]. Before proceeding, please note that we use $\mathbb{Q}$ to represent the rationals, $\mathbb{Z}$ the integers, $\mathbb{N}$ the natural numbers and $\mathbb{P}$ the primes in $\mathbb{Z}$.

It is clear that Gilmer's interest in the rings $\text{Int}(E, D)$ was motivated by his early work on multiplicative ideal theory and the theory of Prüfer domains. In particular, there was a problem open in the early 60's regarding the number of required generators for a finitely generated ideal of a Prüfer domain. It was at the time well known in the Noetherian case (i.e., for a Dedekind domain) that every ideal could be generated by two elements, one of which could be chosen to be an arbitrary non-zero element of the ideal.

* Part of this work was completed while the first author was on an Academic Leave granted by the Trinity University Faculty Development Committee.

Whether or not this property extended to the finitely generated ideals of a Prüfer domain (the non-Noetherian case) was the subject of much inquiry for over two decades. Evidence of this work, often determining cases where the property held, can be found in a 1970 paper of Gilmer and Heinzer [20]. There was much work on this problem during the subsequent years. Notable results were those by Sally and Vasconcelos [42] where it was established that for a one-dimensional Prüfer domain, every finitely generated ideal could be generated by two elements, with one chosen arbitrarily. Later, Heitman and Levy [27] gave an example of a Prüfer domain where the finitely generated ideals were 2-generated, yet one generator could not be chosen arbitrarily. About the same time, Heitman [26] extended the Sally and Vasconcelos results by showing in an n-dimensional Prüfer domain every finitely generated ideal could be generated by $n+1$ elements. Schulting [43] gave the first example of an invertible ideal in a Prüfer domain which required more than 2-generators. Finally, Swan [44] provided a construction for each positive integer $n > 1$ of a Prüfer domain of dimension n which contains a finitely generated ideal requiring $n+1$ generators. In the examples just mentioned, the Prüfer domains are described as intersections of valuation domains and the reasoning used involved geometric techniques beyond those of basic commutative ideal theory. Hence, even after the basic question of generators of finitely generated ideals in Prüfer domains was answered, there were two very general problems of interest to ideal theorists. These were described in the recent work "Non-Noetherian Commutative Ring Theory" [13] [14] and formulated based on work in a paper of Loper [33] as follows:

(A) [14, Problem 76] Let K be a field, and let $\{V_i\}$ be a set of valuation domains all of which have quotient fields equal to K. Give necessary and sufficient conditions on the set $\{V_i\}$, so that $\bigcap V_i$ is a Prüfer domain with quotient field K.

(B) [14, Problem 77] The proof of Swan's result cited above [44] uses tools outside ring theory. Give a construction which yields, for each positive integer n, a Prüfer domain containing a finitely generated ideal requiring n generators, such that the proof of the necessity of n generators can be carried out using elementary ring theoretical techniques without any reliance on geometry.

During the same period that the above work was taking place, considerable work was in progress on rings of integer-valued polynomials. Notable in this regard were the works of Brizolis, Cahen, Chabert and McQuillan. A good general reference for these works can be found in [5]. In particular, it was established in the late 70's that $\text{Int}(\mathbb{Z})$ is a two-dimensional Prüfer domain (for example, see [1]). In particular, in light of the Schulting example and Problem (B) described above, the question arose as to whether or not the simple domain $\text{Int}(\mathbb{Z})$ might represent an example of a Prüfer domain where the finitely generated ideals were not two-generated. Gilmer and Smith considered this problem which resulted in two publications, [23] and [24]. In the first publication, they established the following Theorem (an ideal I in Int$(\mathbb{Z})$ is called *unitary* if $I \cap \mathbb{Z} \neq 0$).

Theorem 1.1 (Gilmer-Smith [23]). *Int($\mathbb{Z}$) has the two-generator property on finitely generated ideals. Moreover, if I is a finitely generated unitary ideal of Int($\mathbb{Z}$) and $0 \neq n \in I \cap \mathbb{Z}$, then there exists $f(X) \in I$ such that $I = (n, f(X))$.*

The main technique in the proof of Theorem 1.1 considers the periodic behavior of the sequence of ideals

$$I(0) + n\mathbb{Z},\ I(1) + n\mathbb{Z},\ I(2) + n\mathbb{Z},\ \ldots$$

where n is a positive integer, I a finitely generated unitary ideal of Int$(\mathbb{Z})$ and

$$I(a) = \{f(a) | f \in I\}.$$

This result was extended to Int(E, D) where D is a Dedekind domain with finite residue fields and E is a "D-fractional" subset of K (i.e., there exists a $d \in D$ such that $dE \subseteq D$) by McQuillan [34, Theorem 5.5] (see also [7]).

Although the above Theorem answered the question regarding the existence of two generators for every finitely generated ideal of Int$(\mathbb{Z})$, it left open the additional question about the arbitrary choice of one of the two generators. The common terminology used is an ideal I is called *strongly two-generated* if it can be generated by two elements and the first of the two generators can be chosen at random from the nonzero elements of I. A ring where every finitely generated ideal is strongly two-generated is said to have the *strong two-generator property.* Finally, an element α of a domain R which can be chosen as one of two generators for every two-generated ideal I in which it is contained is called a *strong two-generator of* R. We note a result of Lantz and Martin [28] yields that the set of strongly two-generated ideals of R forms a subgroup of the set of invertible ideals of R. Hence, we set G_2 to be the subgroup of the class group G of R which is given by the strongly two-generated ideals.

Using the above terminology, Gilmer-Smith established in Theorem 1.1 that every integer $n \neq 0$ is a strong two-generator in Int$(\mathbb{Z})$. However, in a subsequent paper [24], they established that Int$(\mathbb{Z})$ did not have the strong two-generator property. In that paper the following results were established on the presence of elements that were not strong two-generators as well as a description of some elements other than $0 \neq n \in \mathbb{Z}$ which were strong two-generators.

Theorem 1.2. *[24] For Int($\mathbb{Z}$),*

(1) *If d is a square free integer and the class group of $\mathbb{Q}(\sqrt{d})$ is not an elementary abelian 2-group, then $X^2 - d$ is not a strong two-generator of Int($\mathbb{Z}$).*

(2) *For an odd prime p, the ideal (p, X) is not a strongly two-generated ideal of Int($\mathbb{Z}$).*
(3) *If n and a are non-zero elements of $\mathbb{Z}$, and b, c elements of $\mathbb{Z}$ with $c \geq 0$, then every element of the form $n(aX+b)^c$ is a strong two-generator of Int($\mathbb{Z}$).*

Hence, it was established in Int($\mathbb{Z}$) that $G_2 \subsetneq G$ (i.e., Int($\mathbb{Z}$) does not have the strong two-generator property). We outline the main tool used in the proof of Theorem 1.2. Let $f(X) \in \text{Int}(\mathbb{Z})$ be an irreducible polynomial in $\mathbb{Q}[X]$ with root θ. Set $K = \mathbb{Q}(\theta)$ and define the map

$$\varphi : \text{Int}(\mathbb{Z}) \to \mathbb{Q}(\theta)$$

by

$$g(X) \mapsto g(\theta).$$

Now, if

$$\tilde{f} = \text{Ker}(\varphi) = f(X)\mathbb{Q}[X] \cap \text{Int}(\mathbb{Z}),$$

then

$$\text{Int}(\mathbb{Z})/\tilde{f} \cong \{g(\theta) \mid g(X) \in \text{Int}(\mathbb{Z})\}.$$

Set $J(f) = \text{Int}(\mathbb{Z})/\tilde{f}$. $J(f)$ is the homomorphic image of a Prüfer domain and is hence integrally closed. Clearly $J(f) \subseteq \mathbb{Q}(\theta)$ and since $\mathbb{Z} \subseteq J(f)$, we have that $\mathbb{Z}^* \subseteq J(f)$ where $\mathbb{Z}^*$ is the ring of integers of $\mathbb{Q}(\theta)$ over $\mathbb{Z}$. $J(f)$ is an overring of $\mathbb{Z}^*$ and is thus itself a Dedekind domain. Thus, we have a set $S = \{p_\alpha\}_{\alpha \in \mathcal{A}}$ of prime ideals of $\mathbb{Z}^*$ so that

$$J(f) = \bigcap_{p_\alpha \in S} \mathbb{Z}^*_{p_\alpha}.$$

In [24], the authors argue that a nonzero $f \in \text{Int}(\mathbb{Z})$ is a strong two-generator in Int($\mathbb{Z}$) if and only if $J(f)$ is a principal ideal domain. These ideas were extended in a subsequent paper by McQuillan [37].

It remains open to determine the following:

1. Does there exist a non-principal ideal of Int($\mathbb{Z}$) which is strongly two-generated?
2. A characterization of all the strong two-generators of Int($\mathbb{Z}$).

Recalling that $\text{Int}(\mathbb{Z}) = \text{Int}(\mathbb{Z}, \mathbb{Z})$, we summarize the above problems in the context of a general domain D and non-empty subset $E \subseteq D$. Let $\mathcal{R} = \text{Int}(E, D)$, G be the class group of $\mathcal{R}$ and G_2 the subgroup of G represented by the strongly two-generated ideals of R.

General Problems:

1. For a given $\mathcal{R}$, describe G and G_2 (or G/G_2).
2. Describe the strong two-generators of $\mathcal{R}$.

3. Determine those $\mathcal{R}$ for which $G_2 = G$ (that is, $\mathcal{R}$ is strongly two-generated).

We note some additional work of Gilmer in this direction. In [22], Gilmer-Henizer-Lantz-Smith determined the structure of G for Int$(\mathbb{Z})$ as a free abelian group on a countably infinite basis. Additional descriptions of G can be found in [5, Chapter VIII].

Less is known about the subgroup G_2, in particular the strongly two-generated ideals. For the general case Int(E, D), we note that the situation where E is a finite set is very special. A simple description of the domain Int(E, D) where $| E | < \infty$ was given by McQuillan [35]. Using this characterization of Int(E, D), Chapman-Loper-Smith were able to establish the following in [15] and [16].

Theorem 1.3. *Let D be an integral domain and $E = \{e_1, \ldots, e_k\}$ a finite nonempty subset of D. Then Int(E, D) has the strong two-generator property if and only if D is a Bezout domain.*

Theorem 1.4. *If D is a Dedekind domain which is not a principal ideal domain and $E = \{e_1, \ldots, e_k\}$ is a finite nonempty subset of D, then $f(X)$ is a strong two-generator in Int(E, D) if and only if $f(e_i) \neq 0$ for every $1 \leq i \leq k$.*

Theorem 1.3 is of interest, since there is no example in the literature of the form $R = \text{Int}(E, D)$ with the strong two-generator property where the Jacobson radical of R is zero (for other examples of rings of form Int(E, D) with the strong two-generator property, see Rush [41] or Brewer and Klingler [2] [3]). In light of Theorem 1.3, the current authors are unaware of a domain of the form Int(E, D) with zero Jacobson radical with the strong two-generator property where E is infinite. Additional information on the strong two-generator property in Int(E, D) can be found in [5, Chapter VIII].

We briefly mention two other papers of Gilmer on integer-valued polynomials. Following the early observation that Int$(\mathbb{Z})$ was a Prüfer domain, a basic question was raised.

Question: What are necessary and sufficient conditions on a domain D in order that Int(D) be a Prüfer domain?

A necessary condition is that D be an almost Dedekind domain with finite residue fields. In the Noetherian case (for instance when D is a Dedekind domain with finite residue fields) then this condition is also sufficient. In [19] Gilmer considered two related questions:

Q1 If Int(D) is Prüfer, must D be Noetherian?
Q2 Does D almost Dedekind with finite residue fields imply Int(D) is Prüfer?

Exploiting previous work (some joint work with others) Gilmer provides constructions yielding a negative answer to both questions. Additional work on this problem was done by Chabert [7] [8] [9] and Loper [29] [30] [31] and necessary and sufficient conditions were finally given by Loper in [32].

In [21], Gilmer-Heinzer-Lantz discuss the Noetherian properties in the ring $\mathrm{Int}(D)$ and in the prime spectrum of $\mathrm{Int}(D)$. A summary of the results of that work and of subsequent work can be found in [5, Chapter VI].

Finally, there are two other papers of Gilmer regarding a different aspect of study for integer-valued polynomials. In [23], the authors give an argument that if E is a subset of $\mathbb{Z}$ that includes all except a finite number of positive integers, then $f(a) \in \mathbb{Z}$ for all a in E implies $f(b) \in \mathbb{Z}$ for all $b \in \mathbb{Z}$. That is, $\mathrm{Int}(E, \mathbb{Z}) = \mathrm{Int}(\mathbb{Z})$. A question was raised as to exactly what subsets E of $\mathbb{Z}$ have this property. Gilmer answered this question in [18], both for $\mathbb{Z}$ and also in the setting of a Dedekind domain D with finite residue fields. McQuillan [36] extended the Gilmer result describing when $\mathrm{Int}(E_1, D) = \mathrm{Int}(E_2, D)$ in terms of the closures of E_1 and E_2 in the p-adic completions of D. Gilmer's description was more in terms of basic number theory (especially in the case $D = \mathbb{Z}$) and is described as follows in terms of $\mathbb{Z}$ (where the extension to the Dedekind case is straight forward). We require a definition.

Definition 1.5. *For $E \subseteq \mathbb{Z}$ and a prime power p^k where $k \geq 1$,*

$$E \bmod p^k = \{x + (p^k) \mid x \in E\}.$$

Although basically established in [18], the following result is stated in a later Gilmer-Smith paper [25, Theorem 2.2].

Theorem 1.6. *Let E_1 and E_2 be nonempty subsets of $\mathbb{Z}$. Then $\mathrm{Int}(E_1, \mathbb{Z}) = \mathrm{Int}(E_2, \mathbb{Z})$ if and only if $E_1 \bmod p^k = E_2 \bmod p^k$ for every prime p and every integer $k \geq 1$.*

The terminology now used for the property $\mathrm{Int}(E_1, D) = \mathrm{Int}(E_2, D)$ is that E_1 and E_2 are *polynomially equivalent.* Additional work on the notion of polynomial equivlance can be found in [5, Chapter IV].

In this spirit, Gilmer-Smith in [25] consider the following problem:

Question: For $E \subseteq \mathbb{Z}$ and $f(X) \in \mathrm{Int}(\mathbb{Z})$, under what conditions are E and $f(E)$ polynomially equivalent?

In [25], the answer is provided when E is finite and for several special cases when E is infinite. In every case considered where E is infinite, it was shown that E and $f(E)$ polynomially equivalent implied $f(X)$ must be linear (in some cases linear of a special form). These results are of interest in connection with some terminology and questions raised by Narkiewicz. In [39], Narkiewicz defines a property (**P**) for a field as follows.

Definition 1.7. *A field K has property* (**P**) *if the polynomials $f(t)$ in $K[t]$ for which $f(E) = E$ for some infinite set E of K, must be linear.*

Narkiewicz provides several results about property (**P**), including that $\mathbb{Q}$ has the property and poses several interesting questions. Note the problem posed by Gilmer-Smith in [25] is related in that it asked if $f(X)$ must be linear when

"$f(E) = E$" is replaced by "$E \bmod p^k = f(E) \bmod p^k$" (that is, E and $f(E)$ are polynomially equivalent) where E is an infinite subset of $\mathbb{Z}$ and $f(X)$ is in $\mathrm{Int}(\mathbb{Z})$. We note that the Gilmer-Smith question on polynomially equivalence of E and $f(E)$ has been recently answered in the affirmative by Fares [17]. A more general result has been offered by Mulay [38].

All of the above is directly related to the concept of *polynomial closure* of a set E (see [4]). More precisely, if D is a domain and $E \subseteq D$ define

$$\overline{E} = \{d \in D \mid f(d) \in D \text{ for all } f(X) \in \mathrm{Int}(E, D)\}.$$

That is, $\overline{E}$ is the largest subset of E such that $\mathrm{Int}(\overline{E}, D) = \mathrm{Int}(E, D)$. We provide in the next section an alternate description to that given in [36] of $\overline{E}$ using only elementary arithmetical tools available in the domain D.

2 A New Approach to Polynomial Closure

The work in this section is sparked by the papers [10], [11] and [12] which study the ring $\mathrm{Int}(\mathbb{P}, \mathbb{Z})$. In [12, Proposition 2.1] (which is actually a restatement of [11, Propositions 5.1 and 5.2]), the authors argue that $\overline{\mathbb{P}} = \mathbb{P} \cup \{\pm 1\}$ with respect to $\mathbb{Z}$. To show that a polynomial $f(X) \in \mathrm{Int}(\mathbb{P}, \mathbb{Z})$ has the property $f(\pm 1) \in \mathbb{Z}$, they appeal to localizations. In fact, their argument is based on the following result: for any prime p, $\mathrm{Int}(\mathbb{P}, \mathbb{Z})_{(p)} = \mathrm{Int}(X_p, \mathbb{Z}_{(p)})$ where $X_p = (\mathbb{Z} \backslash p\mathbb{Z}) \cup \{p\}$ ([10, Corollaire 3]). We present an elementary proof of this fact which does not rely on localizations or topological closures (see [12, Proposition 2.2]). Our primary result is Propositin 2.5, where we give a new characterization of $\bar{S}$. We then apply this result to give an elementary description of the closure of the set of natural prime integers $\mathbb{P}$ in both $\mathbb{Z}$ and in the larger domain $\mathbb{Z}[\imath]$. We close by applying our method to the set of prime numbers contained in arithmetic sequence in $\mathbb{Z}$. Proposition 2.11 computes the polynomial closure of any such set.

In the following, let S be a subset of D, $a \neq 0$ an element of D and $g(X) \in D[X]$.

Definition 2.1. (1) *We say that* $g(S) \equiv 0 \pmod a$ *if* $s \in S$ *implies that* $g(s) \equiv 0 \pmod a$.
(2) *We say that* $g(S) \not\equiv 0 \pmod a$ *if* $s \in S$ *implies that* $g(s) \not\equiv 0 \pmod a$.

Let S_1 and S_2 be subsets of D.

Definition 2.2. (1) *We say that* $(a, g(X))$ *separates* S_1 *from* S_2 *if* $g(S_1) \equiv 0 \pmod a$ *and* $g(S_2) \not\equiv 0 \pmod a$.
(2) *We say that* a *separates* S_1 *from* S_2 *if there exists a* $g(X) \in D[X]$ *such that* $(a, g(X))$ *separates* S_1 *from* S_2.

In the latter case, we call a a *separator*. Clearly a separator is never a unit of D. We say that S_1 and S_2 are *separable* if there is a separator a separating S_1

from S_2 or S_2 from S_1. We say that S_1 and S_2 are *fully separable* if there is a separator a_1 separating S_1 from S_2 and a separator a_2 separating S_2 from S_1. We note that our definition of *separates* differs sharply from the notion of Int(D) *separating points* in the completion $\hat{D}$ of D as illustrated in [5, Chapter III.4].

Comments.

(1) If there exists $s_1 \in S_1$, $s_2 \in S_2$ and $a \in D$ with $s_1 \equiv s_2 \pmod{a}$, then a cannot separate S_1 from S_2. This clearly holds for all a if $S_1 \cap S_2 \neq \emptyset$.

(2) Set $D = \mathbb{Z}$. If S_1 and S_2 are disjoint and both finite, then S_1 and S_2 are fully separable. To see this, let $p \in \mathbb{P}$ with

$$p > \left(\prod_{s_1 \in S_1} s_1\right)^{|S_2|} \left(\prod_{s_2 \in S_2} s_2\right)^{|S_1|}$$

and set $g(X) = \prod_{s \in S_1}(X - s)$. Clearly $g(S_1) \equiv 0 \pmod{p}$. If $s_2 \in S_2$, then $g(s_2) = \prod_{s \in S_1}(s_2 - s) \not\equiv 0 \pmod{p}$ and so $g(S_2) \not\equiv 0 \pmod{p}$. Similarly, setting $f(X) = \prod_{s \in S_2}(X - s)$ yields that p also separates S_2 from S_1.

(3) If a separates S_1 from S_2 and $0 \neq b \in D$, then ab also separates S_1 from S_2. To see this, suppose $(a, g_1(X))$ separates S_1 from S_2. Set $g(X) = bg_1(X)$. Clearly $g(S_1) \equiv 0 \pmod{ab}$. Suppose for some $s_2 \in S_2$ that $g(s_2) = kab = bg_1(s_2)$ for $0 \neq k \in D$. But then $g_1(s_2) = ka$, so $g_1(s_2) \equiv 0 \pmod{a}$, which is impossible. So $g(S_2) \not\equiv 0 \pmod{ab}$.

(4) Suppose S_1 and S_2 are subsets in D with $S_2 = \{\alpha\}$ a singleton. If we can separate S_1 from S_2, then a separator can be chosen of the form p^k where p is a prime in D. To see this, suppose $(a, g(X))$ separates S_1 from S_2. Write $a = p_1^{m_1} \cdots p_k^{m_k}$. Now, $g(S_1) \equiv 0 \pmod{a}$, hence $g(S_1) \equiv 0 \pmod{p_i^{m_i}}$ for $1 \leq i \leq k$. We cannot have $g(S_2) \equiv 0 \pmod{p_i^{m_i}}$ for $1 \leq i \leq k$, since in that case $g(S_2) \equiv 0 \pmod{a}$. Therefore, one of the $p_i^{m_i}$ must separate S_1 and S_2, using the same $g(X)$.

Example 2.3. If S_1 and S_2 are separable, they may not be fully separable. Let $D = \mathbb{Z}$ and set

$$S_1 = \bigcup_{i=1}^{5} \{i + 6k \mid k \in \mathbb{N} \cup \{0\}\} \text{ and } S_2 = \{0\}.$$

We have $(6, X)$ separating S_2 from S_1. Suppose that $(m, g(X))$ separates S_1 from $\{0\}$. From Comment (4), we may assume m is a prime power p^k, $k > 0$. Whatever this prime power, $p^k \not\equiv 0 \pmod{6}$, thus $p^k \in S_1$. As clearly $p^k \equiv 0 \pmod{p^k}$, we reach a contradiction by Comment (1).

We also find points which cannot be separated from S of interest.

Definition 2.4. *Let S be a nonempty subset of an integral domain D. We say that $x \in D$ sticks to S if there does not exist an $a \in D$ which separates S from $\{x\}$. We denote the set of elements in D which stick to S by $\tilde{S}$.*

Clearly x sticks to S if and only if

$$\forall a \in D,\ a \neq 0,\ g(S) \equiv 0 \pmod{a} \text{ implies } g(x) \equiv 0 \pmod{a}.$$

Proposition 2.5. *Let S be a nonempty subset of the domain D. Then $\tilde{S}$ is equal to the polynomial closure $\overline{S}$ of S.*

Proof. Note that a poplynomial $q[X] \in K[X]$ can be written $q = \frac{f}{a}$, with $f \in D[X]$ and $a \in D \neq 0$. With this in mind for $\alpha \in D$, the following assertions are equivalent.

- $\alpha \notin \overline{S}$,
- there exists a polynomial $q[X] \in K[X]$ such that $q(S) \subseteq D$, but $q(\alpha) \notin D$,
- there exists $a \in D$, $a \neq 0$, and $f(X) \in D[X]$ such that $f(S) \equiv 0 \pmod{a}$, but $f(\alpha) \not\equiv 0 \pmod{a}$,
- $\alpha \notin \tilde{S}$.

□

Notice that our separation method makes verification of the basic properties related to closure relatively simple (see [5, Proposition IV.1.5]). We list these properties here for the convenience of the reader.

Basic Facts: Let D be an integral domain.

(i) If $\emptyset \neq S \subseteq D$, then $\overline{\overline{S}} = \overline{S}$.
(ii) If $\emptyset \neq S \subseteq T \subseteq D$, then $\overline{S} \subseteq \overline{T}$.
(iii) If $\{S_i\}$ is a family of nonempty subsets of D, then

$$\overline{\bigcap_i S_i} \subseteq \bigcap_i \overline{S_i} \text{ and } \bigcup_i \overline{S_i} \subseteq \overline{\bigcup_i S_i}.$$

We now offer the promised applications.

Example 2.6. Let D be any integral domain and b, c nonzero elements of D. If $S = \{b+cz \mid z \in D\}$, then every $\alpha \notin S$ is separated from S by $(c, X-b)$. Thus $\overline{S} = S$ and in general a set which is equal to its closure is called *polynomially closed.* In the above construction, set $D = \mathbb{Z}$ and suppose both b and c are positive integers. If S_1 is a subset of S with $S \backslash S_1$ bounded above or below (or both), then $\overline{S_1} = S$. To see this, first consider $s_2 \not\equiv b \pmod{c}$. Then $(c, X-b)$ separates S_1 from $\{s_2\}$. Now, let $s_2 \equiv b \pmod{c}$. Suppose $(a, g(X))$ separates S_1 from $\{s_2\}$. Now, $s_2 = mc+b$ for some $m \in \mathbb{Z}$ and by hypothesis, $nmc+b \in S_1$ for all n sufficiently large (or small). Choose one such $n \equiv 1 \pmod{a}$. Then $s_2 = mc+b \equiv nmc+b \pmod{a}$. This contradicts the supposition that a separates S_1 from $\{s_2\}$. Therefore, $s_2 \in \overline{S_1}$.

Example 2.7. Let $\mathcal{SF}$ be the set of squarefree integers. Using Theorem 1.6, one can argue that $\mathcal{SF}$ is its own closure. We show how to obtain this result using Proposition 2.5. It suffices to separate $\mathcal{SF}$ from each $y \in \mathbb{Z} \backslash \mathcal{SF}$. Suppose that $p^2 \mid y$ for some prime p. Set $f_1(X) = (X-p)(X-2p)\cdots(X-(p-1)p)$. Set $f_2(X) = (X-1)(X-2)\cdots(X-(p-1))$. Set $f(X) = f_1(X)f_2(X)^p$. Now, consider the pair $(p^p, f(X))$. We claim that this separates $\mathcal{SF}$ from y.

If $X \not\equiv 0 \pmod{p}$, then $f_2(X) \equiv 0 \pmod{p}$, and hence $f(X) \equiv 0 \pmod{p^p}$. If $X \equiv 0 \pmod{p}$, then write $X = kp$. Note that $f_2(X) \not\equiv 0 \pmod{p}$, and hence $v_p(f(X)) = v_p(f_1(X))$. Now, $f_1(X) = (kp-p)(kp-2p)\cdots(kp-(p-1)p) = p^{p-1}(k-1)(k-2)\cdots(k-(p-1))$. If $p \nmid k$, then $k \not\equiv 0 \pmod{p}$ and therefore $p^p \mid f_1(X)$ so $f(X) \equiv 0 \pmod{p^p}$. Therefore, if $p \nmid k$, then $f(\mathcal{SF}) \equiv 0 \pmod{p^p}$. However, if $p \mid k$ then $v_p(f_1(X)) = p-1$ and $f(X) \not\equiv 0 \pmod{p^p}$. Since $p^2 \mid y$ we write $y = pk$, with $p \mid k$. Hence, $f(y) \not\equiv 0 \pmod{p^p}$.

Proposition 2.8. *Let $\mathbb{P}$ be the set of prime numbers. Then $\overline{\mathbb{P}} = \mathbb{P} \cup \{\pm 1\}$.*

Proof. If $x = \pm 1$, then, by Dirichlet's Theorem, for each $a \neq 0$ which is not a unit in $\mathbb{Z}$, there is a prime p congruent to $x \pmod{a}$, hence $\mathbb{P}$ cannot be separated from $\{1\}$ and $\{-1\}$ by Comment (1).

If $x \notin \mathbb{P} \cap \{\pm 1\}$, choose a prime p dividing x and let $r = v_p(x-p)$ (where $v_p(y)$ is the p-adic valuation of y), then $\mathbb{P}$ is separated from $\{x\}$ by

$$(p^{r+1}, (X-p)[(X-1)\dots(X-p+1)]^{r+1}),$$

completing the argument. □

We further demonstrate the versatility of our method by computing $\overline{\mathbb{P}}$ with respect to $\mathbb{Z}[\imath]$. The proof requires a lemma.

Lemma 2.9. *Let $D = \mathbb{Z}[i]$.*

(1) *The closure $\overline{S}$ of a subset S of D such that $S \subseteq \mathbb{Z}$ is itself such that $\overline{S} \subseteq \mathbb{Z}$.*
(2) *If two subsets of D are separable, then a separator m can be chosen in $\mathbb{Z}$.*

Proof. (1) Suppose that α sticks to S (i.e., suppose that $\alpha \in \overline{S}$). Pick a prime p such that $p \equiv 3 \pmod{4}$. Set

$$g(X) = X(X-1)(X-2)\dots(X-p+1).$$

Then $g(S) \equiv 0 \pmod{p}$ and hence, $g(\alpha) \equiv 0 \pmod{p}$. As p is prime in D it divides some $\alpha - j$, $0 \leq j \leq p-1$, in D, and thus clearly divides the imaginary part of α in $\mathbb{Z}$. As there are infinitely many such primes p, it follows that α is real.

(2) If z is a separator, then $m = z\overline{z}$ (where $\overline{z}$ is the conjugate of z) is an integer and, by Comment (3), another separator. □

From the lemma follows this result about the polynomial closure of subsets of $\mathbb{Z}$ in $\mathbb{Z}[i]$.

Proposition 2.10. *Let S be a subset of $D = \mathbb{Z}[i]$ such that $S \subseteq \mathbb{Z}$, then its polynomial closure D is the same as in $\mathbb{Z}$. In particular, $\overline{\mathbb{P}} = \mathbb{P} \cup \{\pm 1\}$.*

Proof. From Lemma 2.9, both closures are contained in $\mathbb{Z}$. Considering $\alpha \in \mathbb{Z}$, we show that it can be separated from S in D if and only if it can be so in $\mathbb{Z}$. Note that if a, b are integers, then a divides b in $\mathbb{Z}$ if and only if this is so in D.

- Suppose α can be separated from S in $\mathbb{Z}$: there exists $(a, g(X))$, with $a \in \mathbb{Z}$, and $g(X) \in \mathbb{Z}[X]$ such that, $\forall s \in S$, a divides $g(s)$ but not $g(\alpha)$ in $\mathbb{Z}$. As a, $g(s)$, and $g(\alpha)$ are integers, the same division properties hold in D. Hence α can be separated from S in D.
- Suppose α can be separated from S in D: From Lemma 2.9, the separated a can be chosen in $\mathbb{Z}$, and there is $g(X) \in D[X]$ such that, $\forall s \in S$, a divides $g(s)$ but not $g(\alpha)$ in D. Write $g(X) = g_1(X) + ig_2(X)$, where both $g_1(X)$ and $g_2(X)$ have their coefficients in $\mathbb{Z}$. Then, for $i = 1$ or $i = 2$, a does not divide both $g_1(s)$ and $g_2(s)$ in D and hence also in $\mathbb{Z}$. Therefore $(a, g_i(X))$ separates α from S in $\mathbb{Z}$. □

Let a and b be relatively prime natural numbers with $1 \leq a < b$ and $b \geq 2$. Further, let $\mathbb{P}_{a,b}$ represent the prime numbers in the arithmetic sequence $A_{a,b} = \{a + bk \mid k \in \mathbb{N} \cup \{0\}\}$. The method we have developed allows us to determine the polynomial closure of the sets $\mathbb{P}_{a,b}$ in $\mathbb{Z}$.

Proposition 2.11. *Let a, b and $\mathbb{P}_{a,b}$ be as above.*

(1) $\overline{\mathbb{P}_{1,2}} = \mathbb{P}_{1,2} \cup \{\pm 1\}$.
(2) *Suppose that $b > 2$.*
 (a) *If $a \neq 1$ or $b - 1$, then $\overline{\mathbb{P}_{a,b}} = \mathbb{P}_{a,b}$.*
 (b) *If $a = 1$, then $\overline{\mathbb{P}_{a,b}} = \mathbb{P}_{a,b} \cup \{1\}$.*
 (c) *If $a = b - 1$, then $\overline{\mathbb{P}_{a,b}} = \mathbb{P}_{a,b} \cup \{-1\}$.*

Proof. Note that $\mathbb{P}_{a,b} = \mathbb{P} \cap A_{a,b}$. By Basic Fact (ii) above, $\overline{\mathbb{P}_{a,b}} \subseteq \overline{\mathbb{P}} \cap \overline{A_{a,b}}$. The results follow immediately taking into account Example 2.6. □

We close with a generalization of the last result.

Corollary 2.12. *Let $a_1, a_2, \ldots, a_k, b \in \mathbb{Z}$ with $1 \leq a_1 < a_2 < \ldots < a_k < b$ and $k > 1$. Suppose that* $\gcd(b, a_i) = 1$ *for each i and set $X = \mathbb{P}_{a_1,b} \cup \mathbb{P}_{a_2,b} \cup \ldots \cup \mathbb{P}_{a_k,b}$. Then:*

(1) $(\overline{X} - X) \subseteq \{-1, 1\}$.
(2) 1 *is in* $(\overline{X} - X)$ *if and only if* $a_1 = 1$.
(3) -1 *is in* $(\overline{X} - X)$ *if and only if* $a_k = b - 1$.

Proof. By our hypothesis, $b > 2$. (2) and (3) follow from the previous proof. Using an argument similar to that used in Example 2.6, the set

$$Y = A_{a_1,b} \cup A_{a_2,b} \cup \cdots \cup A_{a_k,b}$$

is closed. From Basic Fact (iii),

$$\overline{\mathbb{P}_{a_1,b}} \cup \overline{\mathbb{P}_{a_2,b}} \cup \cdots \cup \overline{\mathbb{P}_{a_k,b}} \subseteq \overline{X} \subseteq \overline{Y} = Y.$$

The result now follows immediately. □

Acknowledgement

The authors would like to thank the referee for suggestions which substantially improved the paper.

References

1. Brizolis, D.: A theorem on ideals in Prüfer rings of integer-valued polynomials. *Comm. Algebra*, **7**, 1065-1077 (1979)
2. Brewer, J. and Klinger, L.: The ring of integer-valued polynomials of a semi-local principal-ideal domain. *Linear Algebra Appl.*, **157**, 141–145 (1991)
3. Brewer, J. and Klinger, L.: Rings of integer-valued polynomials and the bcs-property, Commutative ring theory and applications. *Lecture Notes in Pure and Appl. Math.*, **231**, 65–75 (2003)
4. Cahen, P.-J.: Polynomial closure. *J. Number Theory*, **61**, 226–247 (1996)
5. Cahen, P.-J. and Chabert, J.-L.: Integer Valued-Polynomials, *Amer. Math. Soc. Surveys and Monographs* **58**, American Mathematical Society, Providence, (1997)
6. Cahen, P.-J. and Chabert, J.-L.: What's new about integer–valued polynomials on a subset? In: Chapman, S. T. and Glaz, S. (eds) *Non-Noetherian Commutative Ring Theory*, Kluwer Academic Publishers, Boston, 75–96 (2000).
7. Chabert, J.-L.: Un anneau de Prüfer. *J. Algebra*, **107**, 1–16 (1987)
8. Chabert, J.-L.: Anneaux de polynômes à valeurs entières et anneaux de Prüfer. *C. R. Acad. Sci. Paris Sr. I Math.*, **312**, 715–720 (1991)
9. Chabert, J.-L.: Integer-valued polynomials, Prüfer domains, and localization. *Proc. Amer. Math. Soc.*, **118**, 1061–1073 (1993)
10. Chabert, J.-L.: Une caractérisation des polynômes prenant des valeurs entieres sur tous les nombres premiers. *Canad. Math. Bull.*, **99**, 273–282 (1996)
11. Chabert, J.-L., Chapman, S. T. and Smith, W. W.: A basis for the ring of polynomials integer–valued on prime numbers. *Lecture Notes in Pure and Applied Mathematics*, **189**, 271–284 (1997)
12. Chabert, J.-L., Chapman, S. T. and Smith, W. W.: Algebraic properties of the ring of integer–valued polynomials on prime numbers. *Comm. Algebra*, **25**, 1947–1959 (1997)
13. Chapman, S. T. and Glaz, S. (eds): *Non-noetherian commutative ring theory.* Kluwer Academic Publishers, Boston (2000)
14. Chapman, S. T. and Glaz, S.: One hundred problems in commutative ring theory. In: Chapman, S. T. and Glaz, S. (eds) *Non-Noetherian Commutative Ring Theory*, Kluwer Academic Publishers, Boston, 459–476 (2000)

15. Chapman, S. T., Loper, K. A. and Smith, W. W.: The strong two-generator property in rings of integer-valued polynomials determined by finite sets. *Arch. Math*, **78**, 372-377 (2002)
16. Chapman, S. T., Loper, K. A. and Smith, W. W.: Strongly two-generated ideals in rings of integer-valued polynomials determined by finite sets, *C. R. Math. Rep. Acad. Sci. Canada*, **26**, 33-38 (2004)
17. Fares, Y.: Factorial Preservation. *Arch. Math. (Basel)*, **83**, 497-506 (2004)
18. Gilmer, R.: Sets that determine integer–valued polynomials. *J. Number Theory*, **33**, 95–100 (1989)
19. Gilmer, R.: Prüfer domains and rings of integer-valued polynomials. *J. Algebra*, **129**, 502–517 (1990)
20. Gilmer R. and Heinzer, W.: On the number of generators for an invertible ideal. *J. Algebra*, **14**, 139–151 (1970)
21. Gilmer, R., Heinzer, W. and Lantz, D.: The noetherian property in rings of integer-valued polynomials. *Trans. Amer. Math. Soc.*, **338**, 187–199 (1993)
22. Gilmer, R., Heinzer, W., Lantz, D. and Smith, W. W.: The ring of integer-valued polynomials of Dedekind domain. *Proc. Amer. Math. Soc.*, **108**, 673–681 (1990)
23. Gilmer, R. and Smith, W.W.: Finitely generated ideals of the ring of integer-valued polynomials. *J. Algebra*, **81**, 150–164 1983)
24. Gilmer, R. and Smith, W.W.: Integer-valued polynomials and the strong two-generator property. *Houston J. Math.*, **11**, 65–74 (1985)
25. Gilmer, R. and Smith, W. W.: On the polynomial equivalence of subsets E and $f(E)$ of $\mathbb{Z}$. *Arch.Math. (Basel)*, **73**, 355–365 (1999)
26. Heitmann, R. C.: Generating ideals in Prüfer domains. *Pacific J. Math.*, **62**, 117–126 (1976)
27. Heitman, R. C. and Levy, L. S.: $1\frac{1}{2}$ and 2-generator ideals in Prüfer domains. *Rocky Mountain J. Math.*, **5**, 361-373 (1975)
28. Lantz, D. and Martin, M.: Strongly two-generated ideals. *Comm. Algebra*, **16**, 1759-1777 (1988)
29. Loper, K. A.: More almost Dedekind domains and Prüfer domains of polynomials. *Lecture Notes in Pure and Appl. Math.* **171**, 287–298 (1995)
30. Loper, K. A.: Another Prüfer ring of integer-valued polynomials. *J. Algebra*, **187**, 1–6 (1997)
31. Loper, K. A.: Ideals of integer-valued polynomial rings. *Comm. Algebra*, **25**, 833–845 (1997)
32. Loper, K. A.: A classification of all D such that Int(D) is a Prüfer domain. *Proc. Amer. Math. Soc.*, **126**, 657–660 (1998)
33. Loper, K. A.: Constructing examples of integral domains by intersection valuation domains. In: Chapman, S. T. and Glaz, S. (eds) *Non-Noetherian Commutative Ring Theory*, Kluwer Academic Publishers, Boston, 325–340 (2000)
34. McQuillan, D.L.: On Prüfer domains of polynomials. *J. Reine Angew. Math.*, **358**, 162–178 (1985)
35. McQuillan, D. L.: Rings of integer-valued polynomials determined by finite sets. *Proc. Royal Irish Acad.*, **85A**, 177-184 (1985)
36. McQuillan, D. L.: On a theorem of R. Gilmer. *J. Number Theory*, **39**, 245–250 (1991)
37. McQuillan, D. L.: Split Primes and Integer-Valued Polynomials. *J. Number Theory*, **43**, 216–219 (1993)
38. Mulay, S. B.: Polynomial-mappings and M-equivalence. preprint

39. Narkiewicz, W.: Polynomial mappings. Lecture Notes in Mathematics, Springer-Verlag, Berlin, No. 1600 (1995)
40. Rose, H. E.: A Course in Number Theory. Clarendon Press, Oxford (1994)
41. Rush, D.: Generating ideals in rings of integer-valued polynomials. *J. Algebra*, **92**, 389–394 (1985)
42. Sally, J. and Vasconcelos, W.: Stable rings. *J. Pure Appl. Algebra*, **4**, 319-336 (1974)
43. Schülting, H-W.: Über die Erzeugendenanzahl invertierbarer Ideale in Prüferringen. *Comm. Algebra*, **7**, 1331–1349 (1979)
44. Swan, R.: n-generator ideals in Prüfer domains. *Pacific J. Math.*, **111**, 433–446 (1984)

Progress on the dimension question for power series rings

Jim Coykendall

North Dakota State University, Department of Mathematics, Fargo, ND 58105-5075 jim.coykendall@ndsu.edu

1 Introduction and background

The notion of Krull dimension of a ring (commutative with identity) is of great importance in the field of commutative algebra. In particular, the study of dimension behavior in ring extensions is a central issue. More precisely, we may ask if $R \subseteq T$ is an extension, what is the relationship, if any, between $\dim(R)$ and $\dim(T)$?

Two fundamental types of ring extensions are integral extensions and polynomial extensions. It is known that if $R \subseteq T$ is an integral extension, then the extension is going-up (GU) and has the "incomparable" property (INC), which implies that $\dim(R) = \dim(T)$.

The polynomial case is more interesting. It is a classical theorem in commutative algebra that if $\dim(R) = n$ then we have the bounds

$$n + 1 \leq \dim(R[x]) \leq 2n + 1.$$

The lower bound is easy to establish (and, in fact, coincides with $\dim(R[x])$ in many cases, including the case where R is Noetherian). The upper bound of $2n + 1$ follows from the fact that if $\mathfrak{P} \subseteq R$ is a prime ideal, then any chain of primes in $R[x]$ lying over $\mathfrak{P}$ is of length at most two.

It is natural to also consider the relationship between $\dim(T)$ and $\dim(\hat{T})$ where $\hat{T}$ is an $I-$adic completion of T. A particular case of interest is the power series extension $\hat{T} = R[[x]]$. Since it is known that $\dim(R[x])$ is finite (if $\dim(R)$ is finite), we will focus our attention mostly on the work that has been done relating the dimension of R to the dimension of $R[[x]]$. A couple of obvious questions arise concerning the relationship of $\dim(R)$ and $\dim(R[[x]])$. Firstly, if $\dim(R)$ is finite, is $\dim(R[[x]])$ finite? Secondly, if $\dim(R[[x]])$ is finite, does it obey the polynomial upper bound of $2n + 1$)? Finally, if $\dim(R[[x]])$ is not finite in general, under what conditions is it finite (and is $\dim(R[[x, y]])$ finite)?

In the late 1960's some work was begun on the question of the dimension behavior of power series rings by Robert Gilmer and some of his students.

In [F] it was shown that if V is a discrete valuation domain, of dimension n, then $\dim(V[[x]]) = n + 1$. The results of [AB] demonstrated that if V is a nondiscrete valuation domain, then $\dim(V[[x]])$ must be at least 4 (and hence the polynomial bound is violated for a 1-dimensional nondiscrete valuation domain).

In 1973, Jimmy Arnold, one of Gilmer's students, published a stunning paper that showed that if R does not satisfy the strong finite type (SFT) property (a technical, near-Noetherian property) then $\dim(R[[x]]) = \infty$. More precisely, if R is not SFT, then there is an infinitely ascending chain of primes in $R[[x]]$. This paper showed the necessity of the SFT property for any hope of finite dimension behavior of $R[[x]]$, but sufficiency remained open.

In a sequence of papers, Arnold concentrated on dimension formulas for well-behaved SFT domains. It was shown in [A2] that if V is an SFT valuation domain with $\dim(V) = n$, then $\dim(V[[x_1, \cdots, x_m]] = nm + 1$. This result was globalized in [A3] where it was shown that if D is an SFT Prüfer domain of dimension n, then $\dim(D[[x_1, \cdots, x_m]] = nm + 1$ (note that these results imply that $D[[x_1, \cdots, x_m]]$ is SFT for all m).

After a period of relatively little progress, the past few years have given rise to a number of advances in the dimension theory of formal power series rings. It is the main aim of this paper to outline some of this progress and to present some of the recent developments into one "package". Of course this work is a tribute to the work and influence of Robert Gilmer and his students (especially Jimmy Arnold should be noted for his contributions as well as Jim Brewer, whose book [B] is an extremely useful reference).

2 Krull Dimension of Power Series Rings

In this section, one of our main aims will be to give a concrete demonstration of the "bad dimension" phenomenon that can occur in power series extensions. It is a classical result in commutative algebra that if R is any commutative ring with identity then the Krull dimension of the polynomial extension is bounded above. We state more precisely.

Theorem 2.1 *If R is a domain such that dim$(R) = n$, then $n + 1 \leq$ dim$(R[x]) \leq 2n + 1$.*

Although we will not prove this (a standard proof may be found in [K]), we will outline the strategy to highlight some important differences between the behavior of polynomial extensions and power series extensions.

The basic strategy of the above proof is to show that there is no chain of three primes lying over a fixed prime in $\mathfrak{P} \subseteq R$. To this end, one assumes the existence of such a chain and then begins to reduce the problem, first by reducing modulo the ideal $\mathfrak{P}[x]$ to reduce to the domain case, and then by localizing at the set of nonzero elements of $(R/\mathfrak{P})[x]$ to reduce to a polynomial ring over a field.

These two crucial reductions are quite problematic for power series. First of all, it is not true in general that if $\Gamma \subseteq R[[x]]$ is a prime ideal lying over $\mathfrak{P} \subseteq R$ that $\mathfrak{P}[[x]] \subseteq \Gamma$. In fact, although quite easy for the polynomial case, this is (in the general sense) almost never true for power series.

The next hazard is localization. If $S \subseteq R \setminus \{0\}$ is a multiplicative system then $R[x]_S = R_S[x]$ but the inclusion $R[[x]]_S \subseteq R_S[[x]]$ is almost always strict. Gilmer has shown that a necessary and sufficient condition for equality is for

$$\bigcap_{s \in T} (s) \neq 0$$

where T is any (countable) subset of S.

Basically speaking, this is the condition required for "collecting denominators". It is easy to see, for example, that this condition never holds for any subset of $\mathbb{Z}$ that is not a set of units. Put simply, localization in power series rings is not to be taken lightly.

These two failures of passage from "polynomial proof" to "power series proof" are very expensive.

In tackling the dimension question for power series rings, Arnold introduced a near-Noetherian property called the SFT property; we pause at this juncture to formally recall this notion.

Definition 2.2 *We say that an ideal $I \subseteq R$ is an SFT (for "strong finite type") if there is a finitely generated ideal $B \subseteq I$ and a fixed integer N such that $x^N \in B$ for all $x \in I$. We also say that the ring R is SFT if every ideal of R is SFT.*

If R is Noetherian, then globally we can choose $I = B$ and $N = 1$. So SFT rings may be thought of as a generalization of Noetherian rings. Note that in an SFT ring, every ideal is the radical of a finitely generated ideal (but this condition is not equivalent to SFT since, for example, any 1 dimensional valuation domain has this property).

The importance of the SFT property is made clear by the following theorem of J. Arnold [A1].

Theorem 2.3 *If R is a ring that is not SFT then $R[[x]]$ contains an infinite ascending chain of prime ideals. In particular, $dim(R[[x]]) = \infty$.*

The technical difficulty of the proof of this theorem has become somewhat legendary. Although we will not go through the details, we will outline the basic strategy since the technique is quite useful (for example in [CD] this same strategy is used to show that fragmented domains have infinite Krull dimension).

To show that a ring, R, contains an infinitely ascending chain of primes, one might employ the following approach. Beginning with an infinitely descending chain of multiplicatively closed subsets of R

$$S_1 \supsetneq S_2 \supsetneq \cdots S_n \supsetneq S_{n+1} \supsetneq \cdots$$

and a collection of elements $f_n \in S_n$ with the property that $rf_n + s_{n+1} \in S_{n+1}$ for all $r \in R$ and $s_{n+1} \in S_{n+1}$.

Once the existence of these sets and elements are established, the proof goes by induction as follows.

1. Begin with the ideal (f_1) and notice that, if $rf_1 = s_2 \in S_2$, we would have that $rf_1 - s_2 = 0 \in S_2$ which is a contradiction. Hence we expand (f_1) to a prime that is maximal with respect to the exclusion of S_2 [K] and we have constructed our first prime $\mathfrak{P}_1$.
2. Inductively, we assume that we have constructed the nth prime in the chain that is maximal with respect to the property that $\mathfrak{P}_n \bigcap S_{n+1} = \emptyset$.
3. We now consider the ideal $(f_{n+1}, \mathfrak{P}_n)$. If $rf_{n+1} + p_n = s_{n+2}$ with $p_n \in \mathfrak{P}_n$ and $s_{n+2} \in S_{n+2}$ then we have that $rf_{n+1} - s_{n+2} = -p_n \in S_{n+2} \subset S_{n+1}$ which is a contradiction.
4. Since $(f_{n+1}, \mathfrak{P}_n) \bigcap S_{n+2} = \emptyset$ then we expand this ideal to $\mathfrak{P}_{n+1} \supset \mathfrak{P}_n$ and we are done by induction.

Arnold's technique is very clever and straightforward in its implementation. The difficulty (for Arnold's Theorem) arises in establishing the existence of the multiplicative sets and the relevant elements f_n.

If R is not an SFT ring, it can be shown that this implies the existence of a sequence of elements $\{a_n\}$ of elements of R such that

$$a_{n+1}^{n+1} \notin (a_1, a_2, \cdots, a_n).$$

The existence of this sequence allows the construction of sets of power series which are slower and slower to "converge" (in the sense of the above equation). The technical difficulty of the proof of the main theorem in [A1] arises in the construction of this descending chain of multiplicative sets, and the SFT property, in a certain sense, is designed for this construction to be allowed.

To give an intuitive feel for how large the spectrum for even a 1−dimensional nondiscrete valuation domain can be we consider the following.

Let K be a field and let V be a one-dimensional valuation domain of the form

$$V := K[x; S]_{\mathfrak{M}}$$

where S is the monoid consisting of the nonnegative elements of $\mathbb{R}$ and $\mathfrak{M}$ the ideal generated by elements of the form x^α where $\alpha \in S$ is a positive element.

It is well known that this domain is not SFT and hence the dimension of $V[[t]]$ is infinite. More specifically, it is shown in [A1] that in $V[[t]]$ there is an infinitely ascending chain of prime ideals.

For an element $f(t) \in V[[t]]$ we denote by $f_n(t)$ the nth partial sum, and we use the notation

$$vc(f_n)$$

to denote the value of the (generator of the) content ideal of the polynomial f_n.

In the domain $V[[t]]$ we consider the family of ideals

$$I_\alpha = \{f(t) \in V[[t]] | vc(f_{2^n}) \geq \frac{k}{\alpha^n} \text{ for some constant } k > 0\}$$

where $\alpha > 1$ is a real number.

Since the valuation v satisfies $v(a+b) \geq \min(v(a), v(b))$ it is easy to see that the collection $\{I_\alpha\}$ forms a chain of ideals. Intuitively, the ideal I_α consists of the power series in $V[[t]]$ such that the values of the contents of the partial sums converge at a certain rate. In fact, if $\alpha \leq \beta$ then $I_\alpha \subseteq I_\beta$ (and the ideal I_β has terms in the value sequence of faster convergence rate.

The set complement of I_α is the set

$$S_\alpha = \{f \in V[[t]] | \forall k > 0, \exists n \text{ such that } vc(f_{2^n}) < \frac{k}{\alpha^n}\}.$$

The importance to the exposition here is that the set S_α is reminiscent of the sets constructed by Arnold in [A1] to handle the general case. These sets are decreasing and saturated (since they are complements of ideals). It is not clear that they are multiplicatively closed, but they illustrate (via this notion of "fast convergence of values") how the construction in Arnold's seminal paper works.

We make the following conjecture.

Conjecture 2.4 *The sets S_α are multiplicatively closed.*

If settled in the affirmative, this conjecture would give the following interesting result.

Theorem 2.5 *If the sets S_α are multiplicatively closed, then there is a chain of prime ideals in $V[[t]]$ that is order-isomorphic to the positive half-line (in particular, there is an uncountable chain of primes).*

Proof. The ideals I_α form a chain. If their complements are multiplicatively closed, the ideals are prime and hence the theorem follows.

A number of open questions arose after Arnold's paper. The obvious question is one concerning sufficiency. That is, if R is finite dimensional and SFT, does this imply that the dimension of $R[[x]]$ is finite. Another related question concerns stability of the SFT property in power series rings. All known examples where $\dim(R[[x]])$ is finite give a multivariable analog (that is, $\dim(R[[x_1, \cdots, x_n]])$ is also finite for all n). Given the results of [A1] it is clearly necessary in all of these examples for $R[[x_1, \cdots, x_{n-1}]]$ to be SFT. And so the question "if R is SFT is $R[[x]]$ SFT?" arises. Clearly an affirmative answer to the first question would imply an affirmative answer to the latter. Both of these questions were pointed out in [G].

3 The SFT property

In 1999, the paper [CC] studied the SFT property in conjunction with the notion of "super convergence" in an attempt to get a better handle on the SFT/dimension property for power series rings.

The idea behind super convergence is as follows. One often considers a power series

$$a_0 + a_1x + a_2x^2 + \cdots + a_nx^n + \cdots$$

with $a_n \in R$ formally from the point of view of manipulating symbols (for the purposes of forming the ring $R[[x]]$). However, each element a_nx^n is an element of $R[[x]]$ as well and the notation for a formal power series suggests the writing of an infinite sum (rather than just considering this to be a notation). And indeed the Cauchy sequence of polynomials

$$\{a_0, a_0 + a_1x, a_0 + a_1x + a_2x^2, \cdots\}$$

converges $x-$adically. So it makes sense to consider the formal power series alluded to above as an infinite sum and the domain of convergence is all of R (in the sense that if $\{r_i\}_{i=0}^{\infty}$ is any countable collection of elements of R, then the series $\sum_{i=0}^{\infty} r_ix^i$ converges).

Suppose that $I \subseteq R$ is an ideal. It is easy to see that if $\{\alpha_i\}_{i=0}^{\infty}$ is a countable collection of elements in I, then the infinite sum

$$\sum_{i=0}^{\infty} \alpha_ix^i \in I[[x]].$$

Since every term α_ix^i is an element of the extension of I to $R[x]]$, it is natural to ask under what conditions does the sum $\sum_{i=0}^{\infty} \alpha_ix^i$ converge to an element of $IR[[x]]$. We will say that the ideal $\Gamma \subseteq R[[x]]$ has the superconvergence property (or is superconvergent) if given any countable collection $\{\gamma_i\}_{i=0}^{\infty}$ of elements of $R[[x]]$, the sum $\sum_{i=0}^{\infty} \gamma_ix^i$ is an element of Γ.

It is clear that any finitely generated ideal of $R[[x]]$ has this property. It is also easy to see that maximal ideals of $R[[x]]$ must have this property as well (and hence the condition "finitely generated ideal" is not necessary). We recall the following definitions from [CC]

Definition 3.1 *We say that the domain $R[[x]]$ is globally superconvergent if every ideal of $R[[x]]$ has the superconvergence property. We say that $R[[x]]$ is radically superconvergent if each radical ideal has the superconvergence property.*

We note here that it is shown in [CC] that global superconvergence is equivalent to Noetherian. Radical superconvergence is more interesting. In the case of radical superconvergence, it is possible that for a countable collection $\{\gamma_i\}_{i=0}^{\infty}$ of elements of Γ that the infinite sum

$$\sum_{i=0}^{\infty} \gamma_i x^i$$

is not an element of Γ. But the sum must remain in every *prime* ideal containing Γ. For a simple example of a domain that is not radically superconvergent, consider the domain $V[[x]]$ where $(V, \mathfrak{M})$ is a 1 dimensional nondiscrete valuation domain with maximal ideal $\mathfrak{M}$. It is known from [A1] that there is an infinitely ascending chain of prime ideals containing the ideal $\mathfrak{M}V[[x]]$, each of which is contained in $\mathfrak{M}[[x]]$. Since for an arbitrary $\gamma \in \mathfrak{M}[[x]]$, one can select elements of $\mathfrak{M}$ ($\{\gamma_i\}_{i=0}^{\infty}$) such that the sum

$$\sum_{i=0}^{\infty} \gamma_i x^i = \gamma$$

we see that $V[[x]]$ is not radically superconvergent.

The importance of the connection between the notions of superconvergence and the SFT property will be seen soon. We now produce the key connection from [CC] that will help us to answer the question of SFT stability.

Theorem 3.2 *Let $\Gamma \subseteq R[[x]]$ be an ideal that is both SFT and radical. Then Γ has the superconvergence property.*

Proof. Let $\Gamma \subseteq R[[x]]$ be SFT. If $\sum_{i=0}^{\infty} \gamma_i x^i$ is an infinite sum with each $\gamma_i \in \Gamma$, we need to show that this sum is again an element of Γ.

Since Γ is SFT, we know from the work of Arnold ([A1]) that

$$\Gamma[[y]] \subseteq \sqrt{\Gamma R[[x, y]]}.$$

Hence the element

$$\gamma = \sum_{i=0}^{\infty} \gamma_i y^i$$

is in the radical of the ideal generated by Γ in $R[[x, y]]$. Hence for some n we can write

$$\gamma^n = f_1 r_1(x, y) + f_2 r_2(x, y) + \cdots + f_m r_m(x, y)$$

with each $f_i \in \Gamma$ and each $r_i(x, y) \in R[[x, y]]$.

To finish the proof, merely apply the homomorphism

$$\phi_y : R[[x, y]] \longrightarrow R[[x]]$$

defined by $\phi_y(f(x, y)) = f(x, x)$ and note that our equation for γ^n becomes

$$\gamma^n = f_1 r_1(x, x) + f_2 r_2(x, x) + \cdots + f_m r_m(x, x) \in \Gamma.$$

Since Γ is a radical ideal, we have that $\gamma \in \Gamma$ and we are done.

4 Good cases and general counterexamples

The work of Arnold showed that if R is not an SFT ring then this implies the existence of an infinitely increasing chain of prime ideals

$$\Gamma_0 \subseteq \Gamma_1 \subseteq \cdots \subseteq \Gamma_n \subseteq \Gamma_{n+1} \subseteq \cdots$$

It was very natural to ask at this juncture if the SFT condition was sufficient for "good" dimension behavior. That is, if R is an SFT ring of finite Krull dimension, is $\dim(R[[x]])$ finite? In a number of papers in the early 1980's, ([A2] and [A3]) Arnold showed that if R is an SFT valuation domain of dimension n, then the power series ring in m indeterminates is of finite Krull dimension. The following theorem can be found in [A2]

Theorem 4.1 *Let V be an SFT valuation domain with dim$(V) = n$. Then $V[[x_1, x_2, \cdots, x_m]]$ is both SFT and finite dimensional for all m. Additionally, dim$(V[[x_1, x_2, \cdots, x_m]]) = nm + 1$.*

Later this result was broadened from the class of valuation domains to the class of Prüfer domains ([A3]).

Theorem 4.2 *Let R be an SFT Prüfer domain with dim$(R) = n$. Then $R[[x_1, x_2, \cdots, x_m]]$ is both SFT and finite dimensional for all m and dim$(R[[x_1, x_2, \cdots, x_m]]) = nm + 1$.*

The proofs of the previous results depend heavily on the "valuation" and "Prüfer" assumptions respectively. The results were also striking because of the unusual behavior of Krull dimension. Note that in the case where $\dim(R) = 1$, with R an SFT Prüfer domain, we have that for all m, $\dim(R[[x_1, x_2, \cdots, x_m]]) = m + 1$, which parallels the Noetherian case (and indeed, it should be noted that a $1-$dimensional SFT Prüfer domain is Dedekind). However, if the Krull dimension of R exceeds 1, then the dimension of the power series extension in m indeterminates behaves much more strangely.

It is worth observing that, from a "relative size" point of view, the adjunction of the first power series indeterminate creates the biggest jump in the dimension. Note that the jump in dimension from m indeterminates to $m + 1$ indeterminates (with $m > 0$) is always a constant $n = \dim(R)$. In particular, we have that

$$\dim(R[[x_1, \cdots, x_m]][[x_{m+1}]]) = \dim(R[[x_1, \cdots, x_m]]) + n.$$

If we let $R_m := R[[x_1, x_2, \cdots, x_m]]$, we have that $\frac{\dim(R_m[[x]])}{\dim(R_m)} = \frac{nm+n+1}{nm+1}$ and we note that this ratio is much smaller than 2 and is well within the polynomial bounds.

It should also be noted that in [A1] Arnold established that the SFT property was sufficient for 0-dimensional rings. More precisely, if R is $0-$dimensional and SFT, then $\dim(R[[x]]) = 1$.

In 1999, Condo, Coykendall, and Dobbs established the multivariable version of the dimension formula for 0-dimensional rings. The following theorem may be found in [CCD].

Theorem 4.3 *If R is a $0-$dimensional SFT ring then* $dim(R[[x_1, x_2, \cdots, x_n]] = n$.

It is interesting to note that this follows the earlier patterns of discovery in the sense that for all known classes of rings for which $\dim(R[[x]])$ is finite, then $\dim(R[[x_1, \cdots, x_n]])$ is finite for all n.

The general situation, however, is not nearly as placid as the situation for Prüfer domains and 0 dimensional rings. To get a clearer look at some of the obstructions, we produce an important theorem due to B. G. Kang and M. Park. Both of these authors have done important work that has greatly improved the understanding of the spectrum of power series rings over valuation domains. How the content of this theorem affects the dimension question will be made clear presently.

Theorem 4.4 *Let $(V, \mathfrak{M})$ be a $1-$dimensional nondiscrete valuation domain. Then the ring $V[[x_1, x_2, \cdots, x_t]]$ possesses an infinite descending chain of prime ideals inside $\mathfrak{M}[[x_1, x_2, \cdots, x_t]]$ all lying over (0) in V.*

The importance of this result is illustrated by the following construction.

Let K be a field and let V be the one-dimensional valuation domain considered earlier:

$$V := K[x; S]_{\mathfrak{M}}$$

where S is the monoid consisting of the nonnegative elements of $\mathbb{R}$ and $\mathfrak{M}$ the ideal generated by elements of the form x^α where $\alpha \in S$ is a positive element.

As has been pointed out, this domain has the property that $V[[t]]$ is infinite dimensional. Arnold's results give that $V[[t]]$ has an infinite ascending chain of primes and Kang and Park's results show that $V[[t]]$ has an infinite descending chain of primes that lie over (0) (and this is the key fact).

Using the notation from above, we form the $D + M$ construction

$$V_1 = K + xV.$$

As it turns out ([A1]), the SFT property is determined completely by the behavior of the prime ideals. That is, a ring is SFT if and only if all of its prime ideals are SFT. This result makes it easy to check that the domain V_1 is SFT. Since the integral closure of V_1 is V, we only have to check that the maximal ideal of V_1 is SFT, and it is routine to do so. Also note that the element x conducts V to V_1 and hence conducts $V[[t]]$ to $V_1[[t]]$.

The decreasing chain of primes from [KP1]

$$\Gamma_1 \supsetneq \Gamma_2 \supsetneq \cdots$$

with each Γ_n lying over (0), gives rise to a decreasing chain of primes

$$\mathfrak{P}_1 \supseteq \mathfrak{P}_2 \supseteq \cdots .$$

It only remains to see that the chain of primes is strictly decreasing. To this end, we note that if $f \in \Gamma_n \setminus \Gamma_{n+1}$ then $xf \in \mathfrak{P}_n$. If $xf \in \mathfrak{P}_{n+1}$ then $xf \in \Gamma_{n+1}$ and since Γ_{n+1} is a prime lying over (0), we have that $f \in \Gamma_{n+1}$ which is a contradiction.

It can also be shown that V_1 is an example of an SFT domain such that its power series extension $V_1[[t]]$ is not SFT. We will outline the reason for this.

We denote by $\phi{:}V[[t]] \longrightarrow V$ the variable annihilation map $(f(t) \mapsto f(0))$, and for the ease of exposition, we will assume that our valuation domain V has value group $\mathbb{Q}$ and residue field $K = \mathbb{F}_2$. We record the following theorem from [C].

Theorem 4.5 *There is a prime ideal* $\mathfrak{P} \subseteq V[[t]]$ *with the following properties.*

(1) $\mathfrak{P} \bigcap V = 0$.
(2) $\phi(P) = \mathfrak{M}$.
(3) $\mathfrak{P} \subseteq \mathfrak{M}[[t]]$.

Basically, this result gives the existence of an ideal in $V[[t]]$ (inside $\mathfrak{M}[[t]]$, so all coefficients of all elements have positive value) such that for all ϵ there is a series in $\mathfrak{P}$ with value of $f(0) < \epsilon$. Additionally, this ideal $\mathfrak{P}$ contains no nonzero constant terms.

The utility of the existence of this ideal comes from the radical superconvergence property. Recall that if an ideal is prime (radical) and SFT, then it must have the superconvergence property.

Note that the existence of an ideal with the properties listed above cannot have the superconvergence property. Note that properties 2) and 3) above guarantee that we can "construct" any power series in $\mathfrak{M}[[t]]$ with an infinite sum. To construct an arbitrary $f(t) = \sum_{i=0}^{\infty} a_i t^i$, first find $g(t) \in \mathfrak{P}$ such that the value of $g(0)$ is smaller than the value of $f(0)$. We multiply $g(t)$ by some $v_0 \in V$ such that $f(0) = v_0 g(0)$. Since the first nonzero coefficient of $f(t) - v_0 g(t)$ is an element of $\mathfrak{M}$ we can continue this process to express an arbitrary $f(t) \in \mathfrak{M}[[t]]$ as a sum

$$f(t) = \sum_{i=0}^{\infty} t^i g_i(t)$$

with $g_i(t) \in \mathfrak{P}$.

In particular if we take $f(t)$ to be a nonzero constant, condition 1) above shows that $\mathfrak{P}$ is not radically superconvergent.

Of course this is probably no surprise yet, since the ring $V[[t]]$ has an infinitely ascending chain of primes and hence is not SFT. However, the result listed above can be translated by conducting by the element x to the situation concerning V_1 and $V_1[[t]]$. This general strategy shows that V_1 is an SFT domain with the property that $V_1[[t]]$ is not SFT.

To finish this section, we outline how the existence of the prime ideal claimed above is shown. To begin, we denote by $f_n(t)$ the polynomials $x^{\frac{1}{2^n}} + t^{2^n}$ and we let

$$g_n(t) = \prod_{m=n}^{\infty} f_m(t).$$

We note that this infinite product is defined exclusively for use in this situation. The infinite product is an element of $V[[t]]$ and the constant term is x^α with $\alpha = \sum_{m=n}^{\infty} \frac{1}{2^n}$. Note that since all powers of t in the infinite product appear only finitely often (with the exception of t^0), all coefficients of t^k $(k > 0)$ consist of a sum of finitely many well-defined elements of V, and so the product makes sense. It should also be noted that for all $k \geq n$, we have that $f_k(t)$ divides $g_n(t)$ and the ideal generated by each $f_k(t)$ lies over (0) in V.

We now construct the ideal $I = (g_1, g_2, \cdots)$. Showing this ideal is contained in $\mathfrak{M}[[t]]$ is computational. It can also easily be seen that I contains series with arbitrarily small constant value (in fact the value of the constant term of $g_n(t)$ is $\frac{1}{2^{n-1}}$). The fact that I lies over (0) in V follows from a simple computation and the observation that each (f_k) lies over (0). As the final key to the construction, we note that I is an ideal that excludes the multiplicative set $V \setminus \{0\}$ and hence we can expand I to a prime ideal that is maximal with respect to this property.

5 Bounding the dimension of a power series ring

In this section, we give a partial answer to the question concerning the bound for the dimension of $R[[x]]$ when it is finite. As was alluded to earlier, every known example (until recently) of an $n-$dimensional domain with the property that $\dim(R[[x]]) < \infty$ also enjoyed the property that $\dim(R[[x]]) \leq 2n+1$ (in fact the dimension of $R[[x]]$ was usually much smaller than the polynomial bound). The present author asked in [C] if the dimension of $R[[x]]$ is bounded by $2(\dim(R)) + 1$ whenever the dimension of $R[[x]]$ is finite. Recently Kang and Park have shown that this is about as far from true as possible.

In [KP2] Kang and Park used a construction that they termed a *mixed extension* to give arbitrarily bad dimension behavior for power series rings. The definition of a mixed extension is an extension of the form

$$R[x_1]][x_2]] \cdots [x_n]]$$

where the notation $[x_i]]$ means that x_i is adjoined as either a polynomial or a power series indeterminate.

It is shown via some very technical arguments that if R is an $m-$dimensional Prüfer domain and at least one of the adjoined indeterminates is a power series variable, then the spectrum of the mixed extension behaves in a very similar fashion to the behavior of the extension $R[[x_1, \cdots, x_n]]$. From this they obtain the following interesting result.

Theorem 5.1 *If R is an $m-$dimensional SFT Prüfer domain then*

$$dim(R[x_1]][x_2]] \cdots [x_n]]) = \begin{cases} mn+1, & \text{if } [x_i]] = [[x_i]] \text{ for some } 1 \leq i \leq n; \\ m+n, & \text{otherwise.} \end{cases}$$

In other words the formula for the dimension of a mixed extension over a finite-dimensional SFT Prüfer domain corresponds to the multivariable power series formula if the mixed extension is not simply a polynomial extension.

So for example, one could take R to be a $m-$dimensional SFT Prüfer domain. The dimension of $R[x_1, \cdots, x_{n-1}]$ would therefore be $n+m$ and the dimension of $R[x_1, \cdots, x_{n-1}][[x_n]]$ would be $mn+1$. Choosing (for example) $n = m$ allows us to force the ratio $\frac{nm+1}{n+m} = \frac{m^2+1}{2m}$ arbitrarily high by letting m be large.

6 Final questions

We end this overview with a few questions concerning the dimension and spectrum of power series rings.

1. If $R[[x]]$ is SFT, does this imply that $R[[x, y]]$ is SFT?
2. If $R[[x]]$ is finite dimensional, is $R[[x, y]]$ finite dimensional?
3. Is there an uncountable chain of primes in $V[[x]]$ if V is a one-dimensional nondiscrete valuation domain (see Conjecture 2.4 and Theorem 2.5)?
4. Is there a finite dimensional, integrally closed SFT domain R such that $R[[x]]$ is infinite dimensional?

The first two questions are motivated by the quest to understand the multivariable case. Note that an affirmative answer to the first question implies an affirmative answer to the second (and it is the author's guess that these problems may be technically difficult to solve).

The last question is motivated by the fact that the counterexample used to show that $\dim(R[[x]])$ need not be finite when R is SFT, begs some natural questions. Although the example is nice in a certain sense (it is $1-$dimensional and quasilocal) it does have some properties that mirror the valuation domain of which it is a subring. In particular, using the notation of Section 4, the integral closure of V_1 is V and since V is conducted to V_1 by a single nonzero element, the "badness" of V is inherited by V_1. In this spirit (especially given the integral-like nature of the SFT property), it would be nice to know if such an example can exist for an integrally closed domain.

References

[A1] J. Arnold.: Krull dimension in power series rings. Trans. Amer. Math. Soc. **177**, 299–304 (1973)

[A2] J. Arnold.: Power series rings over discrete valuation rings. Pacific J. Math. **93**, 31–33 (1981)

[A3] J. Arnold.: Power series rings with finite Krull dimension. Indiana Univ. Math. J. **31**, 897–911 (1982)

[AB] J. Arnold and J. Brewer.: When $(D[[X]])_{P[[X]]}$ is a valuation ring. Proc. Amer. Math. Soc. **37**, 326–332 (1973)

[B] J. Brewer.: Power series over commutative rings. Dekker, (1981)

[CC] J. Condo, J. Coykendall.: Strong convergence properties of SFT rings. Comm. Algebra **27**, 2073–2085 (1999)

[CCD] J. Condo, J Coykendall, D. Dobbs.: Formal power series rings over zero-dimensional SFT-rings. Comm. Algebra **24**, 2687–2698 (1996)

[C] J. Coykendall.: The SFT property does not imply finite dimension for power series rings. J. Algebra **256**, 85–96 (2002)

[CD] J. Coykendall, D. Dobbs.: Fragmented domains have infinite Krull dimension. Rend. Circ. Mat. Palermo **50**, 377–388 (2001)

[F] D. Fields.: Dimension theory in power series rings. Pacific J. Math. **35**, 326–332 (1970)

[G] R. Gilmer.: Dimension theory of power series rings over a commutative ring. Algebra and logic (Fourteenth Summer Res. Inst., Austral. Math. Soc., Monash Univ., Clayton, 1974), Lecture Notes in Math., Vol. 450. Springer, Berlin Heidelberg New York , 155–162 (1975)

[KP1] B. Kang and M. Park. A localization of a power series ring over a valuation domain. J. Pure Appl. Algebra **140**, 107–124 (1999)

[KP2] B. Kang and M. Park. Krull dimension in power series rings. Preprint.

[K] I. Kaplansky. Commutative Rings. University of Chicago Press, Chicago (1974)

Some research on chains of prime ideals influenced by the writings of Robert Gilmer

D. E. Dobbs

Department of Mathematics, University of Tennessee, Knoxville, TN 37996-1300
dobbs@math.utk.edu

1 Introduction

Robert Gilmer's work has overtly influenced the direction of much of commutative ring theory for the past 40-plus years. Although I have not worked on many of the explicit questions that Robert raised, I feel that Robert's work has definitely influenced my research, both in identifying important topics and themes and in exemplifying the highest standards of exposition. I hope that the next two sections serve to document that influence and to indicate where it has led me and others.

Each of the following sections begins with a theorem of Robert Gilmer and then proceeds to examine a cross-section of subsequent developments. The first of these sections concerns various characterizations of Prüfer domains, beginning with a result of Robert for the one-dimensional case, and features a discussion of going-down domains, treed domains and universally catenarian domains. The final section concerns more recent work. It begins by explaining how an exquisitely finitistic result that Robert obtained jointly with Nashier–Nichols spurred my interest in infinite chains and trees of prime ideals of arbitrary (commutative) rings. The section goes on to discuss several directions of inquiry, including an application to treed domains.

We use the following notation and conventions. R denotes a (commutative integral) domain with integral closure R' and quotient field K. "Dimension" refers to Krull dimension and is denoted by dim; valuative dimension is denoted by $\dim_v$. All rings are assumed commutative with identity. Notation such as $A \subseteq B$ means that A is a (unital) subring of B, while the notation $f{:}A \to B$ means that f is a (unital) ring homomorphism from A to B. If A is a ring, then by an *overring* of A, we mean an (unital) A-subalgebra of the total quotient ring of A. Modifying the usage in [52, page 28], we use GD, GU, LO and INC to refer to the going-down, going-up, lying-over and incomparable properties, respectively, of ring extensions or ring homomorphisms. As usual, the cardinality of a set S is denoted by $|S|$.

2 One-dimensional Prüfer Domains Lead the Way

Two characterizations of one-dimensional Prüfer domains, Theorems 2.1 and 2.7, have stimulated much work in higher dimensions. We begin by stating the first of these seminal results.

Theorem 2.1. ([48, Corollaire 3, page 61], [40, Theorem 6]) *Let* $\dim(R) = 1$. *Then R is a Prüfer domain if and only if R is integrally closed and the polynomial ring $R[X]$ is two-dimensional.*

The above result can easily be reformulated in a way that suggests analogues in higher dimensions and/or for domains that need not be integrally closed. Theorem 2.2 states that reformulation (and, for convenience, includes the earlier statement). We assume familiarity with the definitions of S-domains and strong S-domains [52, page 26], stably strong S-domains [53] and universally catenarian domains [9].

Theorem 2.2. *Let* $\dim(R) = 1$. *Then the following conditions are equivalent.*
(1) *R is a Prüfer domain.*
(2) *R is integrally closed and the polynomial ring $R[X]$ is two-dimensional.*
(3) *R is integrally closed and* $\dim_v(R) = 1$.
(4) *R is an integrally closed S-domain.*

One has the following non-reversible implications: LFD (locally finite dimensional) Prüfer domain $\Rightarrow$ universally catenarian domain $\Rightarrow$ stably strong S-domain $\Rightarrow$ strong S-domain $\Rightarrow$ S-domain. The first of these implications is essentially due to Nagata [56] and was revisited by Malik–Mott [53], Bouvier–Dobbs–Fontana [9] and a number of subsequent authors; the second implication is a consequence of [9, Lemma 2.3]; and the final two implications are immediate from the definitions.

Interest in universally catenarian domains arose, in part, because of a famous example of Nagata [55], who showed that a catenarian Noetherian domain R need not have the property that $R[X]$ is catenarian. Nagata's example R was two-dimensional. No such example could be of smaller dimension, as a consequence of the result of Ratliff [60, (2.6)] that a Noetherian ring A is universally catenarian if (and only if) $A[X]$ is catenarian. Indeed, if R is a one-dimensional Noetherian domain, then $R[X]$ is (like any two-dimensional domain) catenarian. To address the analogous non-Noetherian context, Alain Bouvier conjectured that domains of valuative dimension 1 must be universally catenarian. This was proved by Bouvier, Fontana and me: see the first assertion in the next result. Both assertions in Theorem 2.3 will be generalized in Theorems 2.4–2.6 when we move beyond the (valuative) dimension 1 setting.

Theorem 2.3. [9, Corollaries 6.4 and 3.4] *If* $\dim_v(R) = 1$, *then R is a universally catenarian domain. If* $\dim(R) = 1$, *then R is a Prüfer domain if and only if R is an integrally closed universally catenarian domain.*

To obtain the desired generalizations of Theorem 2.3, we need a definition (cf. [16], [35]): R is said to be a *going-down domain* if $R \subseteq T$ satisfies GD for each extension domain T of R. By [35, Theorem 1], the test domains T can be restricted to be simple overrings of R (that is, $T = R[u]$ with $u \in K$) or valuation overrings of R. The most familiar examples of going-down domains are arbitrary one-dimensional domains and Prüfer domains. The latter family of examples arises because of Richman's characterization [62] of Prüfer domains as the domains R such that each overring of R is R-flat. (One also needs to know that commutative flat algebras satisfy GD: the usual proof of this is somewhat homological, making use of faithful flatness, but Papick and I [36] gave a more computational proof of this fact by using a criterion for GD that had been noted by Kaplansky [52, Exercise 37, page 44].)

We pause for some history that has not been recorded elsewhere. In a seminar talk at UCLA on May 19, 1970, Kaplansky suggested defining a "GD-domain" as a domain R such that $R \subseteq T$ satisfies GD for each *integral* (emphasis mine) extension domain T of R. If memory serves (my notes from the lecture are silent on this point), his motivation for this definition involved the question whether integrally closed Noetherian domains must satisfy the saturated chain condition for prime ideals. I was frankly unimpressed with such a research program at that time (as I still wore the hat of one working in algebraic geometry, algebraic number theory and homological algebra). This reaction was strengthened a year later when Kaplansky [51] answered a question of Krull in the negative by showing that adjacent prime ideals in an integral extension domain need not contract to adjacent primes of the base domain. A year later, when I was teaching Commutative Algebra for the first time, my interest in GD was piqued again by some homework answers of a student named Jeff Dawson. Our discussions in the spring of 1972 led to my first paper on going-down (part of which is discussed following Theorem 2.11) and, soon afterward, to the series of three papers on simple overrings that included the above definition of "going-down domain".

The promised generalizations of Theorem 2.3 include the next three results.

Theorem 2.4. [9, Theorem 8.1] *Let R be a going-down domain in which all maximal ideals have the same height. Then* $\dim_v(R) = \dim(R)$ *if and only if R is a universally catenarian domain.*

Theorem 2.4 is one of the results in [9] that motivated the teams at Lyons, Rome and Knoxville to define Jaffard domains (cf. [2], [11]). Note that Theorem 2.4 generalizes the first assertion in Theorem 2.3. A generalization of the second assertion in Theorem 2.3 will be given in Theorem 2.5. By consulting [9], the reader can check that the proof of Theorem 2.4 depends in part on Theorem 2.5.

Theorem 2.5. [9, Theorem 6.2] *Let R be a going-down domain. Then the following conditions are equivalent.*

(1) *R' is an LFD Prüfer domain.*
(2) $\dim_v(R_M) = \dim(R_M) < \infty$ *for each maximal ideal M of R.*
(3) *R is a universally catenarian domain.*
(4) *R' is a universally catenarian domain.*

We next present a strong companion for Theorem 2.5 that Bouvier–Dobbs–Fontana obtained in their second paper on universal catenarity. Note that the implication (2) $\Rightarrow$ (1) in Theorem 2.6 was known in the earlier paper [9] only in case $\dim(R) = 1$ [9, Corollary 6.3] or R is an LFD locally pseudo-valuation domain [9, Corollaries 6.6 and 6.7].

Theorem 2.6. [10, Theorem 1] *Let R be a going-down domain. Then the following conditions are equivalent.*
(1) *R is a universally catenarian domain.*
(2) *$R[X]$ is a catenarian domain.*
(3) *R is an LFD strong S-domain.*

Theorem 2.6 shows that going-down domains (in particular, one-dimensional domains) R behave like Noetherian domains inasmuch as the catenarity of $R[X]$ implies that of $R[X, Y]$ (the polynomial ring over R in two variables). Bouvier asked if this implication holds for arbitrary domains. This question remained open for about 20 years, until Ben Nasr and Jarboui [6] answered it in the negative by using a clever pullback construction to produce a two-dimensional equidimensional domain R such that $R[X]$ is catenarian but $R[X, Y]$ is not catenarian.

Before leaving the subject of universal catenarity, we note that several papers have been devoted to developing new families of examples of universally catenarian domains. Of these, we mention only two, namely [3, Corollary 2.3] via a $D + M$ construction and [1, Corollary 2.15 and Example 4.1] via semigroup ring constructions. It seems particularly appropriate to focus on these two instances here since the classical $D+M$ construction and semigroup rings owe much of their popularity to Robert Gilmer's two books, [41] and [42], respectively.

Before leaving the subject of S-domains and their relatives, we mention some transfer results that owe their proofs to some new examples of GD-behavior. In a paper submitted for publication in May, 2005, Jay Shapiro and I proved the following results. Let a group G act on a ring A via ring automorphisms in a locally finite way (a situation addressed in [41, Theorem 12.1 (1)]) and let $A^G := \{a \in A \mid ga = a \text{ for each } g \in G\}$, the fixed ring. Then the extension $A^G \subseteq A$ satisfies (universally) GD [38, Theorem 2.2]. As a consequence, it was shown in [38, Corollary 2.13] that if a group G acts on a domain R via ring automorphisms in a locally finite way, then each of the following four properties transfers from R to the fixed ring R^G: S-domain, strong S-domain, stably strong S-domain, and universally catenarian going-down domain.

We turn next to the second seminal characterization of one-dimensional Prüfer domains that has led to many higher-dimensional studies. This result of Quentel was originally stated for (quasi)local R, but the global formulation given below follows easily from the local case. In reporting on this result for Mathematical Reviews in 1968, Gilmer mentioned Ribenboim's example [61] of a one-dimensional completely integrally closed domain that is not a valuation domain as motivating the question of determining when a one-dimensional integrally closed domain must be a Prüfer domain.

Theorem 2.7. [59, Corollaire 2] *Let* $\dim(R) = 1$. *Then R is a Prüfer domain if and only if R is integrally closed and coherent.*

To obtain the desired generalizations of Theorem 2.7, we need the following definition. As in [16], we say that R is a *treed domain* in case $\text{Spec}(R)$, as a poset under inclusion, is a tree, that is, in case no prime (maximal) ideal of R can contain prime ideals of R that are incomparable (with respect to inclusion). Once again, the motivation is Prüferian, as each Prüfer domain is evidently a treed domain. In fact, I showed, more generally, in [16, Theorem 2.2] that each going-down domain is treed. This proof was harder than necessary, as the equivalences in [35, Theorem 1] had not yet been proved and the definition of "going-down domain" that was being used in [16] was the criterion involving simple overrings.

At a meeting in Lincoln, Nebraska in 1974, I mentioned casually in conversation that I hoped to prove that any treed domain must be a going-down domain, if only so that the main section of the paper on it could be titled "Timber!". The next morning, Jim Lewis presented me with a counterexample. The details of Lewis's construction appear in [37, Example 4.4]. By tweaking that construction about ten years later, I was able to show that it is possible for a treed domain that is not a going-down domain to have all its overrings being treed domains [21, Example 2.3].

The next three results were among my earliest efforts to generalize Theorem 2.7 to higher dimensions. These results were obtained prior to [35], that is, while the "simple overring" criterion was being used as the definition of "going-down domain". We begin modestly by examining – in the quasilocal case to avoid issues of equidimensionality – what it would take for an integrally closed two-dimensional treed Jaffard domain to be a Prüfer domain.

Theorem 2.8. [16, Corollary 2.9] *Let (R, M) be a quasilocal two-dimensional domain. Then R is a valuation domain if (and only if) R is an integrally closed treed domain with* $\dim_v(R) = 2$ *such that if $MR[u] \neq R[u]$ for some $u \in K$, then $R[u]$ has a unique height* 1 *prime ideal.*

Theorem 2.8 was actually a consequence of combining Theorem 2.9 with another result [16, Proposition 2.8] that we do not state here. Notice that Theorem 2.9 addresses finite dimensional domains and was another motivation for the later definition of Jaffard domains.

Theorem 2.9. [16, Proposition 2.7] *Let R be a quasilocal domain with $n := \dim(R) < \infty$. Then R is a valuation domain if (and only if) R is an integrally closed going-down domain with* $\dim_v(R) = n$.

It is easy to see that Theorem 2.5 generalizes Theorem 2.9. In fact, I must confess that the proof of the implication (2) $\Rightarrow$ (4) in Theorem 2.5 was motivated by the proof of Theorem 2.9.

Recall that R is said to be a *finite conductor domain* if $Ra \cap Rb$ is a finitely generated R-module for all $a, b \in K$. Any coherent (for instance, Prüfer) domain is an example of a finite conductor domain. There are many other examples of finite conductor domains: consider, for example, any GCD-domain.

We can next state a result from my second paper on going-down. (My work on going-down shared one misfortune of my work on universal catenarity, namely, my first paper on the subject was published one year after my second paper on the subject. Such are the vagaries of publication!) It is clear that Theorem 2.10 generalizes Quentel's seminal result, Theorem 2.7.

Theorem 2.10. [15, Corollary 4] *R is a Prüfer domain if (and only if) R is an integrally closed finite conductor domain that is also a going-down domain.*

Theorem 2.10 should be compared with the following result of Steve McAdam [54]. Like Theorem 2.7, Theorem 2.11 was published only in the quasilocal case, but the global assertion given below is then an easy consequence. Now that we know that going-down domains must be treed [16, Theorem 2.2] but not conversely, it is clear that Theorem 2.11 is a better result than Theorem 2.10. Of course, [54] and [15] were each written before [16] was a glint in my eye.

Theorem 2.11. [54, Theorem 1] *R is a Prüfer domain if (and only if) R is an integrally closed finite conductor domain that is also a treed domain.*

I noted above that McAdam and I had published parallel results (Theorems 2.10 and 2.11) before it was known that going-down domains must be treed. A similar comment could be made in comparing a paper of Sheldon with my collaboration that involved Dawson. Indeed, Sheldon [64, Theorem 3.7] proved that R is a Prüfer domain if and only if R is a GCD-domain that is also a treed domain, while Dawson and I [14, Corollary 4.3] proved that R is a Prüfer domain if and only if R is a GCD-domain that is also a going-down domain. We all drew the corollary that a one-dimensional domain is a Bézout domain if and only if it is a GCD-domain ([64, Corollary 3.9], [14, Corollary 4.4]). (A few years later, I realized that this fact was also a consequence of a result [66, Proposition A] in an earlier paper of Vasconcelos.) Coincidentally, [64] and [14] appeared in the same issue of the Canadian Mathematical Journal. I was amazed that Sheldon had managed to cite my paper with Dawson although I had not been aware of Sheldon's work until I saw the published journal. The reason for this was explained to me several years later by someone who had been on my preprint mailing list: in his capacity as referee of [64], he had shared a copy of the preprint of [14] that I had sent him.

Recall from [18] that a prime ideal P of R is called a *divided prime ideal* of R if $PR_P = P$; and that R is called a *locally divided domain* if, for each prime ideal Q of R, each prime ideal of R_Q is a divided prime ideal of R_Q (that is, if $PR_Q = PR_P$ for all prime ideals $P \subset Q$ of R). Any one-dimensional domain is a locally divided domain, as is any Prüfer domain. Each locally divided domain is a going-down domain, with the converse holding if R is a seminormal domain: cf. [18, Remark 2.7 (b) and Corollary 2.6]. (Note that the "x^2, x^3" criterion appears instead of seminormality in the statement of [18, Corollary 2.6] because I had not heard of seminormality when [18] was being written!)

Of the many results published on the "ascent of going-down domain" theme, I next state one which, in the integrally closed case, gives another generalization of Theorem 2.7.

Theorem 2.12. [19, Corollary 3.3] *If R is a locally divided domain that is also locally coherent and if R' is finitely generated over R, then R' is a Prüfer domain.*

In concert with the valuation-theoretic motivation that began this section of the paper, we next state Badawi's recent characterization of valuation domains that uses the "divided" concept. First, recall the following definition that is due to Heinzer–J. Huckaba–Papick: a nonzero ideal I of R is said to be *m-canonical* if $(I:(I:J)) = J$ for each nonzero ideal J of R. Badawi's result [5, Remark 1.1 and Corollary 2.5] states that R is a valuation domain if and only if R is a quasilocal domain with an m-canonical ideal that is a divided prime ideal of R.

Anyone familiar with "Multiplicative Ideal Theory" [41] knows that one of its many important contributions was to illustrate and popularize the central role that Prüfer domains play in commutative ring theory. We next discuss a class of going-down domains (the i-domains) whose introduction was motivated, in part, by a particularly useful approach to Prüfer domains that was given in [41]. This concerns the characterization of Prüfer domains as the integrally closed domains R such that for each prime ideal P of R, each overring (resp., each chain of prime ideals in any overring) T of R has at most one prime ideal that contracts to P [41, Theorem 26.2]. Dropping the "integrally closed" requirement, Papick [57] was led to the definition of an i-domain (resp., an INC-domain). He showed that R is a quasilocal i-domain if and only if R' is a valuation domain [57, Corollary 2.15] (resp., R is an INC-domain if and only if R' is a Prüfer domain [57, Proposition 2.26]). I had somewhat anticipated part of this result in [17, Corollary 2.5], where I had proved that if R' is a valuation domain, then R and each of its overrings are going-down domains. By permitting the extension domains T to lie inside the algebraic closure of K, I specialized the i-domain concept to the ai-domain concept in [24]. Gabriel

Picavet [58] has recently extended some of that domain-theoretic work to the context of reduced rings.

Apropos of INC and Prüfer domains, we close the section by mentioning two more directions of inquiry. The first of these had its beginnings with the result of Gilmer–Hoffmann [43, Theorem 5] characterizing the domains R such that R' is a Prüfer domain as those R such that each element of K is a root of a polynomial in $R[X]$ having a unit coefficient. I recovered this result in [20, Corollary 5] as a consequence of a characterization of those ring extensions of the form $A \subseteq A[u]$ that satisfy INC [20, Theorem, page 38]. Nearly 20 years later, I returned to a more quantitative study of these issues (focussing on bounds for the degree of the relevant polynomials and the question of which of their coefficients were units): see [25, Theorems 2.3 and 2.5].

Finally, I turn to what I have called Prüfer's Ascent Result, namely, Prüfer's result that the integral closure of a Prüfer domain in any (algebraic) field extension of its quotient field is also a Prüfer domain. The proof of this result given in "Multiplicative Ideal Theory" [41, Theorem 22.3] is so elegant that many, including me, have admired it. (For instance, Kaplansky [52, Theorem 101] simply repeats that proof in his "Commutative Rings". For a survey of other proofs of Prüfer's Ascent Result, see the Introduction of [23].) Mindful of Papick's characterization of INC-domains that was mentioned above, I sought to find an INC-theoretic proof of Prüfer's Ascent Result. This was accomplished in [23, Corollary 4], as a consequence of a result [23, Proposition 3] that generalized work of Uda [65, Corollary 3.12] concerning certain domain extensions that were not necessarily integral or overrings. An analogue of [23, Proposition 3] involving ring extensions with nontrivial zero-divisors was given in [23, Proposition 6], a result whose hypothesis involving GD was shown to be essential in [23, Remark 7 (c)].

3 Infinite Chains of Prime Ideals

For more than ten years, much of my research has concerned infinite chains and trees (of prime ideals, intermediate field extensions, etc.). The immediate stimulus for that inquiry was my reading of a remarkable result of Gilmer–Nashier–Nichols, a significant fragment of which is stated in Theorem 3.1. First, here is some convenient notation: given rings $A \subseteq B$ and a prime ideal P of A, let $F_{A,B}(P) := \{Q \in \mathrm{Spec}(B) \mid Q \cap A = P\}$, the fiber above P in B.

Theorem 3.1. [44, Theorem 1.3] *Let $\{P_i\}$ be a nonempty finite set of pairwise incomparable prime ideals of a ring A; for each index i, let n_i be a positive integer. Then there exists an integral extension B of A such that B is finitely generated over A and, for each i, the cardinality of $F_{A,B}(P_i)$ is n_i.*

Actually, [44, Theorem 1.3] also included the conclusion that it is possible to arrange that each element of $F_{A,B}(P_i)$ has the same height as P_i for each

i. This conclusion is not a feature of our infinitistic analogue of Theorem 3.1 given in Theorem 3.2. However, Theorem 3.2 does extend to infinite cardinals each of the three finitistic aspects in the above statement of Theorem 3.1.

Theorem 3.2. [22, Theorem 2] *Let $\{P_i\}$ be a countable set of pairwise incomparable prime ideals of a ring A; for each index i, let α_i be a nonzero cardinal number such that $\alpha_i \leq \aleph_1$. Then there exists an integral extension B of A such that B is countably generated over A and, for each i, the cardinality of $F_{A,B}(P_i)$ is α_i.*

The proof of Theorem 3.2 involved iterated direct limits. Given the above role of $\aleph_1$, I should also mention that the proof of Theorem 3.2 assumed the Continuum Hypothesis (CH). As explained in [22, Remark 3 (c)], Theorem 3.2 is best-possible (assuming CH). Part (d) of [22, Remark 3] speculates about what may hold (assuming the Generalized Continuum Hypothesis) if one drops the "countably generated" condition and considers cardinals greater than $\aleph_1$.

It seemed natural to me to ask for an analogue of Theorem 3.2 in which the "pairwise incomparable" condition would be removed. Intuitively, I would view such a result on the cardinality of chains within a fiber as an "infinitistic INC-Theorem". Theorem 3.3 records my best-possible answer to this question. This result should be contrasted with the case for a polynomial ring in finitely many variables over a field (cf. [41, Theorem 30.5], [52, page 109, line 8]).

Theorem 3.3. [27, Theorem 2.6 and Remark 2.8 (a)] *Let A be a ring and let B be a ring which is generated as an A-algebra by a set I. Then each chain $\mathcal{C}$ of prime ideals in a given fiber $F_{A,B}(P)$ has at most $2^{|I|}$ elements. Moreover, if A is a field and $B = A[\{X_i \mid i \in I\}]$ is the polynomial ring over A in a set of algebraically independent variables indexed by an infinite set I, then it can be arranged that some such $\mathcal{C}$ (in the only fiber) has cardinality $2^{|I|}$.*

I was first alerted to the phenomenon in the second assertion in Theorem 3.3 by Sylvia and Roger Wiegand for the case of denumerable I, which they handled via Dedekind cuts in the rationals. My proof for the general case in Theorem 3.3 used the following consequence of the Generalized Continuum Hypothesis: each infinite set I has a chain $\mathcal{D}$ of subsets such that $|\mathcal{D}| = 2^{|I|}$.

The above focus on chains has led to the following definitions. Given (a ring homomorphism) $f: A \to B$ and a chain $\{P_i\}$ (of arbitrary cardinality) in Spec(A), we say that a chain $\{Q_i\}$ in Spec(B) *covers* $\{P_i\}$ if ${}^a f(Q_i) = P_i$ for each i. If each chain $\{P_i\}$ in Spec(A) is covered by some chain in Spec(B), we say that f is a *chain morphism.* If the same condition is asserted only in case the chain $\{P_i\}$ consists of at most n elements (where n is a positive integer), we say that f is an *n-chain morphism.* It is clear that any chain morphism is an n-chain morphism for each n; and that if $m > n$ are positive integers, then any m-chain morphism is an n-chain morphism. Note that $f: A \to B$ is a 1-chain morphism if and only if f satisfies LO; and that 2-chain morphisms are the same as the subtrusive morphisms that have been studied in several papers of Gabriel Picavet. It was shown by Dobbs–Fontana–G. Picavet in [32, Theorem 3.26] that " universally 2-chain morphism" and "universally chain morphism" are equivalent concepts. However, as Example 3.4 records, Hetzel and I obtained results with quite a different flavor when the "universally" proviso is dropped. The proof of Example 3.4 was categorical and rather non-constructive, as it applied a realization result of Hochster [46, Theorem 6 (a)] to a surjective spectral map on the left topology on some posets. (A convenient reference for the left topology is [8, Exercice 2, page 89].)

Example 3.4. [34, Example 2.2] For each positive integer n, there exists an n-chain morphism $f{:}A \to B$ that is not an $(n+1)$-chain morphism. In such an example, it can be arranged that A is a domain. It can be further arranged that either B is a reduced ring with exactly two minimal prime ideals or B is an extension domain of A.

In view of the final comment prior to Theorem 3.3, Hetzel and I were initially unsure whether "n-chain morphism for all n" would imply "chain morphism". By using Bergman's work on "templates" [7, Proposition 10 (ii)], we were able to resolve the issue: see Theorem 3.5. It is interesting that the proof of Bergman's result used methods of logic, specifically, the Compactness Theorem of model theory. No doubt, a more penetrating logical analysis would yield additional applications (perhaps via "arrays" in the sense of Bergman, perhaps in a constructive approach via "constructions cachées" in the sense of Coquand–Lombardi [13]).

Theorem 3.5. [34, Proposition 2.5 (a)] *A ring homomorphism $f{:}A \to B$ is an n-chain morphism for each positive integer n (if and) only if f is a chain morphism.*

There has been continuing interest in finding chains that cover given chains of prime ideals. Recently, for any $f{:}A \to B$, Sharma [63] has given a necessary and sufficient condition for a given finite chain of prime ideals of A to be covered by some chain of prime ideals in Spec(B). Surely, among the progenitors for such issues concerning covers of finite chains are the classical Going-up Theorem [41, Corollary 11.6] and the ring-theoretic version of "extension of valuations" [41, Corollary 19.7]. When I first read the latter result in [41], I was struck by its generality (as covers of chains had not been emphasized in versions with which I was already familiar, such as [52, Theorem 56]) and by its extreme clarity. Later, I came to appreciate its proof's subtle and effective use of pullbacks (see the "inverse image" considered in [41, page 230, line 5]). The presentation of this result is typical of Gilmer's high standards of exposition that I referred to in the Introduction. I suspect that others felt the same way, and perhaps that is what led Dan Anderson to ask if there was a "Noetherian" analogue of [41, Corollary 19.7]. This question was answered in the affirmative by Cahen–Houston–Lucas [12]. Among other things, they showed that if $\mathcal{C}$ is a finite chain of n nonzero prime ideals of a Noetherian domain R, then there exists a discrete rank n valuation overring T of R such that some chain in Spec(T) covers $\mathcal{C}$. In reading the proof in [12], I was struck by how well it paralleled the approach in Gilmer's exposition (including the use of a suitable pullback). A similar comment applies to Kang–Oh's proof of the infinitistic generalization of [41, Corollary 19.7] given in Theorem 3.6.

Theorem 3.6. [49, Theorem] *If R is a domain and $\mathcal{C}$ is a chain (of arbitrary cardinality) in Spec(R), then there exists a valuation overring T of R such that some chain in Spec(T) covers $\mathcal{C}$.*

It took little for me to extend the preceding result to the setting of [41, Corollary 19.7], that is, by requiring the extension domain T to be a valuation ring of a given field containing R: see [28, Theorem 3]. I also obtained some zero-divisor generalizations of Theorem 3.6: see Theorem 3.7 below and its consequence [28, Corollary 5]. (The latter result produced a suitable valuation overring of the base ring A in case A is a Marot ring.) For suitable paravaluation-theoretic background, see [47].

Theorem 3.7. [28, Theorem 4] *Let $A \subseteq B$ be rings and $\mathcal{C} = \{P_i\}$ be a chain in Spec(A). Then A is contained in some paravaluation ring V of B such that Spec(V) has a chain that covers $\mathcal{C}$. It can be further arranged that $N \cap A = \cup P_i$, where N is such that (V, N) is a paravaluation pair of B.*

Kang announced Theorem 3.6 in October 1996 in Chattanooga in an invited talk at a Special Session of the American Mathematical Society organized by David Anderson and myself. Immediately after that talk, I mentioned to Kang that during his talk, I had realized how to prove such a result (via transfinite induction) for a general ring-theoretic setting by using the GU property and well ordered chains. When I asked Kang if his chains were general, he replied that they were "of arbitrary ordinal type". I took that to mean that Kang–Oh had used well ordered chains as well and that I had merely found a new proof of their result. When [49] was published a year and a half later, I discovered that Kang and I had failed to communicate in Chattanooga, for, as stated above, Kang–Oh's work is valid for arbitrary chains. At that point, I quickly wrote up [26] and submitted it for publication. There were some problems with that paper and its sequel, but before describing them, let me first state in Theorem 3.8 something that I got right, namely, the Generalized Going-up Theorem for arbitrary integral extensions.

The statement of Theorem 3.8 and other results from [26] will be made easier by a couple of definitions (that were made in later papers, [32] and [45]). A ring homomorphism (possibly the inclusion map of a subring) $f{:}A \to B$ is said to satisfy generalized going-down GGD (resp., generalized going-up GGU) if f satisfies the analogue of the usual GD (resp., GU) property predicated for chains of arbitrary cardinality.

Theorem 3.8. [26, Remark (d)] *Any integral ring extension satisfies GGU.*

What about the "well ordered" results that I mentioned to Kang in 1996? They are in the Corrigendum to [26]. Sadly, just before submitting [26] for publication, I came to believe (incorrectly) that my methods served to reduce GGU (resp., GGD) issues for arbitrary chains to the corresponding issues for chains that are well ordered by inclusion (resp., reverse inclusion). As a result, the general GGU (resp., GGD) result stated in [26, Theorem] (resp., [26, Remark (a)]) needs to be amended by adding the hypothesis that the given chain of prime ideals of the base ring is well ordered by inclusion (resp., reverse inclusion). (Actually, those statements may seem to just be addressing chain maps – a concept which had not been defined yet! – but their proofs are easily modified to establish what was stated above.) This leaves open the question whether the "well ordered" hypotheses in the Corrigendum to [26] were really needed. After much frantic effort, answers were found, by Kang–Oh for GGU and then by Dobbs–Hetzel for GGD: see Theorems 3.9 (stated here using the GGU notation) and 3.10.

Theorem 3.9. [50, Corollary 12] *A ring homomorphism satisfies GGU if (and only if) it satisfies GU.*

We pause for a minor point. Theorem 3.9 was stated in [50] only for ring extensions rather than for arbitrary ring homomorphisms. However, as noted in [33], it is easy to apply standard homomorphism theorems to infer Theorem 3.6 from the case for extensions. The credit for Theorem 3.9 – and for the computational heroics in its proof – belongs entirely to Kang–Oh. By the way, perhaps I should mention that Kang–Oh's notation for GGU was SCLO.

Although one must be careful in speaking of "duality" when dealing with rings (because the category of commutative rings is far from being an Abelian category), Hetzel and I were able to apply the Kang–Oh result on GGU to obtain the GGD result in Theorem 3.10. Perhaps surprisingly, our argument involved a careful analysis of the flat topology and some spectral maps. As noted in [33], our methods actually show that Theorems 3.9 and 3.10 are equivalent.

Theorem 3.10. [33, Theorem 2.2] *A ring homomorphism satisfies GGD if (and only if) it satisfies GD.*

In view of Theorems 3.9 and 3.10, portions of [32] and [45] have become obsolete, namely, those results which gave sufficient conditions for GD (resp., GU) to imply GGD (resp., GGU). However, I believe that much of [32] and [45] remains of interest, especially their results that gave new sufficient conditions for GD and GU, respectively.

Given rings $A \subseteq B$ and subsets $\mathcal{T} \subseteq \mathrm{Spec}(A)$ and $\mathcal{S} \subseteq \mathrm{Spec}(B)$, it is convenient to generalize our earlier usage for chains by saying that $\mathcal{S}$ *covers* (or *contracts to*) $\mathcal{T}$ (*in* A) if $\{Q \cap A \mid Q \in \mathcal{S}\} = \mathcal{T}$. After writing [26], I enlisted the help of Marco Fontana in trying to prove my conjecture that if $A \subseteq B$ satisfies GU, then each tree $\mathcal{T}$ in $\mathrm{Spec}(A)$ could be covered by a suitable tree $\mathcal{S}$ in Spec(B). We quickly produced [31], where we proved my conjecture in case each branch of $\mathcal{T}$ has finite length (which settles matters if A is a Noetherian ring) and also in case $\mathcal{T}$ has only two branches each of which is well ordered under inclusion [31, Theorem 3]; and in [31, Proposition 7], we generalized Theorem 3.6 from chains to trees $\mathcal{T}$ with only finitely many branches (and required the asserted overring T of R to be a Bézout domain). However, settling my conjecture required a formidable combinatorial analysis by Kang–Oh: see Theorem 3.11.

Theorem 3.11. [50, Theorem 7, Corollary 12] *If $A \subseteq B$ satisfies GU and $\mathcal{T}$ is a tree in Spec(A), then there exists a tree $\mathcal{S}$ in Spec(B) such that $\mathcal{S}$ covers $\mathcal{T}$.*

The next three results are motivated by the observation, given $A \subseteq B$, that although chains in Spec(B) must contract to chains in Spec(A), trees in Spec(B) need not contract to trees in Spec(A).

Example 3.12. [29, Example 2.1] There exists an integral (hence GU) extension $A \subseteq B$ of one-dimensional Noetherian rings such that some tree in Spec(B) contracts to a non-tree in Spec(A).

Proof. (Sketch) Let (V, M) be a DVR. Put $B{:} = V \times V$ and $k{:} = V/M$. Consider the surjection $\pi{:}B \to k \times k$, given by $(v_1, v_2) \mapsto (v+1+M, v_2+M)$; and the (injective) diagonal map $\Delta{:}k \to k \times k$, given by $a \mapsto (a, a)$. Let A be the pullback of π and Δ. Then $\mathcal{T}{:} = \text{Spec}(B)$ is a tree with exactly two (disjoint) branches, each of which has cardinality 2; and $\mathcal{T}$ contracts to the non-tree Spec(A) (which consists of two minimal points and a unique maximal point).

Theorem 3.13. [29, Theorem 2.7] *The following conditions are equivalent.*
(1) *R is a treed domain.*
(2) *For all Bézout overrings T of R, each tree in Spec(T) contracts to a tree in Spec(R).*
(3) *For all domains T containing R, each tree in Spec(T) contracts to a tree in Spec(R).*

We close with a domain-theoretic result that retains the cardinality features of the construction in Example 3.12.

Corollary 3.14. [29, Corollary 2.8] *Let R be a non-treed domain. Then there exists a Bézout overring T of R and a tree $\mathcal{T}$ in Spec(T) such that $|\mathcal{T}| = 4$ and $\mathcal{T}$ contracts to a non-tree $\mathcal{S}$ in Spec(R) with $|\mathcal{S}| = 3$. If, in addition, R is LFD (for instance, Noetherian of dimension at least 2), it can be further arranged that $\mathcal{T}$ and $\mathcal{S}$ are saturated in Spec(T) and Spec(R), respectively.*

Because of space limitations, I cannot say much more here about the ways that Robert Gilmer's writings have influenced my research. But the above account is surely not complete. For instance, I could point to [30], where Fontana and I revisited some work of Arnold and Gilmer [4] on dimension sequences, replacing their classical $D+M$ constructions with domains that are locally P^nVDs (in the sense of [39]).

Thank you, Robert, for your example and for your inspiring leadership.

References

1. S. Améziane Hassani, D. E. Dobbs, and S.-E. Kabbaj, On the prime spectrum of commutative semigroup rings, Comm. Algebra 26 (1998), 2559–2589.
2. D. F. Anderson, A. Bouvier, D. E. Dobbs, M. Fontana and S. Kabbaj, On Jaffard domains, Exposition. Math. 6 (1988), 145–175.
3. D. F. Anderson, D. E. Dobbs, S. Kabbaj and S. B. Mulay, Universally catenarian domains of $D+M$ type, Proc. Amer. Math. Soc. 104 (1988), 378–384.
4. J. T. Arnold and R. Gilmer, The dimension sequence of a commutative ring, Amer. J. Math. 96 (1974), 385–408.
5. A. Badawi, A characterization of valuation domains via m-canonical ideals, Comm. Algebra 32 (2004), 4363–4374.
6. M. Ben Nasr and N. Jarboui, A counterexample for a conjecture about the catenarity of polynomial rings, J. Algebra 248 (2002), 785–789.
7. G. M. Bergman, Arrays of prime ideals in commutative rings, J. Algebra 261 (2003), 389–410.
8. N. Bourbaki, Topologie Générale, Chapitres 1–4, Hermann, Paris, 1971.
9. A. Bouvier, D. E. Dobbs and M. Fontana, Universally catenarian integral domains, Adv. in Math. 72 (1988), 211–238.
10. A. Bouvier, D. E. Dobbs and M. Fontana, Two sufficient conditions for universal catenarity, Comm. Algebra 15 (1987), 861–872.
11. A. Bouvier and S. Kabbaj, Examples of Jaffard domains, J. Pure Appl. Algebra 54 (1988), 155–165.
12. P.-J. Cahen, E. G. Houston and T. G. Lucas, Discrete valuation overrings of Noetherian domains, Proc. Amer. Math. Soc. 124 (1996), 1719–1721.
13. T. Coquand and H. Lombardi, Hidden constructions in abstract algebra: Krull dimension of distributive lattices and commutative rings, Commutative Ring Theory and Applications (eds. M. Fontana, S.-E. Kabbaj and S. Wiegand) Vol. 231, Dekker, 2003, 477–499.
14. J. Dawson and D. E. Dobbs, On going-down in polynomial rings, Can. J. Math. 26 (1974), 177–184.
15. D. E. Dobbs, On going-down for simple overrings, Proc. Amer. Math. Soc. 39 (1973), 515–519.
16. D. E. Dobbs, On going-down for simple overrings, II, Comm. Algebra 1 (1974), 439–458.
17. D. E. Dobbs, Ascent and descent of going-down rings for integral extensions, Bull. Austral. Math. Soc. 15 (1976), 253–264.
18. D. E. Dobbs, Divided rings and going-down, Pac. J. Math. 67 (1976), 353–363.
19. D. E. Dobbs, Coherence, ascent of going-down and pseudo-valuation domains, Houston J. Math. 4 (1978), 551–567.

20. D. E. Dobbs, On INC-extensions and polynomials with unit content, Can. Math. Bull. 23 (1980), 37–42.
21. D. E. Dobbs, On treed overrings and going-down domains, Rend. Mat. 7 (1987), 317–322.
22. D. E. Dobbs, Integral extensions with fibers of prescribed cardinality, Zero Dimensional Commutative Rings (eds. D. F. Anderson and D. E. Dobbs) Vol. 171, Dekker, 1995, 143–154.
23. D. E. Dobbs, Prüfer's ascent theorem via INC, Comm. Algebra 23 (1995), 5413–5417.
24. D. E. Dobbs, Absolutely injective integral domains, Houston J. Math. 22 (1996), 485–497.
25. D. E. Dobbs, On characterizations of Prüfer domains using polynomials with unit content, Factorization in Integral Domains (ed. D. D. Anderson) Vol. 189, Dekker, 1997, 295–303.
26. D. E. Dobbs, A going-up theorem for arbitrary chains of prime ideals, Comm. Algebra 27 (1999), 3887–3894. Corrigendum 28 (2000), 1653–1654.
27. D. E. Dobbs, On the prime ideals in a commutative ring, Canad. Math. Bull. 43 (2000), 312–319.
28. D. E. Dobbs, Lifting chains of prime ideals to paravaluation rings, Rend. Circ. Mat. Palermo. Ser. II 49 (2000), 319–324.
29. D. E. Dobbs, Extensions of commutative rings in which trees of prime ideals contract to trees, Rend. Circ. Mat. Palermo. Ser. II 50 (2001), 259–270.
30. D. E. Dobbs and M. Fontana, Sur les suites dimensionnelles et une classe d'anneaux distingués qui les déterminent, C. R. Acad. Sci. Paris Sér. A-B 306 (1988), 11–16.
31. D. E. Dobbs and M. Fontana, Lifting trees of prime ideals to Bézout extension domains, Comm. Algebra 27 (1999), 6243–6252. Corrigendum 28 (2000), 1655–1656.
32. D. E. Dobbs, M. Fontana and G. Picavet, Generalized going-down homomorphisms of commutative rings, Commutative Ring Theory and Applications (eds. M. Fontana, S.-E. Kabbaj and S, Wiegand) Vol. 231, Dekker, 2002, 143–163.
33. D. E. Dobbs and A. J. Hetzel, Going-down implies generalized going-down, Rocky Mountain J. Math. 35 (2005), 479–484.
34. D. E. Dobbs and A. J. Hetzel, On chain morphisms of commutative rings, Rend. Circ. Mat. Palermo 53 (2004), 71–84.
35. D. E. Dobbs and I. J. Papick, On going-down for simple overrings, III, Proc. Amer. Math. Soc. 54 (1976), 35–38.
36. D. E. Dobbs and I. J. Papick, Flat or open implies going-down, Proc. Amer. Math. Soc. 65 (1977), 370–371.
37. D. E. Dobbs and I. J. Papick, Going-down: a survey, Nieuw Arch. v. Wisk. 26 (1978), 255–291.
38. D. E. Dobbs and J. Shapiro, Transfer of Krull dimension, lying-over and going-down to the fixed ring, submitted for publication.
39. M. Fontana, Sur quelques classes d'anneaux divisés, Rend. Sem. Mat. e Fis. di Milano 51 (1981), 179–200.
40. R. Gilmer, Domains in which valuation ideals are prime powers, Arch. Math. 17 (1966), 210–215.
41. R. Gilmer, Multiplicative Ideal Theory, Dekker, New York, 1972.
42. R. Gilmer, Commutative Semigroup Rings, Univ. Chicago Press, Chicago, 1984.

43. R. Gilmer and J. F. Hoffmann, A characterization of Prüfer domains in terms of polynomials, Pac. J. Math. 60 (1975), 81–85.
44. R. Gilmer, B. Nashier and W. Nichols, On the heights of prime ideals under integral extensions, Arch. Math. 52 (1989), 47–52.
45. A. J. Hetzel, Generalized going-up homomorphisms of commutative rings, Commutative Ring Theory and Applications (eds. M. Fontana, S.-E. Kabbaj and S, Wiegand) Vol. 231, Dekker, 2002, 255–265.
46. M. Hochster, Prime ideal structure in commutative rings, Trans. Amer. Math. Soc. 142 (1969), 43–60.
47. J. A. Huckaba, Commutative Rings with Zero Divisors, Dekker, New York, 1988.
48. P. Jaffard, Théorie de la dimension dans les anneaux de polynômes, Mémor. Sci. Math., Fasc. 146, Gauthier-Villars, Paris, 1960.
49. B. G. Kang and D. Y. Oh, Lifting up an infinite chain of prime ideals to a valuation ring, Proc. Amer. Math. Soc. 126 (1998), 645–646.
50. B. G. Kang and D. Y. Oh, Lifting up a tree of prime ideals to a going-up extension, J. Pure Appl. Algebra 182 (2003), 239–252.
51. I. Kaplansky, Adjacent prime ideals, J. Algebra 20 (1972), 94–97.
52. I. Kaplansky, Commutative Rings, rev. ed., Univ. Chicago Press, Chicago, 1974.
53. S. Malik and J. L. Mott, Strong S-domains, J. Pure Appl. Algebra 28 (1983), 249-264.
54. S. McAdam, Two conductor theorems, J. Algebra 23 (1972), 239–240.
55. M. Nagata, On the chain problem of prime ideals, Nagoya Math. J. 10 (1956), 51–64.
56. M. Nagata, Finitely generated rings over a valuation domain, J. Math. Kyoto Univ. 5 (1966), 163–169.
57. I. J. Papick, Topologically defined classes of going-down domains, Trans. Amer. Math. Soc. 219 (1976), 1–37.
58. G. Picavet, Universally going-down rings, 1-split rings and absolute integral closure, J. Pure Appl. Algebra 31 (2003), 4655–4685.
59. Y. Quentel, Sur une caractérisation des anneaux de valuation de hauteur 1, C. R. Acad Sci. Paris Sér. A-B 265 (1967), A659–A661.
60. L. J. Ratliff, Jr., On quasi-unmixed local domains, the altitude formula, and the chain condition for prime ideals, II, Amer. J. Math. 92 (1970), 99–144.
61. P. Ribenboim, Sur une note de Nagata relative à un problème de Krull, Math. Z. 64 (1956), 159–168.
62. F. Richman, Generalized quotient rings, Proc. Amer. Math. Soc. 16 (1965), 794–799.
63. P. K. Sharma, A note on lifting a chain of prime ideals, J. Pure Appl. Algebra 192 (2004), 287–291.
64. P. B. Sheldon, Prime ideals in GCD-domains, Can. J. Math. 26 (1974), 98–107.
65. H. Uda, Incomparabilty in ring extensions, Hiroshima Math. J. 9 (1979), 451–463.
66. W. V. Vasconcelos, The local rings of global dimension two, Proc. Amer. Math. Soc. 35 (1972), 381–386.

Direct-sum decompositions over one-dimensional Cohen-Macaulay local rings

Alberto Facchini[1], Wolfgang Hassler[2], Lee Klingler[3], and Roger Wiegand[4]

[1] Alberto Facchini, Dipartimento di Matematica Pura e Applicata, Università di Padova, Via Belzoni 7, I-35131 Padova, Italy, facchini@math.unipd.it
[2] Wolfgang Hassler, Institut für Mathematik und Wissenschaftliches Rechnen, Karl-Franzens-Universität Graz, Heinrichstraße 36/IV, A-8010 Graz, Austria, wolfgang.hassler@uni-graz.at
[3] Lee Klingler, Department of Mathematical Sciences, Florida Atlantic University, Boca Raton, FL 33431-6498, klingler@fau.edu
[4] Roger Wiegand, Department of Mathematics, University of Nebraska, Lincoln, NE 68588-0323, rwiegand@math.unl.edu

This article is dedicated to Robert Gilmer, a pioneer of multiplicative ideal theory. Thanks to his fundamental contributions and his leadership, the theory has become a major mathematical enterprise, with connections to many areas of mathematics. Here we explore one of these connections.

1 Introduction and Terminology

The theory of commutative cancellative monoids grew out of the multiplicative ideal theory of integral domains. Many problems in multiplicative arithmetic become more clearly focused when one strips away the additive structure of an integral domain and looks only at the multiplicative monoid of non-zero elements. Krull monoids, introduced in [4], and their divisor class groups have provided perhaps the most fertile ground for investigation. It is known [17]

[1] Alberto Facchini's research was partially supported by Gruppo Nazionale Strutture Algebriche e Geometriche e loro Applicazioni of Istituto Nazionale di Alta Matematica, Italy, and by Università di Padova (Progetto di Ateneo CDPA048343 "Decomposition and tilting theory in modules, derived and cluster categories").
[2] The research of W. Hassler was supported by the *Fonds zur Förderung der wissenschaftlichen Forschung*, project number P16770-N12.
[3] Klingler thanks the University of Nebraska–Lincoln, where much of the research was completed.
[4] Wiegand's research was partially supported by a grant from the National Security Agency.

that an integral domain R is a Krull domain if and only if its multiplicative monoid $R - \{0\}$ is a Krull monoid. Moreover, the divisor class group of R agrees with the divisor class group of the monoid $R - \{0\}$. In particular, the divisor class group of a Krull domain depends only on the multiplicative monoid of non-zero elements.

Commutative monoids arise also in representation theory. Given any class $\mathcal{C}$ of finitely generated modules, one can form the set $\mathrm{V}(\mathcal{C})$ consisting of isomorphism classes $[M]$ of modules $M \in \mathcal{C}$. If $\mathcal{C}$ contains (0) and is closed under finite direct sums, the set $\mathrm{V}(\mathcal{C})$ becomes a commutative monoid under the operation $[M]+[N] = [M \oplus N]$. If, moreover, $\mathcal{C}$ is closed under direct summands, the monoid $\mathrm{V}(\mathcal{C})$ carries detailed information about the direct-sum behavior of modules in $\mathcal{C}$, e.g., whether or not the Krull-Remak-Azumaya-Schmidt theorem holds, and, when it does not, how badly it fails.

Our goal in this paper is to give a complete set of invariants for the monoid $\mathrm{V}(R\text{-mod})$, where $(R, \mathfrak{m}, k)$ is a one-dimensional reduced Noetherian local ring and R-mod is the class of all finitely generated R-modules. In order to state our results precisely and place them in context, we recall some terminology.

1.1 Krull monoids

All of our monoids are assumed to be commutative. We always use additive notation, that is, $+$ is the operation in our monoids, and 0 is the identity element. We will always assume that 0 is the only invertible element, that is, $x + y = 0 \implies x = y = 0$. Further, we shall restrict our attention to *cancellative* monoids: $x + z = y + z \implies x = y$. Note that the monoid $\mathrm{V}(R\text{-mod})$, for $(R, \mathfrak{m}, k)$ a commutative Noetherian local ring, is cancellative [5].

If x and y are elements of an additive monoid H we write $x \leq y$ to indicate that $x + z = y$ for some (necessarily unique) $z \in H$. A monoid homomorphism $\phi{:}H_1 \to H_2$ is a *divisor* homomorphism provided $x \leq y$ in $H_1 \iff \phi(x) \leq \phi(y)$ in H_2. In this case, it follows from our standing assumptions that ϕ is injective. A monoid is *free* provided it is isomorphic to the direct sum $\mathbb{N}^{(\Lambda)}$ of copies of the additive monoid $\mathbb{N}{:} = \{0, 1, 2, \dots\}$ for some (possibly infinite) index set Λ. Of course, given elements x and y in the free monoid $\mathbb{N}^{(\Lambda)}$, we have $x \leq y$ if and only if $x_\lambda \leq y_\lambda$ for each $\lambda \in \Lambda$, that is, $\leq$ is the product partial ordering for free monoids. A monoid H is a *Krull* monoid (see, e.g. [11]) provided there exists a divisor homomorphism from H to a free monoid.

It turns out [11, Theorem 23.4] that every Krull monoid H actually has a *divisor theory*, that is, a divisor homomorphism $\phi{:}H \to F$ such that

(1) F is a free monoid, and
(2) every element of F is the greatest lower bound of some finite set of elements in $\phi(H)$.

We note [11, Theorem 20.4] that a divisor theory of a Krull monoid is unique up to canonical isomorphism, that is, if $\phi : H \to F$ and $\phi' : H \to F'$ are two divisor theories, then there exists a unique isomorphism $\psi : F \to F'$ with $\phi' = \psi\phi$.

Given a divisor theory $\phi : H \to F = \mathbb{N}^{(\Lambda)}$, we have an induced homomorphism $\mathcal{Q}(\phi) : \mathcal{Q}(H) \to \mathcal{Q}(F)$, where $\mathcal{Q}(H)$ and $\mathcal{Q}(F) \cong \mathbb{Z}^{(\Lambda)}$ are the quotient groups of H and F, respectively. The *divisor class group* $\mathrm{Cl}(H)$ is, by definition, the cokernel of $\mathcal{Q}(\phi)$. Krull monoids are *atomic*, that is, every non-zero element is a sum of atoms (elements that cannot be written as a sum of two non-zero elements). The representation of an element as a sum of atoms is, in general, highly non-unique (cf. Theorem 4.1).

It is easy to see that the following conditions on a Krull monoid H are equivalent:

(1) H is free.
(2) Every non-zero element of H has a unique (up to a permutation) representation as a sum of atoms (that is, H is *factorial*).
(3) $\mathrm{Cl}(H) = 0$.

Thus the class group is, in some sense, a measure of the deviation of H from being factorial.

Let H be a Krull monoid, and let $\phi : H \to \mathbb{N}^{(\Lambda)}$ be a divisor theory. Given a divisor class $\alpha \in \mathrm{Cl}(H)$, that is, a coset of $\mathrm{im}(\mathcal{Q}(\phi))$ in $\mathbb{Z}^{(\Lambda)}$, we put

$$\mathfrak{A}(\alpha) = \{x \in \alpha \mid x \text{ is an atom of } \mathbb{N}^{(\Lambda)}\}.$$

(Of course the atoms of $\mathbb{N}^{(\Lambda)}$ are just the "unit vectors", with 1 in a single coordinate and 0's elsewhere.) The following result is [11, Theorem 23.5]:

Lemma 1.1 *Let H_1 and H_2 be Krull monoids. Then $H_1 \cong H_2$ if and only if there is a group isomorphism $\theta : \mathrm{Cl}(H_1) \to \mathrm{Cl}(H_2)$ such that $|\mathfrak{A}(\alpha)| = |\mathfrak{A}(\theta(\alpha))|$ for each $\alpha \in \mathrm{Cl}(H_1)$.*

Recall that an arbitrary ring S (not necessarily commutative) with Jacobson radical $\mathrm{J}(S)$ is said to be *semilocal* provided $S/\mathrm{J}(S)$ is semisimple Artinian. Let R be an arbitrary ring (not necessarily commutative or Noetherian), and let $\mathcal{C}$ be a class of right R-modules closed under isomorphism, finite direct sums, and direct summands. Assume that $\mathrm{End}_R(M)$ is semilocal for every $M \in \mathcal{C}$. Then $\mathrm{V}(\mathcal{C})$ is a Krull monoid [7, Theorem 3.4].

The following proposition gives further examples, outside the main context of this paper, in which $\mathrm{V}(\mathcal{C})$ is a Krull monoid. Recall that a *torsion-free R-module* is a module M such that for all $r \in R$, $x \in M$, $rx = 0 \implies r \in \mathfrak{z}(R)$ or $x = 0$, where $\mathfrak{z}(R)$ denotes the set of zerodivisors of R.

Proposition 1.2 *Let R be a commutative ring (not necessarily Noetherian).*

(1) *If R is semilocal, the monoid $\mathrm{V}(R\text{-mod})$ is a Krull monoid.*

(2) *Let R be a one-dimensional commutative Noetherian local ring with no non-zero nilpotents, and let $\mathcal{C}$ be the class of all torsion-free R-modules of finite rank (torsion-free modules M satisfying* $\dim_{R_P}(M_P) < \infty$ *for every minimal prime ideal P of R). Then* $\mathrm{V}(\mathcal{C})$ *is a Krull monoid.*

Proof. (1) If M is a finitely generated R-module, then $\mathrm{End}_R(M)$ is semilocal (cf. [23, Lemma 2.3] or [8, Proposition 3.2]). Therefore it is possible to apply [7, Theorem 3.4].

(2) Let $P_1, \dots, P_s$ be the minimal primes of R. The total quotient ring K of R is the direct product $R_{P_1} \times \cdots \times R_{P_s}$, where the localizations R_{P_i} are the quotient fields of the domains R/P_i. The R-module K is an injective envelope of R [20, Proposition 1.6]. By [19, Theorem 1], the R/P_i-module $R_{P_i}/(R/P_i)$ is Artinian. Therefore every homomorphic image of the R/P_i-module R_{P_i} has finite Goldie dimension. It follows that the same is true for every homomorphic image of the injective R-module K. From [8, Corollary 5.8], we see that every torsion-free R-module M of finite rank is isomorphic to an R-submodule of K^n for some n, and hence $\mathrm{End}_R(M)$ is semilocal [8, Corollary 5.8]. By [7, Theorem 3.4], $\mathrm{V}(\mathcal{C})$ is a Krull monoid.

Now let $(R, \mathfrak{m}, k)$ be a Noetherian local ring with $\mathfrak{m}$-adic completion $\widehat{R}$. The homomorphism $[M] \mapsto [\widehat{M}]$ is a divisor homomorphism from $\mathrm{V}(R\text{-mod})$ to $\mathrm{V}(\widehat{R}\text{-mod})$, [24, Proposition 1.2]. By the Krull-Remak-Azumaya-Schmidt theorem [6] and [22, Lemma 13] (see also [21]), $\mathrm{V}(\widehat{R}\text{-mod})$ is free. Thus we find again that $\mathrm{V}(R\text{-mod})$ is a Krull monoid. Given a finitely generated R-module M one can form the monoid $+(M)$ consisting of isomorphism classes of modules that are direct summands of direct sums of finitely many copies of M. This is always a finitely generated Krull monoid [24], and the main result of [24] is that *every* finitely generated Krull monoid arises in this way. In fact, given any finitely generated Krull monoid H, there exist a one-dimensional Noetherian local domain R (essentially of finite type over the field of rational numbers) and a finitely generated torsion-free R-module M such that $+(M) \cong H$. This result motivated us to seek a characterization of the Krull monoids that arise in the form $\mathrm{V}(\mathfrak{C}(R))$, where $\mathfrak{C}(R)$ is the class of finitely generated torsion-free modules M over a Noetherian local ring R. Alas, we were unsuccessful, for reasons that we discuss briefly in Section 5. We were, however, able to classify the monoids $\mathrm{V}(R\text{-mod})$, for $(R, \mathfrak{m}, k)$ a reduced Noetherian local ring of dimension one. Our main theorem is that these monoids depend, up to isomorphism, only on the *splitting number* $\mathrm{spl}(R) := |\operatorname{Spec}(\widehat{R})| - |\operatorname{Spec}(R)|$, the cardinality of the residue field k, and whether or not R is Dedekind-like (defined below). We show, in all cases, that $\mathrm{Cl}(\mathrm{V}(R\text{-mod})) \cong \mathbb{Z}^{(\mathrm{spl}(R))}$.

1.2 Dedekind-like rings

A commutative Noetherian local ring $(R, \mathfrak{m}, k)$ is *Dedekind-like* [14, Definition 2.5] provided R is one-dimensional and reduced, the integral closure $\overline{R}$ of R

is generated by at most 2 elements as an R-module, and $\mathfrak{m}$ is the Jacobson radical of $\overline{R}$. In a recent series of papers [14, 15, 16] L. Klingler and L. S. Levy classified all finitely generated indecomposable R-modules up to isomorphism over essentially every Dedekind-like ring. The exceptional Dedekind-like rings for which their classification has not yet been worked out are those where $\overline{R}/\mathfrak{m}$ is a purely inseparable field extension of k of degree 2. (An example is the ring $F+XK[[X]]$, where K/F is a purely inseparable field extension of degree 2.) For convenience, we will refer to these rings as *exceptional Dedekind-like rings.*

2 The Main Theorem

Fix a positive integer q, and let B be a $q \times \aleph_0$ matrix whose columns are an enumeration of $\mathbb{Z}^{(q)}$. (The order does not matter. There are $2^{\aleph_0}$ matrices B of this type, one for each permutation of the $\aleph_0$ elements of $\mathbb{Z}^{(q)}$.) Next, let τ be an infinite cardinal, and let $\mathfrak{B} = \mathfrak{B}(q,\tau)$ be the $q \times \tau\aleph_0$ matrix consisting of τ copies of B arranged "horizontally". Of course $\tau\aleph_0 = \tau$, so we can regard $\mathfrak{B}$ as a homomorphism $\mathbb{Z}^{(\tau)} \to \mathbb{Z}^{(q)}$. We define $\mathfrak{H}(q,\tau) := \mathbb{N}^{(\tau)} \cap \ker(\mathfrak{B}(q,\tau))$. Finally, we put $\mathfrak{H}(0,\tau) = \mathbb{N}^{(\tau)}$. These are the monoids we will obtain as $\mathrm{V}(R\text{-mod})$ for the rings that are not Dedekind-like. By the next lemma and Lemma 1.1, it follows that the isomorphism class of $\mathfrak{H}(q,\tau)$ is independent of all choices:

Lemma 2.1 *Let q be a non-negative integer, let τ be an infinite cardinal, and let $\mathfrak{D}$ be a $q \times \tau$ matrix with entries in $\mathbb{Z}$. Assume that, for each $\delta \in \mathbb{Z}^{(q)}$, $\mathfrak{D}$ has τ distinct columns that coincide with δ (equivalently, $\mathfrak{D} = \mathfrak{B}(q,\tau)$ up to column permutations). Let $H = \mathbb{N}^{(\tau)} \cap \ker(\mathfrak{D}{:}\mathbb{Z}^{(\tau)} \to \mathbb{Z}^{(q)}) \cong \mathfrak{H}(q,\tau)$. Then:*

(1) *$\mathfrak{D}{:}\mathbb{Z}^{(\tau)} \to \mathbb{Z}^{(q)}$ is surjective.*
(2) *The inclusion $H \hookrightarrow \mathbb{N}^{(\tau)}$ is a divisor theory.*
(3) $\ker(\mathfrak{D}) = \mathcal{Q}(H)$.
(4) $\mathrm{Cl}(H) = \mathbb{Z}^{(q)}$.
(5) $|\mathfrak{A}(\alpha)| = \tau$ *for each* $\alpha \in \mathrm{Cl}(H)$.

Proof. Since every vector δ of $\mathbb{Z}^{(q)}$ occurs as a column of $\mathfrak{D}$, (1) is clear. To prove (2), let β be an arbitrary element of $\mathbb{N}^{(\tau)}$. Select distinct elements $t, u \in \tau - \mathrm{Supp}(\beta)$ such that the tth and u^{th} columns of $\mathfrak{D}$ both coincide with $-\mathfrak{D}\beta$. Letting e_t and e_u denote the unit vectors of $\mathbb{N}^{(\tau)}$ with supports $\{t\}$ and $\{u\}$, respectively, we see that $\beta + e_t$ and $\beta + e_u$ are both in H. Clearly β is the greatest lower bound of $\beta + e_t$ and $\beta + e_u$ in the monoid $\mathbb{N}^{(\tau)}$.

The inclusion "$\supseteq$" in (3) is clear. For the reverse inclusion, let $\alpha \in \ker(\mathfrak{D})$, and write $\alpha = \beta - \gamma$, with $\beta, \gamma \in \mathbb{N}^{(\tau)}$. Choose, as in the proof of (2), a column index $t \in \tau$ such that $\beta + e_t \in H$. Then $\gamma + e_t = \beta + e_t - \alpha \in \mathbb{N}^{(\tau)} \cap \ker(\mathfrak{D}) = H$. Therefore $\alpha = (\beta + e_t) - (\gamma + e_t) \in \mathcal{Q}(H)$. This proves (3), and assertion (4) follows immediately from (1), (2) and (3).

For (5), let $\alpha \in \mathrm{Cl}(H) = \mathbb{Z}^{(q)}$. By hypothesis, there are τ column indices t such that the tth column coincides with α. Each e_t is then an atom of $\mathbb{N}^{(\tau)}$ in the divisor class α.

For some Dedekind-like rings, we will obtain another monoid. Let E be the $1 \times \aleph_0$ matrix $[1\ -1\ 1\ -1\ 1\ -1\ \cdots]$, and put $\mathfrak{H}_1 := \mathbb{N}^{(\aleph_0)} \cap \ker(E : \mathbb{Z}^{(\aleph_0)} \to \mathbb{Z})$.

In order to include Cohen-Macaulay rings with non-zero nilpotents in our theorem, we introduce the class $\mathfrak{F}(R)$ of *generically free* modules—finitely generated R-modules M such that M_P is R_P-free for every minimal prime ideal P. Of course, if R is reduced, each R_P is a field, so $\mathfrak{F}(R) = R$-mod. Moreover, if R is one-dimensional, Noetherian and reduced, then R is automatically Cohen-Macaulay. If $P_1, \ldots, P_s$ are the minimal prime ideals of R and $M_{P_i} \cong R_{P_i}^{(r_i)}$ for each i, we call $(r_1, \ldots, r_s)$ the *rank* of M.

Theorem 2.2 (Main Theorem) *Suppose $(R, \mathfrak{m}, k)$ is a one-dimensional Noetherian Cohen-Macaulay local ring. Let $q := \mathrm{spl}(R)$ be the splitting number of R, and let $\tau = \tau(R) = \max\{|k|, \aleph_0\}$.*

(1) *If R is not Dedekind-like, then $\mathrm{V}(\mathfrak{F}(R)) \cong \mathfrak{H}(q, \tau)$.*
(2) *If R is a discrete valuation ring, then $\mathrm{V}(\mathfrak{F}(R)) = \mathrm{V}(R\text{-mod}) \cong \mathbb{N}^{(\aleph_0)}$.*
(3) *If R is Dedekind-like but not a discrete valuation ring, and if $q = 0$, then $\mathrm{V}(\mathfrak{F}(R)) = \mathrm{V}(R\text{-mod}) \cong \mathbb{N}^{(\tau)}$.*
(4) *If R is Dedekind-like and $q > 0$, then $q = 1$ and $\mathrm{V}(\mathfrak{F}(R)) = \mathrm{V}(R\text{-mod}) \cong \mathbb{N}^{(\tau)} \oplus \mathfrak{H}_1$.*

In every case, $\mathrm{Cl}(\mathrm{V}(\mathfrak{F}(R))) \cong \mathbb{Z}^{(q)}$.

The proof will be an easy consequence of the following four lemmas.

Lemma 2.3 *Let $(R, \mathfrak{m}, k)$ be a one-dimensional Noetherian Cohen-Macaulay local ring, and let $\tau = \max\{|k|, \aleph_0\}$. Then $|\mathrm{V}(\mathfrak{F}(R))| \leq \tau$.*

Proof. An easy induction argument shows that every finite-length R-module has cardinality at most τ. Now, for each positive integer ℓ, let $\mathfrak{M}_\ell$ be the class of modules of length at most ℓ. We claim that $|\mathrm{V}(\mathfrak{M}_\ell)| \leq \tau$ for each ℓ. Since $|\mathrm{V}(\mathfrak{M}_1)| = 2$, we may assume that $\ell \geq 2$ and proceed by induction. Given a module M of length ℓ, we can choose a short exact sequence $0 \to A \to M \to B \to 0$, in which A and B are strictly shorter than M. By the induction hypothesis there are at most τ choices for A and for B. Moreover, $\mathrm{Ext}_R^1(B, A)$, being a module of finite length, has cardinality at most τ. Thus the claim follows. Letting ℓ vary, we see that there are at most τ non-isomorphic modules of finite length.

Next, let K be the total quotient ring of R, that is, the localization of R with respect to the complement of the union of the associated primes $P_1, \ldots, P_s$ of R. As R is Cohen-Macaulay, the P_i are exactly the minimal primes of R, and thus $K = R_{P_1} \times \cdots \times R_{P_s}$. Let S_i be the image of the natural homomorphism $R \to R_{P_i}$ (inclusion followed by projection).

Now let M be a torsion-free, generically free R-module of rank $(r_1, \dots, r_s)$. Then we have an embedding $M \to K \otimes_R M \cong R_{P_1}^{(r_1)} \times \cdots \times R_{P_s}^{(r_s)}$. Clearing denominators, we see that there is an injective R-homomorphism $j{:}M \to F{:}= S_1^{(r_1)} \times \cdots \times S_s^{(r_s)}$, with $\operatorname{coker}(j)$ a torsion R-module (hence of finite length). Thus M is the kernel of a homomorphism $F \to T$, where T is of finite length. There are only countably many choices for F and only τ choices for T, by the first paragraph of the proof. Since $\operatorname{Hom}_R(F,T)$ has finite length, it has cardinality at most τ, and it follows that there are at most τ choices for M.

Finally, let M be an arbitrary generically free module, and let T be the torsion submodule of M. Then T has finite length and M/T is generically free, so there are at most τ possibilities for T and for M/T. Since $\operatorname{Ext}^1_R(M/T,T)$ has finite length, its cardinality is at most τ, and it follows that there are at most τ possibilities for M, up to isomorphism.

The following theorem is a sharpening the main theorem of [12]. In the next section we will give a sketch of the additional arguments needed to prove the version given here.

Lemma 2.4 *Let $(R, \mathfrak{m}, k)$ be a one-dimensional Noetherian Cohen-Macaulay local ring, and assume that R is not Dedekind-like. Let $P_1, \dots, P_s$ be the minimal prime ideals of R, let $(r_1, \dots, r_s)$ be an s-tuple of non-negative integers, and let $\tau = \max\{|k|, \aleph_0\}$. Then there are τ pairwise non-isomorphic indecomposable generically free R-modules M of rank $(r_1, \dots, r_s)$.*

The next result, which tells us how $\mathrm{V}(\mathfrak{F}(R))$ sits inside $\mathrm{V}(\mathfrak{F}(\widehat{R}))$, is a special case of [18, Theorem 3.4].

Lemma 2.5 *Let $(R, \mathfrak{m}, k)$ be a one-dimensional Noetherian Cohen-Macaulay local ring with $\mathfrak{m}$-adic completion $\widehat{R}$. Let K and L be the total quotient rings of R and $\widehat{R}$, respectively. The following conditions on a generically free $\widehat{R}$-module N are equivalent:*

(1) *There is a finitely generated R-module M (necessarily generically free) such that $N \cong \widehat{R} \otimes_R M$.*
(2) *There is a finitely generated K-module W such that $L \otimes_{\widehat{R}} N \cong L \otimes_K W$.*
(3) $\operatorname{rank}_{\widehat{R}_P}(N_P) = \operatorname{rank}_{\widehat{R}_Q}(N_Q)$ *whenever P and Q are minimal primes of $\widehat{R}$ lying over the same prime of R.*

In [15] Klingler and Levy classify all finitely generated modules over non-exceptional Dedekind-like rings. The next lemma distills the aspects of this classification that we shall need in our proof of the Main Theorem. The key property we need for Dedekind-like *domains*—that there are $\max\{|k|, \aleph_0\}$ non-isomorphic finite-length modules—holds even for exceptional Dedekind-like rings. Since the classification in [15] does not apply to exceptional Dedekind-like rings, we include a sketch of this fact.

Lemma 2.6 *Let $(R, \mathfrak{m}, k)$ be a local Dedekind-like ring. Then R is reduced with at most two minimal prime ideals. Put $\tau := \max\{|k|, \aleph_0\}$.*

(1) *If R is not a discrete valuation ring, then R has τ pairwise non-isomorphic indecomposable finite-length modules.*
(2) *Suppose R is not a domain. For each pair (a, b) of non-negative integers, let $G(a, b)$ be the cardinality of the set of isomorphism classes of finitely generated indecomposable R-modules of rank (a, b). Then $G(0,0) = \tau$; $G(a, b) = \aleph_0$ if $(a, b) \in \{(0,1), (1,0), (1,1)\}$; and $G(a, b) = 0$ otherwise.*

Proof. To prove (1) we may assume that R is $\mathfrak{m}$-adically complete, since R and its completion $\widehat{R}$ have exactly the same finite-length modules. Since the finite-length R-modules $R/\mathfrak{m}^n$ are indecomposable and pairwise non-isomorphic, we may assume that k is infinite. If the normalization $\overline{R}$ is *not* a discrete valuation ring (i.e., R is *strictly split* in the terminology of [15]), we appeal to Section 7 of [15], in particular, [15, Theorem 7.3, (ii)]. Even if we choose the "blocking matrix" L to be 1×1, we still obtain $|k|$ pairwise non-isomorphic "block-cycle" indecomposables, and these are all of finite length. (Cf. [15, Section 6] for the connection between matrices and modules.)

Still assuming that R is complete and k is infinite, we now suppose that $\overline{R}$ is a discrete valuation ring properly containing R. Then $F := \overline{R}/\mathfrak{m}$ is a (possibly inseparable) field extension of degree 2 over k. Write $\mathfrak{m} = \overline{R}\pi$, where $\pi \in \mathfrak{m}$. Given $u \in \overline{R}^\times$ (the group of units of $\overline{R}$), put $M_u := R/R\pi u$, an indecomposable R-module of finite length. Suppose $u, v \in \overline{R}^\times$ and $M_u \cong M_v$. We claim that u and v are in the same coset of $\overline{R}^\times/R^\times = F^\times/k^\times$. Since $|F^\times/k^\times| = |k|$, this will complete the proof of (1).

Choose an R-isomorphism $\phi: M_u \to M_v$. Since $R \twoheadrightarrow M_u$ and $R \twoheadrightarrow M_v$ are projective covers, we can lift ϕ to an automorphism $\Phi: R \to R$. Then Φ is multiplication by some element $t \in R^\times$, with $Rt\pi u = R\pi v$. Write $\pi v = st\pi u$, with $s \in R$. Then $v = stu$, and $st \notin \mathfrak{m}$ (lest $v \in \mathfrak{m}$). Thus $v \in R^\times u$, as desired.

Now we prove (2). By (1) and Lemma 2.3, $G(0,0) = \tau$. Also, if $(a,b) \notin \{(0,0), (0,1), (1,0), (1,1)\}$, then $G(a,b) = 0$ by [16, Lemma 16.2 and Corollary 16.4]. To complete the proof, we need only show that $G(0,1), G(1,0)$ and $G(1,1)$ are all countably infinite. Since both R and $\widehat{R}$ have exactly two minimal primes, Lemma 2.5 implies that every finitely generated $\widehat{R}$ module is extended from a finitely generated R-module. Therefore we may again assume that R is complete. The desired result now follows from the classification given in [15, Section 3].

Proof of the Main Theorem. Suppose first that $q = 0$. Then every minimal prime ideal of R has a unique minimal prime of $\widehat{R}$ lying over it. By Lemma 2.5, the homomorphism $\mathrm{V}(\mathfrak{F}(R)) \to \mathrm{V}(\mathfrak{F}(\widehat{R}))$ is bijective. Therefore $\mathrm{V}(\mathfrak{F}(R))$ is a free monoid, and $\mathrm{Cl}(\mathrm{V}(\mathfrak{F}(R))) = 0$. To complete the proof in cases (2) and (3), and in the case $q = 0$ of (1), we need only show that the number of isomorphism classes of indecomposable generically free modules is $\aleph_0$ if R

is a discrete valuation ring and τ otherwise. If R is a discrete valuation ring, the indecomposable modules are $R/\mathfrak{m}^n, 1 \le n \le \infty$. If R is not a discrete valuation ring, we quote Lemmas 2.3, 2.4 and 2.6 (Part 1).

Next, assume that $q > 0$ and R is Dedekind-like. Then R is a domain and $\widehat{R}$ has exactly two minimal primes. Using part (2) of Lemma 2.6, we let A_t, B_n, C_n, D_n ($t \in \tau$, $n \ge 1$) be the indecomposable finitely generated $\widehat{R}$-modules of rank $(0,0), (0,1), (1,0), (1,1)$, respectively. By Lemma 2.5, an $\widehat{R}$-module $\bigoplus_t A_t^{(a_t)} \oplus \bigoplus_n B_n^{(b_n)} \oplus \bigoplus_n C_n^{(c_n)} \oplus \bigoplus_n D_n^{(d_n)}$ is extended from an R-module if and only if $\sum_n (b_n - c_n) = 0$, so it follows immediately that $\mathrm{V}(\mathfrak{F}(R)) = \mathrm{V}(R\text{-mod}) \cong \mathbb{N}^{(\tau)} \oplus \mathfrak{H}_1$. It is easy to see that the inclusion $\mathrm{V}(R\text{-mod}) \hookrightarrow \mathrm{V}(\widehat{R}\text{-mod})$ is a divisor theory, and it follows that $\mathrm{Cl}(\mathrm{V}(R\text{-mod})) \cong \mathbb{Z}$, completing the proof in case (4).

Finally, assume that $q > 0$ and R is not Dedekind-like. Then $\widehat{R}$ is not Dedekind-like either [16, Lemma 11.8]. Let $P_1, \dots, P_s$ be the minimal prime ideals of R, and let $Q_{i,1}, \dots, Q_{i,t_i}$ be the primes of $\widehat{R}$ lying over P_i for $i = 1, \dots, s$. Then $q{:} = \mathrm{spl}(R) = t_1 + \dots + t_s - s = (t_1 - 1) + \dots + (t_s - 1)$. Let $\mathfrak{L}$ be the $q \times \tau$ matrix whose columns are indexed by the indecomposable finitely generated generically free $\widehat{R}$-modules, and whose columns are given by the following scheme: Let W be an indecomposable generically free $\widehat{R}$-module, of rank $r_{i,j}$ at $Q_{i,j}$. The column indexed by the module W is then the transpose of the array

$$[r_{1,1} - r_{1,2}, \dots, r_{1,1} - r_{1,t_1}, \dots\dots, r_{s,1} - r_{s,2}, \dots, r_{s,1} - r_{s,t_s}].$$

Using Lemma 2.5, we see that $\mathrm{V}(\mathfrak{F}(R)) \cong H{:} = \mathbb{N}^{(\tau)} \cap \ker(\mathfrak{L}{:}\mathbb{Z}^{(\tau)} \to \mathbb{Z}^{(q)})$.

Next, we verify the hypotheses of Lemma 2.1 for the matrix $\mathfrak{L}$. Let α be an arbitrary element of $\mathbb{Z}^{(q)}$. We want to show that $\mathfrak{L}$ has τ columns that coincide with α. Write

$$\alpha{:} = [a_{1,2}, \dots, a_{1,t_1}, \dots\dots, a_{s,2}, \dots, a_{s,t_s}]^{\mathrm{tr}}.$$

Choose a positive integer b greater than each entry of α. Put $r_{i,1}{:} = b$, $i = 1, \dots, s$, and put $r_{i,j}{:} = b - a_{i,j}$, $i = 1, \dots, s$, $j = 2, \dots, t_i$. By Lemma 2.4 there are τ indecomposable generically free $\widehat{R}$-modules W having rank $r_{i,j}$ at the prime $Q_{i,j}$. The column of $\mathfrak{L}$ corresponding to such a W is precisely α, as desired. Now Lemma 2.1 completes the proof of the Main Theorem.

3 Proof of Lemma 2.4

The main theorem of [12] produces infinitely many indecomposable generically free modules of prescribed rank, over any one-dimensional Noetherian Cohen-Macaulay local ring $(R, \mathfrak{m}, k)$ that is not Dedekind-like. If k is uncountable, we need to modify that proof, in order to obtain $|k|$ non-isomorphic indecomposables. The proof of [12, Theorem 1.2] is divided into two cases. We give a careful sketch of the modifications needed in the first case ([12, Theorem 2.3], called the ramified case); we give a brief summary of the modifications needed in the second case ([12, Theorem 2.4], called the unramified case). For consistency with the notation of [12], all matrices and homomorphisms in this section act on the *right.*

Let $(R, \mathfrak{m}, k)$ be a one-dimensional Noetherian Cohen-Macaulay local ring which is not Dedekind-like, and let $\overline{R}$ be the normalization of R. In [12, Theorem 2.3], we suppose that there is a local ring $(\Omega, \mathfrak{n}, k)$ with $R \subsetneq \Omega \subseteq \overline{R}$, such that $\mathfrak{m}$ is the conductor of R in Ω, $\mathfrak{m} \subsetneq \mathfrak{n}$, and Ω is generated by 2 elements as an R-module. Let $P_1, \dots, P_t$ be the minimal prime ideals of R, and suppose that $r_1, \dots, r_t$ are non-negative integers. We want to build a family $(M_\kappa)_{\kappa \in k}$ of indecomposable, pairwise non-isomorphic, finitely generated R-modules, each with rank $(r_1, \dots, r_t)$. Let $Q_1, \dots, Q_t$ be the minimal prime ideals of Ω. Pick $\delta \in \mathfrak{n} - (\mathfrak{m} \cup Q_1 \cup \dots \cup Q_t)$, where $\delta^2 \in \mathfrak{m}$, as described in the first paragraph of the proof of [12, Theorem 2.3], and let $\bar{\delta}$ denote the image of δ in $\Omega/\mathfrak{m}$.

The following construction of the module X is described in detail in the third and fourth paragraphs of the proof of [12, Theorem 2.3]; we recall the main steps of the construction. For each $i, 1 \leq i \leq t$, choose an element $\lambda_i \in \mathcal{Q}(R/P_i)^\times \cap R$. Fixing an integer $n \geq \max\{r_1, \dots, r_t\}$, for each index j, $1 \leq j \leq n$, set $C_j = \{1 \leq i \leq t \mid r_i \geq j\}$, and define $w_{C_j} = \delta^8 \sum_{i \notin C_j} \lambda_i \in \delta^8 \Omega$. Put $X_1 = \Omega/w_{C_1}\Omega \oplus \dots \oplus \Omega/w_{C_n}\Omega$, $X_2 = (\Omega/\delta^4\mathfrak{m})^{(n)}$, $X_3 = (\Omega/\delta^2\mathfrak{m})^{(n)}$, and $X_4 = (\Omega/\mathfrak{m})^{(n)}$, and let $X = X_1 \oplus \dots \oplus X_4$.

The next step is to define an R-module S by the following pullback square (see [12, (2.3.2)]):

$$\begin{array}{ccc} S & \xrightarrow{\subseteq} & X \\ {\scriptstyle \pi}\downarrow & & \downarrow{\scriptstyle \nu} \\ k^{(4n)} & \xrightarrow{A} & (\Omega/\mathfrak{m})^{(4n)} \end{array} \tag{1}$$

Here the elements of $k^{(4n)}$ are viewed as row vectors, subjected to right multiplication by the block matrix

$$A := \begin{pmatrix} I & 0 & 0 & 0 \\ \bar{\delta}I & 0 & I & 0 \\ 0 & I & 0 & 0 \\ 0 & \bar{\delta}I & \bar{\delta}I & I \end{pmatrix},$$

where I is the $n \times n$ identity matrix over k.

As in [12, (2.3.3)], let $\sigma = \left(\begin{smallmatrix} 0 & \sigma_2 & 0 & 0 \\ 0 & 0 & \sigma_3 & 0 \end{smallmatrix}\right) : (\Omega/\mathfrak{m})^{2n} \to X$ denote the injective Ω-homomorphism, where $\sigma_2 : (\Omega/\mathfrak{m})^{(n)} \to X_2$ and $\sigma_3 : (\Omega/\mathfrak{m})^{(n)} \to X_3$ are given by multiplication by δ^4 and δ^2, respectively. Next, let $B_\kappa : k^{(2n)} \to (\Omega/\mathfrak{m}\Omega)^{(2n)}$ be right multiplication by the matrix $B_\kappa = \left(\begin{smallmatrix} I & \bar{\delta}(\kappa I + H) \\ 0 & I \end{smallmatrix}\right)$, where I is the $n \times n$ identity matrix, and H is the $n \times n$ Jordan block

$$H = \begin{pmatrix} 0 & 1 & 0 & \dots & 0 \\ 0 & 0 & 1 & \dots & 0 \\ \vdots & \vdots & \vdots & \ddots & \vdots \\ 0 & 0 & 0 & \dots & 1 \\ 0 & 0 & 0 & \dots & 0 \end{pmatrix}.$$

Finally, we define $M_\kappa := S/\operatorname{im}(\tau_\kappa)$, where $\tau_\kappa = B_\kappa \sigma$. Note that the matrix B in [12, (2.3.4)] is equal to B_0. This is the only modification in the construction of the indecomposable modules we make. By adding κI to H, we "parameterize" the matrix B by the residue field k.

By exactly the same argument as in the paragraph after [12, (2.3.5)] it follows that the rank of M_κ is $(r_1, \dots, r_t)$. To see that the M_κ are indecomposable and pairwise non-isomorphic, suppose $f: M_\kappa \to M_{\kappa'}$ is an R-homomorphism. Note that $\operatorname{im}(\tau_\kappa)$ and $\operatorname{im}(\tau_{\kappa'})$ are contained in $\mathfrak{m}S$, and an easy argument shows that the R-submodules $\operatorname{im}(B_\kappa)$ and $\operatorname{im}(B_{\kappa'})$ of $(\Omega/\mathfrak{m})^{(2n)}$ contain no non-zero Ω-submodules of $(\Omega/\mathfrak{m})^{(2n)}$. It follows that S is a *separated cover* of M_κ and $M_{\kappa'}$ [15, Lemma 4.9]. As described in the proof of [12, Theorem 2.3], f lifts to an R-homomorphism $\theta: S \to S$, and θ extends to an Ω-homomorphism $\theta': X \to X$. Exactly as in the given proof, one shows that θ' modulo the radical of Ω is right multiplication by a block upper-triangular matrix over k with four identical diagonal $n \times n$ blocks Δ.

Since the map $\theta: S \to S$ induces the R-homomorphism f from $M_\kappa = S/\operatorname{im}(\tau_\kappa)$ to $M_{\kappa'} = S/\operatorname{im}(\tau_{\kappa'})$, it follows that $(\operatorname{im}(\tau_\kappa))\theta \subseteq \operatorname{im}(\tau_{\kappa'})$. Therefore θ can be lifted to an R-homomorphism $\tilde{\theta}: k^{(2n)} \to k^{(2n)}$ such that $\tau_\kappa \theta = \tilde{\theta} \tau_{\kappa'}$. Moreover, since B_κ is invertible, $\operatorname{im}(B_\kappa \sigma)$ generates $\operatorname{im}(\sigma)$ as an Ω-submodule of X. Since the map $\theta': X \to X$ extends θ, it follows that $(\operatorname{im}(\sigma))\theta' \subseteq \operatorname{im}(\sigma)$. Therefore θ' lifts to an Ω-homomorphism $\tilde{\theta}': (\Omega/\mathfrak{m})^{(2n)} \to (\Omega/\mathfrak{m})^{(2n)}$ such that $\sigma\theta' = \tilde{\theta}'\sigma$. These maps yield a commutative cube:

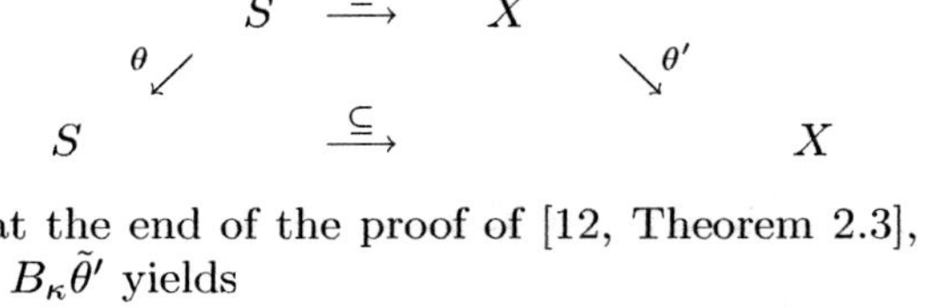

$$\begin{array}{ccccccc}
k^{(2n)} & & & \xrightarrow{B_{\kappa'}} & & & (\Omega/\mathfrak{m})^{(2n)} \\
& \nwarrow^{\tilde{\theta}} & & & & \tilde{\theta}' \nearrow & \\
& & k^{(2n)} & \xrightarrow{B_\kappa} & (\Omega/\mathfrak{m})^{(n)} & & \\
\tau_{\kappa'} \downarrow & & \tau_\kappa \downarrow & & \downarrow \sigma & & \downarrow \sigma \\
& & S & \xrightarrow{\subseteq} & X & & \\
& {}^{\theta}\swarrow & & & & \searrow^{\theta'} & \\
S & & & \xrightarrow{\subseteq} & & & X
\end{array}$$

By arguing as at the end of the proof of [12, Theorem 2.3], we see that the identity $\tilde{\theta} B_{\kappa'} = B_\kappa \tilde{\theta}'$ yields

$$\Delta(\kappa' I + H) = (\kappa I + H)\Delta. \tag{2}$$

On the one hand, if $\kappa = \kappa'$, then (2) implies that $\Delta H = H\Delta$, and the argument proceeds exactly as in the last two paragraphs of the proof of [12,

Theorem 2.3], showing that Δ is upper-triangular with constant diagonal. If, now, f is a non-surjective idempotent endomorphism of M_κ, then $\mathrm{im}(f)$ is contained in $\mathfrak{m}M_\kappa$, and hence f must be zero. This shows that M_κ is indecomposable. On the other hand, if $\kappa \neq \kappa'$, then an inductive argument shows that (2) implies $\Delta = 0$. Then, by Nakayama's Lemma, we can conclude that θ' is not surjective, so θ is not surjective either, since $\theta' = \theta \otimes \mathrm{id}_X$. This implies that f is not surjective, because any surjective f lifts to a surjective θ [15, Theorem 4.12]. Therefore M_κ and $M_{\kappa'}$ are not isomorphic, and the proof of Lemma 2.4 is complete in the ramified case.

If the special hypothesis of [12, Theorem 2.3] does not hold, that is, if there does not exist a local ring $(\Omega, \mathfrak{n}, k)$ with $R \subsetneq \Omega \subseteq \overline{R}$, such that $\mathfrak{m}$ is the conductor of R in Ω, $\mathfrak{m} \subsetneq \mathfrak{n}$, and Ω is generated by 2 elements as an R-module, then [12, Proposition 2.2] proves that R is reduced, its normalization $\overline{R}$ is finitely generated as an R-module, $\mathfrak{m}$ is the Jacobson radical of $\overline{R}$, $\mathfrak{m}$ is the conductor of R in $\overline{R}$, and $\dim_k(\overline{R}/\mathfrak{m}) \geq 3$. This case is handled in [12, Theorem 2.4], where it is further divided into three subcases. The modifications needed to produce, for each possible rank $(r_1, \ldots, r_t)$, a family of $|k|$ indecomposable, pairwise non-isomorphic, finitely generated R-modules are quite similar to those sketched above for the ramified case in [12, Theorem 2.3]. Here we only mention the changes needed from that sketch, leaving the details to the interested reader.

In each of the three subcases of [12, Theorem 2.4], we begin by finding an appropriate ring Ω between R and $\overline{R}$ over which to work. We continue by defining an Ω-module X and then an R-module S defined by the conductor square (1), for the choice of matrix A as given in [12, Theorem 2.4]. In each of the three subcases, we define an injection $\sigma: (\Omega/\mathfrak{m})^{2n} \to X$, parameterize the matrix B in [12, Theorem 2.4] by $\kappa \in k$ to get B_κ, and let $M_\kappa = S/\mathrm{im}(B_\kappa \sigma)$. The matrices B_κ, in the three subcases, are defined as follows:

(1) **Basic case.** Set $B_\kappa = \begin{pmatrix} \kappa I + H & I \\ \bar{u}^2 I & \bar{u} I \end{pmatrix}$, where $1, \bar{u}, \bar{u}^2$ are units of $\Omega/\mathfrak{m}$ linearly independent over k.
(2) **Inseparable case.** Set $B_\kappa = \begin{pmatrix} \kappa I + H & I \\ \bar{v} I & \bar{u} I \end{pmatrix}$, where $1, \bar{u}, \bar{v}, \bar{u}\bar{v}$ are units of $\Omega/\mathfrak{m}$ linearly independent over k.
(3) **Small residue field case.** Set $B_\kappa = \bar{e}_1 \begin{pmatrix} I & 0 \\ 0 & I \end{pmatrix} + \bar{e}_2 \begin{pmatrix} 0 & I \\ I & 0 \end{pmatrix} + \bar{e}_3 \begin{pmatrix} I & \kappa I + H \\ 0 & I \end{pmatrix}$, where $\bar{e}_1, \bar{e}_2, \bar{e}_3$ are primitive idempotents in the ring $\Omega/\mathfrak{m}$.

In each of these cases, the proof proceeds as in [12, Theorem 2.4], with changes similar to those made in the ramified case described above.

4 Sets of lengths

Let H be an atomic monoid and $a \in H$. The set

$$\mathsf{L}(a) = \mathsf{L}_H(a) = \{n \in \mathbb{N} \mid a \text{ is a sum of } n \text{ atoms of } H\}$$

is called the *set of lengths* of a, and $\mathfrak{L}(H) = \{\mathsf{L}(a) \mid a \in H\}$ is called the *system of sets of lengths* of H. Sets of lengths and related invariants (e.g. the elasticity of an atomic monoid) are frequently studied objects in the theory of non-unique factorizations in integral domains and monoids. The reader is referred to [1], [3], [9] and [10] for recent results in this area of research. For certain classes of integral domains and monoids (e.g. for finitely generated monoids, orders in algebraic number fields, congruence monoids in Dedekind domains, and certain higher-dimensional algebras over finite fields), sets of lengths have the following special structure: they are, up to bounded initial and final segments, a union of arithmetical progressions with bounded distance. The interested reader is referred to the monograph [10] for details. In strong contrast to these finiteness results, F. Kainrath proved the following theorem on sets of lengths of Krull monoids with infinite class group:

Theorem 4.1 *Let H be a Krull monoid with infinite class group G such that $|\mathfrak{A}(g)| > 0$ for all $g \in G$. Then, for any non-empty finite subset $L \subseteq \mathbb{N}_{\geq 2} := \{n \in \mathbb{N} \mid n \geq 2\}$, there exists $h \in H$ such that $\mathsf{L}(h) = L$.*

Proof. See [13].

Here we consider the system of sets of lengths of the monoids $\mathrm{V}(\mathfrak{F}(R))$:

Theorem 4.2 *Let $(R, \mathfrak{m}, k)$ be a one-dimensional Noetherian Cohen-Macaulay local ring.*

(1) *If* $\mathrm{spl}(R) = 0$, *then* $\mathrm{V}(\mathfrak{F}(R))$ *is factorial (that is, free).*
(2) *If* $\mathrm{spl}(R) > 0$ *and R is Dedekind-like, then* $\mathrm{V}(\mathfrak{F}(R))$ *is half factorial (that is, $|L(h)| = 1$ for all $h \in \mathrm{V}(\mathfrak{F}(R))$) but not factorial.*
(3) *If* $\mathrm{spl}(R) > 0$ *and R is not Dedekind-like, then*

$$\mathfrak{L}(\mathrm{V}(\mathfrak{F}(R))) = \{L \subseteq \mathbb{N}_{\geq 2} \mid L \textit{ is non-empty and finite}\}.$$

Proof. That $\mathrm{V}(\mathfrak{F}(R))$ is half factorial when $\mathrm{spl}(R) > 0$ and R is Dedekind-like follows, for example, from [16, Corollary 16.4(i)]. The rest of the theorem follows easily from Lemma 2.1, our Main Theorem, and Theorem 4.1.

5 Torsion-free modules

In this section we shall consider the monoid of finitely generated torsion-free modules in the case that $\widehat{R}$ has no non-zero nilpotents.

Setup. Let $(R, \mathfrak{m}, k)$ denote a one-dimensional Noetherian local ring whose completion $\widehat{R}$ has no non-zero nilpotents. We denote by $\mathrm{V}(\mathfrak{C}(R))$ the monoid of all isomorphism classes of finitely generated torsion-free R-modules. Let $p_1, \ldots, p_s$ denote the minimal primes of R. Then the set of minimal primes of $\widehat{R}$ is a disjoint union $\mathfrak{P}_1 \cup \cdots \cup \mathfrak{P}_s$, where each $P \in \mathfrak{P}_i$ contracts to p_i. We denote the rank function by $\mathrm{rank}_R \colon \mathrm{V}(\mathfrak{C}(R)) \longrightarrow \mathbb{N}^{(s)}$.

The main difficulty we encountered in trying to describe the monoid $\mathrm{V}(\mathfrak{C}(R))$ is that the analogue of Lemma 2.4 can fail. For example, let $R := \mathbb{C}[[X,Y]]/(Y^4 - X^5)Y$. Then R has two minimal primes $P_1 := (y^4 - x^5)$ and $P_2 := (y)$. We claim that there is no indecomposable torsion-free R-module of rank $(0,s)$ if $s \geq 2$. For suppose M is a torsion-free R-module with rank $(0,s)$, with $s \geq 2$. Then $M_{P_1} = 0$. Also $(P_2)_{P_2} = 0$, so P_2M vanishes at both minimal primes. Since P_2M is torsion-free, it follows that $P_2M = 0$, that is, M is a module over the discrete valuation ring R/P_2. Since $s \geq 2$, M decomposes. On the other hand, it follows from [24, Lemma 2.2] that for every pair (r,s) of non-negative integers with $0 < r \geq s$ there is an indecomposable torsion-free R-module of with rank (r,s). We have been unable to determine exactly which rank functions can occur for indecomposable finitely generated torsion-free R-modules. For example, is there one of rank $(1,2)$? Are there infinitely many?

A related problem, whose answer seems to depend on which ranks can occur, is to determine when the divisor homomorphism $\mathrm{V}(\mathfrak{C}(R)) \to \mathrm{V}(\mathfrak{C}(\widehat{R}))$ is actually a divisor theory. Suppose, for example, that R is a domain and that $\widehat{R}$ has 2 minimal primes P_1 and P_2. Assume that $\widehat{R}/P_1$ and $\widehat{R}/P_2$ are both discrete valuation rings. Then $S := \widehat{R}/P_1 \times \widehat{R}/P_2$ is the integral closure $\widehat{R}$. Since S is 2-generated as an $\widehat{R}$-module, $(1,0)$, $(0,1)$ and $(1,1)$ are the only possible ranks of indecomposable $\widehat{R}$-modules (by Bass's decomposition theorem, [2, Section 7]). Furthermore, there is only one isomorphism class of $\widehat{R}$-modules having rank $(1,0)$ (and the same for $(0,1)$). Thus it is easy to see that $\mathrm{V}(\mathfrak{C}(R)) \to \mathrm{V}(\mathfrak{C}(\widehat{R}))$ is not a divisor theory in this case.

If R is a domain whose completion $\widehat{R}$ has two minimal primes P_1, P_2, with exactly *one* of the $\widehat{R}/P_i$ a discrete valuation domain (essentially the case discussed in the paragraph after the **Setup**), we have been unable to determine whether or not the map $\mathrm{V}(\mathfrak{C}(R)) \to \mathrm{V}(\mathfrak{C}(\widehat{R}))$ is a divisor theory. The following theorem (in conjunction with the negative result in the preceding paragraph) gives the answer in every other case:

Theorem 5.1 *Suppose R is as in the* **Setup** *and that at least one of the following conditions is satisfied:*

(1) *The number of minimal primes of $\widehat{R}$ is different from* 2.
(2) *R is not a domain, and $\widehat{R}$ has exactly* 2 *minimal primes.*
(3) *R is a domain, $\widehat{R}$ has exactly* 2 *minimal primes P_1 and P_2 and neither $\widehat{R}/P_1$ nor $\widehat{R}/P_2$ is a discrete valuation domain.*

Then the natural homomorphism $\mathrm{V}(\mathfrak{C}(R)) \to \mathrm{V}(\mathfrak{C}(\widehat{R}))$ *is a divisor theory.*

Proof. Denote by $P_1, \ldots, P_t$ the minimal primes of $\widehat{R}$, and let $[A] \in \mathrm{V}(\mathfrak{C}(\widehat{R}))$. We have to display $[A]$ as the greatest lower bound of finitely many torsion-free $\widehat{R}$-modules which are extended from R-modules. If R and $\widehat{R}$ have the same number of minimal primes, then A itself is extended, by Lemma 2.5. Thus the theorem holds in case (2) and also if $t = 1$.

Suppose next that $t \geq 3$. For $1 \leq i \leq t$ let $U_i \in \mathrm{V}(\mathfrak{C}(R))$ denote the indecomposable $\widehat{R}$-module

$$\widehat{R}/(P_1 \cap \cdots \cap P_{i-1} \cap P_{i+1} \cap \cdots \cap P_t).$$

Then the j-th component of $\operatorname{rank}_{\widehat{R}}(U_i)$ is 1 if $j \neq i$ and 0 if $j = i$. Let E_i denote the $\widehat{R}$-module $\widehat{R}/P_i$. The rank of E_i is the i-th unit vector of $\mathbb{N}^{(t)}$. An easy argument shows that there exist $l_i, m_i \in \mathbb{N}$ such that

$$B\colon = A \oplus U_1^{(l_1)} \oplus \cdots \oplus U_t^{(l_t)} \quad \text{and} \quad C\colon = A \oplus E_1^{(m_1)} \oplus \cdots \oplus E_t^{(m_t)}$$

have constant rank. Now the modules B and C are extended, by Lemma 2.5. Since the U_i, E_j are pairwise non-isomorphic, we see that $[A]$ is the greatest lower bound of $[B]$ and $[C]$, as desired. This completes the proof in case (1).

Finally, we suppose that condition (3) is satisfied. For $i = 1, 2$, let U_i be a non-principal ideal of $\widehat{R}/P_i$, and put $E_i = \widehat{R}/P_i$. Exactly as in the case $t \geq 3$, we use the four pairwise non-isomorphic indecomposable torsion-free modules U_i, E_j to represent $[A]$ as the greatest lower bound of two extended modules.

References

1. D. D. Anderson, editor. *Factorization in integral domains*, volume 189 of *Lecture Notes in Pure and Applied Mathematics*, New York, 1997. Marcel Dekker Inc.
2. H. Bass. On the ubiquity of Gorenstein rings. *Math. Z.*, 82:8–28, 1963.
3. D. D. Anderson, D. F. Anderson, and M. Zafrullah. Factorization in integral domains. *J. Pure Appl. Algebra*, 69:1–19, 1990.
4. L. G. Chouinard, II. Krull semigroups and divisor class groups. *Canad. J. Math.*, 33:1459–1468, 1981.
5. E. G. Evans, Jr. Krull-Schmidt and cancellation over local rings. *Pacific J. Math.*, 46:115–121, 1973.
6. A. Facchini. *Module Theory. Endomorphism rings and direct sum decompositions in some classes of modules.* Progress in Math. 167, Birkhäuser Verlag, 1998.
7. A. Facchini. Direct sum decompositions of modules, semilocal endomorphism rings, and Krull monoids. *J. Algebra*, 256:280–307, 2002.
8. A. Facchini, and D. Herbera. Local Morphisms and Modules with a Semilocal Endomorphism Ring. *Algebr. Represent. Theory*, to appear.
9. A. Geroldinger. A structure theorem for sets of lengths. *Colloq. Math.*, 78:225–259, 1998.
10. A. Geroldinger and F. Halter-Koch. *Non-unique factorizations.* Marcel Dekker, to appear.
11. F. Halter-Koch. *Ideal Systems. An Introduction to Multiplicative Ideal Theory.* Marcel Dekker, 1998.

12. W. Hassler, R. Karr, L. Klingler, and R. Wiegand. Indecomposable modules of large rank over Cohen-Macualay local rings. *Trans. Amer. Math. Soc.*, to appear.
13. F. Kainrath. Factorization in Krull monoids with infinite class group. *Colloq. Math.*, 80:23–30, 1999.
14. L. Klingler and L. S. Levy. Representation Type of Commutative Noetherian Rings I: Local Wildness. *Pacific J. Math.*, 200:345–386, 2001.
15. L. Klingler and L. S. Levy. Representation Type of Commutative Noetherian Rings II: Local Tameness. *Pacific J. Math.*, 200:387–483, 2001.
16. L. Klingler and L. S. Levy. Representation Type of Commutative Noetherian Rings III: Global Wildness and Tameness. *Mem. Amer. Math. Soc.*, 832:1–170, 2005.
17. U. Krause. On monoids of finite real character. *Proc. Amer. Math. Soc.*, 105:546–554, 1989.
18. L. S. Levy and C. J. Odenthal. Package deal theorems and splitting orders, in dimension 1. *Trans. Amer. Math. Soc.*, 348:3457–3503, 1996.
19. E. Matlis. Some properties of Noetherian domains of dimension one. *Canad. J. Math.*, 13:569–586, 1961.
20. E. Matlis. The minimal prime spectrum of a reduced ring. *Illinois J. Math.*, 27:353–391, 1983.
21. M. F. Siddoway. On endomorphism rings of modules over Henselian rings. *Comm. Algebra*, 18:1323–1335, 1990.
22. P. Vámos. Decomposition problems for modules over valuation domains. *J. London Math. Soc. (2)*, 41:10–26, 1990.
23. R. B. Warfield Jr. Cancellation of modules and groups and stable range of endomorphism rings. *Pacific J. Math.*, 91:457–485, 1980.
24. R. Wiegand. Direct-sum decompositions over local rings. *J. Algebra*, 240: 83–97, 2001.

An historical overview of Kronecker function rings, Nagata rings, and related star and semistar operations

Marco Fontana[1] and K. Alan Loper[2]

[1] Dipartimento di Matematica Università degli Studi Roma Tre, Largo San Leonardo Murialdo, 1, 00146 Roma, Italy fontana@mat.uniroma3.it

[2] Department of Mathematics, Ohio State University-Newark, Newark, Ohio 43055 lopera@math.ohio-state.edu

The authors would like to express their deep appreciation to Robert Gilmer for the inspiration and friendship he has provided to us and to so many other mathematicians.

1 Introduction: The Genesis

Toward the middle of the XIXth century, E.E. Kummer discovered that the ring of integers of a cyclotomic field does not have the unique factorization property and he introduced the concept of "ideal numbers" to re-establish some of the factorization theory for cyclotomic integers [45, Vol. 1, 203-210, 583-629].

As R. Dedekind wrote in 1877 to his former student E. Selling, *the goal* of a general theory was immediately clear after Kummer's solution in the special case of cyclotomic integers: to extend Kummer's theory to the case of general algebraic integers.

Dedekind admitted to having struggled unsuccessfully for many years before he published the first version of his theory in 1871 [45] (XI supplement to Dirichlet's "Vorlesungen über Zahlentheorie" [12]).

The theory of Dedekind domains, as it is known today, is based on Dedekind's original ideas and results. Dedekind's point of view is based on ideals ("ideal numbers") for generalizing the algebraic numbers; he proved that, *in the ring of the integers of an algebraic number field, each proper ideal factors uniquely into a product of prime ideals.*

L. Kronecker essentially achieved this goal in 1859, but he published nothing until 1882 [41].

Kronecker's theory holds in a larger context than that of rings of integers of algebraic numbers and solves a more general problem. The primary objective of his theory was to extend the set of elements and the concept of divisibility in such a way that any finite set of elements has a GCD (greatest common divisor) in an extension of the original ring which still mirrors as closely as possible the ideal structure of the original ring. It is probably for this reason that the basic objects of Kronecker's theory –corresponding to Dedekind's "ideals"– are called "divisors".

Let D_0 be a PID with quotient field K_0 and let K be a finite field extension of K_0. Kronecker's *divisors* are essentially all the possible GCD's of finite sets of elements of K that are algebraic over K_0; a divisor is *integral* if it is the GCD of a finite set of elements of the integral closure D of D_0 in K.

One of the key points of Kronecker's theory is that it is possible to give an explicit description of the "divisors". The divisors can be represented as equivalence classes of polynomials and a given polynomial in $D[X]$ represents the class of the integral divisor associated with the set of its coefficients.

More precisely, we can give the following definition.

The classical Kronecker function ring. Let D be as above. *The Kronecker function ring of D* is given by:

$$\begin{aligned}\mathrm{Kr}(D) :&= \left\{ \tfrac{f}{g} \mid f, g \in D[X],\ g \neq 0 \text{ and } \boldsymbol{c}(f) \subseteq \boldsymbol{c}(g) \right\} \\ &= \left\{ \tfrac{f'}{g'} \mid f', g' \in D[X] \text{ and } \boldsymbol{c}(g') = D \right\},\end{aligned}$$

(where $\boldsymbol{c}(h)$ denotes *the content* of a polynomial $h \in D[X]$, i.e. the ideal of D generated by the coefficients of h).

Note that we are assuming that D is a Dedekind domain (being the integral closure of D_0, which is a PID, in a finite field extension K of the quotient field K_0 of D_0 [26, Theorem 41.1 and Theorem 37.8]).

In this case, for each polynomial $g \in D[X]$, $\boldsymbol{c}(g)$ is an invertible ideal of D and, by choosing a polynomial $u \in K[X]$ such that $\boldsymbol{c}(u) = (\boldsymbol{c}(g))^{-1} := (D:\boldsymbol{c}(g))$, we have $f/g = uf/ug = f'/g'$, with $f' := uf$, $g' := ug \in D[X]$ and thus $\boldsymbol{c}(g') = D$ (Gauss Lemma).

The fundamental properties of the Kronecker function ring are the following (cf. [63, Chapter II], [26, Theorem 32.6 (for $\star$ equal to the identity star operation)]):

(1) *$\mathrm{Kr}(D)$ is a Bézout domain (i.e. each finite set of elements, not all zero, has a GCD and the GCD can be expressed as linear combination of these elements) and $D[X] \subseteq \mathrm{Kr}(D) \subseteq K(X)$ (in particular, the field of rational functions $K(X)$ is the quotient field of $\mathrm{Kr}(D)$).*

(2) *Let $a_0, a_1, \ldots, a_n \in D$ and set $f := a_0 + a_1 X + \ldots + a_n X^n \in D[X]$, then:*

$$\begin{aligned}&(a_0, a_1, \ldots, a_n)\mathrm{Kr}(D) = f\mathrm{Kr}(D) \ \ (\textit{thus, } \mathrm{GCD}_{\mathrm{Kr}(D)}(a_0, a_1, \ldots, a_n) = f),\\ &f\mathrm{Kr}(D) \cap K = (a_0, a_1, \ldots, a_n)D = \boldsymbol{c}(f)D \ \ (\textit{hence, } \mathrm{Kr}(D) \cap K = D).\end{aligned}$$

Kronecker's classical theory led to two different major extensions:

• Beginning from 1936 [43], W. Krull generalized the Kronecker function ring to the more general context of *integrally closed domains*, by introducing ideal systems associated to particular star operations: the a.b. (*arithmetisch brauchbar*) star operations.

• Beginning from 1956 [51], M. Nagata investigated, *for an arbitrary integral domain D*, the domain $\{f/g \mid f, g \in D[X] \text{ and } \boldsymbol{c}(g) = D\}$, which coincides with $\mathrm{Kr}(D)$ if (and only if) D is a Prüfer domain [26, Theorem 33.4].

We recall these two major extensions of Kronecker's classical theory, but first we fix the general notation that we use in the sequel.

General notation. Let D be an integral domain with quotient field K. Let $\overline{\boldsymbol{F}}(D)$ represent the set of all nonzero D–submodules of K and $\boldsymbol{F}(D)$ the nonzero fractionary ideals of D (i.e. $E \in \overline{\boldsymbol{F}}(D)$ such that $dE \subseteq D$, for some nonzero element $d \in D$). Finally, let $\boldsymbol{f}(D)$ be the finitely generated D-submodules of K. Obviously:

$$\boldsymbol{f}(D) \subseteq \boldsymbol{F}(D) \subseteq \overline{\boldsymbol{F}}(D).$$

One of the major difficulties for generalizing Kronecker's theory is that Gauss Lemma for the content of polynomials holds for Dedekind domains (or, more generally, for Prüfer domains), but not in general:

Gauss Lemma. *Let $f, g \in D[X]$, where D is an integral domain. If D is a Prüfer domain, then:*

$$\boldsymbol{c}(fg) = \boldsymbol{c}(f)\boldsymbol{c}(g),$$

and conversely. Cf. [26, Corollary 28.5].

For general integral domains, we always have the inclusion of ideals $\boldsymbol{c}(fg) \subseteq \boldsymbol{c}(f)\boldsymbol{c}(g)$. We also have the following result which is weaker than the Gauss Lemma but more widely applicable.

Dedekind–Mertens Lemma. *Let D be an integral domain and $f, g \in D[X]$. Let $m := \deg(g)$. Then*

$$\boldsymbol{c}(f)^m \boldsymbol{c}(fg) = \boldsymbol{c}(f)^{m+1}\boldsymbol{c}(g).$$

Cf. [26, Theorem 28.1].

In order to overcome this obstruction to generalizing the definition of Kronecker's function rings, Krull introduced multiplicative ideal systems having a cancellation property which mirrors Gauss's Lemma. These ideal systems can be defined by what are called now the e.a.b. (*endlich arithmetisch brauchbar*) star operations.

Star operations. A mapping $\star: \boldsymbol{F}(D) \to \boldsymbol{F}(D)$, $I \mapsto I^\star$, is called *a star operation of D* if, for all $z \in K$, $z \neq 0$ and for all $I, J \in \boldsymbol{F}(D)$, the following properties hold:

($\star_1$) $(zD)^\star = zD\,, \ \ (zI)^\star = zI^\star\,;$
($\star_2$) $I \subseteq J \Rightarrow I^\star \subseteq J^\star\,;$
($\star_3$) $I \subseteq I^\star$ and $I^{\star\star} := (I^\star)^\star = I^\star\,.$

An *e.a.b. star operation* on D is a star operation $\star$ such that, for all nonzero finitely generated ideals $I,\ J,\ H$ of D:

$$(IJ)^\star \subseteq (IH)^\star \ \Rightarrow \ J^\star \subseteq H^\star\,.$$

Using these notions, Krull recovers a useful identity for the contents of polynomials:

Gauss–Krull Lemma. *Let $\star$ be an e.a.b. star operation on an integral domain D (this condition implies that D is an integrally closed domain [26, Corollary 32.8]) and let $f,\ g \in D[X]$ then:*

$$\boldsymbol{c}(fg)^\star = \boldsymbol{c}(f)^\star \boldsymbol{c}(g)^\star\,.$$

Cf. [26, Lemma 32.6].

Remark 1 Krull introduced the concept of a star operation in his first *Beiträge paper* in 1936 [43]. He used the notation " $'$–Operation " ("Strich–Operation") for his generic operation. [In this paper you can find the terminology " $'$–Operation " in footnote 13 and in the title of Section 6, among other places.]

The notation " $\ast$–operation " ("star–operation") arises from Section 26 of the original version of Gilmer's "Multiplicative Ideal Theory" (1968) [25]. Robert Gilmer gave us this explication ≪ I believe the reason I switched from " $'$–Operation " to " $\ast$–operation " was because " $'$ " was not so generic at the time: I' was frequently used as the notation for the integral closure of an ideal I, just as D' was used to denote the integral closure of the domain D. (Such notation was used, for example, in both Nagata's Local Rings and in Zariski-Samuel's two volumes.) ≫

Moreover, Krull only considered the concept of an "arithmetisch brauchbar (a.b.) $'$–Operation", not of an e.a.b. operation.
Recall that an *a.b.–operation* is a star operation $\star$ such that, if $I \in \boldsymbol{f}(D)$ and $J, K \in \boldsymbol{F}(D)$ and if $(IJ)^\star \subseteq (IH)^\star$ then $J^\star \subseteq H^\star$.

The e.a.b. concept stems from the original version of Gilmer's book [25]. The results of Section 26 show that this (presumably) weaker concept is all that one needs to develop a complete theory of Kronecker function rings.

In this regard, Robert Gilmer gave us this explication ≪ I believe I was influenced to recognize this because during the 1966 calendar year in our graduate algebra seminar (Bill Heinzer, Jimmy Arnold, and Jim Brewer, among others, were in that seminar) we had covered Bourbaki's Chapitres 5 and 7 of *Algèbre Commutative*, and the development in Chapter 7 on the v–operation indicated that e.a.b. would be sufficient.≫

One of the main goals for the classical theory of star operations has been to construct a Kronecker function ring associated to a domain, in a more general context than the original one considered by L. Kronecker in 1882.

More precisely, using star operations, in 1936 W. Krull [43] defined a Kronecker function ring in a more general setting than Kronecker's. (Further references are H. Prüfer (1932) [56], Arnold (1969) [8], Arnold-Brewer (1971) [9], Dobbs-Fontana (1986) [13], D.F. Anderson-Dobbs-Fontana (1987) [6], Okabe-Matsuda (1997) [55].)

Star–Kronecker function ring. Let D be an integrally closed integral domain with quotient field K and let $\star$ be an e.a.b. star operation on D, then:

$$\mathrm{Kr}(D,\star) := \left\{ \frac{f}{g} \mid f, g \in D[X],\ g \neq 0 \text{ and } \boldsymbol{c}(f)^\star \subseteq \boldsymbol{c}(g)^\star \right\}$$

is an integral domain with quotient field $K(X)$, called *the $\star$–Kronecker function ring of D, having the following properties:*

(1) $\mathrm{Kr}(D,\star)$ *is a Bézout domain and* $D[X] \subseteq \mathrm{Kr}(D,\star) \subseteq K(X)$.

(2) *Let* $a_0, a_1, \ldots, a_n \in D$ *and set* $f := a_0 + a_1X + \ldots + a_nX^n \in D[X]$, *then:*

$$(a_0, a_1, \ldots, a_n)\mathrm{Kr}(D,\star) = f\mathrm{Kr}(D,\star),$$
$$(a_0, a_1, \ldots, a_n)\mathrm{Kr}(D,\star) \cap K = ((a_0, a_1, \ldots, a_n)D)^\star$$
$$(\text{i.e. } f\mathrm{Kr}(D,\star) \cap K = (\boldsymbol{c}(f))^\star).$$

In particular, $\mathrm{Kr}(D,\star) \cap K = D^\star = D$.

For the proof cf. [26, Theorem 32.7].

Nagata's generalization of the Kronecker function ring. The following construction is possible for any integral domain D and, even, for an arbitrary ring D.

$$\mathrm{Na}(D) := D(X) := \left\{ \frac{f}{g} \mid f, g \in D[X] \text{ and } \boldsymbol{c}(g) = D \right\},$$

and this ring is called the *Nagata ring of D*. This notion is essentially due to Krull (1943) [44]. Then this ring was studied in Nagata's book (1962) [52, Section 6, page 17], using the notation $D(X)$, and in Samuel's Tata volume (1964) [58, page 27] (where the notation $D(X)_{\mathrm{loc}}$ was used). We introduced the notation $\mathrm{Na}(D)$ that is convenient for generalizations.

In general, $\mathrm{Na}(D)$ is not a Bézout domain. It is not difficult to see that [26, Theorem 33.4] and [1, Theorem 8]

- $\mathrm{Na}(D)$ *is a Bézout domain if and only if D is a Prüfer domain.*
- $\mathrm{Na}(D)$ *coincides with* $\mathrm{Kr}(D)$ *if and only if D is a Prüfer domain.*

- *Every ideal of* $\mathrm{Na}(D)$ *is extended from* D *if and only if* D *is a Prüfer domain.*

The interest in Nagata's ring $D(X)$ is due to the fact that this ring of rational functions has some strong ideal-theoretic properties that D itself need not have, while maintaining a strict relation with the ideal structure of D.

(a) *The map* $P \mapsto PD(X)$ *establishes a 1-1 correspondence between the maximal ideals of* D *and the maximal ideals of* $D(X)$.

(b) *For each ideal* I *of* D,

$$ID(X) \cap D = I\,, \qquad D(X)/ID(X) \cong (D/ID)(X)\,;$$

I is finitely generated if and only if $ID(X)$ is finitely generated.

Among the new properties acquired by $D(X)$ are the following:

(c) *the residue field at each maximal ideal of* $D(X)$ *is infinite;*

(d) *an ideal contained in a finite union of ideals is contained in one of them;*

(e) *each finitely generated locally principal ideal is principal (therefore* $\mathrm{Pic}(D(X)) = 0$*).*

The proofs of the previous results can be found in Arnold (1969) [8] and Gilmer's book [26, Proposition 33.1, Proposition 5.8] (for **(a)**, **(b)**, and **(c)** which is a consequence of **(b)**), Quartararo-Butts (1975) [57] (for **(d)**) and D.D. Anderson (1977) [2, Theorem 2] (for **(e)**).

2 Basic facts on semistar operations

In 1994, Okabe and Matsuda [54] introduced the more flexible notion of semistar operation $\star$ of an integral domain D, as a natural generalization of the notion of star operation, allowing $D \neq D^\star$ (cf. also [53], [48], and [49]).

Semistar operations. A mapping $\star\colon \overline{\boldsymbol{F}}(D) \to \overline{\boldsymbol{F}}(D)$, $E \mapsto E^\star$ is called *a semistar operation of* D if, for all $z \in K$, $z \neq 0$ and for all $E, F \in \overline{\boldsymbol{F}}(D)$, the following properties hold:

$(\star_1)$ $(zE)^\star = zE^\star$;

$(\star_2)$ $E \subseteq F \Rightarrow E^\star \subseteq F^\star$;

$(\star_3)$ $E \subseteq E^\star$ and $E^{\star\star} := (E^\star)^\star = E^\star$.

When $D^\star = D$, we say that $\star$ is *a (semi)star operation* of D, since, restricted to $\boldsymbol{F}(D)$ it is *a star operation of* D.

For star operations, the notion of $\star$–ideal leads to the definition of a canonically associated ideal system.

For semistar operations, we need a more general notion, that coincides with the notion of $\star$–ideal, when $\star$ is a (semi)star operation.

- A nonzero (integral) ideal I of D is a *quasi-$\star$–ideal* [respectively, *$\star$–ideal*] if $I^\star \cap D = I$ [respectively, if $I^\star = I$].

• A *quasi–$\star$–prime* [respectively, *$\star$–prime*] of D is a quasi–$\star$–ideal [respectively, an integral $\star$–ideal] of D which is also a prime.

• A *quasi–$\star$–maximal* [respectively, *$\star$–maximal*] of D is a maximal element in the set of all proper quasi–$\star$–ideals [respectively, integral $\star$–ideals] of D.

We denote by $\mathrm{Spec}^\star(D)$ [respectively, $\mathrm{Max}^\star(D)$, $\mathrm{QSpec}^\star(D)$, $\mathrm{QMax}^\star(D)$] the set of all $\star$–primes [respectively, $\star$–maximals, quasi–$\star$–primes, quasi–$\star$–maximals] of D.

For example, it is easy to see that, *if $I^\star \neq D^\star$, then $I^\star \cap D$ is a quasi–$\star$–ideal that contains I* (in particular, a $\star$–ideal is a quasi–$\star$–ideal).
Note that:

- when $D = D^\star$ the notions of quasi–$\star$–ideal and $\star$–ideal coincide;
- $I^\star \neq D^\star$ is equivalent to $I^\star \cap D \neq D$.

As in the classical star-operation setting, we associate to a *semistar* operation $\star$ of D a new semistar operation $\star_f$ as follows. If $E \in \overline{\boldsymbol{F}}(D)$ we set:

$$E^{\star_f} := \cup\{F^\star \mid F \subseteq E,\, F \in \boldsymbol{f}(D)\}\,.$$

We call $\star_f$ *the semistar operation of finite type of D* associated to $\star$.

• If $\star = \star_f$, we say that $\star$ is *a semistar operation of finite type of* D.
Note that $\star_f \leq \star$ and $(\star_f)_f = \star_f$, so $\star_f$ is of finite type on D.

The following result is in [21, Lemma 2.3].

Lemma 2 *Let $\star$ be a non-trivial semistar operation of finite type on D. Then*

(1) *Each proper quasi–$\star$–ideal is contained in a quasi–$\star$–maximal.*
(2) *Each quasi–$\star$–maximal is a quasi–$\star$–prime.*
(3) *Set*

$$\Pi^\star := \{P \in \mathrm{Spec}(D) \,|\, P \neq 0\,, P^\star \cap D \neq D\}\,.$$

Then $\mathrm{QSpec}^\star(D) \subseteq \Pi^\star$ and the set of maximal elements of $\Pi^\star$, denoted by $\Pi^\star_{\max}$, is nonempty and coincides with $\mathrm{QMax}^\star(D)$.

For the sake of simplicity, when $\star = \star_f$, we will denote simply by $\mathcal{M}(\star)$, the nonempty set $\Pi^\star_{\max} = \mathrm{QMax}^\star(D)$.

3 Nagata semistar domain

A generalization of the classical Nagata ring construction was considered by Kang (1987 [39] and 1989 [40]). We further generalize this construction to the semistar setting.

Nagata semistar function ring. Given *any* integral domain D and *any* semistar operation $\star$ on D, we define *the semistar Nagata ring* as follows:

$$\mathrm{Na}(D,\star) := \left\{ \frac{f}{g} \mid f,g \in D[X]\,,\; g \neq 0\,,\; \boldsymbol{c}(g)^{\star} = D^{\star} \right\}.$$

Note that $\mathrm{Na}(D,\star) = \mathrm{Na}(D,\star_f)$. Therefore, the assumption $\star = \star_f$ is not really restrictive when considering Nagata semistar rings.

If $\star = d$ is the identity (semi)star operation of D, then:

$$\mathrm{Na}(D,d) = D(X)\,.$$

Some results on *star* Nagata rings proved by Kang in 1989 are generalized to the semistar setting in the following:

Proposition 3 *Let $\star$ be a nontrivial semistar operation of an integral domain D. Set:*

$$N(\star) := N_D(\star) := \{h \in D[X] \mid \boldsymbol{c}(h)^{\star} = D^{\star}\}\ .$$

(1) $N(\star) = D[X] \setminus \cup\{Q[X] \mid Q \in \mathcal{M}(\star_f)\}$ *is a saturated multiplicatively closed subset of* $D[X]$ *and* $N(\star) = N(\star_f)$.
(2) $Max(D[X]_{N(\star)}) = \{Q[X]_{N(\star)} \mid Q \in \mathcal{M}(\star_f)\}$.
(3) $\mathrm{Na}(D,\star) = D[X]_{N(\star)} = \cap\{D_Q(X) \mid Q \in \mathcal{M}(\star_f)\}$.
(4) $\mathcal{M}(\star_f)$ *coincides with the canonical image in* $\mathrm{Spec}(D)$ *of the maximal spectrum of* $\mathrm{Na}(D,\star)$; *i.e.* $\mathcal{M}(\star_f) = \{M \cap D \mid M \in \mathrm{Max}(\mathrm{Na}(D,\star))\}$.

For the proof cf. [21, Theorem 3.1]. From the previous Proposition 3 (4) we have:

Corollary 4 *Let D be an integral domain, then:*
Q is a maximal t–ideal of D $\Leftrightarrow$ $Q = M \cap D$, for some $M \in \mathrm{Max}(\mathrm{Na}(D,v))$.

4 The semistar operation associated to $\mathrm{Na}(D,\star)$

We start by recalling some distinguished classes of semistar operations.

• If Δ is a nonempty set of prime ideals of an integral domain D, then the semistar operation $\star_{\Delta}$ defined on D as follows

$$E^{\star_{\Delta}} := \cap\{ED_P \mid P \in \Delta\}, \quad \text{for each } E \in \overline{\boldsymbol{F}}(D)\,,$$

is called *the spectral semistar operation associated to* Δ.

• A semistar operation $\star$ of an integral domain D is called *a spectral semistar operation* if there exists $\emptyset \neq \Delta \subseteq \mathrm{Spec}(D)$ such that $\star = \star_{\Delta}$.

• We say that $\star$ *possesses enough primes* or that $\star$ is *a quasi-spectral semistar operation of* D if, for each nonzero ideal I of D such that $I^{\star} \cap D \neq D$, there exists a quasi–$\star$–prime P of D such that $I \subseteq P$.

• Finally, we say that $\star$ is *a stable semistar operation on* D if

$$(E \cap F)^{\star} = E^{\star} \cap F^{\star}, \quad \text{for all } E,F \in \overline{\boldsymbol{F}}(D)\,.$$

Remark 5 *Mutatis mutandis* the previous notions were considered first in the star settings and, in particular, by D.D. Anderson, D.F. Anderson and S. J. Cook who gave important contributions to the subject [3], [4] and [5]. The general situation was considered among the others by Fontana-Huckaba [17], Fontana-Loper [21] and Halter-Koch [31].

Lemma 6 *Let D be an integral domain and let $\emptyset \neq \Delta \subseteq \mathrm{Spec}(D)$. Then:*

(1) $E^{\star_\Delta} D_P = ED_P$, *for each* $E \in \overline{\boldsymbol{F}}(D)$ *and for each* $P \in \Delta$.
(2) $(E \cap F)^{\star_\Delta} = E^{\star_\Delta} \cap F^{\star_\Delta}$, *for all* $E, F \in \overline{\boldsymbol{F}}(D)$.
(3) $P^{\star_\Delta} \cap D = P$, *for each* $P \in \Delta$.
(4) *If I is a nonzero integral ideal of D and $I^{\star_\Delta} \cap D \neq D$, then there exists $P \in \Delta$ such that $I \subseteq P$.*

For the proof cf. [17, Lemma 4.1].

Lemma 7 *Let $\star$ be a nontrivial semistar operation of an integral domain D. Then:*

(1) *$\star$ is spectral if and only if $\star$ is quasi-spectral and stable.*
(2) *Assume that $\star = \star_f$. Then $\star$ is quasi-spectral and $\mathcal{M}(\star) \neq \emptyset$.*

For the proof cf. [17, Theorem 4.12] and [21, Lemma 2.5].

Theorem 8 *Let $\star$ be a nontrivial semistar operation and let $E \in \overline{\boldsymbol{F}}(D)$. Set*

$$\tilde{\star} := (\star_f)_{sp} := \star_{\mathcal{M}(\star_f)}.$$

[$\tilde{\star}$ is called the spectral semistar operation associated to $\star$.] Then:

(1) $E^{\tilde{\star}} = \cap\{ED_Q \mid Q \in \mathcal{M}(\star_f)\}$ *[and* $E^{\star_f} = \cap\{E^{\star_f} D_Q \mid Q \in \mathcal{M}(\star_f)\}$ *].*
(2) $\tilde{\star} \leq \star_f$.
(3) $E\mathrm{Na}(D, \star) = \cap\{ED_Q(X) \mid Q \in \mathcal{M}(\star_f)\}$, *thus:* $E\mathrm{Na}(D, \star) \cap K = \cap\{ED_Q \mid Q \in \mathcal{M}(\star_f)\}$.
(4) $E^{\tilde{\star}} = E\mathrm{Na}(D, \star) \cap K$.

For the proof cf. [21, Proposition 3.4].

Proposition 3 (4) assures that, when a maximal ideal of $\mathrm{Na}(D, \star)$ is contracted to D, the result is exactly a prime ideal in $\mathcal{M}(\star_f)$. This result can be reversed. Moreover, the semistar operation $\tilde{\star}$ generates the same Nagata ring as $\star$.

Corollary 9 *Let $\star$, $\star_1$, $\star_2$ be semistar operations of an integral domain D. Then:*

(1) $\mathrm{Max}(\mathrm{Na}(D, \star)) = \{QD_Q(X) \cap \mathrm{Na}(D, \star) \mid Q \in \mathcal{M}(\star_f)\}$.
(2) $(\tilde{\star})_f = \tilde{\star} = \tilde{\tilde{\star}}$.
(3) $\mathcal{M}(\star_f) = \mathcal{M}(\tilde{\star})$.
(4) $\mathrm{Na}(D, \star) = \mathrm{Na}(D, \tilde{\star})$.
(5) $\star_1 \leq \star_2 \Rightarrow \mathrm{Na}(D, \star_1) \subseteq \mathrm{Na}(D, \star_2) \Leftrightarrow \widetilde{\star_1} \leq \widetilde{\star_2}$.

For the proof cf. [21, Corollary 3.5 and Theorem 3.8].

Remark 10 Note that, when $\star$ is the (semi)star v–operation, then the (semi)star operation $\tilde{v}$ coincides with *the (semi)star operation* w defined as follows:

$$E^w := \cup\{(E:H) \mid H \in \boldsymbol{f}(D) \text{ and } H^v = D\},$$

for each $E \in \overline{\boldsymbol{F}}(D)$, cf. [17, page 182]. This (semi)star operation was considered by J. Hedstrom and E. Houston in 1980 under the name of F_∞–operation [34].

Later, starting in 1997, this operation was studied by Wang Fanggui and R. McCasland under the name of w–operation [61] (cf. also [59], [60] and [62]). (Unfortunately, the same notation is also used for the star a.b. operations defined by a family of valuation overrings [26, page 398] and the two notions are not related, in general.) Note also that the notion of w–ideal coincides with the notion of semi-divisorial ideal considered by S. Glaz and W. Vasconcelos in 1977 [27].

Finally, in 2000, for each (semi)star operation $\star$, D.D. Anderson and S.J. Cook [5] considered the $\star_w$–operation which can be defined as follows:

$$E^{\star_w} := \cup\{(E:H) \mid H \in \boldsymbol{f}(D) \text{ and } H^\star = D\},$$

for each $E \in \overline{\boldsymbol{F}}(D)$. From their theory (and from the results by Hedstrom and Houston) it follows that:

$$\star_w = \tilde{\star}.$$

The relation between $\tilde{\star}$ and the localizing systems of ideals (in the sense of Gabriel and Popescu) was established by M. Fontana and J. Huckaba in 2000 [17].

5 The Kronecker function ring in a general setting

The problem of the construction of a Kronecker function ring for general integral domains was considered indipendently by F. Halter-Koch (2003) [32] and Fontana-Loper (2001, 2003) [19], [21].

Halter-Koch's approach is axiomatic and makes use of the theory of finitary ideal systems (star operations of finite type) [30]. He also establishes a connection with Krull's theory of Kronecker function rings and introduces the Kronecker function rings for integral domains with an ideal system which does not necessarily verify the cancellation property (e.a.b.). Fontana-Loper's treatment is based on the Okabe-Matsuda's theory of semistar operations [19], [21], [55], and [47].

Halter-Koch [32] gives the following abstract definition which does not rely on semistar operations or valuation overrings.

***K*–function ring.** Let K be a field, R a subring of $K(X)$ and $D := R \cap K$. If

(Kr.1) $X \in \boldsymbol{\mathcal{U}}(R)$ (i.e. X is a unit in R);

(Kr.2) $f(0) \in fR$ for each $f \in K[X]$;
then R is called *a K–function ring of D.*

Using only these two axioms, he proved that R "behaves as a Kronecker function ring":

Theorem 11 *Let R be a K–function ring of $D = R \cap K$, then:*

(1) *R is a Bézout domain with quotient field $K(X)$.*
(2) *D is integrally closed in K.*
(3) *For each polynomial $f := a_0 + a_1X + \ldots + a_nX^n \in K[X]$, we have $(a_0, a_1, \ldots, a_n)R = fR$.*

For the proof cf. [32, Theorem 2.2].

Our next goal is to describe Fontana-Loper's approach and to illustrate the relation with Halter-Koch's K–function rings.

Semistar Kronecker function ring. If $\star$ is *any* semistar operation of *any* integral domain D, then we define *the Kronecker function ring of D with respect to the semistar operation* $\star$ by:

$$\begin{aligned}\mathrm{Kr}(D,\star) := \{f/g \mid f,g \in D[X],\ g \neq 0,\ & \text{and there exists} \\ & h \in D[X]\setminus\{0\} \ \text{with}\ (\boldsymbol{c}(f)\boldsymbol{c}(h))^\star \subseteq (\boldsymbol{c}(g)\boldsymbol{c}(h))^\star\}.\end{aligned}$$

At this point, we need some preliminaries in order:
— to show that this construction leads to a natural extension of the classical Kronecker function ring,
— to investigate the connections between the semistar Kronecker function ring $\mathrm{Kr}(D,\star)$ and the axiomatically defined K-function ring,
— to show that $\mathrm{Kr}(D,\star)$ defines a new semistar operation on D, behaving with respect $\mathrm{Kr}(D,\star)$ in a similar way to $\tilde{\star}$ with respect to $\mathrm{Na}(D,\star)$.

We start by recalling that it is possible to associate to an arbitrary semistar operation an e.a.b. semistar operation.

- Given any semistar operation $\star$ of D, we can define an e.a.b. semistar operation of finite type $\star_a$ of D, called *the e.a.b. semistar operation associated to* $\star$, as follows for each $F \in \boldsymbol{f}(D)$ and for each $E \in \overline{\boldsymbol{F}}(D)$:

$$\begin{aligned} F^{\star_a} &:= \cup\{((FH)^\star : H^\star) \mid H \in \boldsymbol{f}(D)\}, \\ E^{\star_a} &:= \cup\{F^{\star_a} \mid F \subseteq E,\ F \in \boldsymbol{f}(D)\}. \end{aligned}$$

The previous construction is essentially due to P. Jaffard (1960) [38, Chapitre II, §2] and F. Halter-Koch (1997, 1998) [29, Section 6], [30, Chapter 19] (cf. also Lorenzen (1939) [46] and Aubert (1983) [10]).

Obviously $(\star_f)_a = \star_a$. Note that (for instance [19, Proposition 4.3 and 4.5]):
– *when $\star = \star_f$, then $\star$ is e.a.b. if and only if $\star = \star_a$.*
– *$D^{\star_a}$ is integrally closed and contains the integral closure of D.*

When $\star = v$, then D^{v_a} coincides with *the pseudo-integral closure of* D introduced by D.F. Anderson, Houston and Zafrullah (1992) [7].

Remark 12 In the classical context of *star* operations, $\star_a$ is expected to be a star operation too and for this reason is defined on the "star closure" of D (or, on an integral domain which is "star closed"), cf. Okabe-Matsuda (1992) [53], Halter-Koch (1997, 1998, 2003) [29], [30], [32].
More precisely (even if $\star$ is a semistar operation), we call *the $\star$–closure of* D:

$$D^{\mathrm{cl}^\star} := \cup\{(F^\star : F^\star) \mid F \in \boldsymbol{f}(D)\}.$$

It is easy to see that $D^{\mathrm{cl}^\star}$ *is an integrally closed overring of* D and D is said *$\star$–closed* if $D = D^{\mathrm{cl}^\star}$.
We can now define a new (semi)star operation on D if $D = D^{\mathrm{cl}^\star}$ (or, in general, a semistar operation on D), $\mathrm{cl}^\star$ by setting for each $F \in \boldsymbol{f}(D)$, for each $E \in \overline{\boldsymbol{F}}(D)$:

$$\begin{aligned} F^{\mathrm{cl}^\star} &:= \cup\{((H^\star : H^\star)F)^\star \mid H \in \boldsymbol{f}(D)\}, \\ E^{\mathrm{cl}^\star} &:= \cup\{F^{\mathrm{cl}^\star} \mid F \subseteq E,\ F \in \boldsymbol{f}(D)\}. \end{aligned}$$

If we set $\overline{\star} := \mathrm{cl}^\star$, it is not difficult to see that $D^{\mathrm{cl}^{\overline{\star}}} = D^{\mathrm{cl}^\star}$ (and that it coincides with $D^{\star_a}$) and $D^{\mathrm{cl}^\star}$ contains the "classical" integral closure of D. Moreover (as semistar operations on D):

$$\star_{_f} \leq \mathrm{cl}^\star \leq \star_a\,, \quad (\star_{_f})_a = (\mathrm{cl}^\star)_a = (\star_a)_a = \star_a\,.$$

We now turn our attention to the valuation overrings. The notion that we recall next is due to P. Jaffard (1960) [38, page 46] (cf. also Halter-Koch (1997) [30, Chapters 15 and 18]).

- For a domain D and a semistar operation $\star$ on D, we say that a valuation overring V of D is *a $\star$–valuation overring of D* provided $F^\star \subseteq FV$, for each $F \in \boldsymbol{f}(D)$.

Note that, by definition the $\star$–valuation overrings coincide with the $\star_f$–valuation overrings.

Proposition 13 *Let D be a domain and let $\star$ be a semistar operation on D.*

(1) *The $\star$–valuation overrings also coincide with the $\star_a$–valuation overrings.*
(2) $D^{\mathrm{cl}^\star} = \cap\{V \mid V$ *is a $\star$–valuation overring of D*$\}$.
(3) *A valuation overring V of D is a $\tilde{\star}$–valuation overring of D if and only if V is an overring of D_P, for some $P \in \mathcal{M}(\star_f)$.*

For the proof cf. for instance [20, Proposition 3.2, 3.3 and Corollary 3.6] and [21, Theorem 3.9].

Theorem 14 *Let $\star$ be a semistar operation of an integral domain D with quotient field K. Then:*

(1) $\mathrm{Na}(D,\star) \subseteq \mathrm{Kr}(D,\star)$.
(2) *V is a $\star$–valuation overring of D if and only if $V(X)$ is a valuation overring of $\mathrm{Kr}(D,\star)$.*
The map $W \mapsto W \cap K$ establishes a bijection between the set of all valuation overrings of $\mathrm{Kr}(D,\star)$ and the set of all the $\star$–valuation overrings of D.
(3) $\mathrm{Kr}(D,\star) = \mathrm{Kr}(D,\star_f) = \mathrm{Kr}(D,\star_a) = \cap\{V(X) \mid V$ *is a $\star$–valuation overring of D} is a Bézout domain with quotient field $K(X)$.*
(4) $E^{\star_a} = E\mathrm{Kr}(D,\star) \cap K = \cap\{EV \mid V$ *is a $\star$–valuation overring of* $D\}$, *for each $E \in \overline{F}(D)$.*
(5) *$R := \mathrm{Kr}(D,\star)$ is a K–function ring of $R \cap K = D^{\star_a}$ (in the sense of Halter-Koch's axiomatic definition).*

For the proof cf. [19, Theorem 3.11], [20, Theorem 3.5], [21, Proposition 4.1].

6 Some relations between $\mathrm{Na}(D,\star)$, $\mathrm{Kr}(D,\star)$, and the semistar operations $\widetilde{\star}$, $\star_a$

An elementary first question to ask is whether the two semistar operations $\tilde{\star}$ and $\star_a$ are actually the same - or usually the same - or rarely the same.

Proposition 13 indicates that for a semistar operation $\star$ on a domain D, the $\tilde{\star}$–valuation overrings of D are all the valuation overrings of the localizations of D at the primes in $\mathcal{M}(\star_f)$. On the other hand, we know from Theorem 14 that the $\star_a$–valuation overrings (or, equivalently, the $\star$–valuation overrings) of D correspond exactly to the valuation overrings of the Kronecker function ring $\mathrm{Kr}(D,\star)$. In particular, each $\star_a$–valuation overring is also a $\tilde{\star}$–valuation overring.

It is easy to imagine that these two collections of valuation domains can frequently be different and, even when the two collections of valuation domains coincide, it may happen that $\tilde{\star} \neq \star_a$. Fontana-Loper [21] gives some examples which illustrate the different situations that can occur.

It is possible to prove positive statements about the relationship between $\widetilde{(\text{-})}$ and $(\text{-})_a$ under certain conditions [21, Proposition 5.4 and Remark 5.5]. However, we limit ourselves to stating a result that generalizes the fundamental result that is at the basis of Krull's theory of Kronecker function rings, i.e. $\mathrm{Na}(D) = \mathrm{Na}(D,d) = \mathrm{Kr}(D,b) = \mathrm{Kr}(D)$ if and only if D is a Prüfer domain, cf. for instance [26, Theorem 33.4].

We recall the following definition, which generalizes the classical notion of Prüfer domain.

Prüfer semistar multiplication domain. Let $\star$ be a semistar operation on an integral domain D. A *Prüfer $\star$–multiplication domain* (for short, a

P⋆MD) is an integral domain D such that $(FF^{-1})^{\star_f} = D^{\star_f}$ $(= D^{\star})$ (i.e., each F is $\star_f$*–invertible*) for each $F \in \boldsymbol{f}(D)$.

Some of the statements of the following theorem, due to Fontana-Jara-Santos (2003) [18] generalize some of the classical characterizations of the Prüfer v–multiplication domains (for short, PvMD).

Theorem 15 *Let D be an integral domain and $\star$ a semistar operation on D. The following are equivalent:*

(i) *D is a P⋆MD.*
(ii) *$\mathrm{Na}(D,\star)$ is a Prüfer domain.*
(iii) $\mathrm{Na}(D,\star) = \mathrm{Kr}(D,\star)$.
(iv) $\tilde{\star} = \star_a$.
(v) *$\star_f$ is stable and e.a.b..*

In particular, D is a P⋆MD if and only if it is a P$\tilde{\star}$MD.

For the proof cf. [18, Theorem 3.1 and Remark 3.1].

The following gives the converse of the implication PvMD $\Rightarrow$ PwMD proved by Wang-McCasland (1999) [62, Section 2, page 160], cf. also D.D. Anderson-Cook (2000) [5, Theorem 2.18].

Corollary 16 *Let D be an integral domain. The following are equivalent:*

(i) *D is a PvMD.*
(ii) $\mathrm{Na}(D,t) = \mathrm{Kr}(D,t)$.
(iii) $w := \tilde{v} = v_a$.
(iv) *t is stable and e.a.b..*

In particular, D is a PvMD if and only if it is a PwMD.

For the proof cf. [18, Corollary 3.1].

In the star setting the relation between the P⋆MDs and the PvMDs is described by the following:

Corollary 17 *Let D be an integral domain and $\star$ a* star *operation on D.*

D is a P⋆MD $\Leftrightarrow$ D is a PvMD and $t = \tilde{\star}$ (or, equivalently, $t = \star_f$).

For the proof cf. [18, Proposition 3.4].

Remark 18 The PvMDs were studied by Griffin in 1967 [28] under the name of v–multiplication domains, cf. also [42] and [38]. Relevant contributions to the subject were given among the others by Arnold-Brewer (1971) [9], Heinzer-Ohm (1973) [35], Mott-Zafrullah (1981) [50], Zafrullah (1984) [64], Houston (1986) [36], Kang (1989) [40], Dobbs-Houston-Lucas-Zafrullah (1989) [14] and El Baghdadi (2002) [16].

For $\star$ a star operation, P⋆MDs were considered by Houston-Malik-Mott in 1984 [37], introducing a unified setting for studying Krull domains, Prüfer domains and PvMDs. This class of domains was also investigated by Garcia-Jara-Santos (1999) [24] and Halter-Koch (2003) [33]. These papers led naturally to the study of the Prüfer semistar multiplication domains initiated in 2003 by Fontana-Jara-Santos [18].

Related to this study are the questions on the invertibility property of ideals and modules especially in the star and semistar setting, cf. the survey paper by Zafrullah [65], Chang-Park (2003) [11] and Fontana-Picozza (2005) [23].

7 Intersections of local Nagata domains

Given a semistar operation $\star$ on D, the integral domains $\mathrm{Na}(D,\star)$ and $\mathrm{Kr}(D,\star)$ (and the related semistar operations $\tilde{\star}$ and $\star_a$) have in many regards a similar behaviour. The following natural question is the starting point of a recent paper by M. Fontana and K.A. Loper [22]:

Is it possible to find an integral domain of rational functions, denoted by $\mathrm{KN}(D,\star)$ *(*obtained as an intersection of local Nagata domains associated to any semistar operation $\star$*), such that:*

- $\mathrm{Na}(D,\star) \subseteq \mathrm{KN}(D,\star) \subseteq \mathrm{Kr}(D,\star)$ *;*
- $\mathrm{KN}(D,\star)$ *generalizes at the same time* $\mathrm{Na}(D,\star)$ *and* $\mathrm{Kr}(D,\star)$ *and coincides with* $\mathrm{Na}(D,\star) = \mathrm{Na}(D,\tilde{\star})$ *or* $\mathrm{Kr}(D,\star) = \mathrm{Kr}(D,\star_a)$*, when the semistar operation (of finite type)* $\star$ *assumes the extreme values of the interval* $\tilde{\star} \leq \star \leq \star_a$*?*

In order to present the answer to the previous question we need to settle some terminology:

- If F is in $\boldsymbol{f}(D)$, we say that F is $\star$*–e.a.b.* [respectively, $\star$*–a.b.*] if $(FG)^\star \subseteq (FH)^\star$, with $G,\ H \in \boldsymbol{f}(D)$ [respectively, $G,\ H \in \overline{\boldsymbol{F}}(D)$], implies that $G^\star \subseteq H^\star$.

Lemma 19 *Let $\star$ be a semistar operation on an integral domain D, let $F \in \boldsymbol{f}(D)$ be $\star_{_f}$–invertible and let (L,N) be a local $\star$–overring of D. Then FL is a principal fractional ideal of L.*

For the proof cf. [22, Corollary 4.3].

Note that, in general, $\star$–(e.)a.b. does not imply $\star$–invertible, even for finite type semistar operations. However, it is possible to show that, for finite type stable semistar operations $\star$ (i.e. when $\star = \tilde{\star}$), the notions of $\star$–e.a.b., $\star$–a.b. and $\star$–invertible coincide [22, Proposition 5.3(2)].

Our next goal is to generalize Lemma 19 to the case of $\star$–e.a.b. ideals.

Semistar monolocalities. Let $\star$ be a semistar operation on an integral domain D. A *$\star$–monolocality of D* is a local overring L of D such that:

– FL is a principal fractionary ideal of L, for each $\star$–e.a.b. $F \in \boldsymbol{f}(D)$;
– $L = L^{\star_f}$.

Obviously, each $\star$–valuation overring is a $\star$–monolocality. It is not hard to prove that, for each $Q \in \mathcal{M}(\star_f)$, D_Q is a $\tilde{\star}$–monolocality [22, Proposition 5.3(1)]. Set:

$$\begin{aligned}\boldsymbol{\mathcal{L}}(D,\star) &:= \{L \mid L \text{ is a } \star\text{-monolocality of } D\}\\ \mathrm{KN}(D,\star) &:= \cap\{L(X) \mid L \in \boldsymbol{\mathcal{L}}(D,\star)\}.\end{aligned}$$

We are now in condition to state some among the main results proved in Fontana-Loper (2005) [22].

Theorem 20 *Let $\star$ be a semistar operation on an integral domain D with quotient field K.*

(1) $\mathrm{Na}(D,\star) \subseteq \mathrm{KN}(D,\star) \subseteq \mathrm{Kr}(D,\star))$.

(2) $\mathrm{KN}(D,\star) := \{f/g \in K(X) \mid f, g \in D[X],\ g \neq 0,$ *such that* $\mathbf{c}(g)$ *is* $\star$*–e.a.b. and* $\mathbf{c}(f) \subseteq \mathbf{c}(g)^\star\}$.

(3) *For each maximal ideal* $\mathfrak{m}$ *of* $\mathrm{KN}(D,\star)$, *set* $L(\mathfrak{m}) := \mathrm{KN}(D,\star)_{\mathfrak{m}} \cap K$. *Then:*

$L(\mathfrak{m})$ is a $\star$–monolocality of D (with maximal ideal $\mathfrak{M} := \mathfrak{m}\mathrm{KN}(D,\star)_{\mathfrak{m}} \cap L(\mathfrak{m})$),

$\mathrm{KN}(D,\star)_{\mathfrak{m}}$ coincides with the Nagata ring $L(\mathfrak{m})(X)$ and $\mathfrak{m}$ coincides with $\mathfrak{M}(X) \cap \mathrm{KN}(D,\star)$.

(4) *Every $\star$–monolocality of an integral domain D contains a minimal $\star$–monolocality of D. If we denote by $\boldsymbol{\mathcal{L}}(D,\star)_{min}$ the set of all the minimal $\star$–monolocalities of D, then*
$\boldsymbol{\mathcal{L}}(D,\star)_{min} = \{L(\mathfrak{m}) \mid \mathfrak{m} \in \mathrm{Max}(\mathrm{KN}(D,\star))\}$ *and, obviously,*
$\mathrm{KN}(D,\star) = \cap\{L(X) \mid L \in \boldsymbol{\mathcal{L}}(D,\star)_{min}\}$.

(5) *For each $J := (a_0, a_1, \ldots, a_n)D \in \boldsymbol{f}(D)$, with $J \subseteq D$ and J $\star$–e.a.b., let $g := a_0 + a_1X + \ldots + a_nX^n \in D[X]$, then:*

$$J\mathrm{KN}(D,\star) = J^\star\mathrm{KN}(D,\star) = g\mathrm{KN}(D,\star).$$

(6) *Let $\star_\ell$ and $\wedge_{\boldsymbol{\mathcal{L}}}$ be the semistar operations of D defined as follows, for each $E \in \overline{\boldsymbol{F}}(D)$,*

$$\begin{aligned}E^{\star_\ell} &:= E\mathrm{KN}(D,\star) \cap K,\\ E^{\wedge_{\boldsymbol{\mathcal{L}}}} &:= \cap\{EL \mid L \in \boldsymbol{\mathcal{L}}(D,\star)\}.\end{aligned}$$

Then:

$$\tilde{\star} \leq \star_\ell = \wedge_{\boldsymbol{\mathcal{L}}} \leq \star_a.$$

(7) $\mathrm{Na}(D,\star) = \mathrm{Na}(D,\tilde{\star}) = \mathrm{KN}(D,\tilde{\star})$ *and* $\mathrm{KN}(D,\star_a) = \mathrm{Kr}(D,\star_a) = \mathrm{Kr}(D,\star)$.

For the proof cf. [22, Theorem 5.11 ((1),(4),(5),(6), and (7)), Proposition 6.3, and Corollary 6.4 ((1) and (2))].

References

1. D. D. Anderson, *Multiplication ideals, multiplication rings, and the ring $R(X)$*, Canad. J. Math. **28** (1976), 760–768.
2. D. D. Anderson, *Some remarks on the ring $R(X)$*, Comm. Math. Univ. St. Pauli **26** (1977), 137–140.
3. D. D. Anderson, *Star operations induced by overrings*, Comm. Algebra **28** (1988), 2535–2553.
4. D. D. Anderson and D. F. Anderson, *Examples of star operations on integral domains*, Comm. Algebra **18** (1990), 1621–1643.
5. D. D. Anderson and S. J. Cook, *Two star-operations and their induced lattices*, Comm. Algebra **28** (2000), 2461–2475.
6. D. F. Anderson, D. E. Dobbs and M. Fontana, *When is a Bézout domain a Kronecker function ring ?* C.R. Math. Rep. Acad. Sci. Canada **9** (1987), 25–30.
7. D. F. Anderson, E. G. Houston and M. Zafrullah, *Pseudo-integrality*, Canad. Math. Bull. **34** (1991), 15–22.
8. J. Arnold, *On the ideal theory of the Kronecker function ring and the domain $D(X)$*, Canad. J. Math. **21** (1969), 558–563.
9. J. Arnold and J. Brewer, *Kronecker function rings and flat $D[X]$-modules*, Proc. Amer. Math. Soc. **27** (1971), 483–485.
10. K. E. Aubert, *Divisors of finite type*, Ann. Mat. Pura Appl. **133** (1983), 327–361.
11. G. W. Chang and J. Park, *Star-invertible ideals of integral domains*, Boll. Un. Mat. Ital. **6-B** (2003), 141–150.
12. R. Dedekind, *Supplement XI* to *Vorlesungen über Zahlentheorie* von P.G. Lejeune Dirichlet. Gesammelte mathematische Werke (R. Fricke, E. Noether and O. Ore, Editors) Vol. 3, 1-222. Vieweg, 1930-1932.
13. D. E. Dobbs and M. Fontana, *Kronecker function rings and abstract Riemann surfaces*, J. Algebra **99** (1986), 263–284.
14. D. Dobbs, E. Houston, T. Lucas and M. Zafrullah, *t-linked overrings and Prüfer v-multiplication domains*, Comm. Algebra **17** (1989), 2835–2852.
15. H.M. Edwards, *Divisor Theory*, Birkhäuser, 1990.
16. S. El Baghdadi, *On a class of Prüfer v-multiplication domains*, Comm. Algebra **30** (2002), 3723-3742.
17. M. Fontana and J. Huckaba, *Localizing systems and semistar operations*, in "Non Noetherian Commutative Ring Theory" (S. Chapman and S. Glaz, Editors), Kluwer Academic Publishers, Dordrecht, 2000, Chapter 8, 169–187.
18. M. Fontana, P. Jara and E. Santos, *Prüfer $\star$-multiplication domains and semistar operations*, J. Algebra Appl. **2** (2003), 21–50.
19. M. Fontana and K. A. Loper, *Kronecker function rings: a general approach*, in "Ideal Theoretic Methods in Commutative Algebra" (D.D. Anderson and I.J. Papick, Editors), M. Dekker Lecture Notes Pure Appl. Math. **220** (2001), 189–205.
20. M. Fontana and K. A. Loper, *A Krull-type theorem for the semistar integral closure of an integral domain*, ASJE Theme Issue "Commutative Algebra" **26** (2001), 89–95.
21. M. Fontana and K. A. Loper, *Nagata rings, Kronecker function rings and related semistar operations*, Comm. Algebra **31** (2003), 4775–4805.
22. M. Fontana and K. A. Loper, *A generalization of Kronecker function rings and Nagata rings*, Forum Math., to appear.

23. M. Fontana and G. Picozza, *Semistar invertibility on integral domains,* Algebra Colloquium 12 (2005), 645–664.
24. J. M. Garcia, P. Jara and E. Santos, *Prüfer $\star$–multiplication domains and torsion theories,* Comm. Algebra **27** (1999), 1275–1295.
25. R. Gilmer, *Multiplicative Ideal Theory (Part I and II),* Queen's Papers in Pure and Applied Mathematics, Queen's University, Kingston, Ontario, Canada, 1968.
26. R. Gilmer, *Multiplicative Ideal Theory,* M. Dekker, New York, 1972.
27. S. Glaz and W. Vasconcelos, *Flat ideals, II,* Manuscripta Math. **22** (1977), 325–341.
28. M. Griffin, *Some results on v–multiplication rings*, Canad. J. Math. **19** (1967), 710–721.
29. F. Halter-Koch, *Generalized integral closures,* in "Factorization in Integral Domains" (D.D. Anderson, Editor), M. Dekker Lecture Notes Pure Appl. Math. **187** (1997), 349–358.
30. F. Halter-Koch, *Ideal Systems: An Introduction to Multiplicative Ideal Theory,* M. Dekker, New York, 1998.
31. F. Halter-Koch, *Localizing systems, module systems and semistar operations,* J. Algebra **238** (2001), 723–761.
32. F. Halter-Koch, *Kronecker function rings and generalized integral closures,* Comm. Algebra **31** (2003), 45–59.
33. F. Halter-Koch, *Characterization of Prüfer multiplication monoids and domains by means of spectral module systems,* Monatsh. Math. **139** (2003), 19-31.
34. J. R. Hedstrom and E. G. Houston, *Some remarks on star-operations*, J. Pure Appl. Algebra **18** (1980), 37–44.
35. W. Heinzer and J. Ohm, *An essential ring which is not a v-multiplication ring,* Can. J. Math. **25** (1973), 856–861.
36. E. Houston, *On divisorial prime ideals in Prüfer v-multiplication domains,* J. Pure Appl. Algebra **42** (1986), 55–62.
37. E. G. Houston, S. B. Malik, and J. L. Mott, *Characterization of $\star$–multiplication domains*, Canad. Math. Bull. **27** (1984), 48–52.
38. P. Jaffard, *Les Systèmes d'Idéaux,* Dunod, Paris, 1960.
39. B. G. Kang, *$\star$–operations on integral domains*, Ph.D. Dissertation, Univ. Iowa 1987.
40. B. G. Kang, *Prüfer v-multiplication domains and the ring $R[X]_{N_v}$*, J. Algebra **123** (1989), 151–170.
41. L. Kronecker, *Grundzüge einer arithmetischen Theorie der algebraischen Grössen,* J. Reine Angew. Math., **92** (1882), 1–122; *Werke* **2**, 237–387 (K. Hensel, Editor, 5 Volumes published from 1895 to 1930, Teubner, Leipzig) Reprint, Chelsea 1968.
42. W. Krull, *Idealtheorie,* Springer-Verlag, Berlin, 1935 (II edition, 1968).
43. W. Krull, *Beiträge zur Arithmetik kommutativer Integritätsbereiche,* I - II. Math. Z. **41** (1936), 545–577; 665–679.
44. W. Krull, *Beiträge zur Arithmetik kommutativer Integritätsbereiche,* VIII. Math. Z. **48** (1942/43), 533–552.
45. E. E. Kummer, *Collected papers,* (A. Weil, Editor), Springer 1975.
46. P. Lorenzen, *Abstrakte Begründung der multiplikativen Idealtheorie,* Math. Z. **45** (1939), 533–553.

47. R. Matsuda, *Kronecker function rings of semistar operations on rings*, Algebra Colloquium **5** (1998), 241–254.
48. R. Matsuda and I. Sato, *Note on star operations and semistar operations*, Bull. Fac. Sci. Ibaraki Univ. Ser. A **28** (1996), 5–22.
49. R. Matsuda and T. Sugatani, *Semistar operations on integral domains*, II, Math. J. Toyama Univ. **18** (1995), 155–161.
50. J. L. Mott and M. Zafrullah, *On Prüfer v–multiplication domains,* Manuscripta Math. **35** (1981), 1–26.
51. M. Nagata, *A general theory of algebraic geometry over Dedekind domains, I,* Amer. J. Math. **78** (1956), 78–116.
52. M. Nagata, *Local rings,* Interscience, New York, 1962.
53. A. Okabe and R. Matsuda, *Star operations and generalized integral closures,* Bull. Fac. Sci. Ibaraki Univ. Ser. A **24** (1992), 7–13.
54. A. Okabe and R. Matsuda, *Semistar operations on integral domains*, Math. J. Toyama Univ. **17** (1994), 1–21.
55. A. Okabe and R. Matsuda, *Kronecker function rings of semistar operations,* Tsukuba J. Math., **21** (1997), 529-540.
56. H. Prüfer, *Untersuchungen über Teilbarkeitseigenschaften in Körpern,* Journ. Reine Angew. Math. **168** (1932), 1–36.
57. P. Quartaro jr. and H. S. Butts, *Finite unions of ideals and modules*, Proc. Amer. Math. Soc. **52** (1975), 91–96.
58. P. Samuel, *Lectures on unique factorization domains,* Tata Institute, Bombay, 1964.
59. Fanggui Wang, *On w–projective modules and w–flat modules*, Algebra Colloquium **4** (1997), 111–120.
60. Fanggui Wang, *On UMT-domains and w–integral dependence*, Preprint.
61. Fanggui Wang and R. L. McCasland, *On w–modules over strong Mori domains,* Comm. Algebra **25** (1997), 1285–1306.
62. Fanggui Wang and R. L. McCasland, *On strong Mori domains*, J. Pure Appl. Algebra **135** (1999), 155–165.
63. H. Weyl, *Algebraic Theory of Numbers,* Princeton University Press, Princeton, 1940.
64. M. Zafrullah, *Some polynomial characterizations of Prüfer v–multiplication domains,* J. Pure Appl. Algebra **32** (1984), 231–237.
65. M. Zafrullah, *Putting t-invertibility to use,* in "Non Noetherian Commutative Ring Theory" (S. Chapman and S. Glaz, Editors), Kluwer Academic Publishers, Dordrecht, 2000, 429–457.

Generalized Dedekind domains

Stefania Gabelli

Dipartimento di Matematica, Università degli Studi "Roma Tre", Largo S. L. Murialdo, 1, 00146 Roma, Italy gabelli@mat.uniroma3.it

To Robert Gilmer, who paved the way.

1 Introduction

A Dedekind domain is a Noetherian Prüfer domain. By weakening some of the finiteness properties of Dedekind domains, one often obtains classes of Prüfer domains having good multiplicative properties which are interesting in their own right. For example, almost Dedekind domains were defined by R. Gilmer by relaxing the condition that each proper ideal has finitely many minimal primes [21] and the study of these domains led to the introduction of the #-property and of some related conditions [18, 24, 25, 35, 36, 39].

The class of generalized Dedekind domains is complementary to that of almost Dedekind domains, in the sense that a domain which is at the same time almost Dedekind and generalized Dedekind is in fact a Dedekind domain. Generalized Dedekind domains were defined by N. Popescu by means of localizing systems of ideals [39]; but in this context they can be defined more naturally as Prüfer domains with the properties that no nonzero prime ideal is idempotent and each proper ideal has finitely many minimal primes [39, 40]. Thus, roughly speaking, generalized Dedekind domains are "Dedekind domains of dimension greater than one".

These domains behave nicely with respect to several important ring-theoretic and ideal-theoretic conditions introduced in the last few decades, since the appearance of the first version of Gilmer's *Multiplicative Ideal Theory* [23]. Their study gives the opportunity to bring into evidence the connections among all these properties and to provide a unified point of view of various apparently unrelated results.

Generalized Dedekind domains are the subject of [11, Chapter V]. In the present survey paper I report the latest developments and give direct proofs of the main results. However, I do not consider module-theoretic properties; for this aspect of the theory the reader is referred to [14]. Recently the class of generalized Dedekind domains has been enlarged, within Prüfer v-multiplication domains, to the class of the so called generalized Krull domains [5]. Multiplicative properties of generalized Krull domains are studied in [5, 6, 7, 8, 18, 19].

2 Ring-theoretic Properties

2.1 Strongly Discrete Prüfer Domains

A prime ideal P of a domain R is *branched* if there exists a P-primary ideal distinct from P. Branched prime ideals of valuation and Prüfer domains are characterized in [24, Theorems 17.3 and 23.3]. A valuation domain V such that each branched prime ideal is not idempotent is said to be a *discrete valuation domain* [24, p. 192]. If V satisfies the stronger condition that each nonzero prime ideal is not idempotent, then V is called *a strongly discrete valuation domain* [11, p. 145]. V is strongly discrete if and only if it is discrete and satisfies the ascending chain condition on prime ideals [11, Proposition 5.3.1]. A valuation domain is Noetherian if and only if it is a one-dimensional discrete valuation domain, for short a *DVR* [24, Theorem 17.5].

A domain R is a *discrete* (respectively, *strongly discrete*) *Prüfer domain* if the quasilocal ring R_P is a discrete (respectively, strongly discrete) valuation domain, for each nonzero prime ideal P. Thus a Prüfer domain is discrete (respectively, strongly discrete) if and only if $P \neq P^2$ for each branched prime (respectively, each nonzero prime) ideal P [11, Proposition 5.3.2]. In the finite dimensional case, the classes of discrete and strongly discrete Prüfer domains coincide [11, Corollary 5.3.6]. An *almost Dedekind domain* is a one-dimensional (strongly) discrete Prüfer domain [21]. Finally, a *generalized Dedekind domain* is a strongly discrete Prüfer domain such that each nonzero ideal has finitely many minimal primes [39, 40].

Proposition 2.1. *The following conditions are equivalent for a domain R:*

(i) *R is a Dedekind domain;*
(ii) *R is a generalized Dedekind domain of dimension one;*
(iii) *R is a completely integrally closed generalized Dedekind domain;*
(iv) *R is a generalized Dedekind domain and an almost Dedekind domain.*

Proof. It follows from the definitions and the fact that a valuation domain is completely integrally closed if and only if it is one-dimensional [24, Theorem 17.5].

Recall that a domain R has *Noetherian Spectrum* if the closed subsets of Spec(R) in the Zariski topology satisfy the descending chain condition. This property can be expressed in several equivalent ways ([24, Exercise 27.9] and [11, Theorem 3.1.11]).

Proposition 2.2. *The following conditions are equivalent for a domain R:*

(i) *R has Noetherian Spectrum;*

(ii) *R satisfies the ascending chain condition on radical ideals;*

(iii) *R satisfies the ascending chain condition on prime ideals and each proper ideal has finitely many minimal primes;*

(iv) *Each proper ideal of R contains a finitely generated ideal with the same radical;*

(v) *Each prime ideal of R is the radical of a finitely generated ideal.*

Since strongly discrete valuation domains satisfy the ascending chain condition on prime ideals [11, Proposition 5.3.1], from Proposition 2.2 we get:

Theorem 2.3. [12, Théorème 2.7] *The following conditions are equivalent for a domain R:*

(i) *R is a generalized Dedekind domain;*

(ii) *R is a strongly discrete Prüfer domain with Noetherian Spectrum.*

Corollary 2.4. *A strongly discrete Prüfer domain with finite character is a generalized Dedekind domain.*

Proof. If a Prüfer domain has finite character, then each proper ideal has finitely many minimal primes, because the prime Spectrum of a Prüfer domain is treed. Hence we can apply Proposition 2.2 and Theorem 2.3.

Corollary 2.5. *A maximal ideal of a generalized Dedekind domain is invertible.*

Proof. Let R be a generalized Dedekind domain and M a maximal ideal of R. If J is a finitely generated ideal with radical M (Theorem 2.3), then $J = JR_M \cap R = M^e R_M \cap R = M^e$, for some $e \geq 1$ (because JR_M is MR_M-primary and $M \neq M^2$). Hence M is invertible.

2.2 Localizing Systems

A *multiplicative system* of ideals of a domain R is a nonempty family $\mathcal{F}$ of nonzero ideals such that $IJ \in \mathcal{F}$, for each $I, J \in \mathcal{F}$. For each multiplicative system of ideals $\mathcal{F}$, the overring $R_{\mathcal{F}}\colon = \cup\{(R{:}J);\ J \in \mathcal{F}\}$ of R is called the *generalized ring of fractions* of R with respect to $\mathcal{F}$.

We say that a multiplicative system of ideals $\mathcal{F}$ is a *localizing system* of R if the following two conditions hold:

(a) $\mathcal{F}$ is *saturated* (that is, if $J \in \mathcal{F}$ and I is an ideal of R containing J, then $I \in \mathcal{F}$);

(b) If $J \in \mathcal{F}$ and I is an ideal of R such that $(I{:}_R xR) \in \mathcal{F}$ for all $x \in J$, then $I \in \mathcal{F}$.

A multiplicative system of ideals $\mathcal{F}$ is *finitely generated* (respectively, *principal*) if each ideal $I \in \mathcal{F}$ contains a finitely generated (respectively, principal) ideal J which is still in $\mathcal{F}$. If T is an overring of R, it is easy to see that the set of ideals $\mathcal{F}(T){:}=\{I \subseteq R$ such that $IT = T\}$ is a finitely generated localizing system. It is well known that T is R-flat if and only if $T = R_{\mathcal{F}(T)}$. If R is a Prüfer domain, a multiplicative system of ideals $\mathcal{F}$ is finitely generated if and only if $\mathcal{F} = \mathcal{F}(R_{\mathcal{F}})$ [16, Theorem 1.3].

For any nonempty subset Λ of Spec(R), the set of ideals $\mathcal{F}(\Lambda){:}=\{I; I \nsubseteq P$, for each $P \in \Lambda\}$ is a localizing system of R and $R_{\mathcal{F}(\Lambda)} = \cap_{P\in\Lambda} R_P$. A Prüfer domain R has Noetherian Spectrum if and only if the localizing system $\mathcal{F}(\Lambda)$ is finitely generated for each $\Lambda \subseteq$ Spec(R) [16, Theorem 2.6].

Theorem 2.6. ([39, Theorem 2.5] and [40, Theorem 3.1] *The following conditions are equivalent for a Prüfer domain R:*

(i) *R is a generalized Dedekind domain;*
(ii) *Each localizing system of ideals of R is finitely generated;*
(iii) *For each localizing system $\mathcal{F}$ of R, $\mathcal{F} = \mathcal{F}(R_{\mathcal{F}}){:}=\{I \subseteq R; IR_{\mathcal{F}} = R_{\mathcal{F}}\}$;*
(iv) *If $\mathcal{F}$ and $\mathcal{F}'$ are two localizing systems of R and $R_{\mathcal{F}} = R_{\mathcal{F}'}$, then $\mathcal{F} = \mathcal{F}'$.*

Proof. $(i) \Rightarrow (iii)$ Set $T{:}= R_{\mathcal{F}}$. Always $\mathcal{F}(T) \subseteq \mathcal{F}$. Assume that there exists an ideal $I \in \mathcal{F}$ such that $IT \neq T$ and let M be a maximal ideal of T containing IT. By Theorem 2.3, the prime ideal $P{:}= M \cap R$ is the radical of a finitely generated ideal J. Since $I \subseteq P$, by saturation, $P \in \mathcal{F}$. If we show that $J \in \mathcal{F}$, we reach a contradiction, because then $(R{:}J) \subseteq T$ and, since J is invertible, $1 \in R = J(R{:}J) \subseteq JT \subseteq PT = M$.

Since P is the radical of J, then JT is M-primary and, since $M \neq M^2$, $JT = M^n = P^nT$ for some $n \geq 1$ [24, Theorem 23.3]. Let $J{:}= a_1R+\cdots+a_nR$, $a_i \in R$. Then, for each $x \in P^n$, we have $x = a_1y_1 + \cdots + a_ny_n$, with $y_i \in T$. Let $I_i \in \mathcal{F}$ such that $y_iI_i \subseteq R$, $i = 1,\ldots,n$, and set $I{:}= I_1 \ldots I_n$. Then $I \in \mathcal{F}$ and $xI \subseteq J$. Hence $I \subseteq (J{:}_R xR)$ and, by saturation, $(J{:}_R xR) \in \mathcal{F}$, for each $x \in P^n$. Since $P^n \in \mathcal{F}$ and $\mathcal{F}$ is a localizing system, it follows that $J \in \mathcal{F}$.

$(iii) \Rightarrow (ii)$ is clear.

$(ii) \Rightarrow (i)$ For each subset Λ of Spec(R), the localizing system $\mathcal{F}(\Lambda)$ is finitely generated. Thus, by [16, Theorem 2.6], R has Noetherian Spectrum.

By Theorem 2.3, we are left to show that R is a strongly discrete Prüfer domain, that is $P \neq P^2$, for each nonzero prime ideal P.

Let M be a maximal ideal of R containing P. Set $V{:}= R_M$, $Q{:}= PV$ and denote by $\mathcal{F}^*(Q)$ the set of ideals of V containing Q. It is easy to check that, if $Q = Q^2$, then $\mathcal{F}^*(Q)$ is a localizing system of V [11, Theorem 5.1.15]. The set of ideals $\mathcal{F}{:}=\{I \subseteq R; IV \in \mathcal{F}^*(Q)\}$ is a localizing system of R that, by hypothesis, is finitely generated. Thus also $\mathcal{F}^*(Q)$ is finitely generated. This is a contradiction, because $Q \in \mathcal{F}^*(Q)$ and Q is not finitely generated (since $Q = Q^2$). Hence $P \neq P^2$.

$(iii) \Leftrightarrow (iv)$ because, by flatness, $R_{\mathcal{F}} = R_{\mathcal{F}(R_{\mathcal{F}})}$, for each localizing system $\mathcal{F}$ of R.

2.3 The ##-property

The #-property and the ##-property were introduced by R. Gilmer [22] and R. Gilmer and W. Heinzer [25] respectively. A domain R has the *#-property*, or it is a *#-domain*, if $R_{\mathcal{F}(\Lambda_1)} \neq R_{\mathcal{F}(\Lambda_2)}$, for any pair of distinct nonempty subsets Λ_1, Λ_2 of $\mathrm{Max}(R)$. Any overring of a one-dimensional Prüfer #-domain is a #-domain [22, Corollary 2], but in general the #-property is not inherited by overrings [25, Section 2]. We say that R has the *##-property*, or it is a *##-domain*, if each overring of R is a #-domain. A Prüfer domain R is a ##-domain if and only if the localizing system $\mathcal{F}(\Lambda)$ is finitely generated for each subset Λ of $\mathrm{Max}(R)$ [16, Theorem 2.4].

N. Popescu introduced in [39] a condition apparently stronger than the ##-property, which is called the *$\#_P$-property* [11]. A domain R is a *$\#_P$-domain* if $R_{\mathcal{F}(\Lambda_1)} \neq R_{\mathcal{F}(\Lambda_2)}$ for any pair of distinct nonempty subsets Λ_1 and Λ_2 of $\mathrm{Spec}(R)$ with the property that $P + Q = R$ for any two different primes $P \in \Lambda_1$ and $Q \in \Lambda_2$. However the ##-property and the $\#_P$-property coincide in any Prüfer domain [16, Theorem 2.5].

Theorem 2.7. ([39, Proposition 3.2] and [12, Théorème 2.7]) *The following conditions are equivalent for a domain R:*

(i) *R is generalized Dedekind domain;*
(ii) *R is a strongly discrete Prüfer ##-domain;*
(iii) *R is a strongly discrete Prüfer $\#_P$-domain.*

Proof. *(i)* ⇔ *(ii)* Let R be a strongly discrete Prüfer domain. Since R satisfies the ascending chain condition on prime ideals, then R has Noetherian Spectrum if and only if it is a ##-domain [25, Theorem 4]. Thus we can apply Theorem 2.3.

(ii) ⇔ *(iii)* because the ##-property and the $\#_P$-property coincide in every Prüfer domain [16, Theorem 2.5].

In the one-dimensional case, by Proposition 2.2 and Theorem 2.3, we recover the result of R. Gilmer which motivated the introduction of the #-property.

Corollary 2.8. [22, Theorem 3] *The following conditions are equivalent for a domain R:*

(i) *R is Dedekind domain;*
(ii) *R is an almost Dedekind #-domain.*

2.4 The Radical Trace Property

If R is an integral domain and M is a unitary R-module, the *trace* of M is the ideal of R generated by the set $\{f(m)\,;\, f \in \mathrm{Hom}_R(M,R)\,,\, m \in M\}$. An ideal J of R is called a *trace ideal* if it is the trace of some R-module M. This happens if and only if $J = I(R{:}I)$, for some nonzero ideal I of R, equivalently $(J{:}J) = (R{:}J)$ [11, Lemmas 4.2.2. and 4.2.3]. In the literature trace ideals are also called *strong ideals.*

If V is a valuation domain, a trace ideal is either equal to V or it is prime [11, Proposition 4.2.1]; this fact led to the consideration of several conditions related to trace ideals [32]. One of the most useful is the radical trace property introduced by W. Heinzer and I. Papick [30]. R is a domain satisfying the *radical trace property,* or it is an *RTP-domain,* if each proper trace ideal is a radical ideal.

Theorem 2.9. [30, Theorem 2.7] *If R is a Prüfer domain satisfying the ascending chain condition on prime ideals, the following conditions are equivalent:*

(i) *R is an RTP-domain;*
(ii) *R is a ##-domain;*
(iii) *R has Noetherian Spectrum.*

Since a strongly discrete Prüfer domain satisfies the ascending chain condition on prime ideals, by the previous theorem and Theorem 2.3 (or Theorem 2.7), we immediately get:

Theorem 2.10. *The following conditions are equivalent for a domain R:*

(i) *R is generalized Dedekind domain;*
(ii) *R is a strongly discrete Prüfer domain satisfying the radical trace property.*

2.5 The Strong Finite Type Property

Let R be a commutative ring with identity. An ideal I of R is an ideal of *strong finite type,* for short an *SFT-ideal,* if there exists a finitely generated ideal J and a positive integer k such that $J \subseteq I$ and $x^k \in J$ for each $x \in I$ and R is an *SFT-ring* if each ideal of R is an SFT-ideal. This property was introduced by J. T. Arnold for studying the Krull dimension of formal power series rings [1, 2].

The following characterization of generalized Dedekind domains is due to B. G. Kang and M. H. Park [33].

Theorem 2.11. [33, Theorem 2.4] *The following conditions are equivalent for a domain R:*

(i) *R is a generalized Dedekind domain;*
(ii) *R is a Prüfer SFT-ring.*

Proof. *(i) ⇒ (ii)* For a Prüfer domain to be a STF-ring it is sufficient that, for each nonzero prime ideal P, there exists a finitely generated ideal J such that $P^2 \subseteq J \subseteq P$ [2, Proposition 3.1]. Assume that R is a generalized Dedekind domain and let P be a nonzero prime ideal of R. If P is the radical of the finitely generated ideal I (Theorem 2.3) and $x \in P \setminus P^2$, then by checking locally, we see that the ideal $J := I + xR$ has the desired property.

(ii) $\Rightarrow$ *(i)* If R is an SFT-ring, then R has Noetherian Spectrum [2, Proposition 2.5] and each nonzero prime ideal of R is not idempotent [2, Lemma 2.7]. Thus we can apply Theorem 2.3.

3 Ring Extensions

Since generalized Dedekind domains are Prüfer domains, each overring T of a generalized Dedekind domain R is an intersection of localizations of R, that is $T = R_{\mathcal{F}(\Lambda)}$, for some subset Λ of Spec(R) [24, Theorem 26.2].

Theorem 3.1. [39, Corollary 2.3] *Let R be a generalized Dedekind domain and $T := R_{\mathcal{F}(\Lambda)}$, $\Lambda \subseteq$ Spec(R), a proper overring of R. Then T is a generalized Dedekind domain and* Max(T) $= \{PT\,; P$ *maximal in* $\Lambda\}$.

Proof. Each overring of a strongly discrete Prüfer domain is a strongly discrete Prüfer domain and each proper overring of a Prüfer domain with Noetherian Spectrum has Noetherian Spectrum [11, Lemma 5.4.6]. Hence T is a generalized Dedekind domain by Theorem 2.3. In addition, $\mathcal{F}(\Lambda) = \{I \subseteq R\,; IT = T\}$ by Theorem 2.6. Thus $IT \neq T$ if and only if $I \subseteq P$, for some $P \in \Lambda$. Since each ideal of T is extended [24, Theorem 26.1], then Spec(T) $= \{QT\,; Q \subseteq P$ for some $P \in \Lambda\}$. We conclude by observing that Λ has maximal elements by the ascending chain condition on prime ideals.

Corollary 3.2. *The complete integral closure of a generalized Dedekind domain is a Dedekind domain.*

Proof. The complete integral closure of a generalized Dedekind domain is a generalized Dedekind domain by Theorem 3.1 and it is one-dimensional by the ##-property [25, Theorem 10]. Hence we can apply Proposition 2.1.

In [11, Example 8.4.8] an example is given of an infinite-dimensional valuation domain V with the maximal ideal idempotent and each other nonzero prime ideal not idempotent. Thus, each proper overring of V is a generalized Dedekind domain (indeed a strongly discrete valuation domain) but V is not generalized Dedekind (it is a discrete valuation domain which is not strongly discrete).

Theorem 3.1 can be generalized in the following way.

Theorem 3.3. [39, Corollary 2.8] *Let R be a strongly discrete Prüfer domain (resp. a generalized Dedekind domain) with quotient field K and let F be a finite extension of K. Then the integral closure of R in F is a strongly discrete Prüfer domain (resp. a generalized Dedekind domain).*

By using the theory developed by J. T. Arnold and R. Gilmer in [3], M. Fontana and N. Popescu were able to give conditions for an extension of a generalized Dedekind domain in an algebraic extension of its quotient field to be a generalized Dedekind domain [12, Section 3].

Theorem 3.4. [37, Theorem 4.7] *The following conditions are equivalent for a Prüfer domain R:*

(i) *R is a generalized Dedekind domain;*
(ii) *If T is a proper overring of R, each maximal ideal of T is invertible.*

Proof. *(i)* $\Rightarrow$ *(ii)* Since each overring T of R is a generalized Dedekind domain (Theorem 3.1), each maximal ideal of T is invertible by Corollary 2.5.

(ii) $\Rightarrow$ *(i)* For each nonzero prime ideal P of R, PR_P is invertible. Thus $P \neq P^2$ and R is a strongly discrete Prüfer domain. In particular, T is a Prüfer domain with invertible maximal ideals. This implies that T is a #-domain [25, Theorem 2] and so R is a ##-domain. We conclude by applying Theorem 2.7.

4 Pullbacks

Of particular interest in multiplicative ideal theory, mainly because of their success in producing interesting examples, are the *pullback diagrams* of type:

$$\begin{array}{ccc} R:=\pi^{-1}(D) & \longrightarrow & D \\ \downarrow & & \downarrow{\scriptstyle\iota} \\ T & \xrightarrow{\pi} & \frac{T}{M}=:F \end{array}$$

where T is a domain, M is a maximal ideal of T, π is the canonical projection, D is a subring of the field F and ι is the canonical inclusion. When $T = F+M$, we obtain $R = D+M$ and refer to R as a "construction of type $D+M$" [24].

The proof of the following theorem is based on the characterization of generalized Dedekind domains given in Theorem 2.3 and well known properties of ideals in pullbacks [17].

Theorem 4.1. [12, Théorème 4.1] *With the previous notation, R is a (strongly) discrete Prüfer domain (respectively, a generalized Dedekind domain) if and only if T and D are (strongly) discrete Prüfer domains (respectively, generalized Dedekind domains) and F is the quotient field of D.*

If T is a discrete valuation domain of dimension $n \geq 1$ and D is a DVR, then R is a discrete valuation domain of dimension $n+1$. A valuation domain of finite dimension $n \geq 2$ is discrete if and only if it can be obtained by iterating this construction [12, Proposition 2.5].

Corollary 4.2. *Let F be a field and X an indeterminate over F. Then $R:= D+XF[X]$ is a generalized Dedekind domain if and only if D is a generalized Dedekind domain with quotient field F.*

5 Ideal-theoretic Properties

5.1 The Prime Spectrum

We have seen in Section 2.1 that the prime Spectrum of a generalized Dedekind domain is a Noetherian tree with respect to the inclusion. A converse is due to A. Facchini.

Theorem 5.1. [9, Theorem 5.3] *For any ordered set X which is a Noetherian tree with a least element, there exists a generalized Dedekind domain R such that* $\mathrm{Spec}(R)$ *is order isomorphic to X.*

As in [13, 20], if R is a generalized Dedekind domain, for any ordinal number α, we define by transfinite induction a subset of $\mathcal{S} := \mathrm{Spec}(R) \setminus \{0\}$ in the following way:

- $\mathcal{S}_0 := \emptyset$;
- $\mathcal{S}_1 := \mathrm{Max}(R)$;
- Assume that $\mathcal{S}_\beta$ has been defined for all $\beta < \alpha$.
  - If $\alpha = \beta + 1$, denoting by $\mathcal{M}_\alpha$ the set of the maximal elements of $\mathcal{S} \setminus \mathcal{S}_\beta$, we set:
  $$\mathcal{S}_\alpha := \mathcal{S}_\beta \cup \mathcal{M}_\alpha .$$
  Note that $\mathcal{M}_\alpha \neq \emptyset$ whenever $\mathcal{S}_\beta \neq \mathcal{S}$, by the ascending chain condition on prime ideals.
  - If α is a limit ordinal, we set:
  $$\mathcal{S}_\alpha := \cup_{\beta < \alpha} \mathcal{S}_\beta .$$

Clearly $\mathcal{S}_\beta \subseteq \mathcal{S}_\alpha$ when $\beta \leq \alpha$ and there exists an ordinal number δ such that
$$\mathcal{S} = \mathcal{S}_\delta = \cup_\alpha \mathcal{S}_\alpha .$$
The smallest ordinal number with this property, which we denote by $d(R)$, coincides with the classical Krull dimension of R as defined in [28, p. 48].

For each ordinal $\alpha < d(R)$, we can then define the set $\mathcal{H}_\alpha$ of prime ideals of *coheight* α by setting:
$$\mathcal{H}_\alpha := \mathcal{S}_{\alpha+1} \setminus \mathcal{S}_\alpha = \mathcal{M}_{\alpha+1} .$$
We have $\mathcal{H}_0 = \mathrm{Max}(R)$ and, since each nonzero prime ideal P of R is not idempotent, by [24, Theorem 23.2],
$$\mathcal{H}_{\alpha+1} = \{\cap_{n \geq 1} P^n \, ; P \in \mathcal{H}_\alpha\} .$$

5.2 Invertible Ideals

A domain R is a Dedekind domain if and only if the set of nonzero fractional ideals of R is a free abelian group generated by the nonzero prime ideals. We will show in this section that, if R is a generalized Dedekind domain, the group $\mathrm{Inv}(R)$ of the invertible fractional ideals of R has a similar property.

Keeping the notation of Sections 2.2 and 5.1, if R is a generalized Dedekind domain, for each ordinal α, consider the localizing system $\mathcal{F}_\alpha(R) := \mathcal{F}_\alpha$ defined by:

$$\mathcal{F}_\alpha := \mathcal{F}(\mathcal{H}_\alpha) := \{I\,;I \nsubseteq P \text{ for each } P \in \mathcal{H}_\alpha\} = \{I\,;\mathrm{Min}(I) \subseteq \mathcal{S}_\alpha\} \cup \{R\}\,.$$

Thus $\mathcal{F}_0 = \{R\}$ and $\mathcal{F}_\alpha$ is the set of all nonzero ideals of R when $\alpha \geq d(R)$. Since $\mathcal{F}_\alpha$ is finitely generated (Theorem 2.6), then $\mathrm{Inv}(R) \cap \mathcal{F}_\alpha \neq \emptyset$. We set:

$$\mathrm{Inv}_\alpha(R) := \left\{IJ^{-1}\,; I, J \in \mathrm{Inv}(R) \cap \mathcal{F}_\alpha\right\}.$$

For $\beta \leq \alpha$, we have $\mathcal{F}_\beta \subseteq \mathcal{F}_\alpha$. Whence $\mathrm{Inv}_\beta(R) \subseteq \mathrm{Inv}_\alpha(R)$ and

$$\mathrm{Inv}(R) = \cup_\alpha \mathrm{Inv}_\alpha(R) = \mathrm{Inv}_{d(R)}(R).$$

Proposition 5.2. [20, Corollary 1.4] *Let R be a generalized Dedekind domain. Then*

$$\mathcal{F}_1(R) = \{M_1^{e_1} \dots M_n^{e_n}\,; M_i \in \mathrm{Max}(R)\,, e_i \geq 0\} \subseteq \mathrm{Inv}(R).$$

Hence $\mathrm{Inv}_1(R)$ *is the free abelian group generated by* $\mathrm{Max}(R)$.

Proof. We have $\mathcal{F}_1(R) = \{I \subseteq R; \mathrm{Min}(I) \subseteq \mathrm{Max}(R)\}$. If $I \in \mathcal{F}_1(R)$ has minimal primes $\{M_1, \dots, M_n\}$ (Theorem 2.3), then $I = (\cap IR_{M_i}) \cap R = (\cap M^{e_i} R_{M_i}) \cap R = M_1^{e_1} \dots M_n^{e_n}$ for some $e_i \geq 1$ (because IR_{M_i} is $M_i R_{M_i}$-primary and $M_i \neq M_i{}^2$). Finally, $\mathrm{Max}(R) \subseteq \mathrm{Inv}(R)$ by Corollary 2.5.

Now consider the overrings of R:

$$R_\alpha := R_{\mathcal{F}_\alpha} = \cap_{P \in \mathcal{H}_\alpha} R_P.$$

Note that $R_0 = R$ and $R_\alpha = K$ is the quotient field of R for $\alpha \geq d(R)$. By Theorem 3.1, R_α is a generalized Dedekind domain and, for $\alpha < d(R)$, $\mathrm{Max}(R_\alpha) = \{PR_\alpha\,; P \in \mathcal{H}_\alpha\}$. Thus, by Proposition 5.2,

$$\mathcal{F}_1(R_\alpha) = \{P_1^{e_1} \dots P_n^{e_n} R_\alpha\,; P_i \in \mathcal{H}_\alpha\,, e_i \geq 0\}$$

and $\mathrm{Inv}_1(R_\alpha)$ is *isomorphic* to the free abelian group generated by $\mathcal{H}_\alpha$.

If $I \in \mathrm{Inv}(R) \cap \mathcal{F}_{\alpha+1}$ is a proper ideal, then $\mathrm{Min}(I) \subseteq \mathcal{S}_{\alpha+1}$. Hence each minimal prime of IR_α is of type PR_α, for $P \in \mathcal{H}_\alpha$ (because $PR_\alpha = R_\alpha$, for each $P \in \mathcal{S}_\alpha$). It follows that, for each $I \in \mathrm{Inv}_{\alpha+1}(R)$, $IR_\alpha \in \mathrm{Inv}_1(R_\alpha)$.

Proposition 5.3. *Let R be a generalized Dedekind domain. Then, for each ordinal α, there is a splitting exact sequence of groups:*

$$(1) \to \mathrm{Inv}_\alpha(R) \xrightarrow{\iota_\alpha} \mathrm{Inv}_{\alpha+1}(R) \xrightarrow{\pi_\alpha} \mathrm{Inv}_1(R_\alpha) \to (1)$$

where ι_α is the natural inclusion and π_α is the extension.

Proof. Since $\mathrm{Inv}_1(R_\alpha)$ is a free abelian group (Proposition 5.2) and by definition $\mathrm{Ker}(\pi_\alpha) = \mathrm{Inv}_\alpha(R)$, it is enough to check that π_α is surjective. This follows from the fact that each nonzero prime ideal $P \subseteq R$ is the radical of an invertible ideal J such that $PR_P = JR_P$ [20, Corollary 1.13]. Hence, for each $P \in \mathcal{H}_\alpha$, $PR_\alpha = JR_\alpha = \pi_\alpha(J)$, for some $J \in \mathrm{Inv}(R) \cap \mathcal{F}_{\alpha+1}$.

Theorem 5.4. [20, Theorem 2.3] *If R is a generalized Dedekind domain, then* $\mathrm{Inv}(R)$ *is isomorphic to the free abelian group $\mathcal{L}(R)$ generated by* $\mathrm{Spec}(R)$.

Proof. By Proposition 5.3, for each ordinal α,

$$\mathrm{Inv}_{\alpha+1}(R) \cong \mathrm{Inv}_\alpha(R) \oplus \mathrm{Inv}_1(R_\alpha).$$

Since $\mathrm{Inv}_1(R)$ is free generated by $\mathrm{Max}(R) = \mathcal{S}_1 = \mathcal{H}_0$ and $\mathrm{Inv}_1(R_\alpha)$ is isomorphic to the free abelian group generated by $\mathcal{H}_\alpha := \mathcal{S}_{\alpha+1} \setminus \mathcal{S}_\alpha$ (Proposition 5.2), by transfinite induction we conclude that $\mathrm{Inv}_\alpha(R)$ is isomorphic to the free abelian group generated by $\mathcal{S}_\alpha$, for each α. In conclusion $\mathrm{Inv}(R) = \cup_\alpha \mathrm{Inv}_\alpha(R) = \mathrm{Inv}_{d(R)}(R)$ is isomorphic to the free abelian group $\mathcal{L}(R)$ generated by $\mathcal{S}_{d(R)} = \mathcal{S} =: \mathrm{Spec}(R)$.

A way of defining a *specific* isomorphism $\mathrm{Inv}(R) \to \mathcal{L}(R)$ is given in [20, Section 2].

5.3 The Class Group

The *class group* of a Prüfer domain R is the factor group $\mathrm{Cl}(R) := \dfrac{\mathrm{Inv}(R)}{\mathrm{Prin}(R)}$, where $\mathrm{Inv}(R)$ denotes the group of invertible ideals of R and $\mathrm{Prin}(R)$ denotes the group of nonzero principal ideals. A Prüfer domain with trivial class group is called a *Bezout domain*. A Prüfer domain with finitely many maximal ideals is a Bezout domain [24, Proposition 7.4].

If R is a Bezout domain, each overring of R is a ring of quotients [11, p. 50], that is R is a so called *QR-domain*. R. Gilmer and J. Ohm proved that a Dedekind domain is a QR-domain if and only if its class group is a torsion group and asked whether any QR-domain (that is necessarily Prüfer) is forced to have a torsion class group [26]. Even though W. Heinzer answered this question in the negative [29], M. Fontana and N. Popescu extended the result of Gilmer and Ohm to generalized Dedekind domain [13].

Keeping the notation of Section 5.2, if R is a generalized Dedekind domain, for each ordinal α we set:

$$\mathrm{Cl}_{\alpha}(R) := \frac{\mathrm{Inv}_{\alpha}(R)}{\mathrm{Inv}_{\alpha}(R) \cap \mathrm{Prin}(R)}.$$

We have $\mathrm{Cl}_0(R) = (1)$, $\mathrm{Cl}_{\beta}(R) \subseteq \mathrm{Cl}_{\alpha}(R)$ for $\beta \leq \alpha$ and $\mathrm{Cl}(R) = \cup_{\alpha} \mathrm{Cl}_{\alpha}(R) = \mathrm{Cl}_{d(R)}(R)$.

Theorem 5.5. [13, Théorème 1.12] *The following conditions are equivalent for a domain R:*

(i) *Each localizing system of R is principal;*
(ii) *R is a generalized Dedekind domain and a QR-domain;*
(iii) *R is a strongly discrete Prüfer domain and each prime ideal of R is the radical of a principal ideal;*
(iv) *R is generalized Dedekind domain and* $\mathrm{Cl}(R)$ *is a torsion group.*

Proof. $(i) \Rightarrow (ii)$ R is generalized Dedekind domain by Theorem 2.6. If $\mathcal{F}$ is a principal localizing system of R, the set $S := \{x \in R\,; xR \in \mathcal{F}\}$ is not empty and $R_{\mathcal{F}} = R_S$. Thus each overring of R is a ring of quotients.

$(ii) \Rightarrow (iii)$ By Theorem 2.3, R is a strongly discrete Prüfer domain and each prime ideal of R is the radical of a finitely generated ideal. The fact that R is a QR-domain, then implies that each prime ideal of R is the radical of a principal ideal [24, Theorem 27.5].

$(iii) \Rightarrow (iv)$ For $\alpha < d(R)$, $\mathrm{Max}(R_{\alpha}) = \{PR_{\alpha}\,; P \in \mathcal{H}_{\alpha}\}$ (Theorem 3.1). Let $P \in \mathcal{H}_{\alpha}$ be a fixed prime and $x \in R$ be such that $\sqrt{xR} = P$. Then $xR_{\alpha} = xR_P \cap R_{\alpha} = P^n R_P \cap R_{\alpha} = (PR_{\alpha})^n$ for some $n \geq 1$. Since $\mathrm{Inv}_1(R_{\alpha})$ is the free group generated by $\mathrm{Max}(R_{\alpha})$ (Proposition 5.2), it follows that $\mathrm{Cl}_1(R_{\alpha})$ is a torsion group for each α.

By Proposition 5.3, we have an exat sequence of groups:

$$(1) \to \mathrm{Cl}_{\alpha}(R) \to \mathrm{Cl}_{\alpha+1}(R) \to \mathrm{Cl}_1(R_{\alpha}) \to (1).$$

Whence, by transfinite induction, we get that $\mathrm{Cl}_{\alpha}(R)$ is a torsion group for each ordinal α and conclude that $\mathrm{Cl}(R) = \mathrm{Cl}_{d(R)}(R)$ is a torsion group.

$(iv) \Rightarrow (i)$ By Theorem 2.6, each localizing system $\mathcal{F}$ of R is finitely generated. But, if $J \in \mathcal{F}$ is a finitely generated ideal, then a power of J is principal. Hence each ideal in $\mathcal{F}$ contains a principal ideal which is still in $\mathcal{F}$.

If R is a Dedekind domain, $\mathrm{Cl}(R)$ is generated by the classes of the maximal ideals, thus R is a Bezout domain if and only if each maximal ideal of R is principal. According to W. Heinzer, in 1974 M. Boisen informally posed the question whether a Prüfer domain whose maximal ideals are all principal must be a Bezout domain. This question was answered in the negative by A. Loper in 1999 [34]. However, in [10] it was shown that a Prüfer domain has class group generated by the classes of the invertible maximal ideals if and only if a distinguished overring $S(R)$ of R is a Bezout domain. In this case, clearly R itself is Bezout if and only if each invertible maximal ideal is principal. The overring $S(R)$ may be Bezout or not - examples of both types are given in [10].

It turns out that, if R is a generalized Dedekind domain, setting as in Section 5.1 $\mathcal{H}_1 := \{\cap_{n\geq 1}P^n\,; P \in \mathrm{Max}(R)\}$, then $S(R)$ is the overring $R_1 := \cap_{P\in\mathcal{H}_1} R_P$.

Theorem 5.6. [10, Theorem 1.11] *If R is a generalized Dedekind domain, then $\mathrm{Cl}(R)$ is generated by the classes of the maximal ideals of R if and only if the overring R_1 is a Bezout domain. In this case, R is a Bezout domain if and only if each maximal ideal of R is principal.*

Proof. Since, for each ideal $I \subseteq R$, $IR_1 = R_1$ if and only if $I \in \mathcal{F}_1$ (Theorem 2.6), we have the exact sequence of class groups:

$$(1) \to \mathrm{Cl}_1(R) \to \mathrm{Cl}(R) \xrightarrow{\pi} \mathrm{Cl}(R_1) \to (1),$$

where π is the extension. Now observe that $\mathrm{Cl}(R)$ is generated by the classes of the maximal ideals of R if and only if $\mathrm{Cl}(R) = \mathrm{Cl}_1(R)$.

Corollary 5.7. *If R is a generalized Dedekind domain and $\mathcal{H}_1$ is a finite set, then $\mathrm{Cl}(R)$ is generated by the classes of the maximal ideals.*

Example 5.8. By a well-known result of Claborn, for any abelian group G, there exists a Dedekind domain D such that $\mathrm{Cl}(D)$ is isomorphic to G. Let F be the quotient field of D an X an indeterminate over F. Then $R := D+XF[X]$ is a two-dimensional generalized Dedekind domain (Corollary 4.2). Since $\mathrm{Cl}(F[X]) = (1)$, by ideal theory in pullbacks [17], we have that $\mathrm{Cl}(R) \cong \mathrm{Cl}(D) \cong G$. In addition $M := XF[X]$ is the only nonzero nonmaximal prime ideal of R. Thus $R_1 = R_M = F[X]_{(X)}$ is a DVR and $\mathrm{Cl}(R)$ is generated by the classes of the maximal ideals. (Note that the height-one maximal ideals of R are principal.)

By repeated pullbacks, one may obtain a generalized Dedekind domain R (which is not Dedekind) such that $\mathrm{Cl}(R) \cong G$ and R_1 is Bezout semilocal with an arbitrarily large number of maximal ideals. By Theorem 5.5, R is a QR-domain if and only if G is a torsion group.

5.4 Stable ideals

Let R be a commutative ring with identity. J. Sally and W. Vasconcelos say that an ideal I of R is *stable* if I is projective over its ring of endomorphisms. If each nonzero ideal of R is stable, R itself is called *stable* [41, Section 2]. If R is a domain, the endomorphism ring of a nonzero ideal I is isomorphic to the overring $(I{:}I)$ of R [11, p. 34]. Hence a nonzero ideal I of a domain R is stable if it is invertible in $(I{:}I)$. Stable domains have been exhaustively investigated by B. Olberding in a series of recent papers.

Stability of nonzero prime ideals forces a Prüfer domain to be generalized Dedekind. To show this, we need some properties of trace ideals.

Lemma 5.9. *Let R be a Prüfer domain.*

(i) *If* $J = P_1 \dots P_n$*, where* $\{P_1, \dots, P_n\}$ *is a set of pairwise incomparable nonzero nonmaximal prime ideals, then*

$$(J:J) = (R:J) = (\cap_{i=1}^{n} R_{P_i}) \cap R_{\mathcal{F}(\Lambda)},$$

where Λ *denotes the set of maximal ideals of* R *not containing* J.

(ii) *If* R *is a generalized Dedekind domain, then a finite product of nonzero prime ideals of* R *is divisorial and* J *is a proper trace ideal if and only if* $J = P_1 \dots P_n$*, where* $\{P_1, \dots, P_n\}$ *is a set of pairwise incomparable nonzero nonmaximal prime ideals.*

Proof. (1) follows from [31, Theorems 3.2 and 3.8 and Proposition 3.9].

(2) Each maximal ideal of R is invertible (Proposition 2.5). Moreover, by the ##-property (Theorem 2.7), each nonzero nonmaximal prime ideal is divisorial and a finite product of divisorial prime ideals is divisorial [11, Theorem 4.1.21]. A proper trace ideal J is radical by the RTP-property (Theorem 2.10) and has finitely many minimal primes (Theorem 2.2). Let $J = P_1 \dots P_n$. Since maximal ideals are invertible, if P_1 is maximal, then $J = J(R:J) = P_2 \dots P_n$; a contradiction. Thus a minimal prime of a trace ideal cannot be maximal. The converse follows from part (1).

Theorem 5.10. [15, Theorem 5] *The following conditions are equivalent for a Prüfer domain* R*:*

(i) R *is a generalized Dedekind domain;*
(ii) *Each nonzero prime ideal of* R *is stable;*
(iii) *Each nonzero prime ideal is divisorial and each divisorial ideal of* R *is stable.*

Proof. We use Lemma 5.9 repeatedly.

(i) $\Rightarrow$ *(iii)* We know that nonzero prime ideals are divisorial. Let I be a divisorial ideal of R. If I is invertible, clearly it is stable. Otherwise, $H := I(R:I)$ is a proper trace ideal. Thus $H = P_1 \dots P_n$, where P_i is a nonzero nonmaximal prime ideal for $1 \leq i \leq n$. We have $(I:I) = (H:H) = (\cap_{i=1}^{n} R_{P_i}) \cap R_{\mathcal{F}(\Lambda)}$, where the first equality holds because I is divisorial and Λ denotes the set of maximal ideals of R not containing H. Since the overring $T := (I:I) = (H:H)$ of R is a generalized Dedekind domain and P_iT is maximal in T (Theorem 3.1), then P_iT is invertible in T, for $1 \leq i \leq n$. It follows that $H := I(R:I) = P_1 \dots P_n$ is invertible in T and thus I is invertible in T.

(iii) $\Rightarrow$ *(ii)* is clear.

(ii) $\Rightarrow$ *(i)* Let T be an overring of R and $N \in \text{Max}(T)$. Since $P := N \cap R$ is invertible in $(P:P)$ and $(P:P) \subseteq (PT:PT) = (N:N)$, then $N = PT$ is invertible in $(N:N)$. Hence N is invertible in T (otherwise $(N:N) = (T:N) = T$, a contradiction). It follows that each maximal ideal of T is invertible. By Theorem 3.4, R is a generalized Dedekind domain.

Note that the equivalence $(i) \Leftrightarrow (ii)$ in the previous theorem is a sharpened version of Theorem 3.4.

By $(ii) \Rightarrow (i)$, a Prüfer stable domain is a generalized Dedekind domain, but there are generalized Dedekind domains that are not stable. In fact R is an integrally closed stable domain if and only if it is a strongly discrete Prüfer domain with finite character [37, Theorem 4.6]. Whence, from Corollary 2.4, we get:

Theorem 5.11. *The following conditions are equivalent for a domain R:*

(i) *R is a generalized Dedekind domain with finite character;*
(ii) *R is integrally closed and stable.*

Example 5.12. The domain $R := \mathbb{Z} + X\mathbb{Q}[[X]]$ is a two-dimensional generalized Dedekind domain (Theorem 4.1). Since R does not have finite character, R is not stable. As in [15, Example 10], it is possible to check directly that the ideal I of R generated by the set $\{\frac{1}{p}X; p \text{ a prime integer}\}$ is not stable. Of course I is not divisorial; indeed $I \neq I_v = X\mathbb{Q}[[X]]$ (for example $\frac{1}{p^2}X \notin I$).

Since an invertible ideal is a cancellation ideal, if R is a stable domain, each nonzero ideal I of R is a cancellation ideal of $(I{:}I)$. A domain with this last property is called in [27] a domain with *restricted cancellation.* By [27, Proposition 3.2], an integrally closed domain with finite character is stable (equivalently, a generalized Dedekind domain) if and only if it has restricted cancellation. Relaxing the finite character condition, generalized Dedekind domains are characterized by the stronger property of being integrally closed and strongly faithful [27, Theorem 3.7 and Example 5.5].

5.5 Divisorial ideals

Following S. Bazzoni and L. Salce [4], we say that R is a *divisorial domain* if each nonzero ideal of R is divisorial and that R is *totally divisorial* if each overring of R is divisorial. Divisoriality and stability are strictly related; in fact B. Olberding proved that a domain is totally divisorial if and only if it is divisorial and stable [38, Theorem 3.11].

The next theorem gives a characterization of generalized Dedekind domains in terms of divisorial ideals.

Theorem 5.13. [20, Theorem 3.3 and Corollary 3.4] *The following conditions are equivalent for a domain R:*

(i) *R is a generalized Dedekind domain;*
(ii) *R is a strongly discrete Prüfer domain and each finite product of nonzero prime ideals is a divisorial ideal;*
(iii) *R is a Prüfer domain and an ideal I of R is divisorial if and only if $I = JQ_1 \dots Q_n$, where J is a fractional invertible ideal and $\{Q_1, \dots, Q_n\}$ is a nonempty collection of nonzero prime ideals that are pairwise comaximal if $n \geq 2$.*

Proof. $(i) \Rightarrow (iii)$ Each ideal of type $JQ_1 \dots Q_n$, where J is a fractional invertible ideal and Q_i is a nonzero prime ideal, for $1 \leq i \leq n$, is divisorial by Lemma 5.9. Conversely, let I be a divisorial ideal of R. If I is invertible and M is a maximal ideal of R, we have $I = JM$, with $J := I(R{:}M)$, because M is invertible (Proposition 2.5). Otherwise, the trace ideal $H := I(R{:}I)$ is radical and divisorial by Lemma 5.9. In addition $(I{:}I) = (H{:}H)$, because I is divisorial. By Theorem 5.10, both I and H are invertible in the overring $T := (I{:}I) = (H{:}H)$. Thus we can write $H = J_1T$ and $I = J_2T$, with J_1 and J_2 fractional invertible ideals of R. We conclude that $I = JH$, with $J = (R{:}J_1)J_2$.

$(iii) \Rightarrow (ii)$ It is enough to show that $P \neq P^2$, for each nonzero prime ideal P. By hypothesis, each nonzero prime ideal is divisorial; thus maximal ideals are invertible and not idempotent. If P is not maximal, then $(R{:}P) = (P{:}P)$ (Lemma 5.9). Let $p \in P$, $p \neq 0$. Since the ideal $I := p(R{:}P) = p(P{:}P)$ is divisorial, we have $I = JQ_1 \dots Q_n$, where J is a fractional invertible ideal and the Q_i's are nonzero prime ideals that are pairwise comaximal if $n \geq 2$. Since I is not invertible, we can assume that Q_i is not maximal for each i. Setting $H := Q_1 \dots Q_n$, by Lemma 5.9, we have $(R{:}P) = (P{:}P) = (I{:}I) = (H{:}H) = (R{:}H)$. Taking duals, we get $P = H = Q_1 \dots Q_n$. Whence $P = p(R{:}J)(P{:}P)$ is stable. It follows that $P \neq P^2$.

$(ii) \Rightarrow (i)$ For each nonzero prime ideal P of R and $n \geq 1$, P^n is a divisorial ideal. Since $P \neq P^2$, then $(R{:}P) \subsetneq T(P)$, where $T(P) := \cup_{n\geq 1}(R{:}P^n)$ is the Nagata Transform of P [11, Theorem 4.1.19]. This implies that P is stable [11, Theorem 3.3.9] and so by Theorem 5.10, R is a generalized Dedekind domain.

If R is a divisorial generalized Dedekind domain, by the previous theorem, each ideal is a product of a fractional invertible ideal and prime ideals. B. Olberding proved that the converse is also true [37]. Recall that R is an *h-local domain* if it has finite character and each nonzero prime ideal is contained in a unique maximal ideal.

Theorem 5.14. *The following conditions are equivalent for a domain R:*

(i) *R is a Prüfer domain and each ideal of R is of the form $JQ_1 \dots Q_n$, where J is a fractional invertible ideal and $\{Q_1, \dots, Q_n\}$ is a nonempty collection of prime ideals that are pairwise comaximal if $n \geq 2$;*
(ii) *R is an h-local strongly discrete Prüfer domain;*
(iii) *R is a generalized Dedekind domain and each nonzero prime ideal is contained in a unique maximal ideal;*
(iv) *R is integrally closed and totally divisorial;*
(v) *R is integrally closed, stable and divisorial;*
(vi) *R is a divisorial generalized Dedekind domain.*

Proof. $(i) \Leftrightarrow (ii)$ is [37, Theorem 5.2]. $(ii) \Leftrightarrow (iii)$ is clear from the definitions. $(ii) \Leftrightarrow (iv)$ is [37, Proposition 4.2]. $(iv) \Leftrightarrow (v)$ by [38, Theorem 3.11]. $(v) \Rightarrow (vi)$ by Theorem 5.11. $(vi) \Rightarrow (i)$ follows from Theorem 5.13.

A generalized Dedekind domain with finite character is always stable (Theorem 5.11), but by the previous theorem need not be divisorial. An example of a stable non-divisorial generalized Dedekind domain with finitely many maximal ideals is given by $R := D + XF[[X]]$, where D is a semilocal (not local) Dedekind domain with quotient field F (Theorem 4.1).

References

1. J. T. Arnold, *Krull dimension in power series rings*, Trans. Amer. Math. Soc. **177** (1973), 299-304.
2. J. T. Arnold, *Power series rings over Prüfer domains*, Pacific J. Math. **44** (1973), 1-11.
3. J. T. Arnold and R. Gilmer, *Idempotent ideals and unions of nets of Prüfer domains*, J. Sci. Hiroshima Univ. **31** (1967), 131-145.
4. S. Bazzoni and L. Salce, *Warfield Domains*, J. Algebra **185** (1996), 836-868.
5. S. El Baghdadi, *On a class of Prüfer v-multiplication domains*, Comm. Algebra **30** (2002), 3723-3742.
6. S. El Baghdadi, *Factorization of divisorial ideals in a generalized Krull domain*, Rings, Modules, Algebras and Abelian Groups (Venice, 2002), 149-160, Lecture notes in Pure and appl. Math. **236**, Marcel Dekker, New York, 2004.
7. S. El Baghdadi and S. Gabelli, *w-divisorial domains*, J. Algebra **285** (2005), 335-355.
8. S. El Baghdadi and S. Gabelli, *Ring-theoretic properties of PvMDs*, manuscript.
9. A. Facchini, *Generalized Dedekind domains and their injective modules*, J. Pure Appl. Algebra **94** (1994), 159-173.
10. M. Fontana and S. Gabelli, *Prüfer domains with class group generated by the classes of the invertible maximal ideals*, Comm. Algebra **25** (1997), 3993-4008.
11. M. Fontana, J. A. Huckaba and I. J. Papick, *Prüfer domains*, Monographs and Textbooks in Pure and Applied Mathematics **203**, M. Dekker, New York, 1997.
12. M. Fontana and N. Popescu, *Sur une classe d'anneaux qui généralisent les anneaux de Dedekind*, J. Algebra **173** (1995), 44-66.
13. M. Fontana and N. Popescu, *Sur une classe d'anneaux de Prüfer avec groupe de classes de torsion*, Comm. Algebra **23** (1995), 4521-4534.
14. L. Fuchs and L. Salce, *Modules over Non-Noetherian domains*, Math. Surveys and Monographs **84**, Amer. Math. Soc., 2001.
15. S. Gabelli, *A class of Prüfer domains with nice divisorial ideals*, Commutative Ring Theory (Fés, 1995), 313-318, Lecture Notes in Pure and Appl. Math. **185**, M. Dekker, New York, 1997.
16. S. Gabelli, *Prüfer (##)-domains and localizing systems of ideals*, Advances in Commutative Ring Theory (Fés, 1997), 391-409, Lecture Notes in Pure and Appl. Math. **205**, M. Dekker, New York, 1999.
17. S. Gabelli and E. G. Houston, *Ideal theory in pullbacks*, Non-Noetherian Commutative Ring Theory, 199-227, Math. Appl. **520**, Kluwer Acad. Publ., Dordrecht, 2000.
18. S. Gabelli, E. G. Houston and T. G. Lucas, *The t#-property for integral domains*, J. Pure Appl. Algebra **194** (2004), 281-298.
19. S. Gabelli and G. Picozza, *Star stable domains*, manuscript.

20. S. Gabelli and N. Popescu, *Invertible and divisorial ideals of generalized Dedekind domains*, J. Pure Appl. Algebra **135** (1999), 237-251.
21. R. Gilmer, *Integral domains which are almost Dedekind*, Proc. Am. Math. Soc. **15** (1964), 813-818.
22. R. Gilmer, *Overrings of Prüfer domains*, J. Algebra **4** (1966), 331-340.
23. R. Gilmer, *Multiplicative ideal theory*, Queen's Papers in Pure and Applied Mathematics, Queen's University, Kingston, Ontario, Canada, 1968.
24. R. Gilmer, *Multiplicative ideal theory*, M. Dekker, New York, 1972.
25. R. Gilmer and W. J. Heinzer, *Overrings of Prüfer domains. II*, J. Algebra **7** (1967), 281-302.
26. R. Gilmer and J. Ohm, *Integral domains with quotient overrings*, Math. Ann. **153** (1964), 97-103.
27. H. P. Goeters and B. Olberding, *Extension of ideal-theoretic properties of a domain to submodules of its quotient field*, J. Algebra **237** (2001), 14-31.
28. R. Gordon and J. C. Robson, *Krull dimension*, Memoirs of the Amer. Math. Soc. **133**, 1973.
29. W. J. Heinzer, *Quotient overrings of integral domains*, Mathematika **17** (1970), 139-148.
30. W. J. Heinzer and I. Papick, *The radical trace property*, J. Algebra **112** (1988), 110-121.
31. J. A. Huckaba and I. J. Papick, *When the dual of an ideal is a ring*, Manuscripta Math. **37** (1982), 67-85.
32. J. A. Huckaba and I. J. Papick, *Connecting trace properties*, Non-Noetherian Commutative Ring Theory, 313-324, Math. Appl. **520**, Kluwer Acad. Publ., Dordrecht, 2000.
33. B. G. Kang and M. H. Park, *On Mockor's question*, J. Algebra **216** (1999), 481-510.
34. K. A. Loper, *Two Prüfer domain counterexamples*, J. Algebra **221** (1999), 630-643.
35. K. A. Loper, *Almost Dedekind domains,* this volume.
36. K. A. Loper and T. G. Lucas, *Factoring ideals in almost Dedekind domains*, J. Reine Angew. Math. **565** (2003), 61-78.
37. B. Olberding, *Globalizing Local Properties of Prüfer Domains*, J. Algebra **205** (1998), 480-504.
38. B. Olberding, *Stability, duality, 2-generated ideals and a canonical decomposition of modules*, Rend. Sem. Mat. Univ. Padova **106** (2001), 261-290.
39. N. Popescu, *On a class of Prüfer domains*, Rev. Roumanie Math. Pures Appl. **29** (1984), 777-786.
40. E. Popescu and N. Popescu, *A characterization of generalized Dedekind domains*, Bull. Math. Roumanie **35** (1991), 139-141.
41. J. D. Sally and W. V. Vasconcelos, *Stable Rings*, J. Pure Appl. Algebra **4** (1974), 319-336.

Non-unique factorizations: a survey

Alfred Geroldinger[1] and Franz Halter-Koch[2]

[1] Institut für Mathematik, Karl-Franzens-Universität Graz, Heinrichstrasse 36, 8010 Graz, Austria, alfred.geroldinger@uni-graz.at
[2] Institut für Mathematik, Karl-Franzens-Universität Graz, Heinrichstrasse 36, 8010 Graz, Austria, franz.halterkoch@uni-graz.at

1 Introduction

It is well known that the ring of integers of an algebraic number field may fail to have unique factorization. In the development of algebraic number theory in the 19th century, this failure led to Dedekind's ideal theory and to Kronecker's divisor theory. Only in the late 20th century, starting with L. Carlitz' result concerning class number 2, W. Narkiewicz began a systematic combinatorial and analytic investigation of phenomena of non-unique factorizations in rings of integers of algebraic number fields (see Chapter 9 of [24] for a survey of the early history of the subject). In the sequel several authors started to investigate factorization properties of more general integral domains in the spirit of R. Gilmer's book [18] (see for example the series of papers [2], [3], [4] and the survey article [19] by R. Gilmer). It soon turned out that the investigation of factorization problems can successfully be carried out in the setting of commutative cancellative monoids, and this point of view opened the door to further applications of the theory. Among them the most prominent ones are the arithmetic of congruence monoids, the theory of zero-sum sequences over abelian groups and the investigation of Krull monoids describing the deviation from the Krull-Remak-Azumaya-Schmidt Theorem in certain categories of modules.

The proceedings [1] and [6] of two Conferences on Factorization Theory (held 1996 in Iowa City and 2003 in Chapel Hill) and the articles contained in [7] give a good survey on the development of the theory of non-unique factorizations over the past decade. Only recently the authors completed the monograph [16] which contains a thorough presentation of the algebraic, combinatorial and analytic aspects of the theory of non-unique factorizations, together with self-contained introductions to additive group theory, to the theory of v-ideals and to abstract analytic number theory.

The purpose of this survey article is to point out some highlights of the theory of non-unique factorizations (Theorem 5.6.**B.** and Theorem 7.4) with an emphasis on the presentation of the concepts which describe the various phenomena of non-uniqueness in structures of arithmetical relevance. We concentrate on the presentation of the main results and, if at all, we only give the main ideas of the proofs. For more details we refer the reader to the monograph [16] and to the original papers in the volumes cited above.

2 Notations and Preliminaries

Let $\mathbb{N}$ denote the set of positive integers and let $\mathbb{N}_0 = \mathbb{N} \cup \{0\}$. For any set P let $|P| \in \mathbb{N}_0 \cup \{\infty\}$ denote the number of elements in P. For integers $a, b \in \mathbb{Z}$ we define $[a, b] = \{x \in \mathbb{Z} \mid a \le x \le b\}$.

By a semigroup we mean a commutative semigroup with a unit element, and by a monoid we mean a semigroup satisfying the cancellation law. Unless stated otherwise, we use multiplicative notation and denote the unit element by 1. (Semigroup) homomorphisms are always assumed to respect the unit element. For a monoid H, we denote by $H^\times$ the set of invertible elements of H and by $H_{\text{red}} = \{aH^\times \mid a \in H\}$ the associated reduced monoid of H. We say that H is reduced if $H^\times = \{1\}$. We denote by $\mathsf{q}(H)$ a quotient group of H. Let $S \subset H$ be a submonoid. We tacitly assume that $\mathsf{q}(S) \subset \mathsf{q}(H)$, and for $a \in \mathsf{q}(H)$ we denote by $[a] = [a]_{H/S} = a\,\mathsf{q}(S) \in \mathsf{q}(H)/\mathsf{q}(S)$ the class containing a. We set $H/S = \{[a] \mid a \in H\} \subset \mathsf{q}(H)/\mathsf{q}(S)$, and if $aH \cap S \neq \emptyset$ for all $a \in H$, then $H/S = \mathsf{q}(H)/\mathsf{q}(S)$ (this condition is fulfilled throughout the present article).

For $a, b \in H$, we write as usual $a \,|\, b$ (in H) if $b \in aH$, and $a \simeq b$ if $a \,|\, b$ and $b \,|\, a$ (equivalently, $aH^\times = bH^\times$). A submonoid $S \subset H$ is called *divisor-closed* if $a \in S$, $b \in H$ and $b \,|\, a$ implies that $b \in S$.

An element $x \in \mathsf{q}(H)$ is called *almost integral* over H if there exists some $c \in H$ such that $cx^n \in H$ for all $n \in \mathbb{N}$. The set $\widehat{H}$ of all elements of $\mathsf{q}(H)$ which are almost integral over H is a monoid, called the *complete integral closure* of H. The monoid H is called *completely integrally closed* if $\widehat{H} = H$. This definition coincides with the corresponding concept in commutative ring theory (note that the stronger concept of integral elements has no purely multiplicative analog).

For two semigroups H_1, H_2, we denote by $H_1 \times H_2$ their direct product, and we view H_1 and H_2 as subsemigroups of $H_1 \times H_2$. Thus every $a \in H_1 \times H_2$ has a uniquely determined decomposition $a = a_1 a_2$ with $a_1 \in H_1$ and $a_2 \in H_2$.

A monoid F is called *free* (*with basis* $P \subset F$) if every $a \in F$ has a unique representation in the form

$$a = \prod_{p \in P} p^{\mathsf{v}_p(a)} \quad \text{with} \quad \mathsf{v}_p(a) \in \mathbb{N}_0 \text{ and } \mathsf{v}_p(a) = 0 \text{ for almost all } p \in P. \quad (*)$$

In this case F is (up to canonical isomorphism) uniquely determined by P (and conversely P is uniquely determined by F). We set $F = \mathcal{F}(P)$, and if a is as in $(*)$, then we call

$$|a| = \sum_{p \in P} \mathsf{v}_p(a) \quad \text{the } \textit{length} \text{ of } a\,.$$

For every map $\varphi_0\colon P \to S$ into a semigroup S there is a unique homomorphism $\varphi\colon \mathcal{F}(P) \to S$ such that $\varphi \,|\, P = \varphi_0$.

For a monoid H, we consider the v-operation and the theory of v-ideals as explained in [16, Chapter 2] or in [21, Section 11]. A monoid H is v-noetherian if it satisfies the ascending chain condition on v-ideals.

Integral domains are in the center of our interest. If R is an integral domain, then $R^\bullet = R \setminus \{0\}$ is a multiplicative monoid, and a subset $\mathfrak{a} \subset R^\bullet$ is a v-ideal of $R^\bullet$ if and only if $\mathfrak{a} \cup \{0\}$ is a divisorial ideal of R. In particular, $R^\bullet$ is v-noetherian if and only if R is a Mori domain.

We study the (multiplicative) arithmetic of an integral domain R by means of the monoid $R^\bullet$, and we attribute a factorization property or an invariant connected with factorizations to R if and only if it holds for $R^\bullet$.

Throughout this paper, let H be a monoid.

An element $a \in H$ is called

- an *atom* (or an irreducible element) if $a \notin H^\times$ and, for all $b, c \in H$, $a = bc$ implies $b \in H^\times$ or $c \in H^\times$. We denote by $\mathcal{A}(H)$ the set of all atoms of H.
- a *prime* (or a prime element) if $a \notin H^\times$ and, for all $b, c \in H$, $a \,|\, bc$ implies $a \,|\, b$ or $a \,|\, c$.

The monoid H is called

- *atomic* if every $a \in H \setminus H^\times$ is a product of atoms.
- *factorial* if it satisfies one of the following equivalent conditions:
  1. Every $a \in H \setminus H^\times$ is a product of primes.
  2. H is atomic, and every atom is a prime.
  3. Every $a \in H \setminus H^\times$ is a product of atoms, and this factorization is unique up to associates and the order of the factors.
  4. H_{red} is free (in that case H_{red} is free with basis $\{pH^\times \mid p \in P\}$ where P denotes the set of primes of H).
  5. $H = H^\times \times \mathcal{F}(P)$ for some subset $P \subset H$ (in that case P is a maximal set of pairwise non-associated primes of H).

Every prime is an atom, and every factorial monoid is atomic. An element $a \in H$ is an atom [a prime] of H if and only if $aH^\times$ is an atom [a prime] of H_{red}. Thus H_{red} is atomic [factorial] if and only if H has this property. In a factorial monoid, every non-empty subset has a greatest common divisor which is uniquely determined up to associates.

By our convention, an integral domain R is atomic [factorial] if and only if $R^\bullet$ has this property, and these definitions coincide with the usual ones in commutative ring theory.

In the theory of non-unique factorizations we describe the deviation of a monoid from being factorial. For this, we have to formalize the notion of a factorization into irreducibles (see Definition 3.1 below). Having a precise notion of the set of all factorizations, we are able to investigate its structure. Since invertible elements play no role in the theory of factorizations, the reduced monoid H_{red} and not the monoid H itself is the basis for our definitions.

3 Arithmetic of monoids

In this section we present the concepts by which we describe the phenomena of non-unique factorizations.

Sets of Factorizations. The free monoid $\mathsf{Z}(H) = \mathcal{F}\big(\mathcal{A}(H_{\mathrm{red}})\big)$ is called the *factorization monoid* of H, and the unique homomorphism

$$\pi \colon \mathsf{Z}(H) \to H_{\mathrm{red}} \quad \text{satisfying} \quad \pi(u) = u \quad \text{for all} \quad u \in \mathcal{A}(H_{\mathrm{red}})$$

is called the *factorization homomorphism* of H. For $a \in H$, the set

$$\mathsf{Z}(a) = \pi^{-1}(aH^\times) \subset \mathsf{Z}(H) \quad \text{is the } \textit{set of factorizations} \text{ of } a.$$

Note that an element of $\mathsf{Z}(a)$ represents a naive factorization of a into atoms, where the order of the factors and the choice of the factors among associates is disregarded. An element $a \in H$ is said to have unique factorization if $|\mathsf{Z}(a)| = 1$. By definition, we have $\mathsf{Z}(a) = \{1\}$ for all $a \in H^\times$. The monoid H is atomic if and only if $\mathsf{Z}(a) \ne \emptyset$ for all $a \in H$, and H is factorial if and only if $|\mathsf{Z}(a)| = 1$ for all $a \in H$.

Sets of Lengths and distances. For $a \in H$, we call

$$\mathsf{L}(a) = \big\{|z| \bigm| z \in \mathsf{Z}(a)\big\} \subset \mathbb{N}_0 \quad \text{the } \textit{set of lengths} \text{ of } a\,.$$

We denote by $\Delta\big(\mathsf{L}(a)\big)$ the set of all $d \in \mathbb{N}$ for which there exists some $m \in \mathsf{L}(a)$ with $[m, m+d] \cap \mathsf{L}(a) = \{m,\, m+d\}$ (that is, $\Delta\big(\mathsf{L}(a)\big)$ is the set of all successive distances in sets of lengths of factorizations of a).

The system of sets of lengths and the set of distances of H are defined by

$$\mathcal{L}(H) = \big\{\mathsf{L}(a) \bigm| a \in H\big\} \quad \text{and} \quad \Delta(H) = \bigcup_{a \in H} \Delta\big(\mathsf{L}(a)\big)\,.$$

We denote by $\Delta^*(H)$ the set of all $d = \min \Delta(S)$ for some divisor-closed submonoid $S \subset H$ for which $\Delta(S) \ne \emptyset$. The set $\Delta^*(H)$ is a subset of $\Delta(H)$ which (among others) is of importance for the Structure Theorem for Sets of Lengths cited below.

The monoid H is called *half-factorial* if $|\mathsf{L}(a)| = 1$ for all $a \in H$, and it is called a BF-*monoid* if $\mathsf{L}(a)$ is finite and non-empty for all $a \in H$.

By definition, H is atomic if and only if $\mathsf{L}(a) \neq \emptyset$ for all $a \in H$, and H is half-factorial if and only if H is atomic and $\Delta(H) = \emptyset$. Every factorial monoid is half-factorial, every half-factorial monoid is a BF-monoid, and every BF-monoid is atomic. Every v-noetherian monoid is a BF-monoid (this requires some ideal-theoretic effort). In particular, every Mori domain is a BF-domain.

Distance of factorizations. Let $z, z' \in \mathsf{Z}(H)$, say

$$z = u_1 \cdot \ldots \cdot u_l v_1 \cdot \ldots \cdot v_m \quad \text{and} \quad z' = u_1 \cdot \ldots \cdot u_l w_1 \cdot \ldots \cdot w_n ,$$

where $l, m, n \in \mathbb{N}_0$, $u_1, \ldots, u_l, v_1, \ldots, v_m, w_1, \ldots, w_n \in \mathcal{A}(H_{\mathrm{red}})$ and

$$\{v_1, \ldots, v_m\} \cap \{w_1, \ldots, w_n\} = \emptyset .$$

Then we call $\mathsf{d}(z, z') = \max\{m, n\} \in \mathbb{N}_0$ the *distance* of z and z'.

It is easily checked that the distance function $\mathsf{d}\colon \mathsf{Z}(H) \times \mathsf{Z}(H) \to \mathbb{N}_0$ is a metric satisfying $\mathsf{d}(xz, xz') = \mathsf{d}(z, z')$ for all $x, z, z' \in \mathsf{Z}(H)$.

If H fails to be factorial (resp. half-factorial), then there exist elements with arbitrary many distinct factorizations (resp. lengths) as the following lemma shows.

Lemma 3.1 *Let H be atomic.*

(1) *If H is not factorial, then for every $k \in \mathbb{N}$ there exists some $a \in H$ such that $|\mathsf{Z}(a)| \ge k+1$, and there exist factorizations $z, z' \in \mathsf{Z}(a)$ such that $\mathsf{d}(z, z') \ge 2k$.*

(2) *If H is not half-factorial, then for every $k \in \mathbb{N}$ there exists some $a \in H$ such that $|\mathsf{L}(a)| \ge k+1$.*

Proof. We may suppose that H is reduced, and we present a proof of the first assertion (the second one follows by similar simple arguments).

If H is not factorial, then there exists some $c \in H$ such that $|\mathsf{Z}(c)| > 1$. If $z, z' \in \mathsf{Z}(c)$ are distinct and $k \in \mathbb{N}$, then $\mathsf{Z}(c^k) \supset \{z^{k-i} z'^i \mid i \in [0, k]\}$. Hence $|\mathsf{Z}(c^k)| \ge k+1$ and $\mathsf{d}(z^k, z'^k) = k\mathsf{d}(z, z') \ge 2k$. $\square$

Sets of lengths are the best understood invariants of non-unique factorizations. Under suitable finiteness conditions (which are satisfied for orders in algebraic number fields, see Example 5.9) they are almost arithmetical multiprogressions with bounded parameters. We are going to describe this structure.

Almost Arithmetical Multiprogressions. Let $M \in \mathbb{N}$, $d \in \mathbb{N}$ and $\{0, d\} \subset \mathcal{D} \subset [0, d]$. A finite non-empty subset $L \subset \mathbb{Z}$ is called an *almost arithmetical multiprogression* (AAMP *for short*) *with difference d, period $\mathcal{D}$ and bound M* if there exists some $y \in \mathbb{Z}$ such that

$$L = y + (L' \cup L^* \cup L'') \subset y + \mathcal{D} + d\mathbb{Z}$$

with $L^* = (\mathcal{D}+d\mathbb{Z}) \cap [0, \max L^*]$, $L' \subset [-M, -1]$ and $L'' \subset \max L^* + [1, M]$. We call y the shift parameter, L^* the central part, L' the initial part and L'' the end part of the AAMP L.

The Structure Theorem for Sets of Lengths. We say that *the Structure Theorem for Sets of Lengths holds for* H if H is atomic and there exists some $M^* \in \mathbb{N}$ such that every $L \in \mathcal{L}(H)$ is an AAMP with some difference $d \in \Delta^*(H)$ and bound M^*.

For half-factorial monoids, the Structure Theorem for Sets of Lengths holds in a trivial way. If H is not half-factorial and the Structure Theorem for Sets of Lengths holds for H, then H is a BF-monoid and has arbitrarily large sets of lengths (by Lemma 3.1), but all these sets of lengths have bounded initial and end parts, and thus they have arbitrarily large central part.

Catenary degree. Let $z,\ z' \in \mathsf{Z}(H)$ and $N \in \mathbb{N}_0 \cup \{\infty\}$. We say that z and z' can be *concatenated by an* N*-chain* if there exists a finite sequence of factorizations $z = z_0, z_1, \ldots, z_k = z'$ in $\mathsf{Z}(a)$ such that $\mathsf{d}(z_{i-1}, z_i) \le N$ for all $i \in [1, k]$.

For an element $a \in H$, we define its *catenary degree* $\mathsf{c}(a)$ to be the smallest $N \in \mathbb{N}_0 \cup \{\infty\}$ such that any two factorizations of a can be concatenated by an N-chain, and we call

$$\mathsf{c}(H) = \sup\{\mathsf{c}(a) \mid a \in H\} \in \mathbb{N}_0 \cup \{\infty\} \quad \text{the } \textit{catenary degree} \text{ of } H.$$

The catenary degree $\mathsf{c}(a)$ measures how complex the set of factorizations of a is. Note that by definition we have either $\mathsf{c}(a) = 0$ or $\mathsf{c}(a) \ge 2$. In the following Lemma 3.2 (whose proof is straightforward) we list the basic properties of the catenary degree.

Lemma 3.2 *Let* H *be atomic and* $a \in H$.

(1) *a has unique factorization if and only if* $\mathsf{c}(a) = 0$. *In particular, H is factorial if and only if* $\mathsf{c}(H) = 0$.
(2) *If* $\mathsf{c}(a) \le 2$, *then* $|\mathsf{L}(a)| = 1$. *In particular, if* $\mathsf{c}(H) \le 2$, *then H is half-factorial.*
(3) *If* $|\mathsf{L}(a)| \ge 2$, *then* $2 + \sup \Delta(\mathsf{L}(a)) \le \mathsf{c}(a)$. *In particular, if* $\mathsf{c}(H) < \infty$, *then $\Delta(H)$ is finite.*
(4) *If* $\mathsf{c}(a) \le 3$, *then* $\Delta(\mathsf{L}(a)) \subset \{1\}$, *whence* $\mathsf{L}(a) = [y, y+k]$ *for some* $y, k \in \mathbb{N}_0$.

Tame degree and local tameness. For a factorization $x \in \mathsf{Z}(H)$ and $a \in H$ we define the *tame degree* $\mathsf{t}(a, x)$ to be the smallest $N \in \mathbb{N}_0 \cup \{\infty\}$ with the following property:

If $\mathsf{Z}(a) \cap x\mathsf{Z}(H) \neq \emptyset$ and $z \in \mathsf{Z}(a)$, then there exists some factorization $z' \in \mathsf{Z}(a) \cap x\mathsf{Z}(H)$ such that $\mathsf{d}(z, z') \leq N$.

We define $\mathsf{t}(H, x) = \sup\{\mathsf{t}(a, x) \mid a \in H\}$, and for a subset $X \subset \mathsf{Z}(H)$ we define $\mathsf{t}(a, X) = \sup\{\mathsf{t}(a, x) \mid x \in X\}$.

The monoid H is called *locally tame* if $\mathsf{t}(H, u) < \infty$ for all $u \in \mathcal{A}(H_{\text{red}})$.

Local tameness is a basic finiteness property in factorization theory. In most settings, where the finiteness of some arithmetical invariant is derived, local tameness has to be proved first. In particular, local tameness is an essential tool in the proof of the Structure Theorem for Sets of Lengths.

4 Zero-sum sequences over abelian groups

Let G be an additive abelian group, $G_0 \subset G$ and $\mathcal{F}(G_0)$ the free (abelian, multiplicative) monoid with basis G_0. According to the tradition of combinatorial number theory, the elements of $\mathcal{F}(G_0)$ are called *sequences over* G_0. If $S \in \mathcal{F}(G_0)$, then

$$S = g_1 \cdot \ldots \cdot g_l = \prod_{g \in G_0} g^{\mathsf{v}_g(S)},$$

where $\mathsf{v}_g(S)$ is the g-adic value of S (also called the *multiplicity of g in S*), and $\mathsf{v}_g(S) = 0$ for all $g \in G_0 \setminus \{g_1, \ldots, g_l\}$. Then $|S| = l$ is the *length of S*. We call $\operatorname{supp}(S) = \{g_1, \ldots, g_l\}$ the *support* and $\sigma(S) = g_1 + \ldots + g_l$ the *sum* of S. The monoid

$$\mathcal{B}(G_0) = \{S \in \mathcal{F}(G_0) \mid \sigma(S) = 0\} = \mathcal{B}(G) \cap \mathcal{F}(G_0)$$

is called the *block monoid* over G_0. It is a divisor-closed submonoid of $\mathcal{B}(G)$. The elements of $\mathcal{B}(G_0)$ are called *zero-sum sequences* over G_0. The monoid $\mathcal{B}(G_0)$ is a BF-monoid, and the atoms of $\mathcal{B}(G_0)$ are the minimal zero-sum sequences (that is, zero-sum sequences without a proper zero-sum subsequence). If G_0 is finite, then the Structure Theorem for Sets of Lengths holds for $\mathcal{B}(G_0)$ (see Theorem 5.6.**B.**).

For every arithmetical invariant $*(D)$ defined for a monoid D, we write $*(G_0)$ instead of $*(\mathcal{B}(G_0))$. Hence $\mathcal{A}(G_0) = \mathcal{A}(\mathcal{B}(G_0))$, $\mathcal{L}(G_0) = \mathcal{L}(\mathcal{B}(G_0))$, $\Delta(G_0) = \Delta(\mathcal{B}(G_0))$, $\mathsf{c}(G_0) = \mathsf{c}(\mathcal{B}(G_0))$ and so on.

A sequence is called *zero-sumfree* if it contains no non-empty zero-sum subsequence. Clearly, a sequence S is zero-sumfree if and only if the sequence $(-\sigma(S))S$ is a minimal zero-sum sequence. The investigation of the structure and length of zero-sum sequences and of zero-sumfree sequences with extremal properties is a main topic in additive group theory. In this area there is a wealth of classical and still wide open questions (see [12] for a recent survey).

We recall the definition of two central invariants of finite abelian groups and discuss their arithmetical significance (Theorem 4.2). If G is a finite abelian group, then $\mathcal{A}(G)$ is finite and thus $\mathcal{B}(G)$ is a finitely generated monoid.

Definition 4.1 Let G be a finite abelian group. Then

$$\mathsf{D}(G) = \max\{|S| \mid S \in \mathcal{A}(G)\}$$

is called the *Davenport constant* of G. It is the smallest integer $l \in \mathbb{N}$ such that every sequence $S \in \mathcal{F}(G)$ of length $|S| \geq l$ has a non-empty zero-sum subsequence. For a sequence $S \in \mathcal{F}(G)$, we define its *cross number* by

$$\mathsf{k}(S) = \sum_{i=1}^{l} \frac{1}{\operatorname{ord}(g_i)}, \quad \text{and} \quad \mathsf{k}(G) = \max\{\mathsf{k}(S) \mid S \in \mathcal{F}(G) \text{ is zero-sumfree}\}$$

is called the (*little*) *cross number* of G.

In general, the precise values of $\mathsf{D}(G)$ and $\mathsf{k}(G)$ (in terms of the group invariants of G) are unknown. Among others, both $\mathsf{D}(G)$ and $\mathsf{k}(G)$ are known for p-groups, and $\mathsf{D}(G)$ is also known for cyclic groups and for groups of rank 2. For every finite abelian group G we have $\mathsf{D}(G) \leq |G|$, and equality holds if and only if G is cyclic. A straightforward argument shows that

$$\mathsf{D}(G) = 2 \max\Big\{\frac{\max L}{\min L} \;\Big|\; L \in \mathcal{L}(G)\Big\} = \max\{\max L \mid L \in \mathcal{L}(G)\,,\ 2 \in L\}\,.$$

In order to investigate sets of lengths $L \in \mathcal{L}(G)$ with $\{2, \mathsf{D}(G)\} \subset L$ it is necessary to know the structure of minimal zero-sum sequences $S \in \mathcal{A}(G)$ of maximal length $|S| = \mathsf{D}(G)$.

It is conjectured that (apart from some explicitly known exceptions) a finite abelian group G is uniquely determined by the system of sets of lengths $\mathcal{L}(G)$. Up to now, this conjecture could only be verified for special classes of groups (including cyclic groups). Again, any progress concerning this problem heavily depends on the knowledge of the Davenport constant and of the structure of minimal zero-sum sequences of maximal length. It is worth mentioning that, in contrast to this conjecture, for any infinite abelian group G the system $\mathcal{L}(G)$ contains all finite non-empty subsets of $\mathbb{N}_{\geq 2}$ regardless of the structure of G (see [22]).

A subset G_0 of an abelian group G is called half-factorial if the monoid $\mathcal{B}(G_0)$ is half-factorial. Half-factorial subsets (and minimal non-half-factorial subsets as their counterparts) play a crucial role in the investigations of $\Delta^*(G)$ (see Theorem 4.2.3) and in some problems of the analytic theory of non-unique factorizations (see Theorem 7.4). A recent survey of results concerning the structure and cardinality of half-factorial subsets was given by W.A. Schmid [26].

A classical criterion (due to A. Zaks and L. Skula) states that a subset G_0 of a finite abelian group G is half-factorial if and only if $\mathsf{k}(S) = 1$ for all $S \in \mathcal{A}(G_0)$.

Theorem 4.2 *Let G be a finite abelian group with $|G| \geq 3$, say*

$$G = C_{n_1} \oplus \ldots \oplus C_{n_r}, \quad \textit{where} \quad 1 < n_1 \mid \ldots \mid n_r = n, \quad \textit{and} \quad k = \sum_{i=1}^{r} \left\lfloor \frac{n_i}{2} \right\rfloor .$$

(1) *We have* $[1, \max\{n-2, k-1\}] \subset \Delta(G) \subset [1, \mathsf{c}(G)-2]$. *In particular,* $\min \Delta(G) = 1$, *and both* $\Delta(G)$ *and* $\mathsf{c}(G)$ *grow with the exponent and with the rank of* G.

(2) $\mathsf{c}(G) \le \mathsf{D}(G)$, *and equality holds if and only if* G *is cyclic or an elementary 2-group.*

(3) $\max \Delta^*(G) \le \max\{n-2, 2\mathsf{k}(G)-1\}$, *and if* $|G| \le \max\{\mathrm{e}^{n/2}, n^2\}$, *then* $\max \Delta^*(G) = n-2$.

For the proof of Theorem 4.2 we refer to [13], [25] and to [16]. Recall that $\mathsf{c}(G)$ measures how complex the set of factorizations of a zero-sum sequence $S \in \mathcal{B}(G)$ may be. Thus Theorem 4.2 asserts that this set of factorizations may become the more complex the larger G is. In contrast to that, the following Theorem 4.3 shows that (in a precise sense) "almost all" zero-sum sequences over a finite abelian group have catenary degree $\mathsf{c}(S) \le 3$. A similar quantitative result holds for factorizations of algebraic integers (see Theorem 7.4).

Theorem 4.3 *Let G be a finite abelian group.*

(1) *Every sequence* $S \in \mathcal{B}(G)$ *for which* $\mathrm{supp}(S) \cup \{0\} \subset G$ *is a subgroup has catenary degree* $\mathsf{c}(S) \le 3$.

(2) *For every* $A \in \mathcal{B}(G)$ *with* $\mathrm{supp}(A) = G$ *we have*

$$\frac{\left|\{S \in A\mathcal{B}(G) \mid |S| \le N\}\right|}{\left|\{S \in \mathcal{B}(G) \mid |S| \le N\}\right|} = 1 + O\Big(\frac{1}{N}\Big) \quad \textit{for all} \quad N \in \mathbb{N}.$$

In particular,

$$\frac{\left|\{S \in \mathcal{B}(G) \mid \mathsf{c}(S) \le 3,\ |S| \le N\}\right|}{\left|\{S \in \mathcal{B}(G) \mid |S| \le N\}\right|} = 1 + O\Big(\frac{1}{N}\Big) \quad \textit{for all} \quad N \in \mathbb{N}.$$

While the proof of the second assertion in Theorem 4.3 is carried out by a simple counting argument, the proof of the first one needs deep methods from additive group theory and occupies about 30 pages in [16].

5 Krull monoids and C-monoids

Let D be a factorial monoid and $H \subset D$ a submonoid. If H is "not too far" from D, then we can investigate the arithmetic of H by means of the unique factorization in D. The distance between H and D is measured by the notions of class groups and class semigroups (note that this in accordance with the philosophy that "the class group measures the deviation from being factorial"). Krull monoids and C-monoids are the most important classes of monoids which are investigated in this way.

Definition 5.1 (Krull monoids)

1. Let D be a monoid and $H \subset D$ a submonoid. Then $H \subset D$ is called *saturated* if $\mathsf{q}(H) \cap D = H$ (that is, if $a, b \in H$ and a divides b in D, then a divides b in H).
2. H is called a *Krull monoid* if H_{red} is a saturated submonoid of a free monoid.
3. Let H be a Krull monoid and suppose that $H_{\text{red}} \subset D = \mathcal{F}(P)$ is a saturated submonoid of a free monoid such that every $p \in P$ is the greatest common divisor of finitely many elements of H_{red}. Then we call D a monoid of *divisors* and P a set of *prime divisors* of H.
   Every Krull monoid possesses a monoid of divisors, and if D and D' are monoids of divisors of H, then there is a unique isomorphism $\Phi \colon D \to D'$ with $\Phi \,|\, H_{\text{red}} = \text{id}$. Hence the *class group*
   $$\mathcal{C}(H) = D/H_{\text{red}} \quad \text{and the subset} \quad \{[p] \in \mathcal{C}(H) \mid p \in P\}$$
   of all classes containing primes are uniquely determined by H (up to canonical isomorphism).

The arithmetic of a Krull monoid is uniquely determined by its class group and the distribution of primes in the classes (for a precise statement see Theorem 6.6). In particular, a monoid is factorial if and only if it a Krull monoid with trivial class group. By definition, H is a Krull monoid if and only if H_{red} is a Krull monoid, and $\mathcal{C}(H) = \mathcal{C}(H_{\text{red}})$.

In the following proposition we present (without proofs) several ideal-theoretic characterizations of Krull monoids (some of them are in accordance with the well-known characterizations of Krull domains). Full proofs can be found in [21] or [16].

Proposition 5.2 *Then the following statements are equivalent:*

(1) *H is a Krull monoid.*

(2) *H_{red} is a Krull monoid, and $H = H^\times \times H_0$ for some submonoid H_0 of H with $H_0 \cong H_{\text{red}}$.*

(3) *H is v-noetherian and completely integrally closed.*

(4) *H is v-noetherian and every non-empty v-ideal of H is v-invertible.*

From the v-ideal theory of a Krull monoid H a monoid of divisors is obtained as follows. The monoid $\mathcal{I}_v^*(H)$ of all non-empty v-ideals, equipped with the v-multiplication, is a free monoid, and the set $\mathfrak{X}(H)$ of all v-maximal v-ideals is a basis of $\mathcal{I}_v^*(H)$. If we identify H_{red} with the set of all principal ideals of H in the natural way, then $H_{\text{red}} \subset \mathcal{I}_v^*(H)$, the monoid $\mathcal{I}_v^*(H)$ is a monoid of divisors of H, and $\mathfrak{X}(H)$ is a set of prime divisors of H.

Examples 5.3 (Examples of Krull monoids)

1. *Block monoids.* Let G be an abelian group and $G_0 \subset G$ a subset. Then the block monoid $\mathcal{B}(G_0)$ is a saturated submonoid of $\mathcal{F}(G_0)$, and thus it is a Krull monoid. It can even be proved that every reduced Krull monoid is isomorphic to a block monoid over a suitable subset of some abelian group.

   If $|G| \le 2$, then the block monoid $\mathcal{B}(G)$ is factorial, and if $|G| \ge 3$, then $\mathcal{F}(G)$ is a monoid of divisors of $\mathcal{B}(G)$, $\mathcal{C}(\mathcal{B}(G)) \cong G$, and every class contains precisely one prime.

2. *Multiplicative monoids of domains.* The multiplicative monoid $R^\bullet$ of a domain R is a Krull monoid if and only if R is a Krull domain, and in this case $\mathcal{C}(R^\bullet)$ is (canonically isomorphic to) the divisor class group $\mathcal{C}(R)$ of R.

3. *Regular congruence monoids in Krull domains.* Let R be a Krull domain, $\{0\} \ne \mathfrak{f} \triangleleft R$ an ideal and $\Gamma \subset (R/\mathfrak{f})^\times$ a subgroup. Then the monoid

$$H_\Gamma = \{a \in R^\bullet \mid a + \mathfrak{f} \in \Gamma\}$$

is a Krull monoid, called the *(regular) congruence monoid* defined in R modulo $\mathfrak{f}$ by Γ (see also Theorem 5.7 and Remarks 5.8).

4. *Regular Hilbert monoids.* Let $f \in \mathbb{N}_{\ge 2}$ and $\Gamma \subset (\mathbb{Z}/f\mathbb{Z})^\times$ a subgroup. Then the monoid $H_\Gamma = \{a \in \mathbb{N} \mid a + f\mathbb{Z} \in \Gamma\}$ is a Krull monoid with class group $\mathcal{C}(H_\Gamma) \cong (\mathbb{Z}/f\mathbb{Z})^\times/\Gamma$, and (by the Dirichlet Prime Number Theorem) every class contains infinitely many primes (see also Theorem 5.7, Remarks 5.8 and Example 5.9).

5. *Analytic theory.* If $[D, H, |\cdot|]$ is a quasi-formation, then H is a Krull monoid with finite class group and infinitely many primes in every class.

6. *Module theory.* Let R be a ring and $\mathcal{C}$ a class of (right) R-modules, closed under finite direct sums, direct summands and isomorphisms such that $\mathcal{C}$ has a set $V(\mathcal{C})$ of representatives (that is, every $A \in \mathcal{C}$ is isomorphic to a unique $[A] \in V(\mathcal{C})$). Then $V(\mathcal{C})$ becomes a commutative semigroup with multiplication $[A]\cdot[B] = [A \oplus B]$. If every $A \in \mathcal{C}$ has a semilocal endomorphism ring, then $V(\mathcal{C})$ is a Krull monoid (see [8]). Among many other cases, this condition is fulfilled if either R is semilocal (not necessarily commutative) and $\mathcal{C}$ is the class of all finitely generated projective R-modules (see [10]), or if R is commutative local noetherian and $\mathcal{C}$ is the class of all finitely generated R-modules (see [27] and [9]).

Definition 5.4

1. Let D be a monoid and $H \subset D$ a submonoid. Two elements $y,\, y' \in D$ are called *H-equivalent* if $y^{-1}H \cap D = {y'}^{-1}H \cap D$ (that is, for all $a \in D$ we have $ya \in H$ if and only if $y'a \in H$). H-equivalence is a congruence relation on D, and for $y \in D$ we denote by $[y]_H^D$ the congruence class of y. We define the *class semigroup*

$$\mathcal{C}(H,D) = \{[y]_H^D \mid y \in D\}, \quad \text{and} \quad \mathcal{C}^*(H,D) = \{[y]_H^D \mid y \in (D \setminus D^\times) \cup \{1\}\}$$

and is called the *reduced class semigroup.*

2. H is called a C-*monoid* if H is a submonoid of a factorial monoid F such that $F^\times \cap H = H^\times$ and $\mathcal{C}^*(H,F)$ is finite. In this case we say that H is a C-*monoid defined in* F.

If H is a C-monoid defined in a factorial monoid F, then F is far from being unique. However, there is a canonical choice for F which is given by the assertion **A**.3 of Theorem 5.6 below. The most important examples of C-monoids will be presented in Theorem 5.7.

For a saturated submonoid $H \subset D$, the distance between H and D is satisfactory codified in the class group $\mathsf{q}(D)/\mathsf{q}(H)$, but in the general case the more subtle concept of a class semigroup is needed. In Proposition 5.5 below we compare these two concepts and establish the connection between Krull monoids and C-monoids. For full proofs of all assertions concerning C-monoids we refer to [20], [14] and [16].

Proposition 5.5 *Let D be a monoid and $H \subset D$ a submonoid.*

(1) *$\mathcal{C}(H,D)$ is finite if and only if both $\mathcal{C}^*(H,D)$ and $D^\times/H^\times$ are finite.*

(2) *If $aD \cap H \neq \emptyset$ for all $a \in D$, then there are natural epimorphisms*

$$\theta \colon \mathcal{C}(H,D) \to D/H \quad \textit{and} \quad \theta^* \colon \mathcal{C}^*(H,D) \to D/D^\times H\,,$$

and θ is an isomorphism if and only if $H \subset D$ is saturated.

(3) *Every Krull monoid with finite class group is a* C*-monoid. In particular, the block monoid over a finite abelian group is a* C*-monoid.*

(4) *Let H be a* C*-monoid defined in a factorial monoid F such that $\mathcal{C}^*(H,F)$ is a group. Then H is a Krull monoid.*

Theorem 5.6 (Main Theorem on C-monoids) *Let H be a* C*-monoid.*

A. Algebraic Properties.

(1) *H is v-noetherian.*

(2) *$\widehat{H}$ is a Krull monoid with finite class group, and there exists some $a \in H$ such that $a\widehat{H} \subset H$.*

(3) *Suppose that $\widehat{H} = \widehat{H}^\times \times D$ with $D \cong \widehat{H}_{\mathrm{red}}$, let F_0 be a monoid of divisors of D and $F = \widehat{H}^\times \times F_0$. Then H is a* C*-monoid defined in F, and there is an epimorphism $\mathcal{C}^*(H,F) \to \mathcal{C}(\widehat{H})$.*

B. Arithmetical Properties. *H is locally tame, $\mathsf{c}(H) < \infty$, and the Structure Theorem for Sets of Lengths holds for H.*

Structure of the proof of **B.**

1. (Reduction step) By means of a transfer principle (see Proposition 6.4 and Theorem 6.5) we may assume that H is defined in a finitely generated factorial monoid F. Thus let $F = F^\times \times [p_1, \ldots, p_s]$ with pairwise non-associated prime elements $p_1, \ldots, p_s$.

2. (Local tameness) The finiteness of $\mathcal{C}^*(H, F)$ turns out to be equivalent with the following property by means of which local tameness can be verified by explicit calculations:

> There exist some $\alpha \in \mathbb{N}$ and a subgroup $V \subset F^\times$ such that $(F^\times : V) \,|\, \alpha$, $V(H \setminus H^\times) \subset H$, and for all $j \in [1, s]$ and $a \in p_j^\alpha F$ we have $a \in H$ if and only if $p_j^\alpha a \in H$.

3. (Catenary degree) H is a v-noetherian G-monoid and thus it is finitary. Every locally tame finitary monoid has finite catenary degree (of course, this argument uses the definitions and simple properties of finitary monoids and G-monoids, see [17]).

4. (Structure Theorem for Sets of Lengths) The proof splits into an abstract additive part and an ideal-theoretic part. Both steps rest on the concepts of pattern ideals and of tamely generated ideals which are defined as follows.

For a finite non-empty set $A \subset \mathbb{Z}$ the pattern ideal $\Phi(A)$ is the set of all $a \in H$ for which there is some $y \in \mathbb{Z}$ such that $y + A \subset \mathsf{L}(a)$.

A subset $\mathfrak{a} \subset H$ is called tamely generated if there exist a subset $E \subset \mathfrak{a}$ and a bound $N \in \mathbb{N}$ with the following property:

> For every $a \in \mathfrak{a}$ there exists some $e \in E$ such that $e \,|\, a$, $\sup \mathsf{L}(e) \le N$ and $\mathsf{t}(a, \mathsf{Z}(e)) \le N$.

In the additive part one proves that the Structure Theorem for Sets of Lengths holds for every BF-monoid H with finite set $\Delta(H)$ in which all pattern ideals are tamely generated. This is done in the spirit of additive number theory. To apply this additive result to a BF-monoid H, it must be proved that $\Delta(H)$ is finite and that all pattern ideals are tamely generated. For a finitely generated monoid, this is comparatively simple. For a C-monoid, the finiteness of $\Delta(H)$ follows from the finiteness of the catenary degree, while the tame generation of pattern ideals needs deep ideal-theoretic considerations.

Theorem 5.7 (C-monoids in ring theory)

A. *Let R be a Krull domain and $\mathfrak{f}$ an ideal of R such that $\mathcal{C}(R)$ and $R/\mathfrak{f}$ are both finite. Let $\emptyset \ne \Gamma \subset R/\mathfrak{f}$ be a multiplicatively closed subset and $H_\Gamma = \{a \in R^\bullet \mid a + \mathfrak{f} \in \Gamma\} \cup \{1\}$. Suppose that either $\mathfrak{f}$ is divisorial or R is noetherian. Then H_Γ is a C-monoid.*

B. *Let $f \in \mathbb{N}_{\ge 2}$ and $\emptyset \ne \Gamma \subset \mathbb{Z}/f\mathbb{Z}$ a multiplicatively closed subset. Then $H_\Gamma = \{a \in \mathbb{N} \mid a + f\mathbb{Z} \in \Gamma\} \cup \{1\}$ is a C-monoid.*

C. *Let A be a Mori domain, $R = \widehat{A}$ and $\mathfrak{f} = \{a \in R \mid aR \subset A\} \neq \{0\}$. Then R is a Krull domain. If $\mathcal{C}(R)$ and $R/\mathfrak{f}$ are both finite, then $A^\bullet$ is a* C*-monoid.*

Remarks 5.8 The monoid H_Γ considered in **A.** is called the *congruence monoid* defined in R modulo $\mathfrak{f}$ by Γ and that considered in **B.** is called the *Hilbert monoid* defined modulo $\mathfrak{f}$ by Γ (named after D. Hilbert who used such monoids to demonstrate the necessity of a proof for the uniqueness of prime factorizations in the integers).

Regular congruence monoids and Hilbert monoids were already considered in Examples 5.3 (3. and 4.). Using a more general concept of congruence monoids (including sign conditions, see [15] or [16]) it is possible to treat **A.** and **B.** in a uniform way.

Let A be a Mori domain as in **C.** (see [5] for a recent survey on Mori domains). Then $\mathfrak{f}$ is the largest ideal of $\widehat{A}$ lying in A (called the *conductor* of A), and $A = \{a \in \widehat{A} \mid a + \mathfrak{f} \in A/\mathfrak{f}\}$, whence in particular A is the congruence monoid defined in $\widehat{A}$ modulo $\mathfrak{f}$ by $A/\mathfrak{f}$.

We sketch the proof of **A.** (by the above remarks, that of **B.** and **C.** is essentially the same). Let $R^\bullet = R^\times \times R_0$ with a reduced Krull monoid R_0, let F_0 be a monoid of divisors of R_0 and $F = R^\times \times F_0$. Then H_Γ is a C-monoid defined in F (the main task is to deduce the finiteness of $\mathcal{C}^*(H_\Gamma, F)$ from that of $R/\mathfrak{f}$ and $\mathcal{C}(R)$).

Example 5.9 (Orders in Dedekind domains and algebraic number fields)

Let R be a Dedekind domain, $\{0\} \neq \mathfrak{f} \triangleleft R$ and $\Gamma \subset (R/\mathfrak{f})^\times$ a subgroup. Then the regular congruence monoid $H_\Gamma = \{a \in R^\bullet \mid a + \mathfrak{f} \in \Gamma\}$ is a Krull monoid (see Example 5.3.3). We denote by $\mathcal{I}_\mathfrak{f}(R)$ the (multiplicative) monoid of all non-zero ideals $\mathfrak{a} \triangleleft R$ with $\mathfrak{a} + \mathfrak{f} = R$, by $\mathcal{H}_\mathfrak{f}(R)$ its submonoid of principal ideals and by $\mathfrak{X}_\mathfrak{f}(R)$ the set of all prime ideals in $\mathcal{I}_\mathfrak{f}(R)$. Then

$$\mathcal{H}_\Gamma = \{aR \mid a \in H_\Gamma\} \subset \mathcal{H}_\mathfrak{f}(R)$$

is a submonoid (called a *generalized Hilbert monoid*), $\mathcal{I}_\mathfrak{f}(R)$ is a monoid of divisors and $\mathfrak{X}_\mathfrak{f}(R)$ is a set of prime divisors of $\mathcal{H}_\Gamma$. There is a canonical isomorphism $(H_\Gamma)_{\rm red} \cong \mathcal{H}_\Gamma$, and consequently $\mathcal{C}(H_\Gamma) = \mathcal{C}(\mathcal{H}_\Gamma)$. The monoid $\mathcal{S}_\mathfrak{f}(R) = \{aR \mid a \in 1 + \mathfrak{f}\} \subset \mathcal{H}_\Gamma$ is called the *principal ray* and the group $\mathcal{I}_\mathfrak{f}(R)/\mathcal{S}_\mathfrak{f}(R)$ is called the *ray class group* modulo $\mathfrak{f}$. There is a natural exact sequence

$$0 \to \mathcal{H}_\Gamma/\mathcal{S}_\mathfrak{f}(R) \to \mathcal{I}_\mathfrak{f}(R)/\mathcal{S}_\mathfrak{f}(R) \to \mathcal{C}(H_\Gamma) \to 0$$

which shows that every class $C \in \mathcal{C}(H_\Gamma)$ is the union of $|\mathcal{H}_\Gamma/\mathcal{S}_\mathfrak{f}(R)|$ ray classes modulo $\mathfrak{f}$.

Let now $A \subset R$ be an order (that is, A is a subring of R such A and R have the same field of quotients, and R is a finitely generated A-module).

Then A is a one-dimensional noetherian domain with integral closure $\widehat{A} = R$ and conductor $\mathfrak{f} = \{a \in A \mid aR \subset A\}$. Since $A = \{a \in R \mid a + \mathfrak{f} \in A/\mathfrak{f}\}$, the monoid $A^\bullet$ is the congruence monoid defined in R modulo $\mathfrak{f}$ by $A/\mathfrak{f}$, and if both $R/\mathfrak{f}$ and $\mathcal{C}(R)$ are finite, then $A^\bullet$ is a C-monoid (see Theorem 5.7 and the consecutive remarks). The monoid

$$A^* = \{a \in R^\bullet \mid a + \mathfrak{f} \in (A/\mathfrak{f})^\times\} = \{a \in A^\bullet \mid aA + \mathfrak{f} = A\}$$

is a regular congruence monoid in R and thus it is a Krull monoid (see Example 5.3.3). By the above, $\mathcal{I}_\mathfrak{f}(R)$ is a monoid of divisors and $\mathfrak{X}_\mathfrak{f}(R)$ is a set of prime divisors of A^*, and every class $C \in \mathcal{C}(A^*)$ is a union of ray classes modulo $\mathfrak{f}$.

We compare the class group $\mathcal{C}(A^*)$ with the Picard group

$$\mathrm{Pic}(A) = \frac{\{\text{invertible fractional ideals of } A\}}{\{\text{fractional principal ideals of } A\}} \quad \text{of} \quad A\,.$$

It is not difficult to prove that every non-zero ideal $\mathfrak{a} \lhd A$ with $\mathfrak{a} + \mathfrak{f} = A$ is invertible, and that every class $C \in \mathrm{Pic}(A)$ contains such an ideal. Hence there is an epimorphism $\mathcal{I}_\mathfrak{f}(R) \to \mathrm{Pic}(A)$, which maps an ideal $\mathfrak{a} \in \mathcal{I}_\mathfrak{f}(R)$ onto the class $[\mathfrak{a} \cap A] \in \mathrm{Pic}(A)$. This epimorphism induces an isomorphism $\mathcal{C}(A^*) \overset{\sim}{\to} \mathrm{Pic}(A)$.

Let finally R be the ring of integers of an algebraic number field and $A \subset R$ an order with conductor $\mathfrak{f}$. The ray class group $\mathcal{I}_\mathfrak{f}(R)/\mathcal{S}_\mathfrak{f}(R)$ is finite and every ray class contains infinitely many prime ideals (see [24, Theorem 3.7 and Corollary 7 to Proposition 7.16]). Consequently every class of $\mathcal{C}(A^*)$ contains infinitely many primes, and every class of $\mathrm{Pic}(A)$ contains infinitely many prime ideals.

6 Transfer principles

Transfer principles are a central tool in the theory of non-unique factorizations. By means of them it is possible to establish factorization properties in simple auxiliary monoids and then to apply the result to various cases of arithmetical interest. This section is rather technical, and we give no proofs (most of them are simple, see [14] or [16] for details).

Definition 6.1 A monoid homomorphism $\theta\colon H \to B$ is called a *transfer homomorphism* if it has the following properties:

(T 1) $B = \theta(H)B^\times$ and $\theta^{-1}(B^\times) = H^\times$.

(T 2) If $u \in H$, $b, c \in B$ and $\theta(u) = bc$, then there exist $v, w \in H$ such that $u = vw$, $\theta(v) \simeq b$ and $\theta(w) \simeq c$.

Proposition 6.2 *Let $\theta\colon H \to B$ be a transfer homomorphism.*

(1) *If $u \in H$, then $u \in \mathcal{A}(H)$ if and only if $\theta(u) \in \mathcal{A}(B)$.*

(2) *There is a unique homomorphism* $\overline{\theta}\colon \mathsf{Z}(H) \to \mathsf{Z}(B)$ *(referred to as the extension of* θ *to the factorization monoids) satisfying*

$$\overline{\theta}(uH^\times) = \theta(u)B^\times \quad \textit{for all} \quad u \in \mathcal{A}(H)\,.$$

It is surjective and has the following properties:

(a) *If* $z,\ z' \in \mathsf{Z}(H)$*, then* $|\overline{\theta}(z)| = |z|$ *and* $\mathsf{d}\big(\overline{\theta}(z), \overline{\theta}(z')\big) \le \mathsf{d}(z, z')$.

(b) *If* $u \in H$*, then* $\overline{\theta}(\mathsf{Z}_H(u)) = \mathsf{Z}_B(\theta(u))$ *and* $\mathsf{L}_H(u) = \mathsf{L}_B(\theta(u))$*. In particular,* $\mathcal{L}(H) = \mathcal{L}(B)$.

Definition 6.3 Let $\theta\colon H \to B$ be a transfer homomorphism of atomic monoids and $\overline{\theta}\colon \mathsf{Z}(H) \to \mathsf{Z}(B)$ its extension to the factorization monoids.

1. (*Catenary degree in the fibres*) For $a \in H$, we denote by $\mathsf{c}(a,\theta)$ the smallest $N \in \mathbb{N}_0 \cup \{\infty\}$ with the following property:

   If $z,\ z' \in \mathsf{Z}(a)$ and $\overline{\theta}(z) = \overline{\theta}(z')$, then there exists a finite sequence of factorizations $z = z_0, \ldots, z_k = z' \in \mathsf{Z}(a)$ such that $\overline{\theta}(z_i) = \overline{\theta}(z)$ and $\mathsf{d}(z_{i-1}, z_i) \le N$ for all $i \in [1,k]$.

   We call $\mathsf{c}(H,\theta) = \sup\{\mathsf{c}(a,\theta) \mid a \in H\} \in \mathbb{N}_0 \cup \{\infty\}$ the *catenary degree in the fibres* of θ.

2. (*Tame degree in the fibres*) For $a \in H$ and $x \in \mathsf{Z}(H)$, we denote by $\mathsf{t}(a,x,\theta)$ the smallest $N \in \mathbb{N}_0 \cup \{\infty\}$ with the following property:

   If $\mathsf{Z}(a) \cap x\mathsf{Z}(H) \ne \emptyset$, $z \in \mathsf{Z}(a)$ and $\overline{\theta}(z) \in \overline{\theta}(x)\mathsf{Z}(B)$, then there exists some $z' \in \mathsf{Z}(a) \cap x\mathsf{Z}(H)$ such that $\overline{\theta}(z') = \overline{\theta}(z)$ and $\mathsf{d}(z,z') \le N$.

   We call $\mathsf{t}(H,x,\theta) = \sup\{\mathsf{t}(a,x,\theta) \mid a \in H\} \in \mathbb{N}_0 \cup \{\infty\}$ the *tame degree of* x *in the fibres* of θ.

Proposition 6.4 *Let* $\theta\colon H \to B$ *be a transfer homomorphism of atomic monoids.*

(1) *If* $a \in H$*, then* $\mathsf{c}(\theta(a)) \le \mathsf{c}(a) \le \max\{\mathsf{c}(\theta(a)), \mathsf{c}(a,\theta)\}$*. In particular,* $\mathsf{c}(B) \le \mathsf{c}(H) \le \max\{\mathsf{c}(B), \mathsf{c}(H,\theta)\}$.

(2) *If* $\overline{\theta}\colon \mathsf{Z}(H) \to \mathsf{Z}(B)$ *is the extension of* θ *to the factorization monoids and* $x \in \mathsf{Z}(H)$*, then* $\mathsf{t}\big(B,\overline{\theta}(x)\big) \le \mathsf{t}(H,x) \le \mathsf{t}\big(B,\overline{\theta}(x)\big) + \mathsf{t}(H,x,\theta)$*. In particular, if* $\mathsf{t}(H,u,\theta) < \infty$ *for all* $u \in \mathcal{A}(H_{\mathrm{red}})$ *and B is locally tame, then H is also locally tame.*

We apply the Propositions 6.2 and 6.4 to C-monoids and to Krull monoids. The application to C-monoids has already been mentioned and successfully used when we sketched the proof of Theorem 5.6.**B.** The transfer result for Krull monoids (Theorem 6.6) goes back to ideas of W. Narkiewicz (see [23]). It provides the link between factorization theory and additive group theory (see Section 4) and shows that (most) arithmetical invariants of a Krull monoid are in fact combinatorial invariants of the class group.

Theorem 6.5 (A transfer result for C-monoids) *Let H be a C-monoid. Then there is a transfer homomorphism $\theta\colon H \to B$ having the following properties:*

(1) *B is C-monoid defined in a factorial monoid D with only finitely many non-associated primes such that $D^\times$ is finite.*
(2) $\mathsf{c}(H,\theta) \le 2$.
(3) *If H is defined in a factorial monoid F and $u \in \mathcal{A}(H)$ is a product of m primes in F, then $\mathsf{t}(H, uH^\times, \theta) \le m + d$ where $d \in \mathbb{N}$ depends only on $\mathcal{C}^*(H,F)$.*

Theorem 6.6 (A transfer result for Krull monoids) *Let H be a Krull monoid, $D = \mathcal{F}(P)$ a monoid of divisors of H, $G = \mathcal{C}(H)$ its class group and $G_0 \subset G$ the set of all classes containing primes. Let $\widetilde{\boldsymbol{\beta}}\colon D \to \mathcal{F}(G_0)$ be the unique homomorphism satisfying $\widetilde{\boldsymbol{\beta}}(p) = [p]$ for all $p \in P$.*

Then $\widetilde{\boldsymbol{\beta}}^{-1}(\mathcal{B}(G)) = H_{\mathrm{red}}$, and the homomorphism $\boldsymbol{\beta}\colon H \to \mathcal{B}(G_0)$, defined by $\boldsymbol{\beta}(a) = \widetilde{\boldsymbol{\beta}}(aH^\times)$, is a transfer homomorphism satisfying $\mathsf{c}(H,\boldsymbol{\beta}) \le 2$ and $\mathsf{t}(H,u,\boldsymbol{\beta}) \le \mathsf{D}(G) + 1$ for all $u \in \mathcal{A}(H_{\mathrm{red}})$.

Let now H be a Krull monoid with finite class group $G = \mathcal{C}(H)$, and suppose that every class contains primes. This holds true for every Krull monoid fitting into a quasi-formation (see Definition 7.1 and Examples 7.2).

Then Theorem 6.6 together with the Propositions 6.2 and 6.4 implies that $\mathcal{L}(H) = \mathcal{L}(G)$, $\Delta(H) = \Delta(G)$, $\Delta^*(H) = \Delta^*(G)$, and if $|G| \ge 3$, then also $\mathsf{c}(H) = \mathsf{c}(G)$. Hence all these quantities can be described with the methods of Section 4. In particular, we rediscover Carlitz' result that H is half-factorial if and only if $|G| \le 2$. Taking into account that the catenary degree $\mathsf{c}(H)$ measures how complex sets of factorizations of elements of H may be and that $\mathsf{c}(H) = \mathsf{c}(G)$ grows with the size of G (see Theorem 4.2.1) we approve the philosophy of classical algebraic number theory that the class group is a measure for the non-uniqueness of factorizations.

7 Analytic theory

The concept of quasi-formations stems from abstract analytic number theory (as presented in [16]) and allows us to formulate the analytic theory of algebraic numbers and algebraic functions in a uniform way.

Definition 7.1 A *quasi-formation* $[D, H, |\cdot|]$ consists of

- a free monoid $D = \mathcal{F}(P)$,
- a homomorphism $|\cdot|\colon D \to (\mathbb{N}, \cdot)$ such that $|a| = 1$ if and only if $a = 1$, and the Dirichlet series

$$\sum_{p \in P} |p|^{-s} \quad \text{converges for} \quad \Re(s) > 1\,,$$

- a saturated submonoid $H \subset D$ such that $G = D/H$ is finite, and for every $g \in G$ the function ψ_g, defined by

$$\psi_g(s) = \sum_{p \in P \cap g} |p|^{-s} - \frac{1}{|G|} \log \frac{1}{s-1} \quad \text{for} \quad \Re(s) > 1\,,$$

has a holomorphic extension to $s = 1$.

If $[D, H, |\cdot|]$ is a quasi-formation, then H is a Krull monoid, D is a monoid of divisors of H, $G = \mathcal{C}(H)$, and every class contains a denumerable set of primes. We say that the Krull monoid H fits into a quasi-formation.

Examples 7.2 (Examples of quasi-formations)

1. Let R be the ring of integers of an algebraic number field or a holomorphy ring in an algebraic function field over a finite field and $H = \mathcal{H}(R)$ the (multiplicative) monoid of non-zero principal ideals of R (note that $H \cong (R^\bullet)_{\text{red}}$). Let D be the (multiplicative) monoid of all non-zero ideals of R, and for $\mathfrak{a} \in D$ let $|\mathfrak{a}| = (D : \mathfrak{a})$. Then $[D, H, |\cdot|]$ is a quasi-formation. This can be verified using classical analytic number theory (see [24] for the number field case and [11] for the function field case).

More generally, every generalized Hilbert monoid (see Example 5.9) defined in R fits into a quasi-formation (we omit details).

2. Let $f \in \mathbb{N}_{\ge 2}$, $\Gamma \subset (\mathbb{Z}/f\mathbb{Z})^\times$ a subgroup and $H_\Gamma = \{a \in \mathbb{N} \mid a + f\mathbb{Z} \in \Gamma\}$ a regular Hilbert monoid. If $\mathbb{N}_f$ denotes the monoid of all $a \in \mathbb{N}$ which are relatively prime to f and $|a| = a$ for all $a \in \mathbb{N}_f$, then $[\mathbb{N}_f, H_\Gamma, |\cdot|]$ is a quasi-formation. Again this follows by classical analytic number theory.

The following Proposition 7.3 is based on the Tauberian Theorem of Ikehara and Delange. It is the key analytic tool for our arithmetical main result given in the Theorem 7.4. There we show that the set of elements having more than k distinct factorization lengths (provided that $|G| \ge 3$) and the set of all elements having catenary degree at most 3 both have density 1.

Proposition 7.3 *Let $[D, H, |\cdot|]$ be a quasi-formation, $G = D/H$, $G_0 \subset G$, $y \in G$, $S \in \mathcal{F}(G \setminus G_0)$ and $l \in \mathbb{N}_0$. Let $\Omega_y(G_0, S, l)$ denote the set of all sequences $C \in \mathcal{F}(G)$ with $\sigma(C) = y$, $\mathsf{v}_g(C) = \mathsf{v}_g(S)$ for all $g \in G \setminus G_0$ and $\mathsf{v}_g(C) \ge l$ for all $g \in G_0$, and suppose that $\Omega_y(G_0, S, 0) \not\subset \{1\}$. Let $\widetilde{\beta}\colon D \to \mathcal{F}(G)$ be the homomorphism defined in* Theorem 6.6, *and for $x \in \mathbb{R}_{\ge 1}$ let*

$$\Omega_y(G_0, S, l)(x) = \big|\{\mathfrak{a} \in D \mid \widetilde{\beta}(\mathfrak{a}) \in \Omega_y(G_0, S, l)\,,\ |\mathfrak{a}| \le x\,\}\big|\,.$$

Then we have, for $x \to \infty$,

$$\Omega_y(G_0, S, l)(x) \asymp x\,(\log x)^\eta\,(\log\log x)^\delta\,,$$

where

$$\eta = -1 + \frac{|G_0|}{|G|} \quad \text{and} \quad \delta = \begin{cases} |S|\,, & \text{if } G_0 \ne \emptyset\,, \\ |S| - 1\,, & \text{if } G_0 = \emptyset\,. \end{cases}$$

Theorem 7.4 *Let $[D, H, |\cdot|]$ be a quasi-formation and $G = D/H$.*

(1) *If $|G| \ge 3$ and $k \in \mathbb{N}$, then we have (for $x \ge 3$)*

$$\frac{\left|\{a \in H \mid |\mathsf{L}(a)| > k\,, |a| \le x\}\right|}{\left|\{a \in H \mid |a| \le x\}\right|} = 1 + O\left(\frac{(\log\log x)^{\psi_k(G)}}{(\log x)^{1-\mu(G)/|G|}}\right),$$

where $\mu(G)$ is the maximal cardinality of a half-factorial subset of G and $\psi_k(G) \in \mathbb{N}_0$ is a combinatorial invariant depending only on G and k.

(2) *For $x \ge 2$ we have*

$$\frac{\left|\{a \in H \mid \mathsf{c}(a) \le 3\,, |a| \le x\}\right|}{\left|\{a \in H \mid |a| \le x\}\right|} = 1 + O\big((\log x)^{-1/|G|}\big)\,.$$

Idea of the proof. 1. We show that the set $\mathcal{G}_k(H) = \{a \in H \mid |\mathsf{L}(a)| \le k\}$ is a finite union of sets of the form $\Omega_0(G_0, S_0, l_0)$ for some half-factorial subsets $G_0 \subset G$ and sequences $S_0 \in \mathcal{F}(G \setminus G_0)$. Then we apply Proposition 7.3.

2. Let $a \in H$ with $\mathsf{c}(a) > 3$. Then (since $\boldsymbol{\beta} = \widetilde{\boldsymbol{\beta}} \,|\, H \colon H \to \mathcal{B}(G)$ is a transfer homomorphism with $\mathsf{c}(H, \boldsymbol{\beta}) \le 2$, see Theorem 6.6) it follows that $\mathsf{c}(\boldsymbol{\beta}(a)) > 3$ (see Proposition 6.4), and thus Theorem 4.3.1 implies that $\operatorname{supp}(\boldsymbol{\beta}(a)) \cup \{0\} \ne G$, whence $\boldsymbol{\beta}(a) \in \Omega_0(G \setminus \{g\}, 1, 0)$ for some $g \in G$. Thus

$$\left|\{a \in H \mid \mathsf{c}(a) > 3,\ |a| \le x\}\right| \ll x(\log x)^{-1/|G|}$$

by Proposition 7.3, and so the assertion follows. □

References

1. D.D. Anderson (ed.), *Factorization in Integral Domains*, Lect. Notes Pure Appl. Math., vol. 189, Marcel Dekker, 1997.
2. D.D. Anderson, D.F. Anderson, and M. Zafrullah, *Factorizations in integral domains*, J. Pure Appl. Algebra **69** (1990), 1 – 19.
3. D.D. Anderson, D.F. Anderson, and M. Zafrullah, *Factorizations in integral domains II*, J. Algebra **152** (1992), 78 – 93.
4. D.F. Anderson and D.N. El Abidine, *Factorization in integral domains III*, J. Pure Appl. Algebra **135** (1999), 107–127.
5. V. Barucci, *Mori domains*, in [7], pp. 57 – 73.
6. S.T. Chapman (ed.), *Arithmetical Properties of Commutative Rings and Monoids*, Lect. Notes Pure Appl. Math., vol. 241, Chapman & Hall/CRC, 2005.
7. S.T. Chapman and S. Glaz (eds.), *Non-Noetherian Commutative Ring Theory*, Kluwer Academic Publishers, 2000.
8. A. Facchini, *Direct sum decomposition of modules, semilocal endomorphism rings, and Krull monoids*, J. Algebra **256** (2002), 280 – 307.
9. A. Facchini, W. Hassler, L. Klingler, and R. Wiegand, *Direct-sum decompositions over one-dimensional Cohen-Macaulay local rings*, this volume.

10. A. Facchini and D. Herbera, K_0 *of a semilocal ring*, J. Algebra **225** (2000), 47 – 69.
11. M. Fried and M. Jarden, *Field Arithmetic, 2nd ed.*, Springer, 2005.
12. W. Gao and A. Geroldinger, *Zero-sum problems in finite abelian groups: a survey.*
13. W. Gao and A. Geroldinger, *Systems of sets of lengths II*, Abh. Math. Semin. Univ. Hamb. **70** (2000), 31 – 49.
14. A. Geroldinger and F. Halter-Koch, *Transfer principles in the theory of non-unique factorizations*, in [6], pp. 114 – 142.
15. A. Geroldinger and F. Halter-Koch, *Congruence monoids*, Acta Arith. **112** (2004), 263 – 296.
16. A. Geroldinger and F. Halter-Koch, *Non-Unique Factorizations. Algebraic, Combinatorial and Analytic Theory*, Pure and Applied Mathematics, vol. 278, Chapman & Hall/CRC, 2006.
17. A. Geroldinger, F. Halter-Koch, W. Hassler, and F. Kainrath, *Finitary monoids*, Semigroup Forum **67** (2003), 1 – 21.
18. R. Gilmer, *Multiplicative Ideal Theory*, vol. 90, Queen's Papers, 1992.
19. R. Gilmer, *Forty years of commutative ring theory*, Rings, Modules, Algebras, and Abelian Groups, Lect. Notes Pure Appl. Math., vol. 236, Marcel Dekker, 2004, pp. 229 – 256.
20. F. Halter-Koch, C-*monoids and congruence monoids in Krull domains*, in [6], pp. 71 – 98.
21. F. Halter-Koch, *Ideal Systems*, Marcel Dekker, 1998.
22. F. Kainrath, *Factorization in Krull monoids with infinite class group*, Colloq. Math. **80** (1999), 23 – 30.
23. W. Narkiewicz, *Finite abelian groups and factorization problems*, Colloq. Math. **42** (1979), 319 – 330.
24. W. Narkiewicz, *Elementary and Analytic Theory of Algebraic Numbers, 3rd ed.*, Springer, 2004.
25. W.A. Schmid, *Differences in sets of lengths of Krull monoids with finite class group*, J. Théor. Nombres Bordx. **17** (2005), 323 – 345.
26. W.A. Schmid, *Half-factorial sets in finite abelian groups: a survey*, Grazer Math. Ber. **348** (2005), 41–64.
27. R. Wiegand, *Direct-sum decompositions over local rings*, J. Algebra **240** (2001), 83 – 97.

Mixed polynomial/power series rings and relations among their spectra

William Heinzer[1], Christel Rotthaus[2], and Sylvia Wiegand[3]

[1] Purdue University, West Lafayette IN 47907-1395 heinzer@math.purdue.edu
[2] Michigan State University, East Lansing Mi 488824-1024 rotthaus@math.msu.edu
[3] University of Nebraska, Lincoln NE 68588-0130 swiegand@math.unl.edu *

This article is dedicated to Robert Gilmer, an outstanding commutative algebraist, scholar and teacher. It relates to his work in ideal theory.

1 Introduction and Background

In this article we study the nested mixed polynomial/power series rings

$$A:=k[x,y] \hookrightarrow B:=k[[y]]\,[x] \hookrightarrow C:=k[x]\,[[y]] \hookrightarrow E:=k[x,1/x]\,[[y]], \quad (1)$$

where k is a field and x and y are indeterminates over k. In Equation 1 the maps are all flat. Also we consider

$$C \hookrightarrow D_1:=k[x]\,[[y/x]] \hookrightarrow \cdots \hookrightarrow D_n:=k[x]\,[[y/x^n]] \hookrightarrow \cdots \hookrightarrow E. \quad (2)$$

With regard to Equation 2, for n a positive integer, the map $C \hookrightarrow D_n$ is not flat, but $D_n \hookrightarrow E$ is a localization followed by an adic completion of a Noetherian ring and therefore is flat. We discuss the spectra of these rings and consider the maps induced on the spectra by the inclusion maps on the rings. For example, we determine whether there exist nonzero primes of one of the larger rings that intersect a smaller ring in zero. We were led to consider these rings by questions that came up in two contexts.

The first motivation is from the introduction to the paper [AJL] by Alonzo-Tarrio, Jeremias-Lopez and Lipman: If a map between Noetherian formal schemes can be factored as a closed immersion followed by an open one, can this map also be factored as an open immersion followed by a closed one? This is not true in general. As mentioned in [AJL], Brian Conrad observed that a counterexample can be constructed for every triple (R, x, p), where

* The authors are grateful for the hospitality and cooperation of Michigan State, Nebraska and Purdue where several work sessions on this research were conducted. Wiegand thanks the National Security Agency for support.

(1) R is an adic domain, that is, R is a Noetherian domain that is separated and complete with respect to the powers of a proper ideal I.
(2) x is a nonzero element of R such that the completion of $R[1/x]$ with respect to the powers of $IR[1/x]$, denoted $S := R_{\{x\}}$, is an integral domain.
(3) p is a nonzero prime ideal of S that intersects R in (0).

The composition $R \to S \to S/p$ determines a map on formal spectra $\text{Spf}(S/p) \to \text{Spf}(S) \to \text{Spf}(R)$ that is a closed immersion followed by an open one. To see this, recall that a surjection such as $S \to S/p$ of adic rings gives rise to a closed immersion $\text{Spf}(S/p) \to \text{Spf}(S)$ while a localization, such as that of R with respect to the powers of x, followed by the completion of $R[1/x]$ with respect to the powers of $IR[1/x]$ to obtain S gives rise to an open immersion $\text{Spf}(S) \to \text{Spf}(R)$ [EGA, (10.2.2)].

The map $\text{Spf}(S/p) \to \text{Spf}(R)$ cannot be factored, however, as an open immersion followed by a closed one. This is because a closed immersion into $\text{Spf}(R)$ corresponds to a surjective map of adic rings $R \to R/J$, where J is an ideal of R [EGA, page 441]. Thus if the immersion $\text{Spf}(S/p) \to \text{Spf}(R)$ factored as an open immersion followed by a closed one, we would have R-algebra homomorphisms from $R \to R/J \to S/p$, where $\text{Spf}(S/p) \to \text{Spf}(R/J)$ is an open immersion. Since $p \cap R = (0)$, we must have $J = (0)$. This implies $\text{Spf}(S/p) \to \text{Spf}(R)$ is an open immersion, that is, the composite map $\text{Spf}(S/p) \to \text{Spf}(S) \to \text{Spf}(R)$, is an open immersion. But also $\text{Spf}(S) \to \text{Spf}(R)$ is an open immersion. It follows that $\text{Spf}(S/p) \to \text{Spf}(S)$ is both open and closed. Since S is an integral domain this implies $\text{Spf}(S/p) \cong \text{Spf}(S)$. This is a contradiction since p is nonzero.

An example of such a triple (R, x, p) is described in [AJL]: For w, x, y, z indeterminates over a field k, set $R := k[w, x, z]\,[[y]]$, $S := k[w, x, 1/x, z]\,[[y]]$. Notice that R is complete with respect to yR and S is complete with respect to yS. An indirect proof is given in [AJL] that there exist nonzero primes p of S for which $p \cap R = (0)$. In Proposition 4.5 below we give a direct proof of this fact.

A second motivation is from a question raised by Mel Hochster: "Can one describe or somehow classify the local maps $R \hookrightarrow S$ of complete local domains R and S such that every nonzero prime ideal of S has nonzero intersection with R?" In [HRW2] we study this question and define:

Definition 1.1. For R and S integral domains with R a subring of S we say that S is a *trivial generic fiber* extension of R, or a **TGF** extension of R, if every nonzero prime ideal of S has nonzero intersection with R.

In some correspondence to Lipman regarding closed and open immersions, Conrad asked: "Is there a nonzero prime ideal of $E := k[x, 1/x]\,[[y]]$ that intersects $C = k[x]\,[[y]]$ in zero?" If there were such a prime ideal, then $C := k[x]\,[[y]] \hookrightarrow E := k[x, 1/x]\,[[y]]$ would be a simpler counterexample to the assertion that a closed immersion followed by an open one also has a factorization as an open immersion followed by a closed one. In the terminology of Definition 1.1, one can ask:

Question 1.2. Let x and y be indeterminates over a field k. Is $C:=k[x][[y]] \hookrightarrow E:=k[x,1/x][[y]]$ a TGF extension?

We show in Proposition 2.6.2 below that the answer to Question 1.2 is "yes". This is part of our analysis of the prime spectra of A, B, C, D_n and E, and the maps induced on these spectra by the inclusion maps on the rings.

The following example is a local map of the type described in Hochster's question.

Example 1.3. Let x and y be indeterminates over a field k and consider the extension $R:=k[[x,y]] \hookrightarrow S:=k[[x]][[y/x]]$. To see this extension is TGF, it suffices to show $P \cap R \neq (0)$ for each $P \in \operatorname{Spec} S$ with $\operatorname{ht} P = 1$. This is clear if $x \in P$, while if $x \notin P$, then $k[[x]] \cap P = (0)$, and so $k[[x]] \hookrightarrow R/(P \cap R) \hookrightarrow S/P$ and S/P is finite over $k[[x]]$ by [M, Theorem 8.4]. Therefore $\dim R/(P \cap R) = 1$, and so $P \cap R \neq (0)$.

Remarks 1.4. (1) The extension $k[[x,y]] \hookrightarrow k[[x,y/x]]$ is, up to isomorphism, the same as the extension $k[[x,xy]] \hookrightarrow k[[x,y]]$.

(2) We show in [HRW2] that the extension $R:=k[[x,y,xz]] \hookrightarrow S:=k[[x,y,z]]$ is not TGF.

2 Trivial generic fiber (TGF) extensions and prime spectra

The following two propositions about trivial generic fiber extensions from [HRW2] are useful and straightforward to check.

Proposition 2.1. *Let $R \hookrightarrow S$ be an injective map where R and S are integral domains. Then the following are equivalent:*

(1) *S is a TGF extension of R.*
(2) *Every nonzero element of S has a nonzero multiple in R.*
(3) *For $U = R \setminus \{0\}$, $U^{-1}S$ is a field.*

Proposition 2.2. *Let $R \hookrightarrow S$ and $S \hookrightarrow T$ be injective maps where R, S and T are integral domains.*

(1) *If $R \hookrightarrow S$ and $S \hookrightarrow T$ are TGF extensions, then so is $R \hookrightarrow T$. Equivalently if $R \hookrightarrow T$ is not TGF, then at least one of the extensions $R \hookrightarrow S$ or $S \hookrightarrow T$ is not TGF.*
(2) *If $R \hookrightarrow T$ is TGF, then $S \hookrightarrow T$ is TGF.*
(3) *If the map Spec $T \to$ Spec S is surjective, then $R \hookrightarrow T$ is TGF implies $R \hookrightarrow S$ is TGF.*

More information about TGF extensions is in [HRW1] and [HRW2].

Remarks 2.3. Let R be a commutative ring and let $R[[y]]$ denote the formal power series ring in the variable y over R. Then

(1) Each maximal ideal of $R[[y]]$ is of the form $(\mathbf{m}, y)R[[y]]$ where $\mathbf{m}$ is a maximal ideal of R. Thus y is in every maximal ideal of $R[[y]]$.
(2) If R is Noetherian with dim $R[[y]] = n$ and $x_1, \ldots, x_m$ are independent indeterminates over $R[[y]]$, then y is in every height $n+m$ maximal ideal of the polynomial ring $R[[y]]\,[x_1, \ldots, x_m]$.

Proof. Item (1) follows from [N, Theorem 15.1]. For item (2), let $\mathbf{m}$ be a maximal ideal of $R[[y]]\,[x_1, \ldots, x_m]$ with ht $(\mathbf{m}) = n+m$. By [K, Theorem 39], ht $(\mathbf{m} \cap R[[y]]) = n$; thus $\mathbf{m} \cap R[[y]]$ is maximal in $R[[y]]$, and so, by item (1), $y \in \mathbf{m}$. □

Proposition 2.4. *Let n be a positive integer, let R be an n-dimensional Noetherian domain, let y be an indeterminate over R, and let $\mathbf{q}$ be a prime ideal of height n in the power series ring $R[[y]]$. If $y \notin \mathbf{q}$, then $\mathbf{q}$ is contained in a unique maximal ideal of $R[[y]]$.*

Proof. The assertion is clear if $\mathbf{q}$ is maximal. Otherwise $S\colon = R[[y]]/\mathbf{q}$ has dimension one. Moreover, S is complete with respect to the yS-adic topology [M, Theorem 8.7] and every maximal ideal of S is a minimal prime of the principal ideal yS. Hence S is a complete semilocal ring. Since S is also an integral domain, it must be local by [M, Theorem 8.15]. Therefore $\mathbf{q}$ is contained in a unique maximal ideal of $R[[y]]$. □

In Section 3 we use the following corollary to Proposition 2.4.

Corollary 2.5. *Let R be a one-dimensional Noetherian domain and let $\mathbf{q}$ be a height-one prime ideal of the power series ring $R[[y]]$. If $\mathbf{q} \neq yR[[y]]$, then $\mathbf{q}$ is contained in a unique maximal ideal of $R[[y]]$.*

Proposition 2.6. *Consider the nested mixed polynomial/power series rings*

$$\begin{aligned} A\colon = k[x,y] &\hookrightarrow B\colon = k[[y]]\,[x] \hookrightarrow C\colon = k[x]\,[[y]] \\ &\hookrightarrow D_1\colon = k[x]\,[[y/x]] \hookrightarrow D_2\colon = k[x]\,[[y/x^2]] \hookrightarrow \cdots \\ &\hookrightarrow D_n\colon = k[x]\,[[y/x^n]] \hookrightarrow \cdots \hookrightarrow E\colon = k[x, 1/x]\,[[y]], \end{aligned}$$

where k is a field and x and y are indeterminates over k. Then

(1) *If $S \in \{B, C, D_1, D_2, \cdots, D_n, \cdots, E\}$, then $A \hookrightarrow S$ is not TGF.*
(2) *If $\{R, S\} \subset \{B, C, D_1, D_2, \cdots, D_n \cdots, E\}$ are such that $R \subsetneq S$, then $R \hookrightarrow S$ is TGF.*
(3) *Each of the proper associated maps on spectra fails to be surjective.*

Proof. For item (1), let $\sigma(y) \in yk[[y]]$ be such that $\sigma(y)$ and y are algebraically independent over k. Then $(x - \sigma(y))S \cap A = (0)$, and so $A \hookrightarrow S$ is not TGF.

For item (2), observe that every maximal ideal of C, D_n or E is of height two with residue field finite algebraic over k. To show $R \hookrightarrow S$ is TGF, it suffices to show $\mathbf{q} \cap R \neq (0)$ for each height-one prime ideal $\mathbf{q}$ of S. This is clear if $y \in \mathbf{q}$. If $y \notin \mathbf{q}$, then $k[[y]] \cap \mathbf{q} = (0)$, and so $k[[y]] \hookrightarrow R/(\mathbf{q} \cap R) \hookrightarrow S/q$ are injections. By Corollary 2.5, $S/\mathbf{q}$ is a one-dimensional local domain. Since the residue field of $S/\mathbf{q}$ is finite algebraic over k, [M, Theorem 8.4] implies that $S/\mathbf{q}$ is finite over $k[[y]]$. Therefore $S/\mathbf{q}$ is integral over $R/(\mathbf{q} \cap R)$. Hence dim $(R/(\mathbf{q} \cap R) = 1$ and so $\mathbf{q} \cap R \neq (0)$.

For item (3), observe that xD_n is a prime ideal of D_n and x is a unit of E. Thus Spec $E \to$ Spec D_n is not surjective. Now, considering $C = D_0$ and $n > 0$, we have $xD_n \cap D_{n-1} = (x, y/x^{n-1})D_{n-1}$. Therefore xD_{n-1} is not in the image of the map Spec $D_n \to$ Spec D_{n-1}. The map from Spec $C \to$ Spec B is not onto, because $(1 + xy)$ is a prime ideal of B, but $1 + xy$ is a unit in C. Similarly Spec $B \to$ Spec A is not onto, because $(1 + y)$ is a prime ideal of A, but $1 + y$ is a unit in B. This completes the proof. □

Question/Remarks 2.7. Which of the Spec maps of Proposition 2.6 are one-to-one and which are finite-to-one?

(1) For $S \in \{B, C, D_1, D_2, \cdots, D_n, \cdots, E\}$, the generic fiber ring of the map $A \hookrightarrow S$ has infinitely many prime ideals and has dimension one. Every height-two maximal ideal of S contracts in A to a maximal ideal. Every maximal ideal of S containing y has height two. Also $yS \cap A = yA$ and the map Spec $S/yS \to$ Spec A/yA is one-to-one.
(2) Suppose $R \hookrightarrow S$ is as in Proposition 2.6.2. Each height-two prime of S contracts in R to a height-two maximal ideal of R. Each height-one prime of R is the contraction of at most finitely many prime ideals of S and all of these prime ideals have height one. If $R \hookrightarrow S$ is flat, which is true if $S \in \{B, C, E\}$, then "going-down" holds for $R \hookrightarrow S$, and so, for P a height-one prime of S, we have ht $(P \cap R) \leq 1$.
(3) As mentioned in [HW, Remark 1.5], C/P is Henselian for every nonzero prime ideal P of C other than yC.

3 Spectra for two-dimensional mixed polynomial/power series rings

Let k be a field and let x and y be indeterminates over k. We consider the prime spectra, as partially ordered sets, of the mixed polynomial/power series rings A, B, C, $D_1, D_2, \cdots, D_n, \cdots$ and E as given in Equations 1 and 2 of the introduction.

Even for k a countable field there are at least two non-order-isomorphic partially ordered sets that can be the prime spectrum of the polynomial ring

$A := k[x,y]$. Let $\mathbb{Q}$ be the field of rational numbers, let F be a field contained in the algebraic closure of a finite field and let $\mathbb{Z}$ denote the ring of integers. Then, by [rW1] and [rW2], Spec $\mathbb{Q}[x,y] \not\cong$ Spec $F[x,y] \cong$ Spec $\mathbb{Z}[y]$.

The prime spectra of the rings B, C, $D_1, D_2, \cdots, D_n, \cdots$, and E of Equations 1 and 2 are simpler since they involve power series in y. Remark 2.3.2 implies that y is in every maximal ideal of height two of each of these rings.

The partially ordered set Spec B = Spec $k[[y]]\,[x]$ is similar to a prime ideal space studied in [HW] and [Shah]. The difference from [HW] is that here $k[[y]]$ is uncountable, even if k is countable. It follows that Spec B is also uncountable. As a partially ordered set, Spec B can be described uniquely up to isomorphism by the axioms of [Shah] (similar to the CHP axioms of [HW]), since $k[[y]]$ is Henselian and has cardinality at least equal to c, the cardinality of the real numbers $\mathbb{R}$.

The following theorem characterizes $U :=$ Spec B as a *Henselian affine* partially ordered set (where the "$\leq$" relation is "set containment"):

Theorem 3.1. [HW, Theorem 2.7] [Shah, Theorem 2.4] *Let $B = k[[y]]\,[x]$ be as in Equation 1, where k is a field, the cardinality of the set of maximal ideals of $k[x]$ is α and the cardinality of $k[[y]]$ is β. Then the partially ordered set $U :=$ Spec B is called Henselian affine of type (β, α) and is characterized as a partially ordered set by the following axioms:*

(1) $|U| = \beta$.
(2) *U has a unique minimal element.*
(3) *dim $(U) = 2$ and $|\{$ height-two elements of $U\}| = \alpha$.*
(4) *There exists a unique special height-one element $u \in U$ such that u is contained in every height-two element of U.*
(5) *Every nonspecial height-one element of U is in at most one height-two element.*
(6) *Every height-two element $t \in U$ contains cardinality β many height-one elements that are only contained in t. If $t_1, t_2 \in U$ are distinct height-two elements, then the special element from (4) is the unique height-one element less than both.*
(7) *There are cardinality β many height-one elements that are maximal.*

Remarks 3.2. (1) The axioms of Theorem 3.2 are redundant. We feel this redundancy helps in understanding the relationships between the prime ideals.

(2) The theorem applies to the spectrum of B by defining the unique minimal element to be the ideal (0) of B and the special height-one element to be the prime ideal yB. Every height-two maximal ideal $\mathbf{m}$ of B has nonzero intersection with $k[[y]]$. Thus $\mathbf{m}/yB$ is principal and so $\mathbf{m} = (y, f(x))$, for some monic irreducible polynomial $f(x)$ of $k[x]$. Consider $\{f(x) + ay \,|\, a \in k[[y]]\}$. This set has cardinality β and each $f(x)+ay$ is contained in a nonempty finite set of height-one primes contained in $\mathbf{m}$. If $\mathbf{p}$ is a height-one prime contained in $\mathbf{m}$ with $\mathbf{p} \neq yB$, then $\mathbf{p} \cap k[[y]] = (0)$, and so $\mathbf{p}k((y))[x]$ is generated by a monic polynomial in $k((y))[x]$. But for $a, b \in k[[y]]$ with $a \neq b$, we have

$(f(x) + ay, f(x) + by)k((y))[x] = k((y))[x]$. Therefore no height-one prime contained in $\mathbf{m}$ contains both $f(x)+ay$ and $f(x)+by$. Since B is Noetherian and $|B| = \beta$ is an infinite cardinal, we conclude that the cardinality of the set of height-one prime ideals contained in $\mathbf{m}$ is β. Examples of height-one maximals are $(1+xyf(x,y))$, for various $f(x,y) \in k[[y]][x]$. The set of height-one maximal ideals of B also has cardinality β.

(3) These axioms *characterize* Spec B in the sense that every two partially ordered sets satisfying these axioms are order-isomorphic.

The picture of Spec B is shown below:

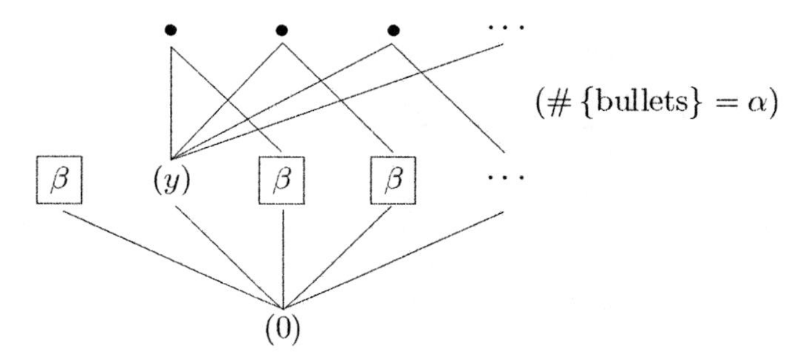

Spec $k[[y]][x]$

In the diagram, β is the cardinality of $k[[y]]$, and α is the cardinality of the set of maximal ideals of $k[x]$ (and also the cardinality of the set of maximal ideals of $k[[y]][x]$); the boxed β means there are cardinality β height-one primes in that position with respect to the partial ordering.

Next we consider Spec $R[[y]]$, for R a Noetherian one-dimensional domain. Then Spec $R[[y]]$ has the following picture by Theorem 3.4 below:

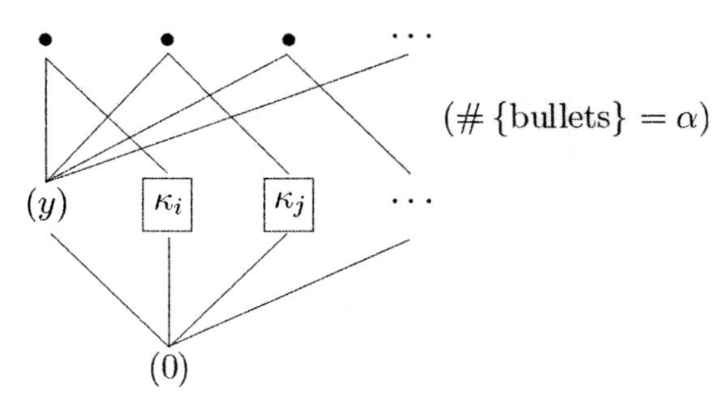

Spec $(R[[y]])$

Here α is the cardinality of the set of maximal ideals of R (and also the cardinality of the set of maximal ideals of $R[[y]]$ by Remarks 2.3.1); the boxed κ_i (one for each maximal ideal of R) means that there are cardinality κ_i prime ideals in that position, where each κ_i is uncountable. If R satisfies certain cardinality conditions described in Remark 3.3.3, for example, if $R = k[x]$, for k a countable field, then each κ_i equals the cardinality of $R[[y]]$. By Remark 3.3.2 below, each κ_i is at least $\gamma^{\aleph_0}$, where γ is the cardinality of $R/\mathbf{m}$ and $\mathbf{m}$ is the maximal ideal of R such that $(\mathbf{m}, y)$ is the maximal ideal of $R[[y]]$ above the κ_i height-one primes in the picture above.

Remarks 3.3. Let $\aleph_0$ denote the cardinality of the set of natural numbers. Suppose that T is a commutative ring of cardinality δ, that $\mathbf{m}$ is a maximal ideal of T and that γ is the cardinality of $T/\mathbf{m}$. Then

(1) The cardinality of $T[[y]]$ is $\delta^{\aleph_0}$, because the elements of $T[[y]]$ are in one-to-one correspondence with $\aleph_0$-tuples having entries in T. If T is Noetherian, then $T[[y]]$ is Noetherian, and so every prime ideal of $T[[y]]$ is finitely generated. Since the cardinality of the finite subsets of $T[[y]]$ is $\delta^{\aleph_0}$, it follows that $T[[y]]$ has at most $\delta^{\aleph_0}$ prime ideals.

(2) If T is Noetherian, there are at least $\gamma^{\aleph_0}$ distinct height-one prime ideals (other than $(y)T[[y]]$) of $T[[y]]$ contained in $(\mathbf{m}, y)T[[y]]$. To see this, choose a set $C = \{c_i \mid i \in I\}$ of elements of T so that $\{c_i + \mathbf{m} \mid i \in I\}$ gives the distinct coset representatives for $T/\mathbf{m}$. Thus there are γ elements of C, and for $c_i, c_j \in C$ with $c_i \neq c_j$, we have $c_i - c_j \notin \mathbf{m}$. Now also let $a \in \mathbf{m}, a \neq 0$. Consider the set

$$G = \{a + \sum_{n \in \mathbb{N}} d_n y^n \mid d_n \in C \, \forall n \in \mathbb{N}\}.$$

Each of the elements of G is in $(\mathbf{m}, y)T[[y]] \setminus yT[[y]]$ and hence is contained in a height-one prime contained in $(\mathbf{m}, y)T[[y]]$ distinct from $yT[[y]]$.

Moreover, $|G| = |C|^{\aleph_0} = \gamma^{\aleph_0}$. Let P be a height-one prime ideal of $T[[y]]$ contained in $(\mathbf{m}, y)T[[y]]$ but such that $y \notin P$. If two distinct elements of G, say $f = a + \sum_{n \in \mathbb{N}} d_n y^n$ and $g = a + \sum_{n \in \mathbb{N}} e_n y^n$, with the $d_n, e_n \in C$, are both in P, then so is their difference; that is

$$f - g = \sum_{n \in \mathbb{N}} d_n y^n - \sum_{n \in \mathbb{N}} e_n y^n = \sum_{n \in \mathbb{N}} (d_n - e_n) y^n \in P.$$

Now let t be the smallest power of y so that $d_t \neq e_t$. Then $(f-g)/y^t \in P$, since P is prime and $y \notin P$, but the constant term, $d_t - e_t \notin \mathbf{m}$, which contradicts the fact that $P \subseteq (\mathbf{m}, y)T[[y]]$. Thus there must be at least $|C|^{\aleph_0} = \gamma^{\aleph_0}$ distinct height-one primes contained in $(\mathbf{m}, y)T[[y]]$.

(3) Using (1) and (2), if T is Noetherian and if $\gamma^{\aleph_0} = \delta^{\aleph_0}$, then there are exactly $\gamma^{\aleph_0} = \delta^{\aleph_0}$ distinct height-one prime ideals (other than $yT[[y]]$) of $T[[y]]$ contained in $(\mathbf{m}, y)T[[y]]$. This is the case, for example, if T is countable, say $T = k[x]$ where k is a countable field, for then $|T| = \aleph_0$ and for every γ with $1 < \gamma \leq \aleph_0$, $\gamma^{\aleph_0} = \aleph_0^{\aleph_0} = c$, the cardinality of the real numbers.

Theorem 3.4. *Suppose that R is a one-dimensional Noetherian domain with cardinality δ: $= |R| =$ and that the cardinality of the set of maximal ideals of*

R is α (α can be finite). Let U = Spec R[[y]], where y is an indeterminate over R. Then
(a) *U as a partially ordered set (where the "≤" relation is "set containment") satisfies the following axioms:*

(1) $|U| \leq \delta^{\aleph_0}$.
(2) *U has a unique minimal element, namely* (0).
(3) *dim* $(U) = 2$ *and* $|\{$ *height-two elements of U* $\}| = \alpha$.
(4) *There exists a unique special height-one element* $u \in U$ *(namely* $u = (y)$*) such that u is contained in every height-two element of U.*
(5) *Every nonspecial height-one element of U is in exactly one height-two element.*
(6) *Every height-two element* $t \in U$ *contains uncountably many height-one elements that are contained only in t. (The number of height-one elements contained only in t is at least* $\gamma^{\aleph_0}$*, where* γ *is the cardinality of the residue field of the corresponding maximal ideal of R.) If* $t_1, t_2 \in U$ *are distinct height-two elements, then the special element from (4) is the unique height-one element less than both.*
(7) *There are no height-one maximal elements in U. Every maximal element has height two.*

(b) *If R is countable, or more generally if* $\delta = |R|$ *satisfies the condition of Remarks 3.3.3, that is,* $\gamma^{\aleph_0} = \delta^{\aleph_0}$*, for every* γ *that occurs as* $|R/\mathbf{m}|$*, where* $\mathbf{m}$ *is a maximal ideal of R, then U = Spec R[[y]] satisfies (1)-(7), with the stronger axioms (1′) and (6′):*

(1′) $|U| = \delta^{\aleph_0}$. *(For R countable, this is c, the cardinality of the real numbers.)*
(6′) *Every height-two element* $t \in U$ *contains* $\delta^{\aleph_0}$ *(uncountably many) height-one elements that are contained only in t.*

(c) *With the additional hypotheses of (b), U is characterized as a partially ordered set by the axioms given in (a) and (b). Every partially ordered set satisfying the axioms (1)-(7) in (a) and (b) is order-isomorphic to every other such partially ordered set.*

Proof. In part(a), item (1) is from Remark 3.3.1. Item (2) and the first part of (3) are clear. The second part of (3) follows immediately from Remark 2.3.1.

For items (4) and (5), suppose that P is a height-one prime of $R[[y]]$. If $P = yR[[y]]$, then P is contained in each maximal ideal of $R[[y]]$ by Remark 2.3.1, and so $yR[[y]]$ is the special element. If $y \notin P$, then, by Corollary 2.5 and Remark 2.3.1, P is contained in a unique maximal ideal of $R[[y]]$ of height two.

For item (6) and items (1′) and (6′) of part (b) use Remarks 3.3.2 and 3.3.3.

For item (c), all partially ordered sets satisfying the axioms of Theorem 3.2 are order-isomorphic, and the partially ordered set U of the present theorem satisfies the same axioms as in Theorem 3.2 except axiom (7) that involves height-one maximals. Since U has no height-one maximals, an order-isomorphism between two partially ordered sets as in item (c) can be deduced by adding on height-one maximals and then deleting them. □

Corollary 3.5. *In the terminology of Equations 1 and 2 of the introduction, we have Spec* $C \cong$ *Spec* $D_n \cong$ *Spec* E*, but Spec* $B \not\cong$ *Spec* C.

Proof. The rings C, D_n, and E are all formal power series rings in one variable over a one-dimensional Noetherian domain R, where R is either $k[x]$ or $k[x, 1/x]$. Thus the domain R satisfies the hypotheses of Theorem 3.4 with the cardinality conditions of parts (b) and (c). If k is finite, then $|R| = |k[x]| = \aleph_0$ and α, the number of maximal ideals of R, is also $\aleph_0$; in this case $|R/\mathbf{m}| = \gamma$ is finite for each maximal ideal $\mathbf{m}$ of R and $\delta = |R| = \gamma \cdot \aleph_0 = \alpha$, and so $\gamma^{\aleph_0} = \delta^{\aleph_0}$. On the other hand, if k is infinite, then $|k| = |k[x]| = |R| = \alpha$, and $|k| = |R/\mathbf{m}| = \gamma$ is the same for every maximal ideal $\mathbf{m}$ of R. Hence also in this case $\delta = |R| = \gamma \cdot \aleph_0 = \alpha$, and so $\gamma^{\aleph_0} = \delta^{\aleph_0}$.

Also the number of maximal ideals is the same for C, D_n, and E, because in each case, it is the same as the number of maximal ideals of R which is $|k[x]| = |k| \cdot \aleph_0$.

Thus in the picture of $R[[y]]$ shown above, for $R[[y]] = C, D_n$ or E, the κ_i are all equal to $|k|^{\aleph_0}$ and $\alpha = |k| \cdot \aleph_0$, and so the spectra are isomorphic. The spectrum of B is not isomorphic to that of C, however, because B contains height-one maximal ideals, such as that generated by $1 + xy$, whereas C has no height-one maximal ideals. □

Remarks 3.6. As mentioned at the beginning of this section, it is shown in [rW1] and [rW2] that Spec $\mathbb{Q}[x,y] \not\cong$ Spec $F[x,y] \cong$ Spec $\mathbb{Z}[y]$, where F is a field contained in the algebraic closure of a finite field. Corollary 3.6 shows that the spectra of power series extensions in y behave differently in that Spec $\mathbb{Z}[[y]] \cong$ Spec $\mathbb{Q}[x]\,[[y]] \cong$ Spec $F[x]\,[[y]]$.

Corollary 3.7. *If* $\mathbb{Z}$ *is the ring of integers,* $\mathbb{Q}$ *is the rational numbers,* F *is a field contained in the algebraic closure of a finite field, and* $\mathbb{R}$ *is the real numbers, then*

$$\textit{Spec}\ \mathbb{Z}[[y]] \cong \textit{Spec}\ \mathbb{Q}[x]\,[[y]] \cong \textit{Spec}\ F[x]\,[[y]] \not\cong \textit{Spec}\ \mathbb{R}[x]\,[[y]].$$

Proof. The rings $\mathbb{Z}, \mathbb{Q}[x]$ and $F[x]$ are all countable with countably infinitely many maximal ideals. Thus if $R = \mathbb{Z}, \mathbb{Q}[x]$ or $F[x]$, then R satisfies the hypotheses of Theorem 3.4 with the cardinality conditions of parts (b) and (c). On the other hand, $\mathbb{R}[x]$ has uncountably many maximal ideals; thus $\mathbb{R}[x]\,[[y]]$ also has uncountably many maximal ideals. □

4 Higher dimensional mixed power series/polynomial rings

In analogy to Equation (1), we display several embeddings involving three variables.

$$(4.0) \qquad k[x,y,z] \overset{\alpha}{\hookrightarrow} k[[z]]\,[x,y] \overset{\beta}{\hookrightarrow} k[x]\,[[z]]\,[y] \overset{\gamma}{\hookrightarrow} k[x,y]\,[[z]] \overset{\delta}{\hookrightarrow} k[x]\,[[y,z]],$$
$$k[[z]]\,[x,y] \overset{\epsilon}{\hookrightarrow} k[[y,z]]\,[x] \overset{\zeta}{\hookrightarrow} k[x]\,[[y,z]] \overset{\eta}{\hookrightarrow} k[[x,y,z]],$$

where k is a field and x, y and z are indeterminates over k.

Remarks 4.1. (1) By Proposition 2.6.2 every nonzero prime ideal of $C = k[x]\,[[y]]$ has nonzero intersection with $B = k[[y]]\,[x]$. In three or more variables, however, the analogous statements fail. We show below that the maps $\alpha, \beta, \gamma, \delta, \epsilon, \zeta, \eta$ in Equation (4.0) fail to be TGF. Thus, by Proposition 2.2.2, no proper inclusion in (4.0) is TGF. The dimensions of the generic fiber rings of the maps in the diagram are either one or two.

(2) For those rings in (4.0) of form $R = S[[z]]$ (ending in a power series variable) where S is a ring, such as $R = k[x,y][[z]]$, we have some information concerning the prime spectra. By Proposition 2.4 every height-two prime ideal not containing z is contained in a unique maximal ideal. By [N, Theorem 15.1] the maximal ideals of $S[[z]]$ are of the form $(\mathbf{m}, z)S[[z]]$, where $\mathbf{m}$ is a maximal ideal of S, and thus the maximal ideals of $S[[z]]$ are in one-to-one correspondence with the maximal ideals of S. As in section 3, using Remarks 2.3, we see that maximal ideals of Spec $k[[z]]\,[x,y]$ can have height two or three, that (z) is contained in every height-three prime ideal, and that every height-two prime ideal not containing (z) is contained in a unique maximal ideal.

(3) It follows by arguments analogous to that in Proposition 2.6.1, that α, δ, ϵ are not TGF. For α, let $\sigma(z) \in zk[[z]]$ be transcendental over $k(z)$; then $(x-\sigma)k[[z]]\,[x,y] \cap k[x,y,z] = (0)$. For δ and ϵ: let $\sigma(y) \in k[[y]]$ be transcendental over $k(y)$; then $(x-\sigma)k[x]\,[[z,y]] \cap k[x]\,[[z]]\,[y] = (0)$, and $(x-\sigma)k[[y,z]]\,[x] \cap k[[z]]\,[x,y] = (0)$.

(4) By [HRW1, Theorem 1.1], η is not TGF and the dimension of the generic fiber ring of η is one.

In order to show in Proposition 4.3 below that the map β is not TGF, we first observe:

Proposition 4.2. *The element $\sigma = \sum_{n=1}^{\infty}(xz)^{n!} \in k[x]\,[[z]]$ is transcendental over $k[[z]]\,[x]$.*

Proof. Consider an expression

$$Z := a_\ell \sigma^\ell + a_{\ell-1}\sigma^{\ell-1} + \cdots + a_1\sigma + a_0,$$

where the $a_i \in k[[z]]\,[x]$ and $a_\ell \neq 0$. Let m be an integer greater than $\ell+1$ and greater than $\deg_x a_i$ for each i such that $0 \leq i \leq \ell$ and $a_i \neq 0$. Regard each $a_i\sigma^i$ as a power series in x with coefficients in $k[[z]]$.

For each i with $0 \leq i \leq \ell$, we have $i(m!) < (m+1)!$. It follows that the coefficient of $x^{i(m!)}$ in σ^i is nonzero, and the coefficient of x^j in σ^i is zero for every j with $i(m!) < j < (m+1)!$. Thus if $a_i \neq 0$ and $j = i(m!) + \deg_x a_i$, then the coefficient of x^j in $a_i\sigma^i$ is nonzero, while for j such that $i(m!) + \deg_x a_i < j < (m+1)!$, the coefficient of x^j in $a_i\sigma^i$ is zero. By our choice of m, for each i such that $0 \leq i < \ell$ and $a_i \neq 0$, we have

$$(m+1)! > \ell(m!) + \deg_x a_\ell \geq i(m!) + m! > i(m!) + \deg_x a_i.$$

Thus in Z, regarded as a power series in x with coefficients in $k[[z]]$, the coefficient of x^j is nonzero for $j = \ell(m!) + \deg_x a_\ell$. Therefore $Z \neq 0$. We conclude that σ is transcendental over $k[[z]]\,[x]$. □

Proposition 4.3. *$k[[z]]\,[x,y] \overset{\beta}{\hookrightarrow} k[x]\,[[z]]\,[y]$ is not TGF.*

Proof. Fix an element $\sigma \in k[x]\,[[z]]$ that is transcendental over $k[[z]]\,[x]$. We define $\pi: k[x]\,[[z]]\,[y] \to k[x]\,[[z]]$ to be the identity map on $k[x]\,[[z]]$ and $\pi(y) = \sigma z$. Let $\mathbf{q} = \ker \pi$. Then $y - \sigma z \in \mathbf{q}$. If $h \in \mathbf{q} \cap (k[[z]]\,[x,y])$, then

$$h = \sum_{j=0}^{s}\sum_{i=0}^{t}\Big(\sum_{\ell \in \mathbb{N}} a_{ij\ell} z^\ell\Big)x^i y^j, \quad \text{for some } \ s,t \in \mathbb{N} \text{ and } a_{ij\ell} \in k, \text{ and so}$$

$$0 = \pi(h) = \sum_{j=0}^{s}\sum_{i=0}^{t}\Big(\sum_{\ell \in \mathbb{N}} a_{ij\ell} z^\ell\Big)x^i (\sigma z)^j = \sum_{j=0}^{s}\sum_{i=0}^{t}\Big(\sum_{\ell \in \mathbb{N}} a_{ij\ell} z^{\ell+j}\Big)x^i \sigma^j.$$

Since σ is transcendental over $k[[z]]\,[x]$, we have that x and σ are algebraically independent over $k((z))$. Thus each of the $a_{ij\ell} = 0$. Therefore $\mathbf{q} \cap (k[[z]][x,y]) = (0)$, and so the embedding β is not TGF. □

Proposition 4.4. *$k[[y,z]]\,[x] \overset{\zeta}{\hookrightarrow} k[x]\,[[y,z]]$ and $k[x]\,[[z]]\,[y] \overset{\gamma}{\hookrightarrow} k[x,y]\,[[z]]$ are not TGF.*

Proof. For ζ, let $t = xy$ and let $\sigma \in k[[t]]$ be algebraically independent over $k(t)$. Define $\pi: k[x]\,[[y,z]] \to k[x]\,[[y]]$ as follows. For

$$f{:} = \sum_{\ell=0}^{\infty} \sum_{m+n=\ell} f_{mn}(x) y^m z^n \in k[x]\,[[y,z]],$$

where $f_{mn}(x) \in k[x]$, define

$$\pi(f){:} = \sum_{\ell=0}^{\infty} \sum_{m+n=\ell} f_{mn}(x) y^m (\sigma y)^n \in k[x]\,[[y]].$$

In particular, $\pi(z) = \sigma y$. Let $\mathbf{p}{:} = \ker \pi$. Then $z - \sigma y \in \mathbf{p}$, and so $\mathbf{p} \neq (0)$. Let $h \in \mathbf{p} \cap k[[y,z]]\,[x]$. We show $h = 0$. Now h is a polynomial with coefficients in $k[[y,z]]$, and we define $g \in k[[y,z]]\,[t]$, by, if $a_i(y,z) \in k[[y,z]]$ and

$$h: = \sum_{i=0}^{r} a_i(y,z)x^i, \text{ then set } g: = y^r h = \sum_{i=0}^{r}(\sum_{\ell=0}^{\infty} \sum_{m+n=\ell} b_{imn} y^m z^n)t^i.$$

The coefficients of g are in $k[[y,z]]$, since $y^r x^i = y^{r-i}t^i$. Thus

$$0 = \pi(g) = \sum_{i=0}^{r}(\sum_{\ell=0}^{\infty} \sum_{m+n=\ell} b_{imn} y^m(\sigma y)^n)t^i = \sum_{i=0}^{r}(\sum_{\ell=0}^{\infty} \sum_{m+n=\ell} b_{imn}\sigma^n y^\ell)t^i$$

$$= \sum_{\ell=0}^{\infty}(\sum_{m+n=\ell} (\sum_{i=0}^{r} b_{imn}t^i)\sigma^n)y^\ell.$$

Now t and y are analytically independent over k, and so the coefficient of each y^ℓ (in $k[[t]]$) is 0; since σ and t are algebraically independent over k, the coefficient of each σ^n is 0. It follows that each $b_{imn} = 0$, that $g = 0$ and hence that $h = 0$. Thus the extension ζ is not TGF.

To see that γ is not TGF, we switch variables in the proof for ζ, so that $t = yz$. Again choose $\sigma \in k[[t]]$ to be algebraically independent over $k(t)$. Define $\psi{:}k[x,y]\,[[z]] \to k[y]\,[[z]]$ by $\psi(x) = \sigma z$ and ψ is the identity on $k[y]\,[[z]]$. Then ψ can be extended to $\pi{:}k[y]\,[[x,z]] \to k[y]\,[[z]]$, which is similar to the π in the proof above. As above, set $\mathbf{p}: = \ker \pi$; then $\mathbf{p} \cap k[[x,z]]\,[y] = (0)$. Thus $\mathbf{p} \cap k[x]\,[[z]]\,[y] = (0)$ and γ is not TGF. □

Proposition 4.5. *Let k be a field and let x and t be indeterminates over k. Then $\sigma = \sum_{n=1}^{\infty} t^{n!}$ is algebraically independent over $k[[x,xt]]$.*

Proof. Let ℓ be a positive integer and consider an expression

$$\gamma: = \gamma_\ell\sigma^\ell + \cdots + \gamma_i\sigma^i + \cdots + \gamma_1\sigma, \text{ where } \gamma_i: = \sum_{j=0}^{\infty} f_{ij}(x)(xt)^j \in k[[x,xt]],$$

that is, each $f_{ij}(x) \in k[[x]]$ and $1 \le i \le \ell$. Assume that $\gamma_\ell \neq 0$. Let a_ℓ be the smallest j such that $f_{\ell j}(x) \neq 0$, and let m_ℓ be the order of $f_{\ell a_\ell}(x)$, that is, $f_{\ell a_\ell}(x) = x^{m_\ell} g_\ell(x)$, where $g_\ell(0) \neq 0$. Let n be a positive integer such that

$$n \ge 2 + \max\{\ell, m_\ell, a_\ell\}.$$

Since $\ell < n$, for each i with $1 \le i \le \ell$, we have

$$\sigma^i = \sigma_{i1}(t) + c_i t^{i(n!)} + t^{(n+1)!}\tau_i(t), \tag{3}$$

where c_i is a nonzero element of k, $\sigma_{i1}(t)$ is a polynomial in $k[t]$ of degree at most $(i-1)n! + (n-1)!$ and $\tau_i(t) \in k[[t]]$.

Claim 4.6. The coefficient of $t^{\ell(n!)+a_\ell}$ in $\sigma^\ell\gamma_\ell = \sigma^\ell(\sum_{j=a_\ell}^{\infty} f_{\ell j}(x)(xt)^j)$ as a power series in $k[[x]]$ has order $m_\ell + a_\ell$, and hence, in particular, is nonzero.

Proof. By the choice of n, $(n+1)! > \ell(n!)+a_\ell$. Hence by the expression for σ^ℓ given in Equation 3, we see that all of the terms in $\sigma^\ell\gamma_\ell$ of the form $bt^{\ell(n!)+a_\ell}$, for some $b \in k[[x]]$, appear in the product

$$(\sigma_{\ell 1}(t) + c_\ell t^{\ell(n!)})(\sum_{j=a_\ell}^{\ell(n!)+a_\ell} f_{\ell j}(x)(xt)^j).$$

One of the terms of the form $bt^{\ell(n!)+a_\ell}$ in this product is

$$c_\ell t^{\ell(n!)} f_{\ell a_\ell}(x)(xt)^{a_\ell} = (c_\ell x^{m_\ell+a_\ell} g_\ell(x))t^{\ell(n!)+a_\ell} = (c_\ell x^{m_\ell+a_\ell} g_\ell(0)+\ldots)t^{\ell(n!)+a_\ell}.$$

Since $c_\ell g_\ell(0)$ is a nonzero element of k, $c_\ell x^{m_\ell+a_\ell} g_\ell(x) \in k[[x]]$ has order $m_\ell + a_\ell$. The other terms in the product $\sigma^\ell\gamma_\ell$ that have the form $bt^{\ell(n!)+a_\ell}$, for some $b \in k[[x]]$, are in the product

$$(\sigma_{\ell 1}(t))(\sum_{j=a_\ell}^{\ell(n!)+a_\ell} f_{\ell j}(x)(xt)^j) = \sum_{j=a_\ell}^{\ell(n!)+a_\ell} f_{\ell j}(x)(xt)^j \sigma_{\ell 1}(t).$$

Since $\deg_t \sigma_{\ell 1} \leq (\ell-1)n! + (n-1)!$ and since, for each j with $f_{\ell j}(x) \neq 0$, we have $\deg_t f_{\ell j}(x)(xt)^j = j$, we see that each term in $f_{\ell j}(xt)^j \sigma_{\ell 1}(t)$ has degree in t less than or equal to $j + (\ell-1)n! + (n-1)!$. Thus each nonzero term in this product of the form $bt^{\ell(n!)+a_\ell}$ has

$$j \geq \ell(n!) + a_\ell - (\ell-1)(n!) - (n-1)! = a_\ell + (n-1)!(n-1) > m_\ell + a_\ell,$$

by choice of n. Moreover, for j such that $f_{\ell j}(x) \neq 0$, the order in x of $f_{\ell j}(x)(xt)^j$ is bigger than or equal to j. This completes the proof of Claim 4.6.

Claim 4.7. For $i < \ell$, the coefficient of $t^{\ell(n!)+a_\ell}$ in $\sigma^i\gamma_i$ as a power series in $k[[x]]$ is either zero or has order greater than $m_\ell + a_\ell$.

Proof. As in the proof of Claim 4.6, all of the terms in $\sigma^i\gamma_i$ of the form $bt^{\ell(n!)+a_\ell}$, for some $b \in k[[x]]$, appear in the product

$$(\sigma_{i1} + c_i t^{i(n!)})(\sum_{j=0}^{\ell(n!)+a_\ell} f_{ij}(x)(xt)^j) = \sum_{j=0}^{\ell(n!)+a_\ell} f_{ij}(x)(xt)^j(\sigma_{i1} + c_i t^{i(n!)}).$$

Since $\deg_t(\sigma_{i1} + c_i t^{i(n!)}) = i(n!)$, each term in $f_{ij}(x)(xt)^j(\sigma_{i1} + c_i t^{i(n!)})$ has degree in t at most $j + i(n!)$. Thus each term in this product of the form $bt^{\ell(n!)+a_\ell}$, for some nonzero $b \in k[[x]]$, has

$$j \geq \ell(n!) + a_\ell - i(n!) \geq n! + a_\ell > m_\ell + a_\ell.$$

Thus $\operatorname{ord}_x b \geq j > m_\ell + a_\ell$. This completes the proof of Claim 4.7. Hence $\gamma \notin k[[x, xz]]$ and so Proposition 4.5 is proved.

Question/Remarks 4.8. (1) As we show in Proposition 2.6, the embeddings from Equation 1 involving two-dimensional mixed power series/polynomial rings over a field k with inverted elements are TGF. So far we have not determined whether the same is true in the three-dimensional case. For example, is θ below TGF?

$$k[x,y]\,[[z]] \overset{\theta}{\hookrightarrow} k[x,y,1/x]\,[[z]]$$

(2) For the four dimensional case, as observed in the introduction, it follows from [HR, p. 364, Theorem 1.12] that the extension $k[x,y,u]\,[[z]] \hookrightarrow k[x,y,u,1/x,]\,[[z]]$ is not TGF. We provide in Proposition 4.9 a direct proof of this fact.

Proposition 4.9. *For k a field and x,y,u and z indeterminates over k, the extension $k[x,y,u]\,[[z]] \hookrightarrow k[x,y,u,1/x,]\,[[z]]$ is not TGF.*

Proof. Let $t = z/x$ and let $\sigma \in k[[t]]$ be algebraically independent over $k[[x,z]]$. (By Proposition 4.5, we may take $\sigma = \sum_{r=1}^{\infty} t^{r!}$.)

Consider

$$\pi{:}k[[x,y,u]]\,[1/x]\,[[z]] \to k[[x,u]]\,[1/x]\,[[z]]$$

defined by mapping

$$\sum_{i=0}^{\infty} a_i(x,y,u,1/x)z^i \mapsto \sum_{i=0}^{\infty} a_i(x,\sigma u,u,1/x)z^i,$$

where $a_i(x,y,u,1/x) \in k[[x,y,u]][1/x]$. Let $\mathbf{p} = \ker\pi$. Then $y - \sigma u \in \mathbf{p}$. We show that $\mathbf{p} \cap k[[x,y,u,z]] = (0)$, and so also $\mathbf{p} \cap k[x,y,u]\,[[z]] = (0)$. Let

$$f{:}= \sum_{\ell=0}^{\infty}\Big(\sum_{i+j=\ell} d_{ij}u^iy^j\Big) \in k[[x,y,u,z]],$$

where $d_{ij} \in k[[x,z]]$. If $f \in \mathbf{p}$, then

$$0 = \pi(f) = \sum_{\ell=0}^{\infty}\Big(\sum_{i+j=\ell} d_{ij}u^i\sigma^ju^j\Big) = \sum_{\ell=0}^{\infty}\Big(\sum_{i+j=\ell} d_{ij}\sigma^j\Big)u^\ell.$$

This is a power series in u, and so, for each ℓ, $\sum_{i+j=\ell} d_{ij}\sigma^j = 0$. Since σ is algebraically independent over $k[[x,z]]$, each $d_{ij} = 0$. Thus $f = 0$. This completes the proof of Proposition 4.9.

References

[AJL] L. Alonso-Tarrio, A. Jeremias-Lopez, and J. Lipman: Correction to the paper "Duality and flat base change on formal schemes". Proc. Amer. Math. Soc., **121, 2**, 351–357 (2002)

[EGA] A. Grothendieck and J. Dieudonné: Eléments de Géométrie Algébrique I. Springer-Verlag, Berlin (1971)

[HR] W. Heinzer and C. Rotthaus: Formal fibers and complete homomorphic images. Proc. Amer. Math. Soc., **120** 359–369 (1994)

[HRW1] W. Heinzer, C. Rotthaus and S. Wiegand: Generic formal fibers of mixed power series/polynomial rings, J. Algebra, to appear.

[HRW2] W. Heinzer, C. Rotthaus and S. Wiegand: Extensions of local domains with trivial generic fiber, preprint

[HW] W. Heinzer and S. Wiegand: Prime ideals in two-dimensional polynomial rings. Proc. Amer. Math. Soc. **107**, 577–586 (1989)

[K] I. Kaplansky: Commutative rings. Allyn and Bacon. Boston (1970)

[M] H. Matsumura: Commutative ring theory. Cambridge Univ. Press. Cambridge (1989)

[N] M. Nagata: Local rings. Interscience. New York (1962)

[Shah] C. Shah: Affine and projective lines over one-dimensional semilocal domains. Proc. Amer. Math. Soc. **124**, 697–705 (1996)

[Shel] P. Sheldon: How changing $D[[x]]$ changes its quotient field. Trans. Amer. Math. Soc. **159**, 223–244 (1971)

[rW1] R. Wiegand: Homeomorphisms of affine surfaces over a finite field. J. London Math. Soc. **18**, 28–32 (1978)

[rW2] R. Wiegand: The prime spectrum of a two-dimensional affine domain. J. Pure & Appl. Algebra **40**, 209-214 (1986)

Uppers to zero in polynomial rings

Evan Houston

Department of Mathematics, University of North Carolina at Charlotte, Charlotte, NC 28223 eghousto@email.uncc.edu

Introduction and preliminaries

Let R be a commutative ring, let X an indeterminate, let Q be a prime ideal of $R[X]$, and let $P = Q \cap R$. If $Q = P[X]$, then Q is called an *extended* prime; otherwise, Q is called an *upper to P*. Observe that in this latter case, $Q/P[X]$ is an *upper to zero* in (the domain) $R[X]/P[X] \cong (R/P)[X]$. The focus of this article in on uppers to zero. Here are several equivalent definitions.

Proposition. *Let R be an integral domain with quotient field K, and let Q be a prime ideal of $R[X]$. The following statements are equivalent.*

(1) $Q = fK[X] \cap R[X]$ *for some irreducible polynomial* $f \in K[X]$.
(2) $Q \neq (0)$ *and* $Q \cap R = (0)$.
(3) *There is an algebraic extension L of K and an element $u \in L$ for which Q is the kernel of the natural homomorphism $R[X] \to R[u]$ sending X to u.*
(4) $Q = (X - u)L[X] \cap R[X]$ *for some algebraic extension L of K.*

When an upper to zero Q arises as in (4), we sometimes refer to it as Q_u. It is useful to observe that the element f in (1) can always be chosen in $R[X]$ if desired. Also, when the generators of $QK[X]$ have degree one, we call Q a *rational* upper to zero. All of this "upper" terminology is due to Stephen McAdam.

Much of what we discuss below can be applied to any contracted ideal $fK[X] \cap R[X]$, but we shall confine ourselves to the prime case for simplicity. For any polynomial $f \in K[X]$, we shall use the notation $c(f)$ for the *content* of f, that is, the ideal of R generated by the coefficients of f; similarly, for any ideal I of $R[X]$, we shall use $c(I)$ for the ideal generated by the coefficients of all the elements of I.

The paper is divided into three sections. The first discusses connections with the work of Robert Gilmer. In Section 2, we give conditions under which an upper to zero has certain desirable ideal-theoretic properties, such as being principal, finitely generated, divisorial, etc. Section 3 is devoted to a sampling of how uppers to zero have been used, primarily in characterizations of such classes of domains as Prüfer v-multiplication domains. Scattered throughout are open questions.

Before proceeding, I want to emphasize that much is left out. In particular, with the exception of a couple of instances at the end, very little is said about the large body of literature concerning uppers to zero in Noetherian polynomial rings.

1 Connections to Gilmer's work

Gilmer and Hoffmann [24] call a unitary extension $R \subseteq S$ of rings a *P-extension* if each element of S satisfies a polynomial g in $R[X]$ one of whose coefficients is a unit of R (or, equivalently, a g with $c(g) = R$ [24, Corollary 1]). They then characterize Prüfer domains as those integrally closed domains R for which the extension $R \subseteq K$, where K is the quotient field of R, is a P-extension [24, Theorem 2]. The connection to uppers to zero is almost immediate. If a domain S is a P-extension of R, then it is clear that for each $u \in S$, the upper to zero $Q_u = (X - u)S[X] \cap R[X]$ satisfies $Q_u \not\subseteq M[X]$ for each maximal ideal M of R. Conversely, if Q_u is not contained in the extension to $R[X]$ of any maximal ideal of R, then $c(Q_u) = R$, and by [24, Corollary 1], there is an element $g \in Q_u$ with a unit coefficient. Hence:

Proposition 1.1. *Let $R \subseteq S$ be an extension of domains. Then $R \subseteq S$ is a P-extension if and only if for each $u \in S$ and each maximal ideal M of R, we have $Q_u \not\subseteq M[X]$. In particular:*

(1) *$R \subseteq K$ is a P-extension if and only if each rational upper to zero avoids extended primes (that is, is not contained in $M[X]$ for any maximal ideal M of R), and*
(2) *$R \subseteq L$ is a P-extension for each algebraic extension L of K if and only if all uppers to zero avoid extended primes.*

This characterization makes it easy to show that the property of being a P-extension is a local one, in the following sense.

Lemma 1.2. *Let $R \subseteq S$ be an extension of domains. Then $R \subseteq S$ is a P-extension if and only if $R_M \subseteq S_{R \setminus M}$ is a P-extension for each maximal ideal M of R.*

Proof. The "only if" implication is a straightforward application of the definition of P-extension. For the "if" half, suppose, contrapositively, that $R \subseteq S$ is not a P-extension. Then by Proposition 1.1 there is an element $u \in S$ and a maximal ideal M of R with $Q_u \subseteq M[X]$. Now $Q_u = fK[X] \cap R[X]$ for some irreducible $f \in K[X]$, and it is easy to see that $fK[X] \cap R_M[X] \subseteq MR_M[X]$. Hence $R_M \subseteq S_{R \setminus M}$ is not a P-extension.

The following result is fundamental and is a nice illustration of the use of the u, u^{-1}-lemma.

Theorem 1.3. *Let (R, M) be a quasi-local integrally closed domain. Then the following statements are equivalent.*

(1) *R is a valuation domain.*
(2) *$R \subseteq K$ is a P-extension.*
(3) *$R \subseteq L$ is a P-extension for each algebraic extension field L of K.*

Proof. (1) $\Rightarrow$ (3): Let Q be an upper to zero in $R[X]$, and let f be a nonzero element of Q. Since one of the coefficients of f divides all the others, we may write $f = ag$ with $a \in R$ and $c(g) = R$. Since $a \notin Q$, we have $g \in Q$ and hence $Q \nsubseteq M[X]$. Now quote Proposition 1.1.

(3) $\Rightarrow$ (2): Trivial.

(2) $\Rightarrow$ (1): Suppose that R is not a valuation domain, and pick $u \in K$ with $u, u^{-1} \notin R$. Set $Q = (X - u)K[X] \cap R[X]$. By the u, u^{-1}-lemma [38, Theorem 67], we must have $Q \subseteq M[X]$. Thus $R \subseteq K$ is not a P-extension.

The proof of (1) $\Rightarrow$ (3) is essentially contained in Seidenberg's proof that if R is a one-dimensional valuation domain, then $R[X]$ has dimension two [51, Theorem 4]. Moreover, I do not think it is stretching the point too far to attribute the proof of (2) $\Rightarrow$ (1) to Seidenberg–see [51, Theorem 6 and its corollary].

Theorem 1.3 (1) $\Leftrightarrow$ (2) is the local version of the Gilmer-Hoffmann characterization of Prüfer domains mentioned above [24, Theorem 2]. By combining Theorem 1.3 and Lemma 1.2, we get (a slight generalization of) the full version:

Corollary 1.4. (cf. [24, Theorem 2]) *Let R be an integrally closed domain with quotient field K. The following statements are equivalent.*

(1) *R is a Prüfer domain.*
(2) *$R \subseteq K$ is a P-extension.*
(3) *$R \subseteq L$ is a P-extension for some algebraic extension field L of K.*
(4) *$R \subseteq L$ is a P-extension for each algebraic extension field L of K.*

The "upper to zero" translation of the equivalence of (2) and (4) above is that in an integrally closed domain, rational uppers to zero avoid extended primes if and only if all uppers to zero avoid extended primes. In fact, the integrally closed hypothesis is unnecessary to obtain this equivalence. To see this, denote the integral closure of R by $\overline{R}$. Then it is easy to see by standard arguments involving the integral extension $R[X] \subseteq \overline{R}[X]$ that rational uppers to zero in $R[X]$ avoid extended primes if and only if rational uppers to zero in $\overline{R}[X]$ avoid extended primes. That is, $R \subseteq K$ is a P-extension if and only if $\overline{R} \subseteq K$ is a P-extension. (This also follows from [24, Theorem 4].) But this latter statement is equivalent to $\overline{R}$ being a Prüfer domain and is therefore also equivalent to having all uppers to zero in $\overline{R}[X]$ avoid extended primes, and this, in turn, is equivalent to having $\overline{R} \subseteq L$ be a P-extension for each algebraic extension L of K. This yields:

Corollary 1.5. [24, Theorem 6] *Let R be a domain with quotient field K. Then the integral closure $\overline{R}$ of R is a Prüfer domain if and only if $R \subseteq L$ is a P-extension for each algebraic extension L of K (if and only if $R \subseteq K$ is a P-extension if and only if $R \subseteq L$ is a P-extension for some algebraic extension L of K).*

At this point, we insert a variation due to Joe Mott, Budh Nashier, and Muhammad Zafrullah.

Theorem 1.6. [43, Theorem 1.7] *A domain R has Prüfer integral closure if and only if each rational upper to zero contains a polynomial with invertible content.*

The following pretty result, due to David Dobbs, yields another way to view P-extensions.

Theorem 1.7. [15, Theorem] *For any rings $R \subseteq T$ and any element $u \in T$, u is primitive over R (that is, u satisfies a polynomial in $R[X]$ one of whose coefficients is a unit) if and only if the extension $R \subseteq R[u]$ satisfies incomparability (INC).*

As a corollary of Theorem 1.7 and Corollary 1.5, we have the following.

Corollary 1.8. *Let R be a domain with quotient field K. Then $\overline{R}$ is Prüfer if and only if $R \subseteq R[u]$ satisfies INC for each $u \in K$ (if and only if $R \subseteq R[u]$ satisfies INC for each element u in each algebraic extension of K).*

It is then not difficult to obtain the following, which is due to Ira Papick.

Theorem 1.9. [48, Proposition 2.26] *A domain R with quotient field K has Prüfer integral closure if and only if $R \subseteq T$ satisfies INC for each overring T of R (if and only if $R \subseteq T$ satisfies INC for each algebraic extension T of R).*

Gilmer deserves the credit for synthesizing most of the ideas presented so far; indeed, much of it follows from Theorem 19.15 of his book [23]. Here is the one-indeterminate version of (part of) this theorem.

Theorem 1.10. ([23, Theorem 19.15] and [22, Theorem 16.10]) *Let P be a prime ideal of the integrally closed domain R. The following conditions are equivalent:*

(1) *R_P is a valuation domain.*
(2) *If Q is an upper to zero, then $Q \nsubseteq P[X]$.*
(3) *For each overring T of R, there is at most one prime ideal of T lying over P.*

(4) *For each overring T of R, there is at most one member in a chain of prime ideals lying over P.*

I do not think this result has received the attention it deserves. This may be partly due to the fact that it is stated "locally". Here is a globalized version.

Theorem 1.11. *Let R be an integrally closed domain. Then the following statements are equivalent.*

(1) *R is a Prüfer domain.*
(2) *If Q is an upper to zero in $R[X]$, then $Q \not\subseteq M[X]$ for each maximal ideal M of R.*
(3) *For each prime ideal P of R and each overring T of R, there is at most one prime ideal of T lying over P.*
(4) *For each prime ideal P of R and each overring T of R, there is at most one member in a chain of prime ideals of T lying over P.*

In the sequel, we will discuss extensions of some of the results above to the non-Prüfer context. We close this section by mentioning that one can axiomatize the condition that uppers are not contained in extended primes and study it in rings which are not necessarily domains–see [6] by Ahmed Ayache, Paul-Jean Cahen, and Othman Echi.

2 Properties of uppers to zero

We will need some terminology for this section. Let R be a domain with quotient field K. For a nonzero ideal I of R, we write $I^{-1} = \{x \in K \mid xI \subseteq R\}$, $I_v = (I^{-1})^{-1}$, and $I_t = \bigcup J_v$, where the union is taken over all nonzero finitely generated subideals J of I. The ideal I is *divisorial* if $I = I_v$, and I is a *t-ideal* if $I = I_t$. We shall often be interested in the case $I_v = R$, which is equivalent to $I^{-1} = R$. An ideal maximal among all proper divisorial ideals (t-ideals) is called a maximal divisorial (maximal t-)ideal. Finally we say that I is v-invertible (t-invertible) if $(II^{-1})_v = R$ ($(II^{-1})_t = R$). For information on the v- and t-operations, see Gilmer's book [23]. We observe that uppers to zero are t-ideals (since they have height one).

We begin by discussing finitely generated uppers to zero. Our first two results are taken from Sarah Glaz's book on coherent rings [26]; the results are also in her thesis [25].

Theorem 2.1. [26, Lemma 7.4.6] *If R is an integrally closed domain, and Q is an upper to zero with $c(Q) = R$, then Q is invertible.*

Theorem 2.2. [26, Theorem 7.4.9] *If R is a coherent domain and Q is an upper to zero with $c(Q)_t = R$, then Q is finitely generated.*

Actually, in the two preceding results, Glaz phrases the condition "$c(Q) = R$" ("$c(Q)_t = R$") as "Q contains f with $c(f) = R$" ("$c(f)^{-1} = R$"). In case $c(Q) = R$, we have already seen that by [24, Corollary 1] we have $c(f) = R$ for some $f \in Q$. In case $c(Q)_t = R$, here is an argument (see [1, proof of Theorem 2]) which produces the desired f: First, there is a finitely generated subideal A of $c(Q)$ with $A_t = R$. Hence there are finitely many polynomials $f_1, \ldots, f_n \in Q$ with $(c(f_1) + \cdots + c(f_n))_v = R$, and for sufficiently large k, we have that the polynomial $f = f_1 + X^k f_2 + \cdots + X^{(n-1)k} f_n$ satisfies $f \in Q$ and $c(f) = c(f_1) + \cdots + c(f_n)$ (and hence that $c(f)_v = R = c(f)^{-1}$).

What about uppers to zero being principal? Hwa Tsang Tang characterized this property (in a slightly different form):

Theorem 2.3. (cf. [53, Theorem A]) *Let R be a domain with quotient field K, and let Q be an upper to zero. Then Q is principal if and only if $Q = fK[X] \cap R[X]$ for some $f \in R[X]$ with $c(f)^{-1} = R$.*

We next describe the maximal t-ideals of a polynomial ring.

Theorem 2.4. [35, Proposition 1.1] *If R is a domain and N is a maximal t-ideal of $R[X]$, then N is either the extension to $R[X]$ of a maximal t-ideal of R or N is an upper to zero.*

We emphasize that the preceding result applies only to *maximal* t-ideals– see the discussion following Example 2.12 below.

Theorem 2.5. [35, Theorem 1.4] *Let R be a domain, and let Q be an upper to zero. The following are equivalent.*

(1) *Q is a maximal t-ideal.*
(2) *Q is t-invertible.*
(3) $c(Q)_t = R$.

Moreover, if $Q = fK[X] \cap R[X]$ is a maximal t-ideal, then there is an element $g \in Q$ with $Q = (f, g)_v$ (and so Q is divisorial).

We pause for some more terminology. Let R be a domain with quotient field K, and let $Q = fK[X] \cap R[X]$ be an upper to zero. We say that Q is *almost principal* if $aQ \subseteq fR[X]$ for some nonzero element $a \in R$. Such uppers to zero were discussed by Eloise Hamann, Jon Johnson, and this author in [29].

Theorem 2.6. *Let R, K, Q be as above. Then Q is almost principal under any of the following conditions.*

(1) *Q is finitely generated (or, more generally, Q is generated by a set of polynomials of bounded degree).*
(2) $c(Q)_t = R$.
(3) *Q contains an element h with $\bigcap_{n=1}^{\infty} (c(h))^n \neq (0)$.*

(4) $Q = (X - u)K[X] \cap R[X]$, *where* $u \in K$ *is almost integral over* R.

Proof. (1) is clear, (2) is [29, Propostion 1.8], (3) is [29, Propositin 1.9] (but is due to Jimmy Arnold), and (4) is [29, Theorem 2.4].

We next state a theorem from [36] (most of which is proved in [29]) which gives several equivalent forms of the definition of "almost principal".

Theorem 2.7. [36, Proposition 2.3] *Let* R *be a domain with quotient field* K, *and let* $Q = fK[X] \cap R[X]$ *be an upper to zero, where* f *is irreducible in* $K[X]$. *The following statements are equivalent.*

(1) Q *is almost principal.*
(2) $Q^{-1} \not\subseteq K[X]$.
(3) $Q^{-1} \neq (Q{:}Q)$.
(4) *There is an element* $g \in R[X] \setminus Q$ *with* $gQ \subseteq fR[X]$.
(5) $Q = (f{:}_{R[X]}a)$ *for some* $a \in R$.
(6) $Q = (R[X]{:}_{R[X]}\psi)$ *for some* $\psi \in K(X)$.

Next, we give implications among some of the aforementioned conditions on uppers to zero.

Theorem 2.8. [36, Proposition 2.5] *Let* R *be a domain, and let* Q *be an upper to zero. Consider the following conditions on* Q.

(1) Q *is a maximal divisorial ideal of* $R[X]$.
(2) Q *is* v*-invertible.*
(3) $(Q{:}Q) = R[X]$
(4) $Q^{-1} \cap K = R$.
(5) $c(Q)_v = R$.
(6) Q *is almost principal.*
(7) Q *is divisorial.*

Then (1) $\Rightarrow$ *(2)* $\Leftrightarrow$ *(3)* $\Rightarrow$ *(4)* $\Leftrightarrow$ *(5), and (1)* $\Rightarrow$ *(6)* $\Rightarrow$ *(7).*

Question 2.9. In Theorem 2.8, do we have (4) $\Rightarrow$ (3)? More interestingly, do we have (7) $\Rightarrow$ (6)?

It is shown in [36] that, other than the two possibilities in this question, no implications other than those given in the theorem exist.

Conditions (1)-(3) above are "almost" equivalent:

Theorem 2.10. [36, Proposition 2.7] *With the notation of Theorem 2.8, assume that* $Q^{-1} \neq R[X]$. *Then conditions (1)-(3) are equivalent.*

If we assume that R is integrally closed, then we can do even better:

Theorem 2.11. [36, Proposition 2.7] *Assume that* R *is an integally closed domain. Then conditions (1)-(5) of Theorem 2.8 are equivalent. Moreover, every upper to zero in* $R[X]$ *is almost principal (and therefore divisorial).*

Examples of uppers to zero which are not divisorial (and therefore not almost principal) are not easy to come by. The following example is due to Jimmy Arnold and has been discussed in [29, 33, 36]; also, see Question 20 on Zafrullah's help desk webpage www.lohar.com/mithelpdesk.

Example 2.12. (Arnold) Let k be a field, let s, t be indeterminates, and set $R = k[s, \{st^{2^n} \mid n \geq 0\}]$. Let $Q = (X - t)K[X] \cap R[X]$. Then Q is not divisorial.

It is shown in [36] that we actually have $Q^{-1} = R[X]$.

Arnold's example was used in [33] to produce prime t-ideals in polynomial rings which are not extended and are not uppers to zero. However, as explained in [33], this does not produce in $R[X]$ long chains of prime t-ideals. We close this section with a couple of related questions.

Question 2.13. Let us say that a domain R has t-dimension n if there is a chain of primes $(0) \subsetneq P_1 \subsetneq \cdots \subsetneq P_n$, where each P_i is a prime t-ideal but there is no longer such chain. It is easy to show that for a domain R, we have $t\text{-dim}(R) \leq t\text{-dim}(R[X]) \leq 2(t\text{-dim}(R))$. It is also easy to produce examples with $t\text{-dim}(R[X]) = t\text{-dim}(R) + 1$. What other possibilities for $t\text{-dim}(R[X])$ are there? For that matter what are the possibilities for the t-dimension sequence of a domain? That is, which sequences of positive integers are realizable as $t\text{-dim}(R), t\text{-dim}(R[X_1]), t\text{-dim}(R[X_1, X_2]), \ldots$ for some domain R? (Arnold and Gilmer answered this question for ordinary (Krull) dimension [8, 9].)

Question 2.14. Arnold's example has dimension two. Is there a one-dimensional domain R which allows a non-almost principal upper to zero in $R[X]$?

3 Other uses of uppers to zero

In [39] Krull gave an example of a one-dimensional, quasi-local integrally closed domain which is not a valuation domain. His example was a specific instance of the following. Let $F \subseteq K$ be fields, with F algebraically closed in K, let V be a one-dimensional valuation domain of the form $K + M$, and let $R = F + M$. Since for $u \in K$ with u transcendental over F, we clearly have $u, u^{-1} \notin R$, it follows from standard facts about the $D+M$-construction that R is a one-dimensional, quasi-local integrally closed domain which is not a valuation domain. We *also* have that the upper to zero $(X-u)K[X] \cap R[X]$ is contained in $M[X]$, which shows that the height of $M[X]$ is (at least) two and thus that the dimension of $R[X]$ is (at least) three. This is exactly the idea that Seidenberg uses to produce his famous examples of domains D with $\dim D = n$ and $\dim D[X] = m$ for m chosen arbitrarily in $\{n+1, n+2, \ldots, 2n+1\}$ [52]. Gilmer gives a very efficient presentation of Seidenberg's construction in Proposition 30.15 of [23] (and Gilmer's proof uses the already-mentioned

Theorem 19.15 of [23]). A somewhat different method of constructing such examples is presented in [46].

Now recall that Kaplansky called a domain R an *S-domain* if height-one primes of R extend to height-one primes of $R[X]$ (equivalently, if uppers to zero in $R[X]$ avoid extensions of height-one primes); a ring R is a *strong S-ring* if R/P is an S-domain for each prime P of R. (See [40] for a study of strong S-domains.) The construction mentioned in the preceding paragraph produces pseudo-valuation domains. A quasi-local domain (R, M) is a *pseudo-valuation domain* (PVD) if M^{-1} is a valuation domain with maximal ideal M [30]. Note that in the construction above, the PVD R is not an S-domain. Here is a characterization of PVDs which are strong S-domains.

Theorem 3.1. [31, Remark 2.3 and Proposition 2.4] *Let (R, M) be a PVD. If the integral closure $\overline{R}$ of R satisfies $\overline{R} = M^{-1}$, then R is a strong S-domain. If $\overline{R} \neq M^{-1}$, then R is a strong S-domain if and only if M is unbranched in the valuation domain M^{-1} (that is, M is the union of the primes properly contained in M).*

We now turn to "upper to zero" characterizations of certain desirable ring-theoretic properties. The following is a characterization of integrally closed domains, which I have often found useful. It appears in a paper by Julien Querré [47].

Theorem 3.2. [47, Lemme 1] *Let R be a domain with quotient field K. The following are equivalent:*

(1) *R is integrally closed.*
(2) *For nonzero elements $f, g \in K[X]$, $c(fg)_v = (c(f)c(g))_v$.*
(3) *For $f \in K[X]$, f irreducible, and $Q = fK[X] \cap R[X]$, we have $Q = fc(f)^{-1}R[X]$.*

We note that (1) $\Rightarrow$ (2) and (2) $\Rightarrow$ (3) are proved in [23, Section 34]. Querré proves (2) $\Rightarrow$ (1) and (3) $\Rightarrow$ (2). In fact, Dan Anderson pointed out to me that (2) $\Rightarrow$ (1) was proved much earlier by H. Flanders [17]. Here is a simple direct proof of (3) $\Rightarrow$ (1): Let $u \in K$ be integral over R, and let $Q = (X - u)K[X] \cap R[X]$. Then $Q = (X - u)(1, u)^{-1}R[X]$. Since u is integral over R, we may pick $g \in Q$ with g monic. Write $g = (X - u)h$. Then $h \in (1, u)^{-1}R[X]$, whence $hu \in R[X]$. Since h is monic, this yields $u \in R$. Note that this proof of (3) $\Rightarrow$ (1) makes it clear that we can add a fourth equivalence: For $u \in K$, we have $(X-u)K[X] \cap R[X] = (X-u)(1, u)^{-1}R[X]$. Moreover, a slight variation of this argument yields an amusing proof of the u, u^{-1}-lemma. (See also the proof given by Dan Anderson and Dong Kwak [4].)

Theorem 3.3. *Let (R, M) be a quasi-local integrally closed domain with quotient field K, let $u \in K$, and suppose that $g(u) = 0$ for some $g \in R[X]$ with $c(g) = R$. Then either u or u^{-1} is in R.*

Proof. As above, let $Q = (X - u)K[X] \cap R[X] = (X - u)(1, u)^{-1}R[X]$. Then $g \in Q$, and we may write $g = (X - u)h$ with $h \in (1, u)^{-1}R[X]$. Thus $h, hu \in R[X]$. If $c(h) = R$, then the fact that $hu \in R[X]$ implies that $u \in R$. Hence we suppose that $c(h) \subseteq M$. We then have $R = c(g) \subseteq (1, u)c(h) \subseteq M + Mu$. Thus there are elements $a, b \in M$ with $1 = a + bu$, and this equation shows that $u^{-1} \in M$.

Dan Anderson and Muhammad Zafrullah gave another characterization of integrally closed domains; it sharpens Theorem 2.1.

Theorem 3.4. [5, Theorem 3.2] *For an integral domain R with quotient field K, the following statements are equivalent.*

(1) *For each nonunit $f \in R[X]$ with $c(f) = R$, $fR[X]$ is a product of prime ideals.*
(2) *For each nonunit $f \in R[X]$ with f monic, f is a product of principal primes.*
(3) *R is integrally closed.*
(4) *If Q is an upper to zero with $c(Q) = R$, then Q is invertible.*
(5) *If Q is an upper to zero which contains a monic polynomial, then Q is principal.*

Dan Anderson and Zafrullah also note that the following equivalence can be added: each irreducible monic polynomial of $R[X]$ is prime. They then show that the condition that each $fR[X]$ with $c(f)_v = R$ is a t-product of prime ideals is not only not another equivalence but that it always holds:

Theorem 3.5. [5, Proposition 3.3] *Let R be any integral domain and $f \in R[X]$ a nonunit with $c(f)_v = R$. Then $fR[X] = (Q_1 \cdots Q_n)_t$, where the Q_i are uppers to zero. Conversely, if $fR[X]$ is such a product, then $c(f)_v = R$.*

We next mention a characterization of the GCD-property, due to Tang. It is essentially a corollary of her characterization of principal uppers to zero (Theorem 2.3).

Theorem 3.6. [53, Theorem I (I) $\Leftrightarrow$ (IV) $\Leftrightarrow$ (V)] *The following are equivalent for a domain R with quotient field K.*

(1) *R is a GCD-domain.*
(2) *Each upper to zero in $R[X]$ is principal.*
(3) *Each rational upper to zero in $R[X]$ is principal.*

According to Dan and David Anderson [2] a domain R is a generalized GCD domain (GGCD domain) if the intersection of each pair of invertible ideals of R is invertible. Recently, Dan Anderson, Tiberiu Dumitrescu, and Muhammad Zafrullah gave the following characterization of GGCD domains; it may be viewed as a companion result to Theorem 3.6.

Theorem 3.7. [3, Theorem 15] *The following statements are equivalent for a domain R.*

(1) *R is a GGCD domain.*
(2) *Each upper to zero in $R[X]$ is invertible.*
(3) *Each rational upper to zero in $R[X]$ is invertible.*

It is well known (and easy to verify) that if R is a Prüfer domain which is not a field, then $R[X]$ is *not* a Prüfer domain. This "defect" may be remedied by working in the class of Prüfer v-multiplication domains (PVMDs). We recall the (modern) definition: A domain R is a PVMD if each nonzero finitely generated ideal is t-invertible, equivalently, if R_M is a valuation domain for each maximal t-ideal M of R. Although these rings were studied by Krull, more recent interest in them was sparked by Malcolm Griffin's paper [28].

Analogous to the characterizations of Prüfer domains given above, we have the following.

Theorem 3.8. *Let R be a domain with quotient field K. Then the following statements are equivalent.*

(1) *R is a PVMD.*
(2) *R is integrally closed, and each rational upper to zero in $R[X]$ contains an element f with $c(f)_v = R$.*
(3) *R is integrally closed, and each upper to zero in $R[X]$ contains an element f with $c(f)_v = R$.*
(4) *R is integrally closed and each upper to zero in $R[X]$ is a maximal t-ideal of $R[X]$.*
(5) *$R[X]$ is a PVMD.*

The equivalence (1) $\Leftrightarrow$ (3) was proved by Papick in [49, Theorem] and in [50, Proposition 2.5] and, as pointed out in [50], by Joe Mott and Muhammad Zafrullah (in a different form) in [42, Theorem 3.4]. The entire theorem is proved in [34]. A key ingredient in the proof in [34] is the so-called content formula, or Hilfssatz von Dedekind-Mertens. Not surprisingly, Gilmer had a good deal to say about this too; it appears as Theorem 28.1 in [23], he gave some uses of it in [21], and he and Jimmy Arnold proved some generalizations and gave some more applications of it in [7].

This seems an appropriate time to mention a conjecture of Glaz and Vasconcelos: if R is a one-dimensional coherent domain, then $\overline{R}$ is Prüfer. Of course, this is equivalent to showing that for such a domain, we have $\dim (R[X]) = 2$, that is, that uppers to zero avoid extensions to $R[X]$ of maximal ideals of R. A more general conjecture is that for any coherent domain, the integral closure is a PVMD. (A one-dimensional PVMD is a Prüfer domain.) In case the domain is integrally closed and coherent, then it is a PVMD–see [23, Exercise 21, p. 432] and [54]. (In fact, "coherence" can be weakened to "finite conductor"; that is, it is enough to suppose that each intersection of two principal ideals is finitely generated.)

Now recall the two definitions given for PVMD above. We briefly consider what happens when each of these is weakened. First, it is well known that for any domain R we have $R = \bigcap R_P$, where the intersection is taken over all associated primes of principal ideals (that is, primes minimal over an ideal of the form $(a{:}b)$) [10, Proposition 4]. If we weaken the requirement that localizations at maximal t-ideals be valuation domains to the condition that localizations at associated primes of principal ideals be valuation domains, we obtain the class of P-domains (not to be confused with the P-extensions discussed above!). Thus a P-domain R is *essential*, that is, we can express $R = \bigcap R_P$, where each R_P is a valuation domain. In [32] Bill Heinzer and Jack Ohm give an example of an essential domain which is not a PVMD, and in [42] Mott and Zafrullah show that the Heinzer-Ohm example is a P-domain (and also prove several interesting results about P-domains). Papick gave an "upper to zero" characterization of P-domains:

Theorem 3.9. (cf. [50, Proposition 2.0 and Corollary 2.3]) *The following are equivalent for a domain R with quotient field K.*

(1) *R is a P-domain.*
(2) *R is integrally closed, and $Q \not\subseteq P[X]$ for each upper to zero Q in $R[X]$ and each associated prime of a principal ideal P.*
(3) *R is integrally closed, and $Q \not\subseteq P[X]$ for each rational upper to zero Q and each associated prime of a principal ideal P.*

If we weaken the condition "each nonzero finitely generated is t-invertible" in the definition of PVMD to "each nonzero finitely generated ideal is v-invertible", we obtain the class of v-domains. It is (fairly) well known that essential domains are v-domains. (In fact, we have the implications PVMD $\Rightarrow$ P-domain $\Rightarrow$ essential domain $\Rightarrow$ v-domain, and, in general, none of these is reversible.) The Heinzer-Ohm example mentioned above is therefore a v-domain which is not a PVMD. (The first example of a v-domain which is not a PVMD was given by Dieudonné [13].) We shall give "upper to zero" characterizations of v-domains below.

The question arises: Are there extensions of the characterizations in Theorem 3.8 analogous to those of domains with Prüfer integral closure? Indeed there are. In view of the characterization of PVMDs as integrally closed domains in which each upper to zero is a maximal t-ideal (Theorem 3.8), Zafrullah and I in [35] called a domain R a *UMT-domain* if each upper to zero in $R[X]$ is a maximal t-ideal. Thus a PVMD is an integrally closed UMT domain. We proceed to give several results about this class of domains.

Theorem 3.10. *The following are equivalent for a domain R with quotient field K.*

(1) *R is a UMT-domain.*
(2) *Each rational upper to zero is a maximal t-ideal.*

(3) $c(Q)_t = R$ *for each upper to zero* Q *in* $R[X]$.
(4) *For each upper to zero* Q *in* $R[X]$, *there is an element* $f \in Q$ *with* $c(f)_v = R$.
(5) *Each upper to zero in* $R[X]$ *is t-invertible.*
(6) $Q \not\subseteq M[X]$ *for each upper to zero* Q *and each maximal t-ideal* M *of* R.
(7) *Each prime ideal of* $R[X]_{N_v}$ *is extended from* R. *(Here,* $N_v = \{f \in R[X] \mid c(f)^{-1} = R\}$.*)*

With the exception of condition (2), Theorem 3.10 is stated as Theorem 1.1 in [19] (but the proof is in [35]). The equivalence (1) $\Leftrightarrow$ (2) follows from [36, Proposition 4.6]. The equivalence of (1) and (7) was motivated by B.G. Kang's result that a domain R is a PVMD if and only if each ideal of $R[X]_{N_v}$ is extended from R [37, Theorem 3.1].

Noetherian UMT-domains are easily characterized:

Theorem 3.11. [35, Theorem 3.7] *A Noetherian domain* R *is a UMT-domain if and only if every prime t-ideal of* R *has height one.*

The UMT-property leads to another characterization of domains with Prüfer integral closure:

Theorem 3.12. [16, Theorem 2.4] *A domain* R *has Prüfer integral closure if and only if* R *is a UMT-domain and each maximal ideal of* R *is a t-ideal.*

The equivalence (1) $\Leftrightarrow$ (2) in our next result shows that UMT-domains are precisely those domains having Prüfer integral closure "t-locally".

Theorem 3.13. [19, Theorem 1.5 and Theorem 2.4] *The following statements are equivalent for a domain* R.

(1) R *is a UMT-domain.*
(2) R_M *has Prüfer integral closure for each maximal t-ideal* M *of* R.
(3) *For each maximal t-ideal* M *of* R, R_M *is a UMT-domain and* MR_M *is a maximal t-ideal of* R_M.
(4) $R[\{X_\alpha\}]$ *is a UMT-domain for any set* $\{X_\alpha\}$ *of indeterminates.*

Because of their use in constructing interesting examples, it has become fashionable to find results on the transfer of desirable properties in pullback diagrams. The PVMD-version of our next theorem was first proved (in this generality) by Marco Fontana and Stefania Gabelli in [18].

Theorem 3.14. [19, Theorem 3.7] *Let* T *be a domain,* M *a maximal ideal of* T, D *a subring of* T/M, *and let* R *be defined by the following pullback diagram of canonical homomorphisms:*

$$\begin{array}{ccc} R & \longrightarrow & D \\ \downarrow & & \downarrow \\ T & \xrightarrow{\varphi} & k = T/M. \end{array}$$

Then R is a UMT-domain (PVMD) if and only if D and T are UMT-domains (PVMDs), M is a maximal t-ideal of T, and k is algebraic over (is equal to) the quotient field of D.

These results combine to yield the following corollaries.

Corollary 3.15. [19, Corollary 3.10] *In a pullback diagram as described above, R has Prüfer integral closure if and only if D and T have Prüfer integral closure and k is algebraic over the quotient field of D.*

Corollary 3.16. [19, Corollary 3.11] *A domain R has Prüfer integral closure if and only if each overring of R is a UMT-domain.*

Corollary 3.17. [19, Corollary 3.12]) *If R is a domain with Prüfer integral closure, then R/P also has Prüfer integral closure for each prime ideal P of R.*

The preceding result is due to Ahmed Ayache, Paul-Jean Cahen, and Othman Echi [6].

In [36] Zafrullah and I called a domain R a *UMV-domain* if each upper to zero in $R[X]$ is a maximal divisorial (v-) ideal. We have the following characterization.

Theorem 3.18. [36, Theorem 3.2] *Let R be a domain. Then the following statements are equivalent.*

(1) *R is a UMV-domain.*
(2) *Each upper to zero Q satisfies $(Q:Q) = R[X] \neq Q^{-1}$.*
(3) *Each upper to zero Q satisfies $Q^{-1} \neq R[X]$ and $c(Q)_v = R$.*

Just as a PVMD is an integrally closed UMT-domain, a v-domain is an integrally closed UMV-domain. This and a couple of other equivalences are given in our next result, the proof of which uses Theorems 2.11, 3.18, and 3.2 as well as a few related ideas.

Theorem 3.19. [36, Theorem 3.3] *The following statements are equivalent for a domain R.*

(1) *R is a v-domain.*
(2) *R is an integrally closed UMV-domain.*
(3) *R is integrally closed, and each upper to zero in $R[X]$ is v-invertible.*
(4) *R is integrally closed, and each rational upper to zero in $R[X]$ is v-invertible.*

Mott, Nashier, and Zafrullah proved the following companion result to their characterization of Prüfer domains (Theorem 1.6):

Theorem 3.20. [43, Theorem 2.5] *A domain R with quotient field K is a v-domain if and only if it is integrally closed and each element of K satisfies a polynomial over R with v-invertible content.*

It follows from Theorem 3.19 (1) $\Leftrightarrow$ (2) that, since there are v-domains which are not PVMDs, there are UMV-domains which are not UMT-domains. (Also, see [20].) A method for constructing non-integrally closed examples is given in [36].

One should not expect completely parallel behavior between UMV-domains and UMT-domains. After all, the t-operation has finite type and the v-operation does not. This means, for example, that each element of a domain is contained in a maximal t-ideal (which is necessarily prime), while there may be no maximal divisorial ideals at all (but when they do exist they are prime). We now explore some of the ramifications of this.

Theorem 3.21. [36, Theorem 3.6 and Corollary 3.7] *If R is a UMV-domain, then R_P has Prüfer integral closure for each divisorial prime ideal P of R. Hence if R is a v-domain, then R_P is a valuation domain for each divisorial prime P of R.*

This result is as close as we can come to a companion result to Theorem 3.12. Thus we pose

Question 3.22. [36, Question 4.1] If a domain R is such that R_P has Prüfer integral closure for each divisorial prime P of R, is R necessarily a UMV-domain?

One problem with this question is that the hypothesis could be satisfied vacuously. This motivates another question.

Question 3.23. If (R, M) is a one-dimensional, quasi-local, integrally closed domain with M *not* divisorial, must R be a v-domain?

We remark that Nagata gave an example of a one-dimensional, quasi-local, completely integrally closed domain (and hence a v-domain) (R, M) which is not a valuation domain [44, 45]. In any such example, we must have M not divisorial. To see this, note that M certainly cannot be principal. Hence if M is divisorial, then $M = (R:_R u)$ for any $u \in M^{-1} \setminus R$. Non-principality of M then forces $uM \subseteq M$. However, this implies that u is almost integral over R, whence $u \in R$, a contradiction.

One theme running throughout this work is that when something is true for all *rational* uppers to zero, then it is often true for *all* uppers to zero. In particular, according to Theorem 3.10, a domain R is a UMT-domain if and only if all rational uppers to zero are maximal t-ideals. This motivates:

Question 3.24. [36, Question 4.5] If R is a domain and every rational upper to zero in $R[X]$ is a maximal divisorial ideal, must R be a UMV-domain? (The answer is yes if R is integrally closed by Theorem 3.19.)

Chevalley showed [12] that if R is a Noetherian domain and P is any nonzero prime of R, then there is a discrete rank one valuation overring centered on P. On the other hand, for any domain R and any chain of primes $(0) \subseteq P_1 \subseteq \cdots \subseteq P_n$ in R, there is a valuation overring of R having a chain of primes contracting term by term to the P_i. Dan Anderson asked whether in the Noetherian case, this valuation overring could be taken to be discrete. Cahen, Lucas, and I answered this question with the following result.

Theorem 3.25. [11, Theorem] *Let $(0) = P_0 \subsetneqq P_1 \subsetneqq \cdots \subsetneqq P_n$ be a chain of primes in a Noetherian domain R, and let $s_1, \ldots, s_n$ be a sequence of integers with $1 \leq s_i \leq ht(P_i/P_{i-1})$. Then there exist an overring T of R and a chain of primes $(0) = P'_0 \subsetneqq P'_1 \subsetneqq \cdots \subsetneqq P'_n$ in T such that $P'_i \cap R = P_i$ and $ht(P'_i/P'_{i-1}) = s_i$ for $i = 1, \ldots, n$. Moreover, T can be taken to be either a finitely generated extension of R or a discrete valuation ring of rank $s = \sum s_i$.*

The key to the construction of the finitely generated extension is to show that a reduction of one in $\mathrm{ht}(P_i/P_{i-1})$ can be achieved in a simple extension of R, and this simple extension is, of course, realized as $R[X]/Q$ for an appropriate upper to zero Q.

We close by briefly touching upon the use of uppers to zero in the study of $\mathrm{Spec}(R[X])$. One nice tool is the following lovely result of Ada Maria de Souza Doering and Yves Lequain, called the principal ideal theorem for polynomial rings.

Theorem 3.26. [14, Theorem A] *Let R be a domain, let P be prime in $R[X]$ with $P \neq (P \cap R)[X]$ (so that P is an upper to $P \cap R$), and let $f \in P \setminus (P \cap R)[X]$. Then there is an upper to zero Q with $f \in Q \subseteq P$.*

Since the term "upper to zero" is due to McAdam, it is perhaps fitting to end with a discussion of his celebrated example showing that it is possible for two height-two prime ideals in a Noetherian domain to fail to contain a common (nonzero) prime. His solution makes double use of uppers to zero: the first step avoids them, and the second step produces an appropriate one.

Example 3.27. [41] For (V, P) a complete DVR, McAdam first shows that in $V[X]$ the intersection of the primes (P, X) and $(P, X+1)$ contains no uppers to zero and that (consequently) the only nonzero prime of $R[X]$ contained in the intersection is $P[X]$. By localizing, he thus obtains a Noetherian domain R with exactly two maximal ideals M and N, both of height two, and such that $M \cap N$ contains a unique nonzero prime p. He then shows that for $b \in p$ and $c \in M \cap N$ with c not in any of the height one primes containing b, the upper to zero $Q = (bX - c)K[X] \cap R[X]$ is such that in the Noetherian domain $R[X]/Q$ the primes $(M, X)R[X]/Q$ and $(N, X)R[X]/Q$ have height two and there is no nonzero prime in their intersection.

I would like to thank Muhammad Zafrullah for some helpful comments on an earlier draft of this paper.

References

1. D.D. Anderson, *Some remarks on the ring R(X)*, Comment. Math. Univ. St. Pauli **XXVI-2** (1977), 137-140.
2. D.D. Anderson and D.F. Anderson, *Generalized GCD domains*, Comment. Math. Univ. St. Pauli **28** (1979), 215-221.
3. D.D. Anderson, D. Dumitrescu, and M. Zafrullah, *Quasi-Schreier domains II*, manuscript.
4. D.D. Anderson and D. Kwak, *The u, u^{-1}-lemma revisited*, Comm. Algebra **24** (1996), 2447-2454.
5. D.D. Anderson and M. Zafrullah, *On t-invertibility III*, Comm. Algebra **21** (1993), 1189-1201.
6. A. Ayache, P.-J. Cahen, and O. Echi, *Anneaux quasi-prüferiens et P-anneaux*, Boll. Un. Mat. Ital. B(7) **10** (1996), 1-24.
7. J. Arnold and R. Gilmer, *On the contents of polynomials*, Proc. Amer. Math. Soc. **24** (1970), 556-562.
8. J. Arnold and R. Gilmer, *The dimension sequence of a commutative ring*, Amer. J. Math. **96** (1974), 385-408.
9. J. Arnold and R. Gilmer, *Two questions concerning dimension sequences*, Arch. Math. (Basel) **29** (1977), 497-503.
10. J. Brewer and W. Heinzer, *Associated primes of principal ideals*, Duke Math. J. **41** (1974), 1-7.
11. P.-J. Cahen, E. Houston, and T. Lucas, *Discrete valuation overrings of Noetherian domains*, Proc. Amer. Math. Soc. **124** (1996), 1719-1721.
12. C. Chevalley, *La notion d'anneau de décomposition*, Nagoya Math. J. **7** (1954), 21-33.
13. J. Dieudonné, *Sur la théorie de la divisibilité*, Bull. Soc. Math. France **69** (1941), 133-144.
14. A. de Souza Doering and Y. Lequain, *Chains of prime ideals in polynomial rings*, J. Algebra **78** (1982), 163-180.
15. D. Dobbs, *On INC-extensions and polynomials with unit content*, Canad. Math. Bull. **23** (1980), 37-42.
16. D. Dobbs, E. Houston, T. Lucas, M. Roitman, and M. Zafrullah, *On t-linked overrings*, Comm. Algebra **20** (1992), 1463-1488.
17. H. Flanders, *A remark on Kronecker's theorem on forms*, Proc. Amer. Math. Soc. **3** (1952), 197.
18. M. Fontana and S. Gabelli, *On the class group and the local class group of a pullback*, J. Algebra **181** (1996), 803-835.
19. M. Fontana, S. Gabelli, and E. Houston, *UMT-domains and domains with Prüfer integral closure*, Comm. Algebra **26** (1998), 1017-1039.
20. S. Gabelli and M. Roitman, *Maximal divisorial ideals and t-maximal ideals*, JP J. Algebra Number Theory Appl. **4** (2004), 323-336.
21. R. Gilmer, *Some applications of the Hilfssatz von Dedekind-Mertens*, Math. Scan. **20** (1967), 240-244.
22. R. Gilmer, *Multiplicative ideal theory*, Queen's Papers on Pure and Applied Mathematics, No. 12, Queen's University, Kingston, Ontario, 1968.
23. R. Gilmer, *Multiplicative ideal theory*, Dekker, New York, 1972.
24. R. Gilmer and J. Hoffmann, *A characterization of Prüfer domains in terms of polynomials*, Pac. Math. J. **60** (1975), 81-85.

25. S. Glaz, *Finiteness and differential properties of ideals*, Ph.D. thesis, Rutgers University, 1977.
26. S. Glaz, *Commutative coherent rings*, Lecture Notes in Mathematics **1371**, Springer, Berlin, 1989.
27. S. Glaz and W. Vasconcelos, *Flat ideals III*, Comm. Algebra **12** (1984), 199-227.
28. M. Griffin, *Some results on Prüfer v-multiplication rings*, Canad. J. Math. **19** (1967), 710-722.
29. E. Hamann, E. Houston, and J. Johnson, *Properties of uppers to zero in* $R[x]$, Pac. J. Math. **135** (1988), 65-79.
30. J. Hedstrom and E. Houston, *Pseudo-valuation domains*, Pac. J. Math. **75** (1978), 137-147.
31. J. Hedstrom and E. Houston, *Pseudo-valuation domains (II)*, Houston J. Math. **4** (1978), 199-207.
32. W. Heinzer and J. Ohm, *An essential ring which is not a v-multiplication ring*, Canad. J. Math. **25** (1973), 856-861.
33. E. Houston, *Prime t-ideals in* $R[X]$, in *Commutative ring theory*, P.-J. Cahen, D. Costa, M. Fontana, S. Kabbaj, eds., Lecture Notes in Pure and Applied Mathematics **153**, Dekker, New York, 1994, pp. 163-170.
34. E. Houston, S. Malik, and J. Mott, *Characterizations of $*$-multiplication domains*, Canad. Math. Bull. **27** (1984), 48-52.
35. E. Houston and M. Zafrullah, *On t-invertibility II*, Comm. Algebra **17** (1989), 1955-1969.
36. E. Houston and M. Zafrullah, *UMV-domains*, in *Arithmetical properties of commutative rings and modules*, S. Chapman, ed., Lecture Notes in Pure and Applied Mathematics **241**, Chapman & Hall/CRC, London, 2005, pp. 304-315.
37. B.G. Kang, *Prüfer v-multiplication domains and the ring* $R[X]_{N_v}$, J. Algebra **123** (1989), 151-170.
38. I. Kaplansky, *Commutative rings*, Allyn and Bacon, Boston, 1970.
39. W. Krull, *Beiträge zur Arithmetik kommutativer Integritätsbereiche II*, Math. Z. **41** (1936), 665-679.
40. S. Malik and J. Mott, *Strong S-domains*, J. Pure. Appl. Algebra textbf28 (1983), 249-264.
41. S. McAdam, *A Noetherian example*, Comm. Algebra **4** (1976), 245-247.
42. J. Mott and M. Zafrullah, *On Prüfer v-multiplication domains*, Manuscripta Math. **35** (1981), 1-26.
43. J. Mott, B. Nashier, and M. Zafrullah, *Contents of polynomials and invertibility*, Comm. Algebra **18** (1990), 1569-1583.
44. M. Nagata, *On Krull's conjecture concerning valuation rings*, Nagoya Math. J. **4** (1952), 29-33.
45. M. Nagata, *Corrections to my paper "On Krull's conjecture concerning valuation rings"*, Nagoya Math. J. **9** (1955), 209-212.
46. A. Ouertani, *Exemples de dimension de Krull d'anneaux de polynômes*, Canad. Math. Bull. **33** (1990), 135-138.
47. J. Querré, *Idéaux divisoriels d'un anneau de polynômes*, J. Algebra **64** (1980), 270-284.
48. I. Papick, *Topologically defined classes of going-down domains*, Trans. Amer. Math. Soc. **219** (1976), 1-37.
49. I. Papick, *A note on Prüfer v-multiplication domains*, Boll. Un. Mat. Ital. A(6) **1** (1982), 133-136.

50. I. Papick, *Super-primitive elements*, Pac. J. Math. **105** (1983), 217-226.
51. A. Seidenberg, *On the dimension theory of rings*, Pac. J. Math **3** (1953), 505-512.
52. A. Seidenberg, *On the dimension theory of rings II*, Pac. J. Math. **4** (1954), 603-614.
53. H. Tang, *Gauss' lemma*, Proc. Amer. Math. Soc. **35** (1972), 372-376.
54. M. Zafrullah, *On finite conductor domains*, Manuscripta Math. **24** (1978), 191-204.
55. M. Zafrullah, *Various facets of rings between $D[X]$ and $K[X]$*, Comm. Algebra **31** (2003), 2497-2540.

On the dimension theory of polynomial rings over pullbacks

S. Kabbaj

Department of Mathematics, P.O. Box 5046, King Fahd University of Petroleum & Minerals, Dhahran 31261, Saudi Arabia kabbaj@kfupm.edu.sa

1 Introduction

Since Seidenberg's (1953-54) papers [35, 36] and Jaffard's (1960) pamphlet [28] on the dimension theory of commutative rings, the literature abounds in works exploring the prime ideal structure of polynomial rings, including four pioneering articles by Arnold and Gilmer on dimension sequences [3, 4, 5, 6]. Of particular interest is Bastida-Gilmer's (1973) precursory article [8] which established a formula for the Krull dimension of a polynomial ring over a $D+M$ issued from a valuation domain. During the last three decades, numerous papers provided in-depth treatments of dimension theory and other related notions (such as the S-property, strong S-property, and catenarity) in polynomial rings over various pullback constructions. All rings considered in this paper are assumed to be integral domains.

A polynomial ring over an arbitrary domain R is subject to Seidenberg's inequalities: $n + \dim(R) \leq \dim(R[X_1, ..., X_n]) \leq n + (n+1)\dim(R)$, $\forall\ n \geq 1$. A finite-dimensional domain R is said to be Jaffard if $\dim(R[X_1, ..., X_n]) = n + \dim(R)$ for all $n \geq 1$; equivalently, if $\dim(R) = \dim_v(R)$, where $\dim(R)$ denotes the Krull dimension of R and $\dim_v(R)$ its valuative dimension (i.e., the supremum of dimensions of the valuation overrings of R). The study of this class was initiated by Jaffard [28]. For the convenience of the reader, recall that, in general, for a domain R with $\dim_v(R) < \infty$ we have: $\dim(R) \leq \dim_v(R)$, $\dim_v(R[X_1, ..., X_n]) = n + \dim_v(R)$ for all $n \geq 1$, and $\dim(R[X_1, ..., X_n]) = n + \dim_v(R)$ for all $n \geq \dim_v(R) - 1$ (Cf. [2, 11, 18, 26, 28]).

As the Jaffard property does not carry over to localizations (see Example 3.5 below), R is said to be locally Jaffard if R_p is a Jaffard domain for each prime ideal p of R; equivalently, $S^{-1}R$ is a Jaffard domain for each multiplicative subset S of R. A locally Jaffard domain is Jaffard [2]. The class of (locally) Jaffard domains contains most classes involved in dimension theory, including Noetherian domains [31], Prüfer domains [26], and universally catenarian domains [10].

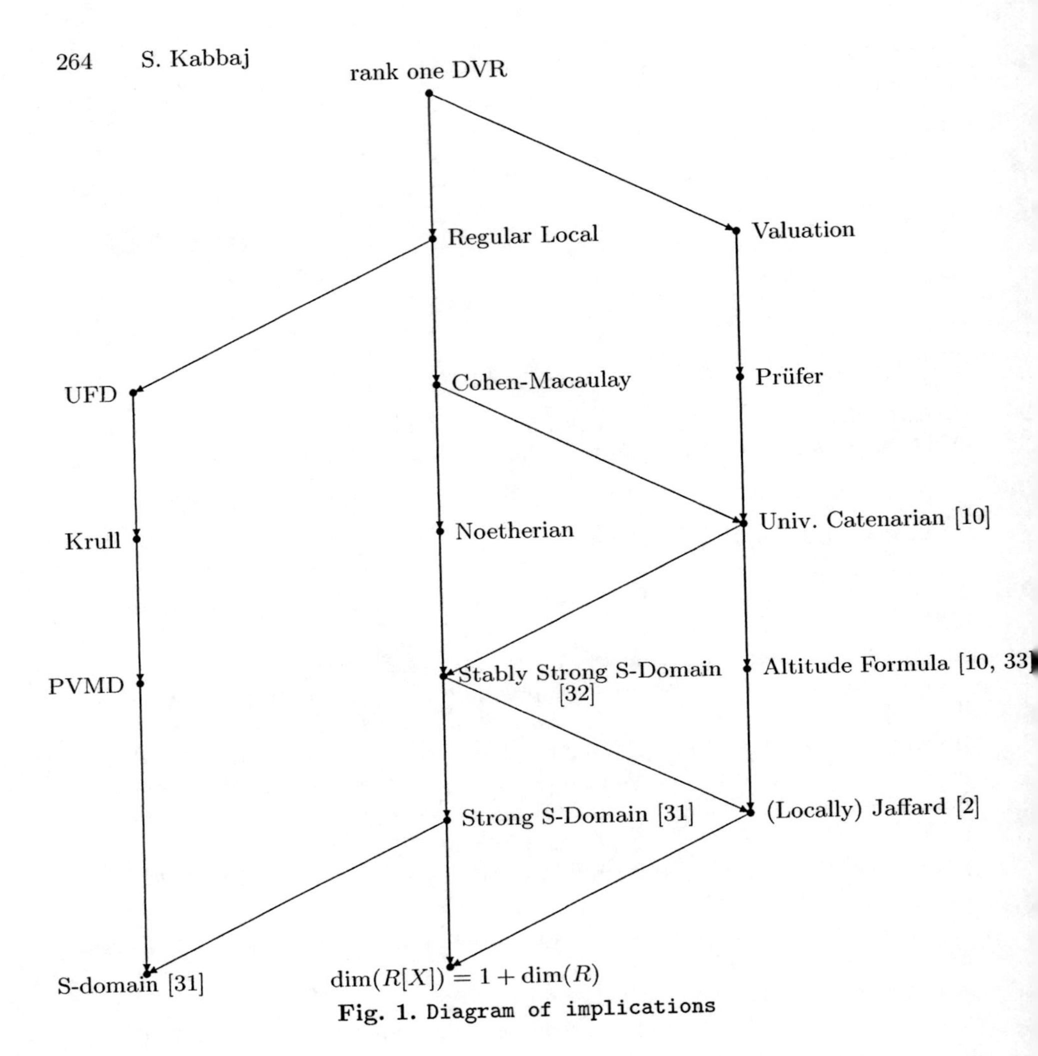

Fig. 1. Diagram of implications

In order to treat Noetherian domains and Prüfer domains in a unified manner, Kaplansky [31] introduced the following concepts: A domain R is called an S-domain if, for each height-one prime ideal p of R, the extension $pR[X]$ in $R[X]$ has height 1 too; and R is said to be a strong S-domain if $\frac{R}{p}$ is an S-domain for each prime ideal p of R. A strong S-domain R satisfies $\dim(R[X]) = \dim(R) + 1$. Notice that while $R[X]$ is always an S-domain for any domain R [24], $R[X]$ need not be a strong S-domain even when R is a strong S-domain [12]. Thus R is called a stably strong S-domain (also called a universally strong S-domain) if the polynomial ring $R[X_1, ..., X_n]$ is a strong S-domain for each positive integer n. A stably strong S-domain is locally Jaffard [2, 29, 32].

This review paper deals with dimension theory of polynomial rings over certain families of pullbacks. While the literature is plentiful, this field is still developing and many contexts are yet to be explored. I will thus restrict the scope of the present survey, mainly, to topics I have worked on over the last decade. The set of pullback constructions studied includes $D+M$, $D+(X_1,...,X_n)D_S[X_1,...,X_n]$, $A+XB[X]$, and $D+I$.

Any unreferenced material is standard, as in [9, 26, 28, 31, 33]. In Figure 1, a diagram of implications summarizes the relations between some spectral notions and well-known classes of integral domains (some of which should be either finite-dimensional or locally finite dimensional).

2 Preliminaries on Pullbacks

Pullbacks have proven to be useful for the construction of original examples and counter-examples in Commutative Ring Theory. The oldest in date is due to Krull (Cf. [8, page 1]). However, the first systematic investigation of a particular family of pullbacks; namely, $D+M$ issued from valuation domains, was carried out by Gilmer [25, Appendix 2] and [26]. Later, during the 1970s, six ground-breaking papers [8, 27, 19, 16, 13, 20] provided further development in various pullback contexts and paved the path for most subsequent works on these constructions. In Figure 2, a diagram provides more details on the contexts studied in these works.

Let's recall some results on the classical $D+M$ constructions (i.e., those issued from valuation domains). We shall use $\mathrm{qf}(R)$ to denote the quotient field of a domain R.

Theorem 2.1. ([25] and [19]) *Let V be a valuation domain of the form $K+M$, where K is a field and M is the maximal ideal of V. Let D be a proper subring of K with $k:=\mathrm{qf}(D)$. Set $R:=D+M$. Then:*
(1) $\dim(R)=\dim(V)+\dim(D)$.
(2) $\dim_v(R)=\dim(V)+\max\{\dim(W) \mid W$ *is valuation on K containing D*$\}$.
(3) *The integral closure of R is $D'+M$, where D' is the integral closure of D.*
(4) *R is a valuation domain $\Leftrightarrow$ D is a valuation domain and $k=K$.*
(5) *R is Prüfer $\Leftrightarrow$ D is Prüfer and $k=K$.*
(6) *R is Bezout $\Leftrightarrow$ D is Bezout and $k=K$.*
(7) *R is Noetherian $\Leftrightarrow$ V is a DVR, $D=k$, and $[K:k]<\infty$.*
(8) *R is coherent $\Leftrightarrow$ either "$k=K$ and D is coherent" or "M is a finitely generated ideal of R." The latter condition yields $D=k$ and $[K:k]<\infty$.*

In [16], the authors established several results, similar to the statements (1-6) and (8) above, for rings of the form $D+XK[X]$ where $K:=qf(D)$; particularly, $\dim(D+XK[X])=1+\dim(D)$ and $\dim_v(D+XK[X])=1+\dim_v(D)$. The next result handles the general context of $D+XD_S[X]$ rings.

Theorem 2.2. [16] *Let D be an integral domain and S a multiplicative subset of D. Set $R^{(S)}:=D+XD_S[X]$. Then:*
(1) *$R^{(S)}$ is GCD $\Leftrightarrow$ D is GCD and GCD(d,X) exists in $R^{(S)}, \forall\ d\in D^*$.*

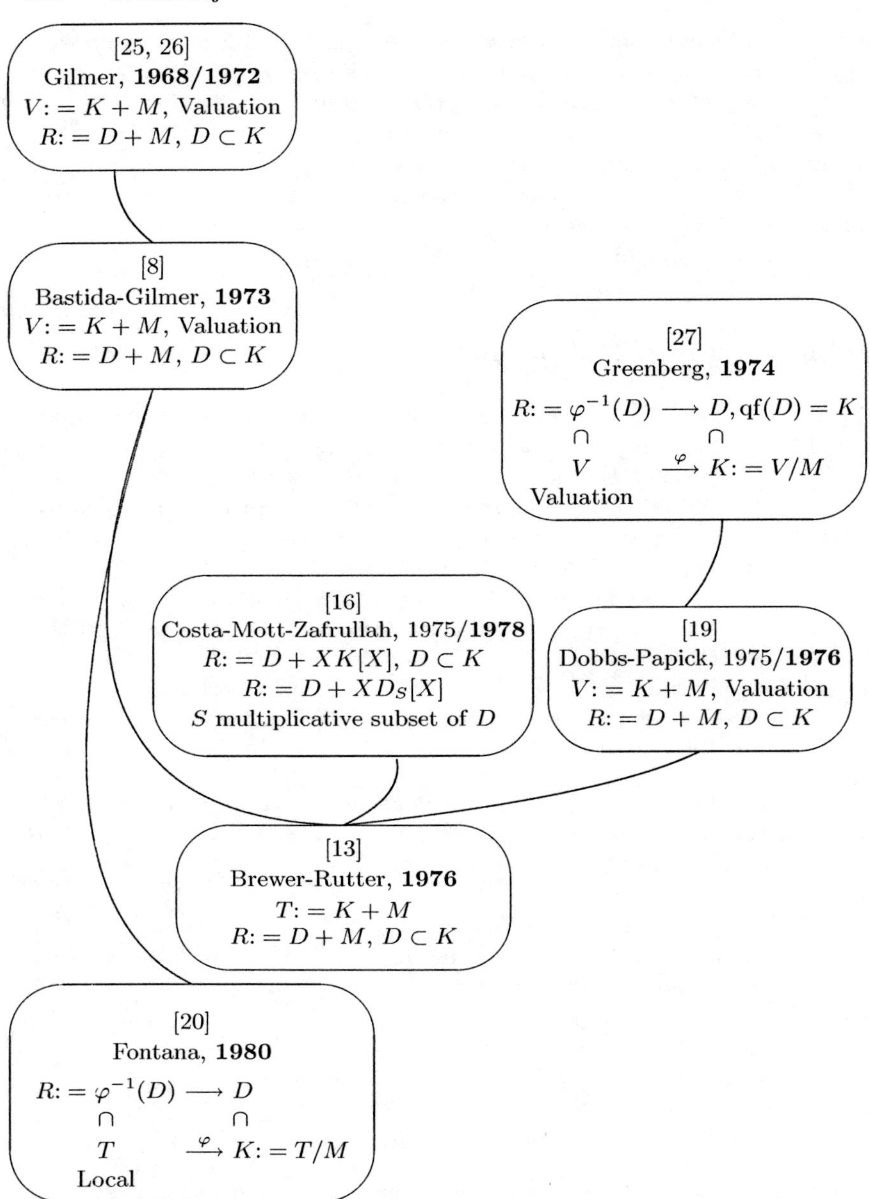

Fig. 2. Diagram of various pullback contexts studied in the 1970s

(2) $\dim(D_S[X]) \leq \dim(R^{(S)}) \leq \dim(D[X])$.
(3) *If D is a valuation domain, then* $\dim(R^{(S)}) = 1 + \dim(D)$. □

in [13], Brewer and Rutter investigated general $D+M$ constructions (i.e., issued from an integral domain not necessarily valuation) and gave unified proofs of most results known on classical $D+M$ and $D+XK[X]$ rings. Their result on the Krull dimension reads as follows:

Theorem 2.3. [13] *Let T be an integral domain of the form $K+M$, where K is a field and M is a maximal ideal of T. Let D be a proper subring of K with $k{:}= \mathrm{qf}(D)$. Set $R{:}= D+M$.*
If $k=K$, then $\dim(R) = \max\{\mathrm{ht}_T(M) + \dim(D), \dim(T)\}$. □

Later, Fontana [20] used topological methods (particularly, his study of amalgamated sums of two spectral spaces) to extend most of these results to pullbacks (issued from local domains). We close this section by citing some basic facts connected with the prime ideal structure of a pullback. These will be used frequently in the sequel without explicit mention. We shall use Spec(R) to denote the set of prime ideals of a ring R.

Theorem 2.4. ([20] and [2, Lemma 2.1]) *Let T be an integral domain, M a maximal ideal of T, K its residue field, $\varphi{:}T \longrightarrow K$ the canonical surjection, D a proper subring of K, and $k{:}= \mathrm{qf}(D)$. Let $R{:}= \varphi^{-1}(D)$ be the pullback issued from the following diagram of canonical homomorphisms:*

$$\begin{array}{ccc} R & \longrightarrow & D \\ \downarrow & & \downarrow \\ T & \xrightarrow{\varphi} & K = T/M \end{array}$$

(1) *$M = (R{:}T)$ and $R/M \cong D$.*
(2) $\mathrm{Spec}(R) \simeq \mathrm{Spec}(D) \coprod_{\mathrm{Spec}(K)} \mathrm{Spec}(T)$ *(i.e., topological amalgamated sum)*
(3) *Assume T is local. Then M is a divided prime and so every prime ideal of R compares with M under inclusion. If, in addition, $k=K$ then $R_M = T$.*
(4) *Assume T is local. Then* $\dim(R) = \dim(T) + \dim(D)$.
(5) *For each prime ideal P of R such that $M \not\subseteq P$, there exists a unique prime ideal Q of T such that $Q \cap R = P$, and hence $T_Q = R_P$.*
(6) *For each prime ideal P of R such that $M \subseteq P$, there exists a unique prime ideal p of D such that $P = \varphi^{-1}(p)$, and hence R_P can be viewed as the pullback of T_M and D_p over K.*
(7) *T is integral over $R \Leftrightarrow D = k$ and K is algebraic over k.* □

3 Dimension Theory

This section studies the Krull dimension and valuative dimension of polynomial rings over various families of pullbacks. It also examines the transfer of the Jaffard property to these constructions.

In 1969, Arnold established a fundamental theorem, [3, Theorem 5], on the dimension of a polynomial ring over an arbitrary integral domain; namely, for any integral domain R with quotient field K and for any positive integer n, $\dim(R[X_1, ..., X_n]) = n + \max\{\dim(R[t_1, ..., t_n]) \mid \{t_i\}_{1 \leq i \leq n} \subseteq K\}$. In [8], Bastida and Gilmer generalized this result to the case where $\{t_i\}_{1 \leq i \leq n}$ is a subset of an extension field of K. It allowed them to establish a formula for the Krull dimension of a polynomial ring over a classical $D + M$ as stated below:

Theorem 3.1. [8, Theorem 5.4] *Let V be a valuation domain of the form $K+M$, where K is a field and M is the maximal ideal of V. Let D be a proper subring of K with $k{:} = \mathrm{qf}(D)$ and let $\mathrm{t.d.}(K{:}\ k)$ denote the transcendence degree of K over k. Let n be a positive integer. Set $R{:} = D + M$. Then:*

$$\dim(R[X_1, ..., X_n]) = \dim(V) + \dim(D[X_1, ..., X_n]) + \min\{n, \mathrm{t.d.}(K{:}\ k)\}. \quad \square$$

In [11], we refined Gilmer's statement on the valuative dimension of a classical $D + M$ in order to build a family of examples of Jaffard domains which are neither Noetherian nor Prüfer domains.

Proposition 3.2. [11, Proposition 2.1] *Under the same notation of Theorem 3.1, we have:*
(1) $\dim_v(R) = \dim_v(D) + \dim(V) + \mathrm{t.d.}(K{:}\ k)$.
(2) *R is a Jaffard domain $\Leftrightarrow$ D is a Jaffard domain and* $\mathrm{t.d.}(K{:}\ k) = 0$. $\square$

From this result stems a first family of Jaffard domains A_n with dimension $n + 3$ which are neither Noetherian nor Prüfer, for every $n \geq 1$. Indeed, the ring $B{:} = \mathbb{Z} + Y\mathbb{Q}(X)[Y]_{(Y)}$ is not a Jaffard domain since $\dim(B) = 2$ and $\dim_v(B) = 3$ by Proposition 3.2. For each $n \geq 1$, set $A_n{:} = B[X_1, ..., X_n]$. For $n = 1$, $A_1 = B[X_1]$ is a 4-dimensional Jaffard domain, since, by Theorem 3.1, $\dim(B[X_1]) = 4 = \dim_v(B) + 1 = \dim_v(B[X_1])$. Clearly, for each $n \geq 2$, A_n is an $(n+3)$-dimensional Jaffard domain. Further, A_1 is not a strong S-domain, otherwise B would be so and hence we would have $5 = \dim(B[X_1, X_2]) = 1 + \dim(B[X_1]) = 2 + \dim(B) = 4$, which is absurd. Consequently, none of the rings A_n is a strong S-domain (hence it is neither Noetherian nor Prüfer), as desired.

We now proceed to explore a general context. Let T be an integral domain, M a maximal ideal of T, K its residue field, $\varphi{:}T \longrightarrow K$ the canonical surjection, D a proper subring of K, and $k{:} = \mathrm{qf}(D)$. Let $R{:} = \varphi^{-1}(D)$ be the pullback issued from the following diagram of canonical homomorphisms:

$$\begin{array}{ccc} R & \longrightarrow & D \\ \downarrow & & \downarrow \\ T & \xrightarrow{\varphi} & K = T/M. \end{array}$$

Theorem 3.3. [2, Theorem 2.6] *Assume T is local. Then:*

(1) $\dim_v(R) = \dim_v(D) + \dim_v(T) + \text{t.d.}(K\colon k)$.
(2) *R is Jaffard $\Leftrightarrow$ D and T are Jaffard and* $\text{t.d.}(K\colon k) = 0$. □

The next result generalizes Theorem 2.1(1), Theorem 2.4(4), and Theorem 3.3.

Theorem 3.4. [2, Theorem 2.11 and Corollary 2.12] *Assume T is an arbitrary domain (i.e., not necessarily local). Then:*

(1) $\dim(R) = \max\{\dim(T), \dim(D) + \text{ht}_T(M)\}$.
(2) $\dim_v(R) = \max\{\dim_v(T), \dim_v(D) + \dim_v(T_M) + \text{t.d.}(K\colon k)\}$.
(3) *R is locally Jaffard $\Leftrightarrow$ D and T are locally Jaffard and* $\text{t.d.}(K\colon k) = 0$.
(4) *If T is locally Jaffard with* $\dim_v(T) < \infty$, *D is Jaffard, and* $\text{t.d.}(K\colon k) = 0$, *then R is a Jaffard domain.* □

There are examples which show that none of the hypotheses in Theorem 3.4(4) is a necessary condition for R to be Jaffard. Indeed, let V and W be two incomparable valuation domains of a suitable field K with $n\colon = \dim(V) \geq 3$ and $\dim(W) = 1$. By [34, Theorem 11.11], $T\colon = V \cap W$ is an n-dimensional Prüfer domain with two maximal ideals, say M_1 and M, $T_{M_1} = V$, and $T_M = W$. Let $\varphi\colon T \longrightarrow T/M \cong K$ be the canonical surjection. We further require that K has a subfield k and a subring D such that $\dim(D) = \dim_v(D) = 1$, $\text{qf}(D) = k$, and $\text{t.d.}(K\colon k) = 1$. Set $R\colon = \varphi^{-1}(D)$. By Theorem 3.4(1) & (2), $\dim(R) = \dim_v(R) = n$. So that R is Jaffard though K is not algebraic over k. Now, alter the above construction by taking $n \geq 4$ and $\dim_v(D) = 2$, so that D is not Jaffard anymore, but one can easily check that R is Jaffard.

Next we proceed to the construction of the first example of a Jaffard domain which is not locally Jaffard.

Example 3.5. [2, Example 3.2] Let k be a field and X_1, X_2, Y indeterminates over k. Set $V_1\colon = k(X_1, X_2)[Y]_{(Y)} = k(X_1, X_2) + M_1$ and $A\colon = k(X_1) + M_1$, where $M_1 = YV_1$. Let (V, M) be a one-dimensional valuation domain of the form $V = k(Y) + M$ such that $k(Y)[X_1, X_2] \subset V \subset k(X_1, X_2, Y)$ (In order to build such a ring, consider the valuation $v\colon k(Y)[X_1, X_2] \longrightarrow \mathbb{Z}^2$ defined by $v(X_1) = (1, 0)$ and $v(X_2) = (0, 1)$, where $\mathbb{Z}^2$ is endowed with the order induced by the group isomorphism $i\colon \mathbb{Z}^2 \longrightarrow \mathbb{Z}[\sqrt{2}]$ defined by $i(a, b) = a + b\sqrt{2}$). Consider the two-dimensional valuation ring $V_2\colon = k[Y]_{(Y)} + M = k + M_2$ with maximal ideal $M_2 = Yk[Y]_{(Y)} + M$. One can easily check that V_1 and V_2 are incomparable. By [34, Theorem 11.11], $B\colon = V_1 \cap V_2$ is a 2-dimensional Prüfer domain with two maximal ideals, say N_1 and N_2, $B_{N_1} = V_1$, and $B_{N_2} = V_2$. Finally, put $R\colon = A \cap V_2$. One can show that R is semi-local with two maximal ideals $\mathcal{M}_1 = N_1 \cap R$ and $\mathcal{M}_2 = N_2 \cap R$ with $R_{\mathcal{M}_1} = A$ and $R_{\mathcal{M}_2} = V_2$ (Cf. [17, Example 2.5]). Via Theorem 3.4, we obtain $\dim(R) = \max\{\dim(R_{\mathcal{M}_1}), \dim(R_{\mathcal{M}_2})\} = 2$ and $\dim_v(R) = \max\{\dim_v(R_{\mathcal{M}_1}), \dim_v(R_{\mathcal{M}_2})\} = 2$. Thus R is Jaffard but not locally Jaffard, since $\dim(R_{\mathcal{M}_1}) = \dim(A) = 1 \neq \dim_v(R_{\mathcal{M}_1}) = \dim_v(A) = 2$. □

The next result examines the possibility of extending Bastida-Gilmer's result (Theorem 3.1) on the classical $D+M$ ring to a general context.

Theorem 3.6. [2, Proposition 2.3 and Proposition 2.7] *Under the same notation as above, the following statements hold.*

(1) *Assume* $k = K$. *Then:* $\dim(R[X_1, ..., X_n]) = \dim(D[X_1, ..., X_n]) + \dim(T[X_1, ..., X_n]) - \dim(K[X_1, ..., X_n])$, *for each positive integer* n.
(2) *Assume* $D = k$ *and set* $d := \text{t.d.}(K : k)$. *Then, for each* $n \geq 0$, *we have:* $n + \dim(T) + \min\{n, d\} \leq \dim(R[X_1, ..., X_n]) \leq n + \dim_v(T) + d$. □

Now, one should design an example to show that the above can be strict.

Example 3.7 ([2, Example 3.9]). Let Y_1, Y_2, U, V, Z, W be indeterminates over a field k. Define $K := k(Y_1, Y_2)$, $S := K(U)[V]_{(V)}$, $R_1 := K(U, V, Z)[W]_{(W)}$, $A := K(U, V) + WR_1$, $B := K + VS$, $R_2 := S + WR_1$, and $T := K + VS + WR_1$. Thus, we have the following pullbacks (with canonical homomorphisms):

$$\begin{array}{ccccc}
T & \longrightarrow & B & \longrightarrow & K \\
\downarrow & & \downarrow & & \downarrow \\
R_2 & \longrightarrow & S & \longrightarrow & K(U) \\
\downarrow & & \downarrow & & \\
A & \longrightarrow & K(U,V) & & \\
\downarrow & & \downarrow & & \\
R_1 & \longrightarrow & K(U,V,Z) & &
\end{array}$$

R_1 and S are discrete valuation rings. Further, by applying Theorem 2.4(4) and Theorem 3.3, we obtain:

$\dim(A) = 1$; $\dim_v(A) = 2$
$\dim(R_2) = \dim(S) + \dim(R_1) = 2$; $\dim_v(R_2) = 3$
$\dim(B) = 1$; $\dim_v(B) = 2$
$\dim(T) = \dim(k) + \dim(R_2) = 2$; $\dim_v(T) = 4$.

Let $\varphi : T \longrightarrow K$ be the canonical surjection and $R := \varphi^{-1}(k)$. The pullback R has Krull dimension 2 and valuative dimension 6. Further, $\dim(R[X]) = 5$ by [21, Theorem 2.1]. Set $d := \text{t.d.}(K : k) = 2$. The desired strict inequalities follow: $1 + \dim(T) + \min\{1, d\} \lneqq \dim(R[X]) \lneqq 1 + \dim_v(T) + d$. □

Next, we explore Costa-Mott-Zafrullah's $D + XD_S[X]$ construction under a slight generalization. Let D be a domain, S a multiplicative subset of D, and r an integer ≥ 1. Put $R^{(S,r)} := D + (X_1, ..., X_r)D_S[X_1, ..., X_r]$. Let $p \in \text{Spec}(D)$. The S-coheight of p, denoted $S\text{-coht}(p)$, is defined as the supremum of the lengths of all chains $p \subset p_1 \subset p_2 \subset ... \subset p_n$ of prime ideals of D with $p_1 \cap S \neq \emptyset$. Set $S\text{-dim}(D) := \max\{S\text{-coht}(p) \mid p \in \text{Spec}(D)\}$.

Theorem 3.8. ([16] and [24]) *Under the above notation, the following statements hold.*

(1) $\max\{\dim(D_S[X_1, ..., X_r]), r + \dim(D)\} \leq \dim(R^{(S,r)})$

$$\leq \min\{\dim(D[X_1, ..., X_r]), \dim(D_S[X_1, ..., X_r]) + S\text{-}\dim(D)\}.$$

(2) $\dim_v(R^{(S,r)}) = r + \dim_v(D)$.

(3) *D is Jaffard $\Leftrightarrow$ $R^{(S,r)}$ is Jaffard and* $\dim(R^{(S,r)}) = r + \dim(D)$.

(4) *$R^{(S,r)}$ is Jaffard $\Leftrightarrow$ so is $D[X_1, ..., X_r]$ with the same dimension as $R^{(S,r)}$.*

□

Now, we provide an example to show that the Jaffard property of $R^{(S,r)}$ does not force D to be Jaffard. Here too we appeal to pullbacks. Let k be a field and X, Y two indeterminates over k. Put $V := k(X) + Yk(X)[Y]_{(Y)}$ and $D := k + Yk(X)[Y]_{(Y)}$. Clearly, D is a local domain with maximal ideal $M := Yk(X)[Y]_{(Y)}$, $\dim(D) = 1$, and $\dim_v(D) = 2$ by Theorem 2.1(1) and Proposition 3.2. Set $S := D \setminus M$ and $R^{(S,1)} := D + XD_S[X]$. So $R^{(S,1)} \cong D[X]$ since $D_M \cong D$. It follows that $\dim(R^{(S,1)}) = \dim(D[X]) = 1 + \dim_v(D) = 3 = \dim_v(R^{(S,1)})$, as desired.

Next we move to a general context. Let $A \subseteq B$ an extension of integral domains and X an indeterminate over B. Put $R := A + XB[X] = \{f \in B[X] \mid f(0) \in A\}$. This construction was introduced by D.D. Anderson-D.F. Anderson-Zafrullah in [1]. Also, R is a particular case of the constructions B, I, D introduced by P.-J. Cahen [15]. Also, $\text{Int}(A) \cap B[X] = \{f \in B[X] \mid f(A) \subseteq A\}$ is a subring of R and hence a deeper knowledge of $A + XB[X]$ constructions may have some interesting impact on the integer-valued polynomial rings.

As a consequence of some general properties of the spectrum of a pullback [20], we state the following: First, $XB[X]$ is a prime ideal of $R := A + XB[X]$ with $R/XB[X] \cong A$ and hence we have an order-isomorphism $\text{Spec}(A) \longrightarrow \{P \in \text{Spec}(R) \mid XB[X] \subseteq P\}$, $p \longmapsto p + XB[X]$. Second, $S := \{X^n \mid n \geq 0\}$ is a multiplicatively closed subset of R and $B[X]$ with $S^{-1}R = S^{-1}B[X] = B[X, X^{-1}]$; by contraction, we obtain an order-isomorphism $\{Q \in \text{Spec}(B[X]) \mid X \notin Q\} \longrightarrow \{P \in \text{Spec}(R) \mid X \notin P\}$. Finally, the spectral space $\text{Spec}(R)$ is canonically homeomorphic to the amalgamated sum of $\text{Spec}(A)$ and $\text{Spec}(B[X])$ over $\text{Spec}(B)$.

For the subfamilies $D + XK[X]$ and $D + XD_S[X]$, it is known that $\text{ht}(XK[X]) = \text{ht}(XD_S[X]) = 1$. The next result probes the situation of $XB[X]$ inside $\text{Spec}(R)$.

Theorem 3.9. [22, Theorem 1.2] *Let $R := A + XB[X]$ and $N := A \setminus \{0\}$. Then:*

(1) $\text{ht}_R(XB[X]) = \dim(N^{-1}B[X]) = \dim(B[X] \otimes_A \text{qf}(A))$.

(2) $1 \leq \text{ht}_R(XB[X]) \leq 1 + \text{t.d.}(B : A)$. □

Thus, if $\text{qf}(A) \subseteq B$, then $\text{ht}_R(XB[X]) = \dim(B[X])$; and if $A \subseteq B$ is an algebraic extension, then $\text{ht}_R(XB[X]) = 1$. In general, $\text{ht}_R(XB[X])$ can describe all integers between 1 and $1 + \text{t.d.}(B : A)$, as shown by the following example: Let d be an integer, $t \in \{1, ..., d+1\}$, K a field, and $X, X_1, ..., X_{d+1}, Y_1, ..., Y_d$ indeterminates over K. Set $A := K$ and $B :=$

$K(X_1,...,X_{d-t+1})[Y_1,...,Y_{t-1}]$. Hence $\text{t.d.}(B\colon A) = d$ and $\text{ht}_R(XB[X]) = \dim(B[X]) = t$.

The next result studies the Krull and valuative dimensions as well as the transfer of the Jaffard property.

Theorem 3.10. [22, Theorems 2.1 & 2.3] *Let* $R\colon = A + XB[X]$ *and set* $k\colon = \text{qf}(A)$ *and* $d\colon = \text{t.d.}(B\colon A)$. *Then:*
(1) $\max\{\dim(A) + \text{ht}_R(XB[X]), \dim(B[X])\} \leq \dim(R)$
$\leq \dim(A) + \dim(B[X])$.
(2) *If* $k \subseteq B$, *then* $\dim(R) = \dim(A) + \dim(B[X])$.
(3) $\dim_v(R) = \dim_v(A) + d + 1$.
(4) *R is Jaffard and* $\dim(R) = \dim(A) + 1 \Leftrightarrow$ *A is Jaffard and* $d = 0$.
(5) *If* $k \subseteq B$, *then: R is Jaffard $\Leftrightarrow$ so is A and* $\dim(B[X]) = 1 + d$. □

Now, one can easily construct new classes of Jaffard domains. For instance, $\mathbb{R} + X\mathbb{C}[X,Y]$ and $\mathbb{Z} + X\overline{\mathbb{Z}}[X]$ both are 2-dimensional Jaffard domains, where $\overline{\mathbb{Z}}$ denotes the integral closure of $\mathbb{Z}$ inside an algebraic extension of $\mathbb{Q}$.

The next result handles the locally Jaffard property.

Theorem 3.11. [22, Theorem 2.8] *Let* $R\colon = A + XB[X]$ *and suppose that A is a locally Jaffard domain. Then R is locally Jaffard $\Leftrightarrow$ $B[X]$ is locally Jaffard and* $\text{ht}_R(XB[X]) = 1 + \text{t.d.}(B\colon A)$. □

We cannot knock down the hypothesis "A is locally Jaffard" to "A is Jaffard." For, assume A is Jaffard but not locally Jaffard (Example 3.5). Set $B\colon = \text{qf}(A)$ and $R\colon = A + XB[X] = A + X\text{qf}(A)[X]$. In this situation $B[X]$ is locally Jaffard and $\text{ht}_R(XB[X]) = 1 = 1 + \text{t.d.}(B\colon A)$; whereas, R is not locally Jaffard by Theorem 3.4(3). Notice, however, that the hypothesis "A is locally Jaffard" is not necessary as shown below.

While several results concerning $D + XK[X]$ and $D + XD_S[X]$ are recovered, some known results on these rings do not carry over to the general context of $A+XB[X]$ constructions. Next, an example provides some of these pathologies and, also, shows that the double inequality established in Theorem 3.10(1) can be strict.

Example 3.12. [22, Example 3.1] Let K be a field and let X, X_1, X_2, X_3, X_4 be indeterminates over K. Set:

$$\begin{array}{lll} L := K(X_1, X_2, X_3) & ; & V_1 := k + N \\ k := K(X_1, X_2) & ; & D := K(X_1)[X_2]_{(X_2)} + N \\ M := X_4 L[X_4]_{(X_4)} & ; & A := K[X_1]_{(X_1)} + M \\ N := X_3 k[X_3]_{(X_3)} & ; & B := D + M \\ V := L + M & ; & R := A + XB[X] \end{array}$$

Then:
(1) $\max\{\dim(A)+\text{ht}_R(XB[X]), \dim(B[X])\} \lneqq \dim(R) \lneqq \dim(A)+\dim(B[X])$.
(2) $\dim(A[X]) \lneqq \dim(R)$ (in contrast with Theorem 3.8(1)).
(3) R is Jaffard and $A[X]$ is not Jaffard (in contrast with Theorem 3.8(4)).

(4) R is locally Jaffard and A is not locally Jaffard (in contrast with Theorem 3.4(3) applied to $D + XK[X]$).

Indeed, by Theorems 2.1 & 3.1 & 3.3, V, V_1, D, and B are valuation domains of dimensions 1, 1, 2, and 3, respectively; moreover, we have:

- $\dim(B[X]) = \dim(B) + 1 = 4$,
- $\dim(A) = \dim(K[X_1]_{(X_1)}) + \dim(V) = 2$,
- $\dim_v(A) = \dim_v(K[X_1]_{(X_1)}) + \dim(V) + \text{t.d.}(L\colon K(X_1)) = 4$,
- $\dim(A[X]) = \dim(K[X_1]_{(X_1)}[X]) + \dim(V) + \min\{1, \text{t.d.}(L\colon K(X_1))\} = 4$,
- $\text{Spec}(B) = \{(0), M, P_1\colon= N + M, P_2\colon= X_2K(X_1)[X_2]_{(X_2)} + P_1\}$,
- $\text{Spec}(A) = \{(0), M, Q\colon= X_1K[X_1]_{(X_1)} + M\}$,
- $M \cap A = P_1 \cap A = P_2 \cap A = M$.

Notice first that $\text{qf}(A) = \text{qf}(B) = \text{qf}(V)$. Now, inside $\text{Spec}(R)$ we have the following chain of prime ideals (in view of the discussion in the paragraph right before Theorem 3.9):

$$(0) \subsetneqq M[X] \cap R \subsetneqq P_1[X] \cap R \subsetneqq P_2[X] \cap R \subsetneqq M + XB[X] \subsetneqq Q + XB[X].$$

Therefore $\dim(R) \geq 5$, and hence R is a 5-dimensional Jaffard domain since $\dim_v(R) = \dim_v(A) + \text{t.d.}(B\colon A) + 1 = 5$ by Theorem 3.10. Consequently, (1) and (2) hold, and so does (3) since $\dim_v(A[X]) = \dim_v(A) + 1 = 5$. It remains to deal with (4). The domain A is not locally Jaffard (since it is not Jaffard). Let $P \in \text{Spec}(R)$ with $X \notin P$. Then $R_P = B[X, X^{-1}]_{PB[X,X^{-1}]}$ is a universally strong S-domain (Cf. [10, 32]) and hence Jaffard (since B is a valuation domain). So, in order to show that R is locally Jaffard, it suffices to consider the localizations with respect to the prime ideals that contain X. Let $P\colon= p + XB[X] \in \text{Spec}(R)$ with $p \in \text{Spec}(A)$. One can check that $R_P = A_p + XB[X]_P$ and thus $A_p + XB_p[X] \subseteq R_P \subseteq A_p + XL[X]_{(X)}$. We obtain, via Theorems 3.3 & 3.10, that $\dim_v(R_P) = \dim_v(A_p + XB_p[X]) = \dim_v(A_p + XL[X]_{(X)}) = \dim_v(A_p) + \text{t.d.}(B\colon A) + 1 = \dim_v(A_p) + 1$. We claim that R_P is Jaffard for all $p \in \text{Spec}(A)$:

- Let $p\colon= (0)$. Then $\dim(R_P) = \text{ht}_R(XB[X]) = 1 = \dim_v(A_{(0)}) + 1$.
- Let $p\colon= M$. Then the above maximal chain yields $\text{ht}(P) = 4$. Hence $\dim(R_P) = 4 = \dim_v(K(X_1)) + \dim(V) + \text{t.d.}(L\colon K(X_1)) + 1 = \dim_v(A_M) + 1$. Here we view A_M as a pullback of V and $K(X_1)$ over L.
- Let $p\colon= Q$. Then $\dim(R_P) = 5 = \dim_v(A) + 1 = \dim_v(A_Q) + 1 = \dim_v(R_P)$ (since $A_Q = A$). □

Next we move to a more general context. let T be a domain, I an nonzero ideal of T, and D a subring of T such that $D \cap I = (0)$. Throughout, D will be identified with its image in T/I. Also $\text{ht}_T(I)$ will be assumed to be finite (though it's not always indispensable). Let $R\colon= D + I$; it is a pullback determined by the following diagram of canonical homomorphisms:

$$\begin{array}{ccc} R: = D + I & \longrightarrow & D \\ \downarrow & & \downarrow \\ T & \longrightarrow & T/I. \end{array}$$

So Spec(R) is canonically homeomorphic to the amalgamated sum of Spec(D) and Spec(T) over Spec(T/I). Precisely, I is a prime ideal of R and we have the order isomorphisms: Spec(D) $\longrightarrow \{P \in \text{Spec}(R) \mid I \subseteq P\}$, $p \longmapsto p + I$; and $\{Q \in \text{Spec}(T) \mid I \nsubseteq Q\} \longrightarrow \{P \in \text{Spec}(R) \mid I \nsubseteq P\}$, $Q \longmapsto Q \cap R$.

This construction was introduced and developed by Cahen [14, 15]. Since its study has proven to be difficult in its generality, the scope was mainly limited to the so-called $(T = B, I, D)$ almost-simple constructions (i.e., every ideal of T containing I is maximal). The following results -due to Cahen- approximate $\text{ht}_R(I)$ and $\dim(R)$ with respect to $\text{ht}_T(I)$, $\dim(D)$, and $\dim(T)$ in the general context.

Theorem 3.13. [14, Proposition 5, Théorème 1, and Corollaire 1]
(1) $\text{ht}_T(I) \leq \text{ht}_R(I) \leq \dim(T)$.
(2) $\dim(D) + \text{ht}_R(I) \leq \dim(R) \leq \dim(D) + \dim(T)$.
(3) $\dim(R) \geq \max\{\text{ht}_T(Q) + \dim(R/Q \cap R) \mid Q \in \text{Spec}(T), I \subseteq Q\}$. □

Later, Ayache devoted his paper [7] to the special case where T is either a finitely generated K-algebra or a quotient of a power series ring in a finite number of indeterminates. He established the following results:

Theorem 3.14. [7] *Let K be a field, T a finitely generated K-algebra or a quotient of a power series ring in a finite number of indeterminates, I a proper non-zero ideal of T, D a subring of K with $k: = \text{qf}(D)$, and $R: = D+I$. Then:*
(1) $\dim(R) = \dim(D) + \dim(T)$.
(2) *Assume either T is a finitely generated K-algebra or* $\text{ht}_T(I) = \dim(T)$. *Then:* $\dim_v(R) = \dim_v(D) + \dim_v(T) + \text{t.d.}(K: k)$, *and hence R is Jaffard if and only if D is Jaffard and* $\text{t.d.}(K: k) = 0$. □

We return to the general context. The next result shades more light on I within the spectrum of R.

Lemma 3.15. [23, Lemme 1.2] *Set $\mathcal{X}: = \{Q \in \text{Spec}(T) \mid Q \cap R = I\}$ and $\mathcal{Y}: = \{Q \in \text{Spec}(T) \mid I \nsubseteq Q, \exists\, Q' \in \mathcal{X}, (0) \subset Q \subset Q'\}$. Then:*
(1) $\mathcal{X} \neq \emptyset$.
(2) $\mathcal{Y} = \emptyset$ *if and only if* $\text{ht}_R(I) = 1$.
(3) $\text{ht}_R(I) = 1 + \max\{\text{ht}_T(Q) \mid Q \in \mathcal{Y}\}$.
(4) *If* $\text{ht}_{R[X]}(I[X]) = 1$, *then* $\text{t.d.}(T/Q: D) = 0, \forall\, Q \in \mathcal{X}$. □

Next we show how the S-domain property is reflected on $\text{ht}_R(I)$.

Theorem 3.16. [23, Théorème 1.3] *Assume T is an S-domain. Then R is an S-domain if and only if* $\text{ht}_R(I) > 1$ *or* $\text{t.d.}(\frac{T}{Q}: D) = 0$, $\forall\, Q \in \text{Spec}(T)$ *such that $Q \cap R = I$.* □

In the special case where $T := V$ is a valuation domain, one can easily check that $\mathrm{ht}_R(I) = \mathrm{ht}_V(I)$ and $\dim(R) = \dim(D) + \mathrm{ht}_V(I)$. Moreover, we have the following:

Theorem 3.17. [23, Théorème 1.13] *Let V be a valuation domain, I an nonzero ideal of V, D a subring of V with $D \cap I = (0)$, and $R := D + I$. Let P_0 denote the prime ideal of V that is minimal over I and let n be a positive integer. Then:*
(1) $\dim_v(R) = \dim_v(D) + \dim_v(V_{P_0}) + \mathrm{t.d.}(\frac{V}{P_0} : D)$.
(2) $\dim(R[X_1, ..., X_n]) = \dim(V_{P_0}) + \dim(D[X_1, ..., X_n]) + \min\{n, \mathrm{t.d.}(\frac{V}{P_0} : D)\}$.
(3) *R is a Jaffard domain $\Leftrightarrow$ D is a Jaffard domain and* $\mathrm{t.d.}(\frac{V}{P_0} : D) = 0$. □

Another special case is when the $D+I$ ring arises from a polynomial ring. Namely, let B be a domain, X an indeterminate over B, D a subring of B, and I an ideal of $B[X]$ with $I \cap B = 0$. Put $R := D+I$. We have the following pullbacks (with canonical homomorphisms):

$$\begin{array}{ccc} R := D + I & \longrightarrow & D \\ \downarrow & & \downarrow \\ B + I & \longrightarrow & B \\ \downarrow & & \downarrow \\ B[X] & \longrightarrow & B[X]/I. \end{array}$$

Theorem 3.18. [23, Théorème 2.1] *Under the above notation, set* $d := \mathrm{t.d.}(B : D)$. *We have:*
(1) $\dim_v(R) = \dim_v(D) + d + 1$.
(2) *R is Jaffard and* $\dim(R) = \dim(D) + 1 \Leftrightarrow$ *D is Jaffard and $d = 0$.* □

The above result applies to the particular context of $A + X^n B[X]$ constructions. Specifically, Let $A \subseteq B$ an extension of integral domains, X an indeterminate over B, and n an integer ≥ 1. Put $R_n := A + X^n B[X]$. Then $\dim_v(R_n) = \dim_v(A) + \mathrm{t.d.}(B : A) + 1$; and R_n is Jaffard and $\dim(R_n) = \dim(A) + 1$ if and only if A is Jaffard and $\mathrm{t.d.}(B : A) = 0$. Here the effect of the S-property appears as follows: R_n is an S-domain if and only if $\mathrm{ht}_{R_1}(XB[X]) > 1$ or $\mathrm{t.d.}(B : A) = 0$. (Since B[X] is always an S-domain.)

In this vein, the ring $R := \mathbb{Z}[(XY^i)_{i \geq 0}] = \mathbb{Z} + X\mathbb{Z}[X, Y]$ was shown by Ayache in [7] to be a 3-dimensional totally Jaffard domain [15]. In [23], we improved this result by stating that $R_n := \mathbb{Z}[(X^n Y^i)_{i \geq 0}] = \mathbb{Z} + X^n \mathbb{Z}[X, Y]$ is a universally strong S-domain, for each integer $n \geq 1$.

References

1. Anderson, D.D., Anderson, D.F., Zafrullah, M.: Rings between $D[X]$ and $K[X]$. Houston J. Math., **17**, 109–129 (1991)
2. Anderson, D.F., Bouvier, A., Dobbs, D.E., Fontana, M., Kabbaj, S.: On Jaffard domains. Exposition. Math., **6**, 145–175 (1988)

3. Arnold, J.T.: On the dimension theory of overrings of an integral domain. Trans. Amer. Math. Soc., **138**, 313–326 (1969)
4. Arnold, J.T., Gilmer, R.: Dimension sequences for commutative rings. Bull. Amer. Math. Soc., **79**, 407–409 (1973)
5. Arnold, J.T., Gilmer, R.: The dimension sequence of a commutative ring. Amer. J. Math., **96**, 385–408 (1974)
6. Arnold, J.T., Gilmer, R.: Two questions concerning dimension sequences. Arch. Math. (Basel), **29**, 497–503 (1977)
7. Ayache, A.: Sous-anneaux de la forme $D+I$ d'une K-algèbre intègre. Portugal. Math., **50**, 139–149 (1993)
8. Bastida, E., Gilmer, R.: Overrings and divisorial ideals of rings of the form $D+M$. Michigan Math. J., **20**, 79–95 (1973)
9. Bourbaki, N.: Commutative Algebra, Chapters 1-7. Springer-Verlag, Berlin (1998)
10. Bouvier, A., Dobbs, D.E., Fontana, M.: Universally catenarian integral domains. Advances in Math., **72**, 211–238 (1988)
11. Bouvier, A., Kabbaj, S.: Examples of Jaffard domains. J. Pure Appl. Algebra, **54**, 155–165 (1988)
12. Brewer, J.W., Montgomery, P.R., Rutter, E.A., Heinzer, W.J.: Krull dimension of polynomial rings. Lecture Notes in Math., Springer, **311**, 26–45 (1973)
13. Brewer, J. W., Rutter, E. A.: $D+M$ constructions with general overrings. Michigan Math. J., **23**, 33–42 (1976)
14. Cahen, P.-J.: Couples d'anneaux partageant un idéal. Arch. Math. (Basel), **51**, 505–514 (1988)
15. Cahen, P.-J.: Construction B, I, D et anneaux localement ou résiduellement de Jaffard. Arch. Math. (Basel), **54**, 125–141 (1990)
16. Costa, D., Mott, J.L., Zafrullah, M.: The construction $D+XD_S[X]$. J. Algebra, **53**, 423–439 (1978)
17. Dobbs, D.E., Fontana, M.: Locally pseudovaluation domains. Ann. Mat. Pura Appl., **134**, 147–168 (1983)
18. Dobbs, D.E., Fontana, M., Kabbaj, S.: Direct limits of Jaffard domains and S-domains. Comment. Math. Univ. St. Paul., **39**, 143–155 (1990).
19. Dobbs, D.E., Papick, I.J.: When is $D+M$ coherent?. Proc. Amer. Math. Soc., **56**, 51–54 (1976)
20. Fontana, M: Topologically defined classes of commutative rings. Ann. Mat. Pura Appl., **123**, 331–355 (1980)
21. Fontana, M: Sur quelques classes d'anneaux divisés. Rend. Sem. Mat. Fis. Milano, **51**, 179–200 (1981)
22. Fontana, M., Izelgue, L., Kabbaj, S.: Krull and valuative dimension of the rings of the form A+XB[X]. Lect. Notes Pure Appl. Math., Dekker, **153**, 111–130 (1993)
23. Fontana, M., Izelgue, L., Kabbaj, S.: Sur quelques propriétés des sous-anneaux de la forme $D+I$ d'un anneau intègre. Comm. Algebra, **23**, 4189–4210 (1995)
24. Fontana, M., Kabbaj, S.: On the Krull and valuative dimension of $D+XD_S[X]$ domains. J. Pure Appl. Algebra, **63**, 231–245 (1990)
25. Gilmer, R.: Multiplicative Ideal Theory. Queen's Papers in Pure and Applied Mathematics, No. 12. Queen's University, Kingston, Ontario (1968)
26. Gilmer, R.: Multiplicative Ideal Theory. Pure and Applied Mathematics, No. 12. Marcel Dekker, Inc., New York (1972)

27. Greenberg, B.: Global dimension of cartesian squares. J. Algebra, **32**, 31–43 (1974)
28. Jaffard, P.: Théorie de la Dimension dans les Anneaux de Polynômes. Mém. Sc. Math. 146, Gauthier-Villars, Paris (1960)
29. Kabbaj, S.: La formule de la dimension pour les S-domaines forts universels, Boll. Un. Mat. Ital. D (6), **5**, 145–161 (1986)
30. Kabbaj, S.: Sur les S-domaines forts de Kaplansky. J. Algebra, **137**, 400–415 (1991)
31. Kaplansky, I.: Commutative Rings. The University of Chicago Press, Chicago (1974)
32. Malik, S., Mott, J.L.: Strong S-domains. J. Pure Appl. Algebra, **28**, 249–264 (1983)
33. Matsumura, H.: Commutative Ring Theory. Cambridge University Press, Cambridge (1989)
34. Nagata, M.: Local Rings. Interscience, New York (1962)
35. Seidenberg, A.: A note on the dimension theory of rings. Pacific J. Math., **3**, 505–512 (1953)
36. Seidenberg, A.: On the dimension theory of rings. II. Pacific J. Math., **4**, 603–614 (1954)

Almost Dedekind domains which are not Dedekind

K. Alan Loper

Department of Mathematics, Ohio State University-Newark, Newark, Ohio 43055 USA lopera@math.ohio-state.edu

1 Introduction

A domain D is a Prüfer domain if D_M is a valuation domain for each maximal ideal M of D. The Noetherian Prüfer domains are the Dedekind domains. It follows that if D is a Dedekind domain then D_M is a Noetherian valuation domain for each maximal ideal M of D. A domain D is almost Dedekind if D_M is a Noetherian valuation domain for each maximal ideal M of D. Clearly, Dedekind domains are almost Dedekind. The point of the designation is that almost Dedekind domains satisfy the characterization given above of Dedekind domains, except that they are not assumed to be Noetherian.

Dedekind domains have been very extensively studied and are very well understood. The class of almost Dedekind domains, as the name suggests, share many of the strong properties of Dedekind domains that make them an attractive object of study. There are many ways in which a nonNoetherian almost Dedekind domain is one of the simplest nonNoetherian rings one could choose to study, especially if the desired ring is to be finite dimensional.

[10] contains a section on almost Dedekind domains. We give below a result from [10] which demonstrates some of the elegant properties of this class of domains.

Theorem 21 [10, Theorems 36.4 and 36.5] *Let D be a domain. The following are equivalent.*

(1) *D is an almost Dedekind domain.*
(2) *D is one-dimensional and every primary ideal of D is a prime power.*
(3) *If I, J, H are nonzero ideals of D and $IH = JH$ then $I = J$. We say in this case that D satisfies the ideal cancellation law.*

[10] contains very little, however, in terms of discussion of application of the theory of almost Dedekind domains or of methods for constructing nontrivial examples. Indeed, many of the results along these lines have appeared since

the 1972 publication of [10]. The goal of this paper then is to give an account of the study of this beautiful class of rings.

In the second section we discuss some of the motivation for studying and applications for almost Dedekind domains. In the third section we discuss methods for constructing nontrivial (i.e. nonNoetherian) almost Dedekind domains.

2 Motivation and application

The first example of an almost Dedekind domain which was not Dedekind was given by Nakano in [20]. Nakano's focus was properties of the rings of algebraic integers in infinite degree extensions of the rational numbers. Such domains are always one-dimensional Prüfer domains, but they are not generally almost Dedekind. In this setting he states our characterization of an almost Dedekind domain and proves that the ring of integers in the field obtained by adjoining the pth root of unity for all primes p satisfies this characterization. I have heard plausible speculation that Nakano could actually have accomplished much more in terms of construction and properties of almost Dedekind domains but saw no need. He seems to have been content to simply demonstrate that an example of such a domain existed.

Gilmer [7] then gave an abstract definition of an almost Dedekind domain apart from rings of integers and coined the term *almost Dedekind.* The goal of Gilmer's work is to demonstrate that various properties which characterize Dedekind domains amongst the class of Noetherian domains also hold true of almost Dedekind domains.

Before we press on to more substantial results, we make some simple observations about almost Dedekind domains which make them an attractive object of study.

- An almost Dedekind domain which is not Noetherian provides an example of a domain which is locally Noetherian but not Noetherian.
- Arnold [1] says that an ideal I of a domain D satisfies the SFT-property provided there exists a finitely generated ideal J and a positive integer n such that $J \subseteq I$ and for any $d \in I$ we have $d^n \in J$. A domain satisfies the SFT-property provided each ideal is an SFT-ideal. He then proves that if D is a finite dimensional domain which is not SFT then the ring $D[[x]]$ of formal power series over D is infinite dimensional. The relevance here is that almost Dedekind domains which are not Dedekind provide especially nice examples of one-dimensional domains which are not SFT. (Note: they are one of the two types of examples given by Arnold.)

2.1 Integer-valued polynomials

The most prominent use for almost Dedekind domains has been in the area of integer-valued polynomials. For a domain D with quotient field K the ring of integer-valued polynomials $Int(D)$ is defined by $Int(D) = \{f(x) \in K[x] \mid f(D) \subseteq D\}$. This ring was first considered in separate papers, both published in 1919 by Polya and Ostrowski. It achieved significant prominence though in the 1970's. Brizolis proved [2] that $Int(Z)$ is a two dimensional Prüfer domain. Subsequently, Chabert [5] and McQuillan [19] each proved that if D is a Noetherian domain then $Int(D)$ is a Prüfer domain if and only if D is a Dedekind domain with all residue fields finite.

Chabert also proved in [5] that if $Int(D)$ is a Prüfer domain then D is an almost Dedekind domain with all residue fields finite. At the time no examples of almost Dedekind domains with all residue fields finite which were not Dedekind were known to exist. Gilmer produced such examples in [11] and proved that for some examples $Int(D)$ was a Prüfer domain and for some it was not. Chabert then gave a somewhat different class of examples in [6]. Gilmer and Chabert both used infinite degree integral extensions in their constructions. Loper provided yet more examples by using infinite degree transcendental extensions in [15] and by intersecting valuation domains in [14]. He then gave the following classification theorem in [17] which is essentially stated but unproven in [6].

Theorem 22 *Let D be an almost Dedekind domain with all residue fields finite and with quotient field K. Suppose that K has characteristic 0. (See the note following the theorem concerning positive characteristic.) For each maximal ideal P of D let v_P denote the valuation associated with the Noetherian valuation domain D_P, normalized so that the value group is the group Z of integers. For each prime number p which is not a unit in D define the sets*

$$F_p = \{|D/P| \mid p \in P\}$$

$$E_p = \{v_P(p) \mid p \in P\}$$

Then $Int(D)$ is a Prufer domain if and only if both of the sets F_p and E_p are bounded for each prime p which is a nonunit in D.

Note: If K has positive characteritic then D must contain a finite field F and must contain an element t which is transcendental over F. We can then let the irreducible polynomials of $F[t]$ play the role of the prime numbers in the statement of the theorem above.

2.2 Factorization of ideals into radical ideals

A distinct application of almost Dedekind domain theory began with a 1978 paper of Vaughan and Yeagy [23]. In this paper, they define a domain to be an SP domain provided each proper ideal can be expressed as a product of radical ideals. They prove that an SP domain must be an almost Dedekind domain. They also demonstrate that a nonNoetherian almost Dedekind domain constructed by Heinzer and Ohm in [13] is an SP domain. Yeagy in [24] proved a partial classification theorem which demonstrated that Nakano's original example is also an SP domain. In the opposite direction, Butts and Yeagy gave an example in [3] of a nonNoetherian almost Dedekind domain which is not an SP domain. This question was completely resolved in a 2005 paper of Olberding [21] who proved the following classification theorem and made clever use of a theorem of Jaffard-Kaplansky-Ohm to construct a wealth of examples which are often quite different from those of Gilmer-Chabert-Loper mentioned above.

We say that a maximal ideal M of an almost Dedekind domain D is critical if every finitely generated subideal of M is contained in the square of some maximal ideal of D.

Recall that any SP domain has been proven to be an almost Dedekind domain. We characterize the SP domains within the class of almost Dedekind domain.

Theorem 23 *Let D be an almost Dedekind domain. Then the following are equivalent.*

(1) *D is an SP domain*
(2) *D contains no critical maximal ideals.*
(3) *If I is a proper finitely generated ideal of D, then $\sqrt{I}$ is also a finitely generated ideal of D.*
(4) *Every proper nonzero ideal of I of D can be represented uniquely as a product $I = J_1 J_2 \cdots J_n$ of radical ideals J_i such that $J_1 \subseteq J_2 \subseteq \cdots \subseteq J_n$*

2.3 Factorization of finitely generated ideals

Another recent study of almost Dedekind domains is in a 2005 paper of Loper and Lucas [18]. In some ways this paper picks up on the original themes of Gilmer's 1964 paper in that it expands some of the classical properties of Dedekind domains.

In [8] Gilmer defines a domain D to have property # provided whenever Ω and Λ are subsets of $Max(D)$ such that $\bigcap_{M_\alpha \in \Omega} D_{M_\alpha} = \bigcap_{M_\beta \in \Lambda} D_{M_\beta}$ then $\Omega = \Lambda$. It is easy to demonstrate that a Dedekind domain is a # domain. Gilmer proves in [8] that an almost Dedekind domain which is not Dedekind is not a # domain. This then demonstrates a clear difference between Dedekind domains and nonNoetherian almost Dedekind domain.

In [18] the concept of # domain is expanded to provide a way of measuring how close an almost Dedekind domain which is not Dedekind is to being Dedekind. In particular, [8, Theorem 2] implies that if D is an almost Dedekind domain then D is a # domain if and only if each maximal ideal contains a finitely generated ideal which is contained in no other maximal ideal. If D is a nonNoetherian almost Dedekind domain then some maximal ideals will not satisfy this property. But perhaps some will. So we say that a maximal ideal M of an almost Dedekind domain D is a sharp prime of D provided M contains a finitely generated ideal which is not contained in any other maximal ideal. We then define a sequence of overrings of a given almost Dedekind domain D as follows:

1. Let D be an almost Dedekind domain with quotient field K.
2. Let $M_{\#}(D)$ be the collection of all sharp primes of D.
3. Let $D_1 = D$.
4. Let $D_n = \bigcap_{M \notin M_{\#}(D)} D_M$ (but let $D_n = K$ if $M_{\#}(D) = Max(D)$). Note that each D_n which is not equal to K is again an almost Dedekind domain.

The idea is that all of the sharp primes in D_1 do not extend to proper primes in D_2 but all of the dull (i.e. not sharp) primes do extend to prime ideals in D_2. Moreover, some of the dull primes of D_1 may extend to sharp primes in D_2. If a maximal ideal M of D extends to a proper sharp prime in D_n and then blows up in the extension to D_{n+1} then we say that M has sharp degree n. If a maximal ideal M of D has sharp degree n for some n then we say that M has finite sharp degree. It is possible that an almost Dedekind domain could have only dull primes and that $D_1 = D_2 = \cdots$. Or it could happen that the chain stabilizes at some later point. Or it could be that the chain never stabilizes. If the chain never stabilizes then most any conceivable situation could apply to the union of the chain of D_i's. [18] contains abundant examples, constructed using yet another construction technique, demonstrating the various sharp and dull possibilities.

The goal of the generalization of the sharp property is to attempt to generalize some of the classical Dedekind domain properties to almost Dedekind domains that are especially close to being Dedekind, which in this case would mean that every maximal ideal has finite sharp degree.

Theorem 24 *Let D be an almost Dedekind domain such that every maximal ideal of D has finite sharp degree. Then there exists a set $S = \{J_\alpha \mid \alpha \in \Omega\}$ of finitely generated ideals of D such that each J_α is either a finitely generated maximal ideal or J_α is a finitely generated ideal associated with a maximal ideal M_α of sharp degree k such that $M_\alpha D_{M_\alpha} = J_\alpha D_{M_\alpha}$ and $M_\alpha D_k$ is the only prime ideal of D_k which contains J_α. Then every finitely generated ideal of D can be factored uniquely into a product of ideals from S. (Note: Some of the exponents may be negative.)*

So the unique factorization of ideals into products of prime ideals which holds for Dedekind domains has a good generalization when the almost Dedekind domain is sufficiently nice. [10, Proposition 37.5], which states that in any almost Dedekind domain an ideal which is contained in only finitely many prime ideals can be factored uniquely into a product of those primes hints at this result. The point is that sharp primes tend to be well behaved.

2.4 Glad domains

Glad domains were defined and studied somewhat in [14]. Rush gave the following slightly broader definition in [22]. Rush then applies the concept to integer-valued polynomials and gives strong hints of other applications.

Definition 25 *A domain D with quotient field K (different from D) is a glad domain provided:*

(1) $D = \bigcap_{i \in \Lambda} V_i$ *where* $\{V_i \mid i \in \Lambda\}$ *is a family of Noetherian valuation overrings of a domain R. Let v_i be the normed additive valuation associated with V_i, and let M_i be the maximal ideal of V_i.*
(2) *There is a monic polynomial $f \in D[x]$ of degree > 1 such that for each $i \in \Lambda$ and each $a \in V_i$, $f(a)$ is a unit of V_i.*
(3) *For each $a \in D$ the set $\{v_i(a) \mid i \in \Lambda\}$ is bounded.*
(4) *There exists $t \in D$ such that $tV_i = M_i$ for each $i \in \Lambda$.*
(5) *There exists a finite subset T of D which is a set of representatives for V_i/M_i for each $i \in \Lambda$.*

Rush then gives the following alternate characterization.

Proposition 26 *Let D be an almost Dedekind domain with quotient field K (different from D). Then D is a glad domain if and only if the following statements hold.*

(1) *Each principal ideal of D contains a power of its radical.*
(2) *The Jacobson radical J of D is a nonzero principal ideal.*
(3) *There exists a finite subset T of D which is a set of representatives for D/P for each maximal ideal P of D.*

The main application to integer-valued polynomial theory has to do with localization. In particular, when D is a Dedekind domain with all residue fields finite then for each maximal ideal P of D we have $Int(D_P) = Int(D)_{D-P}$. It was thought for a time that perhaps this condition (which we call *good behavior under localization*) in the context of almost Dedekind domains with finite residue fields would be equivalent to $Int(D)$ being a Prüfer domain. Rush gives a complete characterization of when we have good behavior under localization.

Note: We make reference here to the standard Zariski topology on the spectrum of a ring. The notation $V(aD)$ refers to the closed set consisting of all prime ideals of D which contain a.

Theorem 27 *Let D be an almost Dedekind domain with finite residue fields. Then $Int(D_P) = Int(D)_{D-P}$ for each $P \in Spec(D)$ if and only if for each nonzero $a \in D$, there is a partition $\{F_1, F_2, \ldots, F_m\}$ of $V(aD)$ into open sets such that the ring $D_i = \bigcap\{D_Q \mid Q \in F_i\}$ is a glad domain for $i = 1, \ldots, m$.*

Rush indicates that glad domains are almost Dedekind domains which seem to have properties reminiscent of the properties of Noetherian valuation domains. For example, he proves the following result.

Proposition 28 *Let $D = D_1 \bigcap \ldots \bigcap D_m$ with each D_i a glad overring of D. Then the following hold.*

(1) $D_i = D_S$ *where* $S = D - \bigcup\{Q \bigcap D \mid Q a maximal ideal of D_i\}$.
(2) D *is an almost Dedekind domain with finite residue fields.*
(3) D *is a Bézout domain.*

Rush also proves some results concerning strong-two generators in integer-valued polynomial rings in which glad domains mirror the behavior of Noetherian valuation domains. It would be interesting to see if this similarity can be exploited to any great benefit. A place to begin might be to see if a domain which is a locally finite intersection of glad domains behaves in a manner similar to Krull domains.

To close this section we remark that the notion of a glad domain appeared first in a paper written the present author. In fact, the definition was essentially posed by Robert Gilmer in response to some questions concerning an example that I had asked him to consider.

2.5 Sequence domains

This subsection concerns a very special class of almost Dedekind domains for which I am not aware of any grand theorems. The reason for including them is that the remarkable simplicity of these domains make them a potentially useful source of examples and intuition concerning almost Dedekind domains and nonNoetherian domains in general. The notion of a sequence domain was defined in [16] and used there to great effect in the study of integer-valued polynomials. The definition here is somewhat more general in that we omit restrictions on the residue fields.

Definition 29 *Let D be an almost Dedekind domain. We call D a sequence domain provided the following hold.*

(1) *The maximal ideals of D are $\{M_i \mid i \in Z^+\}$ and M_∞.*
(2) *Each M_i is principal and M_∞ is not finitely generated.*
(3) *The Jacobson radical J of D is nonzero.*

In the subsection above concerning glad domains we made reference to the proposal that was floated for some time that perhaps good behavior under localization for $Int(D)$ was equivalent to $Int(D)$ being a Prüfer domain for an almost Dedekind domain D with finite residue fields. It was known that good behavior implied Prüfer. So we want to find a domain D such that $Int(D)$ is a Prüfer domain, but we do not have good behavior under localization. Suppose we wanted to look for a counterexample to this in the context of sequence domains. In [16] the counterexample is constructed. The key is to construct a sequence domain such that $|D/M_i| = p^2$ for each i and $|D/M_\infty| = p$. Choose a nonzero element d in M_∞. It is easy to demonstrate that d will also be contained in all but a finite number of the M_i's. Then to satisfy Rush's criterion for good behavior under localization we would have to partition the corresponding collection of valuation domains so that each set in the partition is open in Spec(D) and gives rise to a glad overring. However, any overring of D in which M_∞ survives along with an infinite number of M_i's is obviously not a glad domain and M_∞ alone is not an open set in Spec(D). So this domain does not satisfy good behavior under localization. However, such a domain can be constructed so that Int(D) is a Prüfer domain. Sequence domains are not necessary to construct a counterexample here, but it is very easy to see how the domain should be constructed inside the realm of sequence domains.

We close by noting that after the 1964 paper in which he coined the term almost Dedekind, Gilmer's next reference to almost Dedekind domains was in the paper concerning overrings of Prüfer domains in which he defined the # property. One of his principal objectives in this paper was to answer a conjecture made in the original almost Dedekind paper that almost Dedekind domains do not necessarily have the # property. His example [8, Example 2] given to demonstrate that an almost Dedekind domain need not have the # property is a sequence domain.

3 Construction of almost Dedekind domains

On several occasions, I have beeen asked to give intuitive examples of Prüfer domains. The simplest examples to give are rings of algebraic integers in finite degree extensions of the rational numbers. Of course these are perhaps a little too nice because they are Dedekind domains, and hence, Noetherian. The ring of entire functions and the ring of all algebraic integers can then be used as intuitive examples of Prüfer domains which are not Noetherian.

I have also been asked to give a similarly intuitive example of a nonNoetherian almost Dedekind domain. This is harder. In fact, it may not be possible to give as natural an example as the ring of all algebraic integers and the ring of entire functions are of Prüfer domains.

Although the construction techniques are not natural in general, there are a substantial number of different such techniques in the literature.

3.1 Technigue 1: Infinite extensions

There have been at least three somewhat different constructions involving infinite extensions, all of which followed the same general scheme.

Theorem 30 [18, Theorem 2.10] *Let $R_1 \subset R_2 \subset \cdots$ be a chain of Dedekind domains which satisfy all of the following:*

(1) *For $i < j$, each maximal ideal of R_i survives in R_j.*
(2) *Each maximal ideal of R_j contracts to a maximal ideal of R_1.*
(3) *If M' is a maximal ideal of R_j and $M = M' \cap R_1$, then $MR_{jM'} = M'R_{jM'}$.*

Then $D = \bigcup R_n$ is an almost Dedekind domain.

Most of the early constructions (including Nakano's first construction) of nonNoetherian almost Dedekind domains were built using the infinite chain method of Theorem 10 with the extensions of Dedekind domains being integral. Note that every integral extension of Z or any localization of it is a one dimensional Prüfer domain. The difficult part is to insure that the localizations are discrete. When working with infinite degree integral extensions of Dedekind domains then, the key is to place very sharp control on ramification. Nakano's example is then quite natural because when adjoining the pth roots of unity the only primes that will ramify at all will be those which lie above p. So ramification is made very strongly finite, and the goal of constructing an almost Dedekind domain is achieved. To show that the almost Dedekind domain is not Noetherian, it is enough to show that some prime q does not remain inert (i.e. splits into at least two primes) in the pth cyclotomic field for infinitely many primes p.

More delicate control is necessary if a nonNoetherian almost Dedekind domain with all residue fields finite is desired. For this we need both ramification and inertial behavior to be strongly controlled. Gilmer [11] and Chabert [6] gave examples of infinite integral extension constructions with various nice properties by making use of powerful theorems of Krull (cited in [11] as Theorem K) and of Hasse (cited as 6.1 in [6]).

In [15] Loper gave an infinite extension construction by using the method of Theorem 10 with R_{i+1} being an overring of $R_i[x_i]$ (where x_i is a variable). The idea is that for each maximal ideal $M \subseteq R_i$ we find a finite number of Noetherian valuation overrings of $R_i[x]$ with maximal ideals centered on M. We then let R_{i+1} be the intersection of the Noetherian valuation overrings chosen corresponding to all of the maximal ideals of R_i. We do need to exercise some care to insure that R_{i+1} is a Dedekind domain. For example we can insist that each R_i is an intersection of fintely many Noetherian valuation domains - and hence Dedekind.

The method of [15] involves an infinite number of indeterminates which have no relation to each other. In [18] Loper and Lucas use an infinite number of indeterminates introduced in stages, but the indeterminates are very strongly related to each other. The point of the construction is to begin with a single variable Y. We call this stage 0 and designate $Y = Y_{0,1}$. For stage m we have the variables $\{Y_{m,1}, Y_{m,2}, \ldots, Y_{m,n_m}\}$. We then choose a field K and at each stage we define $V_{m,k} = K[\{Y_{m,1}, Y_{m,2}, \ldots, Y_{m,n_m}\}]_{(Y_{m,k})}$. Then we define $R_m = \bigcap_{1 \leq k \leq n_m} V_{m,k}$. So R_m is an intersection of finitely many Noetherian valuation domains and hence Dedekind. The idea of the factoring is that we attain the variables for stage $m+1$ by factoring the variables of stage m. So for example let $Y_{m,1}$ be a variable at stage m. Then perhaps $Y_{m,1}$ is still a variable for stage $m+1$. Or perhaps, we designate that

$$Y_{m,1} = (Y_{m+1,1})(Y_{m+1,2})(Y_{m+1,1}) \cdots (Y_{m+1,t})$$

with each $Y_{m+1,i}$ being a distinct variable for R_{m+1}. The result is a tree structure. Each level of the tree corresponds to a semilocal Dedekind domain R_m and the number of maximal ideals increases as the tree goes upwards. By varying the branching behavior, we can exert very strong control over the sharp and dull behavior.

3.2 Technique 2: Group of divisibility construction

In [21] Olberding makes use of the following theorem of Jaffard-Kaplansky-Ohm to construct almost Dedekind domains.

Let D be a domain with quotient field K. Let $U(D)$ represent the group of units of D and let K^* represent the nonzero elements of K. We then call the multiplicative quotient group $K^*/U(D)$ the group of divisibility of D.

Theorem 31 [10, Theorem 18.6] *Let G be a lattice ordered abelian group. Then there exists a Bezout domain D with G as its group of divisibility.*

The domain constructed is actually a particular overring of the group ring of G over a field K. In particular, let $d = a_1x^{g_1} + a_2x^{g_2} + \cdots + a_nx^{g_n} \in K[G]$ We define a function w from $K[G]$ to G by $w(d) = inf\{g_i\}_{1 \leq n}$. We extend w additively to the quotient field of $K[G]$ and let D be the set of all positive elements.

This theorem only guarantees that the domain constructed will be a Bezout domain. Olberding assures that it will be an almost Dedekind domain by means of the following steps.

1. Choose a Boolean space X
2. Note that X is homemorphic to the space of ultrafilters on the Boolean algebra $L(X)$ of clopen sets of X.
3. Construct a lattice-ordered abelian group G such that the space of proper prime filters of G^+ is homeomorphic to the space of ultrafilters of $L(X)$.
4. Use the Jaffard-Kaplansky-Ohm theorem to construct a Bezout domain which has G as group of divisibility.
5. Show that the spectral topology on the nonzero primes of D is homeomorphic to the space of proper prime filters of G^+.
6. Show that the Boolean property on X guarantees that the domain D obtained is an almost Dedekind domain (in fact, an SP domain).

Note that the above sequence produces an almost Dedekind domain with the topology on $Max(D)$ homeomorphic to the arbitrarily chosen Boolean space X. Olberding remarks that he has very little control over the residue fields in this construction. On the other hand this construction does answer some questions that the other known constructions do not. In particular all of the other constructions have a strong finite/countability component to them. It is easy to construct an almost Dedekind domain with as many maximal ideals as desired; simply take the ring of polynomials over a very large field. To construct an almost Dedekind domain with nonzero Jacobson radical with a large number of maximal ideals is not so easy. Olberding's construction demonstrates that even with the nonzero Jacobson radical assumption, the maximal spectrum of an almost Dedekind domain can be as large as desired. To achieve a large maximal spectrum simply start with a large Boolean space and use Olberding's construction.

3.3 Technique 3: Intersection of valuation domains:

This is a technique used by Loper in [14] and [16]. The technique involves several stages.

1) Choose an infinite collection of Noetherian valuation domains which share a common quotient field.

2) Intersect the chosen Noetherian valuation domains to obtain a domain D.

3) Prove that D is a Prüfer domain.

4) Prove that D is an almost Dedekind domain.

It is often easy to choose a collection of Noetherian valuation domains which share a given quotient field. To prove that the intersection is a Prüfer domain is a nontrivial task. The following theorem of Gilmer from [9] has been used to great effect in this regard.

Theorem 32 *Let D be a domain with quotient field K. Let $f(x)$ be a monic polynomial in $D[x]$ which has no roots in K. Let $S = \{V_\alpha \mid \alpha \in \Omega\}$ be a collection of valuation overrings of D such that f does not have a root in any of the residue fields of the V_α's. Let $R = \bigcap_{\alpha\in\Omega} V_\alpha$. Then R is a Prüfer domain.*

So once we have intersected an infinite collection of Noetherian valuation domains and obtained a Prüfer domain, how do we then demonstrate that the intersection is almost Dedekind? The ring of entire functions is an example of a Prüfer domain which can be expressed as an intersection of Noetherian valuation overrings and yet is not almost Dedekind. In fact, it is infinite dimensional.

For some collections of Noetherian valuation domains the job is made easier by topology. For example, choose a prime p and consider the valuation overrings of the Prüfer domain $Int(Z_p)$ in which p is a nonunit. These valuation overrings are indexed in a natural way by the ring $\widehat{Z_p}$ of p-adic integers. Moreover, for a given p-adic integer α the corresponding valuation domain V_α is a Noetherian valuation domains if and only if α is transcendental over Q. So the first step is to choose a collection of transcendental p-adic integers and intersect the corresponding valuation domains. The question then is whether when we intersect, we obtain additional valuation overrings which are not Noetherian as in the entire function case. The following theorem addresses this concern.

Theorem 33 [14] *Let $S \subseteq \widehat{Z_p}$ be a collection of p-adic integers. Let $D = \bigcap_{\alpha \in S} V_\alpha$. Then the valuation overrings of D in which p is a nonunit exactly correspond to the elements of the closure of S in the p-adic topology.*

So the key to obtaining an almost Dedekind domain in this case is to choose a collection of p-adic integers which is closed under the p-adic metric and which consists entirely of transcendental numbers.

Example 34 *Choose a p-adic integer α which is transcendental over Q. Then the set $S = \{\alpha, \alpha + p, \alpha + p^2, \alpha + p^3, \cdots\}$ is a closed set of p-adic integers which consists entirely of transcendental numbers.*

It is also possible (but harder) to give examples with an uncountable number of transcendental numbers [14].

3.4 Technique 4: Kronecker function ring

Another construction which appears in the literature is the use of the Kronecker function ring. In particular, if D is an almost Dedekind domain which is not Noetherian then $Kr(D)$, the Kronecker function ring of D is also. This technique is a little unsatisfying because one needs to have a nonNoetherian almost Dedekind domain in hand in order to produce one. One can alter the construction a little however to give a method with more potential.

Let V be a valuation domain with quotient field K. Let $D = V(x)$ be the trivial extension of V to the field of rational functions $K(x)$. Consider the polynomial $f(t) = t^2 - x$ in the ring $D[t]$ of polynomials over D. Note that f is monic and will not have any roots in the residue field of $V(x)$. Then Theorem 12 can be applied to prove that the intersection of a collection of such domains $V(x)$ can be intersected to produce a Prüfer domain with quotient field $K(x)$. (Note: This demonstrates in a manner other than the standard manner that the Kronecker function ring is a Prüfer domain.) So, if we wish to produce an almost Dedekind domain with quotient field $K(x)$ we should begin by choosing a collection of Noetherian valuation overrings of a domain T with quotient field K and then intersecting the corresponding $V(x)$'s. This will always yield a Prüfer domain with quotient field $K(x)$ by Theorem 12. The key is to choose the valuation domains carefully enough that the resulting domain is actually almost Dedekind.

Example 35 *Choose a field K and consider the domain $K[y, z]$ of polynomials in the two variables y and z. Localize at the maximal ideal generated by y and z. We say $D = K[y, z]_{(y,z)}$ - the described local ring. Let $D_1 = D[z/y]$ and let $D_2 = D[y/z]$. Let $\{M_\alpha \mid \alpha \in \Lambda\}$ be the collection of maximal ideals* *of D_1. For each $\alpha \in \Lambda$ choose a DVR overring V_α of D_1 with maximal ideal centered on M_α. Then choose a DVR overring V_∞ of D_2 with maximal ideal centered on the maximal ideal of D_2 which contains y/z. Then the domain $T = \bigcap_{\alpha\in\Lambda} V_\alpha(x) \bigcap V_\infty(x)$ is an almost Dedekind domain. As described above, we already know that T is a Prüfer domain. To prove that T is an almost Dedekind domain it is only necessary to show that the D_i's are the only valuation overrings of T. The principal tool one would use to show this is [4, Proposition 2.8]. The full details are beyond the scope of this article. Note that since this domain has an infinite number of maximal ideals containing y and z it is certainly not Noetherian.*

3.5 Technique 5: Semigroup ring

In [12] Gilmer and Parker found conditions under which the semigroup ring $D[S]$ is an almost Dedekind domain.

Theorem 36 *Let D be a domain and let S be an additive, cancellative, torsion-free semigroup. Then the semigroup ring $D[S]$ is an almost Dedekind domain if and only if D is a field and one of the following holds.*

(1) *S is isomorphic to the additive monoid of nonnegative integers.*
(2) *D has characteristic zero and S is isomorphic to an additive subgroup of the field Q of rational numbers which contains the integers.*
(3) *D has positive characteristic q and S is isomorphic to an additive subgroup of the field Q of rational numbers which contains the integers such that $1/q^k \notin S$ for some k.*

Construction 1 in this list obviously gives a domain which is isomorphic to the ring of polynomials over the field D. The second and third constructions produce almost Dedekind domains which are not Noetherian provided the subgroup S is not finitely generated.

This construction essentially begins with the domain $K[x, 1/x]$ where K is a field and x is a variable and then takes remarkably simple integral extensions, simply adding in roots of the variable x. This construction has not been exploited in any works of which we are aware, but the simplicity of the construction would seem to give it great potential.

References

1. J.T. Arnold, *Krull dimension in power series rings*, Trans. Amer. Math. Soc. **177** (1973), 299-304.
2. D. Brizolis, *A theorem on ideals in Prüfer rings of integral-valued polynomials*, Comm. Algebra **7** (1979), 1065-1077.
3. H.S. Butts and R.W. Yeagy, *Finite bases for integral closures*, J. reine angew. Math. **282** (1976), 114-125.

4. P.J. Cahen, A. Loper and F. Tartarone, *Integer-valued polynomials and Prüfer-v-multiplication domains*, J. Algebra **226**, (2000), 765-787.
5. J.L. Chabert, *Un anneaux de Prüfer domains*, J. Algebra **107** (1987), 1-16.
6. J.L. Chabert, *Integer-valued polynomials, Prüfer domains, and localization*, Proc. Amer. Math. Soc. **118** (1993), 1061-1073.
7. R. Gilmer, *Integral domains which are almost Dedekind*, Proc. Amer. Math. Soc. **15** (1964), 813-818.
8. R. Gilmer, *Overrings of Prüfer domains*, J. Algebra **4** (1966), 331-340.
9. R. Gilmer, *Two constructions of Prüfer domains*, J. reine angew. Math. **129** (1990), 502-517.
10. R. Gilmer, *Multiplicative Ideal Theory*, M. Dekker, New York, 1972.
11. R. Gilmer, *Prüfer domains and rings of integer-valued polynomials*, J. Algebra **129** (1990), 502-517.
12. R. Gilmer and T. Parker, *Semigroup rings as Prüfer rings*, Duke Math. J. **41** (1974), 219-230.
13. W. Heinzer and J. Ohm, *Locally Noehterian commutative rings*, Trans. Amer. Math. Soc. **158** (1971), 273-284.
14. K.A. Loper, *More almost Dedekind domains and Prüfer domains of polynomials*, in Zero-dimensional commutative rings (Knoxville, TN, 1994), 287-298, Lecture Notes in Pure and Appl. Math., **171** (1995), Dekker, New York.
15. A. Loper, *Another Prüfer ring of integer-valued polynomials*, J. Algebra **187**, (1997), 1-6.
16. A. Loper, *Sequence domains and integer-valued polynomials*, Journal of Pure and Applied Algebra **119**, (1997), 185-210.
17. K.A. Loper, *A classification of all D such that Int(D) is a Prufer domain*, Proc. Amer. Math. Soc., *126*, (1998), 657-660.
18. K.A. Loper and T.G. Lucas, *Factoring ideals in almost Dedekind domains*, J. reine angew. Math., **565**, (2003), 61-78.
19. D.L. McQuillan, *On Prüfer domains of polynomials*, J. reine angew. Math., **565**, (2003), 61-78.
20. N. Nakano, *Ideal theorie in einem speziellen undindlichen algebraischen Zahlkörper*, J. Sci. Hiroshima Univ. Ser. A **16** (1953), 425-439.
21. B. Olberding, *Factorization into radical ideals*, Lec. Notes Pure Appl. **241** (2005), 363-377.
22. D. Rush, *The conditions* $Int(R) \subseteq R_S[X]$ *and* $Int(R_s) = Int(R)_S$ *for integer-valued polynomials*, Journal of Pure and Applied Algebra **125**, (1998), 287-303.
23. N. Vaughan and R.W. Yeagy, *Factoring ideals into semiprime ideals*, Can. J. Math. **30** (1978), 1313-1318.
24. R.W. Yeagy, *Semiprime factorizations in unions of Dedekind domains*, J. reine angew. Math. **310** (1979), 182-186.

Integrality properties of polynomial rings and semigroup rings

Thomas G. Lucas

University of North Carolina Charlotte, Charlotte, NC 28223, USA
tglucas@email.uncc.edu

1 Introduction

For a set of indeterminates $\{x_\lambda\}$, an exercise in Robert Gilmer's book *Multiplicative Ideal Theory* assigns to the reader the following task [Gi92, Exercise 11, page 100].

> Suppose each regular element of R is a unit of R. Determine necessary and sufficient conditions in order that $R[\{x_\lambda\}]$ be integrally closed.

Obviously, if R contains a nonzero nilpotent n, then n/x is integral over $R[x]$ but certainly does not belong to $R[x]$. Thus a necessary condition is that R must be a reduced ring.

It is well-known that if R is an integral domain, then $R[x]$ is integrally closed if and only if R is integrally closed. Heinz Prüfer gave a proof in Section 10 of his paper [Pr32]. The result also appears in [Le47], which Wolfgang Krull reviews in Zentralblatt by simply providing a proof using valuations [Zbl 0035.01902]. The problem can be linked to both Gilmer and Joe Mott through their thesis advisor Hubert Butts. In 1954, Butts together with Marshall Hall and Henry Mann gave yet another proof in [BHM54].

Related questions can be formed with regard to semigroup rings and other integrality properties. Of the latter, the ones we will consider are: n-root closed, root closed, $(2,3)$-closed, u-closed, t-closed and completely integrally closed. We will recall the relevant definitions at the appropriate time.

Throughout the article, R denotes a commutative ring with nonzero identity and $T(R)$ denotes its corresponding total quotient ring and $Z(R)$ its set of zero divisors. For an ideal I, we let $Ann(I)$ denote its annhilator.

A few definitions are in order before we go any further. First an ideal I is said to be *regular* if it contains an element that is not a zero divisor, such an element is said to be *regular*. A regular ideal obviously has the property that it has no nonzero annihilators, but this can happen for other ideals as well. An ideal J is *dense* if it has no nonzero annihilators and it is *semiregular* if it contains a finitely generated dense ideal. Recall that McCoy's Theorem states that a polynomial $f(\mathrm{x}) \in R[\mathrm{x}]$ is a zero divisor if and only if there is a nonzero element $r \in R$ such that $rf(\mathrm{x}) = 0$. Thus an ideal J of R is semiregular if and only if its extension to the polynomial ring $R[\mathrm{x}]$ is regular. Following the lead of Carl Faith [F91], we say that a ring R is a *McCoy ring* if each (finitely generated) semiregular ideal is regular. Earlier authors had used "Condition (C)" and "Property A" for this (see, for example, [Q71] and [Ak80] for the former and [HK79] and [H88] for the latter).

For a pair of dense ideals I and J and a pair of R-module homomorphisms, $f \in Hom_R(I, R)$ and $g \in Hom_R(J, R)$, both the sum, $f+g$, and the product, fg, can be defined as R-module homomorphisms on the dense ideal IJ. Moreover, one can define an equivalence relation by setting $f \equiv g$ if $f(r) = g(r)$ for each $r \in I \bigcap J$. In [La86], Joachim Lambek uses the resulting equivalence classes to build $Q(R)$, the complete ring of quotients of R.

Given a pair of rings $V \subseteq T$, V is said to be a *paravaluation ring of* T if there is a totally ordered Abelian group G and a map (called a *paravaluation*) $\nu{:}T \to G \bigcup \{\infty\}$ such that for all $a, b \in T$

(i) $\nu(ab) = \nu(a) + \nu(b)$, with $g < g + \infty = \infty + \infty = \infty$ for each $g \in G$,
(ii) $\nu(a+b) \geq min\{\nu(a), \nu(b)\}$ (with equality if $\nu(a) \neq \nu(b)$),
(iii) $\nu(1) = 0$ and $\nu(0) = \infty$, and
(iv) $V = \{t \in T \mid \nu(t) \geq 0\}$.

Their value lies in the fact that each paravaluation ring of T is integrally closed in T and for $R \subseteq T$, the integral closure of R in T is the intersection of the paravaluation rings of T that contain R. The latter is due to a combination of results of Pierre Samuel [Sa57, Théorème 8] and Malcolm Griffin [Gri70, Proposition 2]. For a relatively simple proof see [Gra82] (reproduced in [H88, Theorem 9.1]). Also, it is easy to show that if V is a paravaluation ring of T with corresponding paravaluation ν, then $V[\mathrm{x}]$ is a paravaluation ring of $T[\mathrm{x}]$ using $\nu(f(\mathrm{x})) = min\{\nu(f_0), \ldots, \nu(f_n)\}$ for $f(\mathrm{x}) = f_n\mathrm{x}^n + \cdots + f_0 \in T[\mathrm{x}]$ (as in the proof given for [Gi92, Proposition 18.4], for polynomials $f(\mathrm{x})$ and $g(\mathrm{x})$ and integers i and j such that $\nu(f(\mathrm{x})) = \nu(f_i) < \nu(f_r)$ for all $r < i$ and $\nu(g(\mathrm{x})) = \nu(g_j) < \nu(g_s)$ for all $s < j$, the value of the coefficient on x^{i+j} in $f(\mathrm{x})g(\mathrm{x})$ is $\nu(f_i) + \nu(g_j)$ and no other power of x has a smaller value). Clearly if $S = \bigcap V_\alpha$, then $S[\mathrm{x}] = \bigcap V_\alpha[\mathrm{x}]$ which establishes Theorem 2.2 below.

Finally, for a pair of rings $R \subseteq S$ and a polynomial $f \in S[\mathrm{x}]$, the R-content of f is the R-submodule of S generated by the coefficients of f. We will use $C_R(f)$ to denote this submodule. The notation is extended to ideals by letting $C_R(I)$ be the R-submodule of S generated by the coefficients of the polynomials in I. A handy result to have is the Dedekind-Mertens content formula (see, for example, [Gi92, Theorem 28.3]). This result is often referred to as the Dedekind-Mertens Lemma. In this article, we will invoke its use by simply saying "by the content formula...".

Theorem 1.1. *Let $R \subseteq S$ be a pair of rings and let f and g be polynomials in $S[\mathrm{x}]$. Then $C_R(fg)C_R(g)^k = C_R(f)C_R(g)^{k+1}$ for some integer $k \geq 0$.*

2 When $R[\mathrm{x}]$ is Integrally Closed

We start with a bit of the history leading up to the complete solution of Gilmer's exercise. Then in subsequent sections we look at the other problems.

Theorem 2.1. (cf. [Pr32, Section 10]) *Let R be an integral domain. Then $R[\mathrm{x}]$ is integrally closed if and only if R is integrally closed.*

If $R[\mathrm{x}]$ is integrally closed, then R is integrally closed. One way to prove the converse for R an integral domain with quotient field K is to make use of the content formula, Cramer's Rule and the fact that $K[\mathrm{x}]$ is integrally closed. Start with $h(\mathrm{x}) \in K(\mathrm{x})$ integral over $R[\mathrm{x}]$ and let $g_0(\mathrm{x}), \ldots, g_n(\mathrm{x}) \in R[\mathrm{x}]$ be such that $h(\mathrm{x})^{n+1} + g_n(\mathrm{x})h(\mathrm{x})^n + \cdots + g_0(\mathrm{x}) = 0$. Since $K[\mathrm{x}]$ is integrally closed, $h(\mathrm{x})$ is a polynomial with coefficients in K. Write $h(\mathrm{x}) = f(\mathrm{x})/a$ where $a \in R$ and $f(\mathrm{x}) \in R[\mathrm{x}]$. Next consider the ideal $J = (f(\mathrm{x})^n, af(\mathrm{x})^{n-1}, \ldots, a^n)$ of $R[\mathrm{x}]$. Using the integrality equation, we have $hf(\mathrm{x})^n \in J$. Hence $h(\mathrm{x})J \subseteq J$. For $m = 1 + n \cdot deg(f(\mathrm{x}))$, the polynomial $s(\mathrm{x}) = a^n + a^{n-1}f(\mathrm{x})\mathrm{x}^m + \cdots + f(\mathrm{x})^n\mathrm{x}^{nm}$ has the same content in R as the ideal J. Moreover, we have $h(\mathrm{x})s(\mathrm{x}) \in J$. By the content formula, $C_R(hs)C_R(s)^k = C_R(h)C_R(s)^{k+1}$ for some integer $k \geq 0$. As $C_R(s) = C_R(J)$ and $h(\mathrm{x})s(\mathrm{x}) \in J$, $C_R(hs) \subseteq C_R(s)$ and therefore $C_R(h)C_R(s)^{k+1} \subseteq C_R(s)^{k+1}$. Let $\{a_1, a_2, \ldots, a_q\}$ be a set of generators for $C_R(s)^{k+1}$. By Cramer's Rule, each $b \in C_R(h)$ is a zero of the corresponding monic polynomial $det(\mathrm{x}I - B)$ where $B = [b_{i,j}]$ is the $q \times q$ matrix corresponding to the equations $ba_i = \sum b_{i,j}a_j$. Thus $h(\mathrm{x}) \in R[\mathrm{x}]$ since R is integrally closed.

An alternate way to prove it is to again use that $K[\mathrm{x}]$ is integrally closed. Then finish the proof by applying the following general result.

Theorem 2.2. (cf. [Gi92, Theorem 10.7] *Let $R \subseteq T$ be a pair of rings and let S be the integral closure of R in T. Then for an indeterminate* x *(over T), $S[\mathrm{x}]$ is the integral closure of $R[\mathrm{x}]$ in $T[\mathrm{x}]$.*

If $T(R)$ is von Neumann regular, then $T(R)[\mathrm{x}]$ is integrally closed since locally $T(R)[\mathrm{x}]$ is a discrete rank one valuation domain. Thus by Theorem 2.2, if $T(R)$ is von Neumann regular, then $R[\mathrm{x}]$ is integrally closed if and only if R is integrally closed. In 1979, Jim Brewer (a student of Gilmer), Doug Costa and Kevin McCrimmon gave an example of a reduced total quotient ring R such that $R[\mathrm{x}]$ is not integrally closed [BCM79, Example 1]. In a 1980 paper, Tomoharu Akiba made several contributions, including completely settling the case of a ring whose set of minimal primes is compact in the Zariski topology [Ak80].

Theorem 2.3. [Ak80, Theorem 2.1] *Let R be a reduced ring for which $Min(R)$ is compact. Then the following are equivalent.*

(1) *$R[\mathrm{x}]$ is integrally closed.*
(2) *R is an integrally closed McCoy ring.*

(3) *R is integrally closed and $T(R)$ is von Neumann regular.*

In 1971, Yann Quentel proved that for a reduced ring R, the total quotient ring of R is von Neumann regular if and only if R is a McCoy ring and $Min(R)$ is compact [Q71, Propostion 9]. He also showed how to construct a reduced ring R where $Min(R)$ is compact but $T(R)$ is not von Neumann regular [Q71, Section 4]. Akiba uses this example to illustrate that it need not be the case that $R[\mathrm{x}]$ is integrally closed when $R = T(R)$ is reduced. The rest of Akiba's paper deals with different sufficient conditions for $R[\mathrm{x}]$ to be integrally closed. Each is in some way connected to the McCoy property.

Theorem 2.4. [Ak80, Corollary 1.2] *Let R be a ring. If R_M is an integrally closed integral domain for each maximal ideal M, then $R[x]$ is integrally closed.*

In a footnote, he acknowledges that Gilmer also knew this but had not published it. Note that having R_M be an integral domain for each maximal ideal M implies that R is reduced. Akiba includes an example of a reduced ring R which is locally an integrally closed integral domain but is not a McCoy ring. He attributes the example to Masayoshi Nagata.

Theorem 2.5. [Ak80, Theorem 3.2] *Let R be a McCoy ring. Then $R[\mathrm{x}]$ is integrally closed if and only if R is reduced and integrally closed.*

Quentel showed that if R is a reduced ring, then $R[\mathrm{x}]$ is a McCoy ring [Q71, Proposition 6]. Thus as a corollary we have a complete solution of Gilmer's original exercise in the special case that R is a McCoy ring.

Theorem 2.6. [Ak80, Corollary 3.4] *Let R be a reduced ring. Then $R[\mathrm{x}]$ is integrally closed if and only if $R[\{\mathrm{x}_\lambda\}]$ is integrally closed for each nonempty set of indeterminates $\{\mathrm{x}_\lambda\}$. Moreover, if R is a McCoy ring, then the following are equivalent.*

(1) *R is integrally closed.*
(2) *$R[\mathrm{x}]$ is integrally closed.*
(3) *$R[\{\mathrm{x}_\lambda\}]$ is integrally closed for each nonempty set of indeterminates $\{\mathrm{x}_\lambda\}$.*

It is still not known whether my first contribution to the solution of Gilmer's exercise was a contribution at all–but at least no one else had ever stated it (see [H88, Theorem 16.10]).

Theorem 2.7. *Let R be a reduced ring. If R_M is an integrally closed McCoy ring for each maximal ideal M, then $R[\mathrm{x}]$ is integrally closed.*

An open question (perhaps justifiably so) is whether there exists a reduced (integrally closed) ring R that is not a McCoy ring, but is locally an integrally closed McCoy ring with at least one such localization not an integral domain.

As noted at the beginning, a necessary condition for $R[\mathrm{x}]$ to be integrally closed is that R have no nonzero nilpotents. When R is reduced, the complete ring of quotients of R, $Q(R)$, is von Neumann regular. So the next thing to observe for the question at hand is the following.

Theorem 2.8. (cf. [Lu89a] and [Lu89b]) *Let R be a reduced ring and let S be its integral closure in $Q(R)$. Then*

(a) *$S[\mathrm{x}]$ is the integral closure of $R[\mathrm{x}]$ in $Q(R)[\mathrm{x}]$.*
(b) *Since $Q(R)$ is von Neumann regular, $Q(R)[\mathrm{x}]$ is integrally closed.*
(c) *$T(R[\mathrm{x}])$ embeds naturally in $T(Q(R)[\mathrm{x}])$. Thus $S[\mathrm{x}] \bigcap T(R[\mathrm{x}])$ is the integral closure of $R[\mathrm{x}]$ in $T(R[\mathrm{x}])$.*

Some examples are in order. First we consider the case that $R = T(R)$ is a reduced ring with $Min(R)$ compact. In 1964, Shizuo Endo proved that if $T(R)$ is von Neumann regular, then R is integrally closed if and only if R_M is an integrally closed integral domain for each maximal ideal M [E61, Propositions 5 and 6]. Since $R[\mathrm{x}]$ is a McCoy ring, having $Min(R)$ compact and R reduced implies $T(R[\mathrm{x}])$ is von Neumann regular [Q71, Proposition 9]. Thus if R is reduced with $Min(R)$ compact, $R[\mathrm{x}]$ is integrally closed if and only if it is locally an integrally closed domain. This is the approach Akiba used to show that having $R[\mathrm{x}]$ integrally closed when $Min(R)$ is compact is equivalent to having R integrally closed and $T(R)$ von Neumann regular. While useful in that case, the approach is not of much use when $Min(R)$ is not compact. So as first example we will take a direct approach of showing that having $R = T(R)$ be a reduced ring with $Min(R)$ compact and $R[\mathrm{x}]$ integrally closed is enough to guarantee that each principal ideal of R is idempotent.

Example 2.9. Let $R = T(R)$ be a reduced ring with $Min(R)$ compact and let $a \in Z(R)$ be a nonzero zero divisor. Since $Min(R)$ is compact, there are finitely many elements $b_0, b_1, \ldots, b_m \in Ann(a)$ such that $I = (a, b_0, \ldots, b_m)$ is semiregular [Q71, Proposition 4]. Set $b(x) = b_0 + b_1\mathrm{x} + \cdots + b_m\mathrm{x}^m$ and $a(\mathrm{x}) = a + \mathrm{x}b(\mathrm{x})$. Since I is semiregular, $a(\mathrm{x})$ is regular. Thus $f = a/a(\mathrm{x}) \in T(R[\mathrm{x}])$. Obviously, $fb_i = 0$ for each i. Moreover, by cross multiplying we see that $fa = fa(\mathrm{x}) = a$. Since $a(\mathrm{x})$ is not a zero divisor and $fa(\mathrm{x}) = f^2a(\mathrm{x})$, it must be that $f = f^2$ is idempotent, and therefore integral over both R and $R[\mathrm{x}]$. Note that multiplication by f defines an R-module homomorphism from the semiregular ideal I into R with $fb_i = 0$ for each i. Since R is a reduced ring, f cannot be a polynomial of positive degree. So the only way to have $R[\mathrm{x}]$ be integrally closed is to have $f \in R$. Assume this is the case and consider the ideal $J = aR + (1-f)R$. As $fb_i = 0$ for each i, $(1-f)b_i = b_i$ for each i and we have that J is semiregular since it contains I. Moreover, if $d(a + (1-f)) = 0$ for some $d \in R$, then $0 = fd(a + (1-f)) = fda = da$, hence $dJ = (0)$. It follows that $d = 0$ and therefore $a + (1-f)$ is a unit since $R = T(R)$. Set $u = a + (1-f)$. Clearly $fu = fa = a$. Hence $f = u^{-1}a \in aR$ and we have that $aR = fR$ is idempotent. It follows that R must be von Neumann regular when $Min(R)$ is compact and $R[\mathrm{x}]$ is integrally closed.

For a second example we use the so-called "$A + B$ construction" (see, for example [H88, Section 26] and [Lu94a]). Let D be an integral domain and let $\mathcal{P}$ be a nonempty set of prime ideals of D. For each $P_\alpha \in \mathcal{P}$ let $D_\alpha = D/P_\alpha$

and let K_α be the quotient field of D_α. Next let $\mathcal{I} = \mathcal{A} \times \mathbb{N}$ where $\mathcal{A}$ is an index set for $\mathcal{P}$. For each $i = (\alpha, n) \in \mathcal{I}$, let $K_i = K_\alpha$ and then let $B = \sum K_i$. Form the ring $R = D + B$ from $D \times B$ by setting $(r, b) + (s, c) = (r + s, b + c)$ and $(r, b)(s, c) = (rs, rc + sb + bc)$. One may identify B with $\{0\} \times B$. For a finitely generated ideal I of D, IR is semiregular if and only if no P_α contains I. In this case $IR = I + B$. Also we may identify the polynomial ring $R[\mathrm{x}]$ with $D[\mathrm{x}] + B[\mathrm{x}]$.

Example 2.10. [Lu89a, Example 3] Let K be a field and let $D = K[\mathrm{y}^2, \mathrm{y}^3, \mathrm{yz}, \mathrm{z}]$. Let $\mathcal{P} = Max(D) \backslash \{M\}$ where $M = (\mathrm{y}^2, \mathrm{y}^3, \mathrm{yz}, \mathrm{z})D$. Form the corresponding $A + B$ ring $R = D + B$.

(a) R is a reduced ring with $R = T(R)$.
(b) The ideal MR is a finitely generated semiregular ideal of R. Thus R is not a McCoy ring.
(c) The element $t = (\mathrm{y}^3\mathrm{x}^3 + \mathrm{y}^4\mathrm{x}^2 + \mathrm{y}^2\mathrm{zx} + \mathrm{yz}, 0)/(\mathrm{y}^2\mathrm{x}^3 + \mathrm{y}^3\mathrm{x}^2 + \mathrm{yzx} + \mathrm{z}, 0) \in T(R[\mathrm{x}])$ is integral over $R[\mathrm{x}]$ but is not in $R[\mathrm{x}]$. Specifically, t is a zero of $\mathrm{x}^2 - (\mathrm{y}^2, 0)$. Thus $R[\mathrm{x}]$ is not integrally closed.
(d) Multiplication by the element t in statement (c) defines an R-module homomorphism from the semiregular ideal MR into R which maps MR into itself. Specifically, $t(\mathrm{y}^2, 0) = (\mathrm{y}^3, 0)$, $t(\mathrm{y}^3, 0) = (\mathrm{y}^4, 0)$, $t(\mathrm{yz}, 0) = (\mathrm{y}^2\mathrm{z}, 0)$ and $t(\mathrm{z}, 0) = (\mathrm{yz}, 0)$.

The element t in statement (c) above is an example of a "finite fraction".
A relatively simple exercise is to show that if $I = (a_0, a_1, \ldots, a_n)$ is a finitely generated semiregular ideal of R, then each R-module homomorphism from I into R can be represented by multiplication by a suitably chosen member of $T(R[\mathrm{x}])$. For $f \in Hom_R(I, R)$, let $f(a_i) = b_i$ for each i. Next let $a(\mathrm{x}) = \sum a_i \mathrm{x}^i$ and $b(\mathrm{x}) = \sum b_i \mathrm{x}^i$. Since I has no nonzero annihilator, $a(\mathrm{x})$ is a regular element of $R[\mathrm{x}]$. Thus $b(\mathrm{x})/a(\mathrm{x}) \in T(R[\mathrm{x}]$ and for each i, $a_i b(\mathrm{x})/a(\mathrm{x}) = b_i a(\mathrm{x})/a(\mathrm{x}) = b_i$ since $a_i b_j = a_i f(a_j) = f(a_i a_j) = f(a_i) a_j = b_i a_j$ for each i and j. Hence we may identify f with multiplication by the "rational function" $b(\mathrm{x})/a(\mathrm{x})$.

One way to construct the ring of finite fractions corresponding to a given ring R is to simply take the subring of $T(R[\mathrm{x}])$ consisting of those fractions $b(\mathrm{x})/a(\mathrm{x})$ with $a(\mathrm{x}), b(\mathrm{x}) \in R[\mathrm{x}]$ where $a_i b_j = b_i a_j$ for each pair of integers i and j. Another is to follow the method Lambek uses to construct the complete ring of quotients of R, but use only those homomorphisms defined on semiregular ideals. The basic steps are to first note that the product of two semiregular ideals is semiregular. Thus given a pair of semiregular ideals I and J and homomorphisms $f \in Hom_R(I, R)$ and $g \in Hom_R(J, R)$, both $f + g$ and fg can be defined as homomorphisms on IJ. With this method, the ring of finite fractions consists of the equivalence classes of homomorphisms where $f \in Hom_R(I, R)$ and $g \in Hom_R(J, R)$ are equivalent if $f(a) = g(a)$ for each $a \in I \cap J$. For more details on the two constructions see [Lu94b]. We let $Q_0(R)$ denote the ring of finite fractions over R.

We now have enough to completely describe the integral closure of $R[\mathrm{x}]$ when R is a reduced ring. For this we let $\mathcal{S}(R)$ denote the set of semiregular ideals of R and we let $\mathcal{S}_f(R)$ denote the set of finitely generated semiregular ideals of R. The former will be used when we deal with complete integral closure. For a semiregular ideal I, $\langle I{:}I\rangle = \{t \in Q_0(R) \mid tI \subseteq I\}$. Note that $\langle I{:}I\rangle$ can be identified with $Hom_R(I, I)$.

Theorem 2.11. [Lu92, Theorems 3.6 and 3.5] *Let R be a reduced ring. Then the integral closure of R in $Q_0(R)$ is the ring $R^\dagger = \bigcup\{\langle I{:}I\rangle \mid I \in \mathcal{S}_f(R)\}$. Moreover, $R^\dagger[\mathrm{x}]$ is the integral closure of $R[\mathrm{x}]$ in $T(R[\mathrm{x}])$.*

Proof. For the first statement, we first show that each element of $R^\dagger$ is integral over R. For this, let $t \in \langle I{:}I\rangle$ where $I = (a_1, a_2, \ldots, a_n)$ is a finitely generated semiregular ideal of R. Thus there are elements $b_{i,j} \in R$ such that $ta_i = \sum b_{i,j}a_j$. Convert to the matrix equation $(tI - B)A = 0$ with B the $n \times n$ matrix $[b_{i,j}]$ and A the column vector $[a_i]$. Next apply Cramer's Rule and use the fact that I has no nonzero annihilator to conclude that $det(tI - B) = 0$. Clearly, t is a zero of the monic polynomial $det(\mathrm{x}I - B) \in R[\mathrm{x}]$.

As noted earlier, if S is the integral closure of R in $Q(R)$, then the integral closure of $R[\mathrm{x}]$ in $T(R[\mathrm{x}])$ is simply the intersection of $S[\mathrm{x}]$ and $T(R[\mathrm{x}])$. In particular, each element of $T(R[\mathrm{x}])$ that is integral over $R[\mathrm{x}]$ can be written in the form of a polynomial with coefficients in $Q(R)$. Thus $S[\mathrm{x}]$ contains $R^\dagger[\mathrm{x}]$. To finish the proof we show that $S[\mathrm{x}] \bigcap T(R[\mathrm{x}]) = R^\dagger[\mathrm{x}]$. The technique we use is similar to the approach we used above for integral domains.

Suppose $h \in Q(R)[\mathrm{x}] \bigcap T(R[\mathrm{x}])$ is integral over $R[\mathrm{x}]$. Then there are polynomials $f_0, \ldots, f_{n-1} \in R[\mathrm{x}]$ such that $h^{n+1} + f_n h^n + \cdots + f_0 = 0$ and there are polynomials $a, b \in R[\mathrm{x}]$ such that $h = b/a$ with a not a zero divisor. Consider the ideal $J = (b^n, ab^{n-1}, \ldots, a^n)$ of $R[\mathrm{x}]$. As above, $hJ \subseteq J$ where we again use the integrality equation to take care of the only difficult case of showing $hb^n \in J$. Clearly, $ha^ib^j = a^{i-1}b^{j+1} \in J$ for $1 \leq i \leq n$ and $i + j = n$. For m sufficiently large, the polynomial $g(\mathrm{x}) = b^n + x^m ab^{n-1} + \cdots + x^{nm}a^n$ is such that there is no overlap of terms in the sum. Thus the single polynomial g has the same content in R as the ideal J has. Also $hg \in J$. Next apply the content formula to get $C_R(hg)C_R(g)^k = C_R(h)C_R(g)^{k+1}$ for some integer $k \geq 0$. Since $hg \in J$, $C_R(hg) \subseteq C_R(J) = C_R(g)$. Thus $C_R(h)C_R(g)^{k+1} \subseteq C_R(g)^{k+1}$. So $h \in R^\dagger[\mathrm{x}]$.

With this we have enough to state necessary and sufficient conditions for $R[\mathrm{x}]$ to be integrally closed.

Theorem 2.12. [Lu89b, Corollary 4] *Let R be a commutative ring. Then $R[\mathrm{x}]$ is integrally closed if and only if R is reduced and integrally closed in $Q_0(R)$, the ring of finite fractions over R.*

Combining this with Akiba's work we have a complete solution to Gilmer's exercise plus a little more–no restriction to having R be its own total quotient ring.

Theorem 2.13. *Let R be a ring and let $\{x_\lambda\}$ be a nonempty set of indeterminates. Then $R[\{x_\lambda\}]$ is integrally closed if and only if R is reduced and integrally closed in $Q_0(R)$.*

3 When $R[x]$ is n-Root Closed, Root Closed, etc.

Given a positive integer $n \geq 2$ and a pair of rings $R \subseteq S$, R is said to be n-root closed in S if $b^n \in R$ for some $b \in S$ implies $b \in R$. If R is n-root closed in S for each $n \geq 2$, it is said to be root closed in S.

A related notion is $(2,3)$-closed. The ring R is $(2,3)$-closed in S if for $b \in S$, $b \in R$ whenever both b^2 and b^3 are in R. An alternate name for this is to say that R is *seminormal in S* (see, for example, [PP93]). This terminology comes from the fact that the following are equivalent for an integral domain D with corresponding quotient field K ([BC79, Theorem 1], [GH80, Theorem 1.1], [R80, Theorem 1], [Sw80, Theorem 1]): (1) D is $(2,3)$-closed in K, (2) $Pic(D) = Pic(D[x_1, x_2, \ldots, x_m])$ for each $m \geq 1$, (3) $Pic(D) = Pic(D[x])$, (4) for $a, b \in D$, if $a^2 = b^3$, then there is a $c \in D$ such that $a = c^3$ and $b = c^2$, and (4) D is the intersection of the rings $D_P + \mathcal{J}(D'_P)$ where D' is the integral closure of D (in K) and the P's range over the prime ideals of D with $\mathcal{J}(D'_P)$ the Jacobson radical of D' localized at P. Richard Swan proved the equivalence of (3) and (4) for reduced rings, and the condition in (4) is often taken as the definition of a seminormal ring. An example in [GH80] shows that a reduced total quotient ring need not be seminormal in the sense of (2), but is trivially $(2,3)$-closed. The ring in this particular example is a McCoy ring, so the corresponding polynomial ring is integrally closed.

For the pair $R \subset S$, R is *u-closed in S* if for $b \in S$, having both $b^2 - b$ and $b^3 - b^2$ in R implies b is in R. Note that both $b^2 - b$ and $b^3 - b^2$ are in R if and only if $b^2 - b$ and $b^3 - b$ are in R. Closely related is the notion of R being *t-closed in S*. For R to be t-closed in S it must be that simply having $b^2 - rb$ and $b^3 - rb^2$ in R for some $r \in R$ (rather than just $r = 1$) must imply b is in R. The "r" can be 0, so if R is t-closed in S, it is also $(2,3)$-closed in S. The terms u-closed and t-closed first appeared in [O85].

Let $***$-closed represent any one of n-root closed, root closed, (2,3)-closed, u-closed and t-closed. The problem at hand is the following.

> For a reduced ring R, give necessary and sufficient conditions in order that $R[x]$ is $***$-closed in $T(R[x])$.

A first step is to simply note that all of the properties above involve knowing the zeros of some particular type of polynomial with coefficients in R (or $R[x]$). Thus we at least know that if $f \in T(R[x])$ fits the role of the "b in S", it must be in $R^\dagger[x]$.

The easiest of the properties to deal with is u-closed. The proof of the following is nearly identical to one given for [O85, Proposition 7], the only real difference is that in [O85] the rings in question are assumed to be integral domains.

Theorem 3.1. *Let $R \subseteq S$ be a pair of rings and let $f \in S[\mathrm{x}]$ be such that $f^2 - f$ and $f^3 - f^2$ are in $R[\mathrm{x}]$. If f_0 is in R, then $f \in R[\mathrm{x}]$.*

Proof. Assume $f_0 \in R$ and let $f = f_n\mathrm{x}^n + \cdots + f_0$. Since both $f^3 - f^2$ and $f^2 - f$ are in $R[\mathrm{x}]$, so is $f^3 - f = f^3 - f^2 + (f^2 - f)$. The rest of the proof is by induction. Assume we have $f_0, f_1, \ldots, f_i \in R$ and consider the coefficients on x^{i+1} in both $f^2 - f$ and $f^3 - f$. Since $f_0, \ldots, f_i \in R$ and both $f^2 - f$ and $f^3 - f$ are in $R[\mathrm{x}]$, both $2f_{i+1}f_0 - f_{i+1}$ and $3f_{i+1}f_0^2 - f_{i+1}$ are in R. We have $6f_0 + 3 \in R$ and thus $(6f_0 + 3)(2f_{i+1}f_0 - f_{i+1}) - 4(3f_{i+1}f_0^2 - f_{i+1}) = 12f_{i+1}f_0^2 - 3f_{i+1} - 12f_{i+1}f_0^2 + 4f_{i+1} = f_{i+1} \in R$. Continue to the end to see that $f(\mathrm{x})$ is in $R[\mathrm{x}]$.

The immediate consequence is the following.

Theorem 3.2. *Let R be a reduced ring. Then $R[\mathrm{x}]$ is u-closed in $T(R[\mathrm{x}])$ if and only if R is u-closed in $Q_0(R)$.*

Proof. Since $R[\mathrm{x}] \bigcap Q_0(R) = R$ and $Q_0(R)$ is contained in $T(R[\mathrm{x}])$, if $R[\mathrm{x}]$ is u-closed in $T(R[\mathrm{x}])$, then R is u-closed in $Q_0(R)$. For the converse, simply note that if $f \in T(R[\mathrm{x}])$ is such that $f^2 - f$ and $f^3 - f^2$ are in $R[\mathrm{x}]$, then $f \in R^\dagger[\mathrm{x}] \subseteq Q_0(R)[\mathrm{x}]$. Clearly, f_0, the constant term of f, satisfies $f_0^2 - f_0, f_0^3 - f_0^2 \in R$. Hence if R is u-closed in $Q_0(R)$, then $f_0 \in R$ and by the theorem above $f \in R[\mathrm{x}]$.

For a pair of integral domains $R \subseteq S$, it is possible to do a relatively straightforward induction proof to show that $R[\mathrm{x}]$ is t-closed in $S[\mathrm{x}]$ if and only if R is t-closed in S (see [O85, Proposition 8]). More complicated proofs are needed for the t-closed property for arbitrary pairs of rings and for the other three properties: n-root closed, root closed and $(2, 3)$-closed. The basic idea is to establish a theorem like Theorem 2.2 above for each of these four and then use that $R^\dagger[\mathrm{x}]$ is the integral closure of $R[\mathrm{x}]$ in $T(R[\mathrm{x}])$.

With regard to n-root closed, root closed and $(2, 3)$-closed, the desired extension of Theorem 2.2 is due to Brewer, Costa and McCrimmon.

Theorem 3.3. [BCM79, Theorem 1] *Let $R \subseteq T$ be a pair of rings and let* x *be an indeterminate over T.*

(a) *If R is n-root closed in T, then $R[\mathrm{x}]$ is n-root closed in $T[\mathrm{x}]$.*
(b) *If R is root closed in T, then $R[\mathrm{x}]$ is root closed in $T[\mathrm{x}]$.*
(c) *If R is $(2, 3)$-closed in T, then $R[\mathrm{x}]$ is $(2, 3)$-closed in $T[\mathrm{x}]$.*

First note if $s \in R[\mathrm{x}]$ is such that $s^n \in R[\mathrm{x}]$ or both $s^2 - rs$ and $s^3 - rs^2$ are in $R[\mathrm{x}]$ for some $r \in R[\mathrm{x}]$, then for each $b \in R[\mathrm{x}]$, $(bs)^n \in R[\mathrm{x}]$ or both $(bs)^2 - br(bs)$ and $(bs)^3 - br(bs)^2$ are in $R[\mathrm{x}]$. By way of contradiction, declare $s \in T[\mathrm{x}] \backslash R[\mathrm{x}]$ to be a "minimal counterexample" if it satisfies the appropriate condition with (i) minimal degree among all members of $T[\mathrm{x}]\backslash R[\mathrm{x}]$ that satisfy the same condtion and (ii) the integer i for which $s_i \in T\backslash R$ with $s_j \in R$ for all $j < i$ is as large as possible among all polynomials that satisfy (i). The most complicated part of the proof is to show that such an s exists with constant term s_0 such that $s_0 s^k \in R[\mathrm{x}]$ for all k in some approprite finite set of positive integers $\{1, \ldots, m\}$ (note that ns^k may be what is required for the n-root closed case instead of s_0). For such a minimal s, $(s - s_0)\mathrm{x}^{-1}$ will be a (contradictory) counterexample of smaller degree. The most complicated case is for n-root closed, the scheme in this case takes up nearly three pages in [BCM79].

The basic scheme used in [BCM79] was successfully adapted to the t-closed property by Gabriel Picavet and Martine Picavet-L'Hermite in [PP93].

Theorem 3.4. [PP93, Théorème 3.8] *Let $R \subseteq T$ be a pair of rings and let* x *be an indeterminate over T. Then $R[\mathrm{x}]$ is t-closed in $T[\mathrm{x}]$ if and only if R is t-closed in T.*

Combining the last two theorems with Theorem 2.11 above we have the following. As far as I know, statement (d) in the next theorem is new. The other three are from [Lu89b].

Theorem 3.5. (cf. [Lu89b, Corollaries 5 and 6]) *Let R be a reduced ring.*

(a) *$R[\mathrm{x}]$ is n-root closed in $T(R[\mathrm{x}])$ if and only if R is n-root closed in $Q_0(R)$.*
(b) *$R[\mathrm{x}]$ is root closed in $T(R[\mathrm{x}])$ if and only if R is root closed in $Q_0(R)$.*
(c) *$R[\mathrm{x}]$ is $(2,3)$-closed in $T(R[\mathrm{x}])$ if and only if R is $(2,3)$-closed in $Q_0(R)$.*
(c) *$R[\mathrm{x}]$ is t-closed in $T(R[\mathrm{x}])$ if and only if R is t-closed in $Q_0(R)$.*

If $s \in T(R)$ is such that $s^2 - rs = a$ and $s^3 - rs^2 = b$ are in R for some $r \in R$, then the resolvent of the polynomials $\mathrm{x}^2 - r\mathrm{x} - a$ and $\mathrm{x}^3 - r\mathrm{x}^2 - b$ is $a^3 + rab - b^2$ with $a^3 + rab - b^2 = 0$ since the polynomials have a common zero. Reversing the view, Picavet and Picavet-L'Hermite defined a ring R to be *t-closed* if for $a, b, r \in R$ with $a^3 + rab - b^2 = 0$, there is an element $c \in R$ such that $a = c^2 - rc$ and $b = c^3 - rc^2$. A ring R is said to be a *weak Baer ring* if the annihilator of each element is idempotent (see, for example, [PP93]). If R is a weak Baer ring, then R is t-closed in $T(R)$ if and only if R is t-closed [PP93, Théorème 2.3]. Moreover, if R is a weak Baer ring, then $R[\mathrm{x}]$ is t-closed if and only if R is t-closed [PP93, Théorème 2.19]. A similar definition is valid for a ring being u-closed: for all $a, b \in R$, if $a^3 + ab - b^2 = 0$, then there is an element $c \in R$ such that $a = c^2 - c$ and $b = c^3 - c^2$. The following are equivalent for a weak Baer ring R: (1) R is u-closed, (2) R is u-closed in $T(R)$, (3) $R[\mathrm{x}]$ is u-closed in $T(R[\mathrm{x}])$ and (4) $R[\mathrm{x}]$ is u-closed [Pi98, Theorem 1.12 and Proposition 1.34].

4 Complete Integral Closure

Lastly for polynomial rings we have the notion of R being completely integrally closed in S. An element $s \in S$ is said to be almost integral over R (as an element of S) if $R[s]$ is contained in a finitely generated R-submodule of S. If each such element of S is in fact in R, then R is said to be completely integrally closed in S. The definitions leave open the possibility that $s \in S$ may fail to be almost integral over R based on the ring S but then be almost integral over R based on some larger ring T that contains S. Perhaps worse is that R can be completely integrally closed in S and S completely integrally closed in T, but with R not completely integrally closed in T. For example, given a field K and indeterminates x and y, the ring $R = K[\{\mathrm{xy}^n \mid n \geq o\}]$ is completely integrally closed in $S = K[\mathrm{x}, \mathrm{y}]$ and S is completely integrally closed in $T = K(\mathrm{x}, \mathrm{y})$. However, y is almost integral over R when considered as an element of T since $\mathrm{y}^n \in R + (1/\mathrm{x})R$, a finitely generated R-submodule of T (but not a submodule of S).

Recall that with regard to an integral domain D, an element t of its corresponding quotient field K, is almost integral over D if and only if there is a nonzero $r \in D$ such that $rt^n \in D$ for each positive integer n. This makes it relatively easy to show that $D[\mathrm{x}]$ is completely integrally closed in $K(\mathrm{x})$ if and only if D is completely integrally closed in K. Since $K[\mathrm{x}]$ is completely integrally closed in $K(\mathrm{x})$, if $t \in K(\mathrm{x})$ is almost integral over $D[\mathrm{x}]$, it must be in $K[\mathrm{x}]$. Also for $s \in K[\mathrm{x}]$, s is almost integral over $D[\mathrm{x}]$ (as an element of $K(\mathrm{x})$) if and only if $\mathrm{x}^m s$ is almost integral over $D[\mathrm{x}]$ for some (hence all) positive integer(s) m. The proof from this point on is a straightforward argument based on recursion/induction. First show that the constant term of s is in D, subtract the constant term and show that $s - s_0 = \mathrm{x}r$ is almost integral so that s_1, the constant term of r, is in D. Continue until each coefficient of s is in D.

Based on the results above, one may be tempted to think that $R[\mathrm{x}]$ is completely integrally closed in $T(R[\mathrm{x}])$ if and only if R is reduced and completely integrally closed in $Q_0(R)$. This is not quite right. An odd thing can happen, R can be reduced and completely integrally closed in $Q_0(R)$ but there can be an element $t \in Q_0(R)$ that is almost integral over R when considered as an element of $T(R[\mathrm{x}])$. An example of such behavior is given in [Lu94b, Example 10]. There is a way around this problem. For a trio of rings $R \subseteq S \subseteq T$, we say that R is *completely integrally closed in S as a subring of T* if each element of $s \in S$ that is almost integral over R as an element of T, is in R [Lu97].

Theorem 4.1. [Lu92, Theorem 3.4] *Let R be a reduced ring and let $R^{\#} = \bigcup\{\langle I{:}I\rangle \mid I \in \mathcal{S}(R)\}$.*

(a) *$R^{\#}$ is the complete integral closure of R in $Q_0(R)$ as a subring of $T(R[\mathrm{x}])$ and $R^{\#}[\mathrm{x}]$ is the complete integral closure of $R[\mathrm{x}]$ in $T(R[\mathrm{x}])$.*
(b) *$R[\mathrm{x}]$ is completely integrally closed in $T(R[\mathrm{x}])$ if and only if R is completely integrally closed in $Q_0(R)$ as a subring of $T(R[\mathrm{x}])$.*

5 Semigroup Rings

In this section we consider semigroup rings of the form $R[S]$ where $\langle S,+\rangle$ is a torsion-free commutative cancellative monoid. Such an S is sometimes referred to as a *torsion-free grading monoid* (see, for example [AA82]). The elements of $R[S]$ can be thought of as finite sums of the form $\sum r_i\mathrm{x}^{\alpha_i}$ with each $r_i \in R$ and each $\alpha_i \in S$. An alternate notation for $R[S]$ is $R[X;S]$ which is used in David Anderson's article in this same volume dealing with Gilmer's contributions in the area of semigroup rings [An06]. Since S is cancellative, it is a submonoid of a group G where $G = \{\alpha - \beta \mid \alpha,\beta \in S\}$. The group G is referred to as the *quotient group of* S. An element γ of G is said to be *integral over* S if there is a positive integer n such that $n\gamma \in S$, and S is *n-root closed in* G if $n\gamma \in S$ implies $\gamma \in S$. The set of elements that are integral over S forms a submonoid of G, called the *integral closure of* S. Note that S is root closed in G if and only if it is integrally closed in G. Also S is $(2,3)$-*closed in* G, if $\gamma^2, \gamma^3 \in S$ implies $\gamma \in S$. It is also possible to extend the notion of being almost integral by saying that $\beta \in G$ is *almost integral over* S it there is an element $\alpha \in S$ such that $\alpha + n\beta \in S$ for each positive integer n. Finally, the set of elements that are almost integral over S forms a submonoid of G, called the *complete integral closure*, and S is *completely integrally closed* if it equals its complete integral closure. If R is an integral domain with quotient field K, $R[S]$ is an integral domain with quotient field $K(G)$.

We begin by recalling a theorem from Gilmer's book Commutative Semigroup Rings [Gi84]. For semigroup rings, this theorem plays a role similar to that played by the one of Prüfer dealing with polynomial rings.

Theorem 5.1. [Gi84, Corollary 12.11] *Let R be an integral domain with quotient field K and let S be a torsion-free commutative cancellative monoid with quotient group G.*

(a) *The integral closure of $R[S]$ is the ring $R'[S']$ where R' is the integral closure of R and S' is the integral closure of S in G.*
(b) *$R[S]$ is integrally closed if and only if R is integrally closed and S is integrally closed.*

Since the finite sums and products of integral elements are again integral, it is clear that each element of $R'[S']$ is integral over $R[S]$. Thus all one needs prove is statement (b). Half of this is easy. If $R[S]$ is integrally closed, then R must be integrally closed in K and S must be integrally closed in G. For the converse, first show that $R[G]$ is integrally closed when R is integrally closed. For this the key is to note that if f is integral over $R[G]$, then there is a finitely generated subgroup H of G such that f is integral over $R[H]$ and in the quotient field of $R[H]$. Since G is Abelian and torsion-free, H is isomorphic to $\mathbb{Z}^n$ for some integer n. With this $R[H]$ is isomorphic to $R[\mathrm{x}_1,\ldots,\mathrm{x}_n][\mathrm{x}_1^{-1},\ldots,\mathrm{x}_n^{-1}]$. As R is an integrally closed integral domain, so is $R[\mathrm{x}_1,\ldots,\mathrm{x}_n][\mathrm{x}_1^{-1},\ldots,\mathrm{x}_n^{-1}]$. It follows that $f \in R[H]$ and we have $R[G]$

integrally closed. Now assume $f \in R[G]$ is integral over $R[S]$. Proving $f \in R[S]$ is not as straightforward as one might think. The appoach that Gilmer takes is to first show that S is the intersection of the valuation monoids of G that contain $S \bigcap H$. Then he shows that if V is a valuation monoid, then $R[V]$ is integrally closed. A valuation on G is a map ν from G onto a totally ordered Abelian group Γ such that (i) $\nu(a+b) = \nu(a) + \nu(b)$ with $\nu(0) = 0$. The corresponding valuation monoid is the set $V = \{a \in G \mid \nu(a) \geq 0\}$. Clearly V is integrally closed in G.

In the case of an integral domain R, to show $R[V]$ is integrally closed in $R[G]$ it suffices to show that ν can be extended to a paravaluation on $R[G]$ with corresponding paravaluation ring $R[V]$. This is done using the same notion used earlier to a extend a paravaluation on a ring T to the polynomial ring $T[\mathrm{x}]$. It is trivial to extend ν to $R[G]$, simply set $\nu(f) = min\{\nu(\alpha_i) \mid f_i \neq 0$ in $f = \sum f_i \mathrm{x}^{\alpha_i} \neq 0\}$ and $\nu(0) = \infty$. It is clear that $\nu(f+g) \geq min\{\nu(f), \nu(g)\}$ and $\nu(r) = 0$ for each nonzero $r \in R$. For a nonzero product fg, first isolate the α_i's of $f = \sum f_i \mathrm{x}^{\alpha_i}$ with minimum ν value and the β_j's of $g = \sum g_j \mathrm{x}^{\beta_j}$ with minimum ν. Since G is Abelian and torsion-free, there is a total order compatible with the operation. So among all of the α_i's with minimum ν value there is a unique α_n with $\alpha_n < \alpha_i$ for all α_i with minimum ν value. Similarly there is a unique β_m with minimum ν value where $\beta_m < \beta_j$ for all other β_j with minimum ν value. In the product fg, the term $f_n g_m \mathrm{x}^{\alpha_n + \beta_m}$ is unique in that for all other $f_i g_j \mathrm{x}^{\alpha_i + \beta_j}$, $\alpha_i + \beta_j \neq \alpha_n + \beta_m$ either because at least one of α_i or β_j does not have minimal ν value or because both do but $\alpha_n + \beta_m < \alpha_i + \beta_j$ under the total order on G. Since R is an integral domain, $f_n g_m \neq 0$. Thus $\nu(fg) = \nu(\alpha_n) + \nu(\beta_m) = \nu(f) + \nu(g)$. Clearly $R[V]$ is the corresponding valuation ring. Thus $R[V]$ is integrally closed.

Note that for a ring with nonzero zero divisors, we might have $f_n g_m = 0$ with neither f_n nor $g_m = 0$. In fact we could have $fg = 0$ with neither f nor g the zero polynomial. Thus, the simple extension we have used for ν will not work for rings with nonzero zero divisors. In fact, $R[V]$ cannot be a paravaluation ring of $R[G]$ except in the trivial case that $V = G$. If R is reduced, select a minimal prime P and set $\nu(r) = \infty$ when $r \in P$ and $\nu(r) = 0$ for $R \in R \backslash P$. Then for $\nu(r\mathrm{x}^{\alpha}) = \nu(r) + \nu(\alpha)$, and extend to $R[G]$ using the minimum of $\nu(f_i \mathrm{x}^{\alpha_i})$. With this definition for the extension ν, $R[V] + P[G]$ will be a paravaluation ring of $R[G]$. Since R is reduced, the intersection of its minimal primes is (0). Thus $R[V] = \bigcap(R[V] + P[G])$ where the P's range over the minimal primes of R. Therefore $R[V]$ is integrally closed [Gi84, Theorem 12.9].

Gilmer also handles the problem of determining the complete integral closure of $R[S]$ when R is an integral domain.

Theorem 5.2. [Gi84, Theorem 12.5, Corollaries 12.6 and 12.7] *Let R be an integral domain with quotient field K and let S be a torsion-free commutative cancellative monoid with quotient group G.*

(a) *The complete integral closure of $R[S]$ in its quotient field is the ring $R^*[S^*]$ where R^* is the complete integral closure of R in K and S^* is the complete integral closure of S in G.*
(b) *$R[S]$ is completely integrally closed in its quotient field if and only if R is completely integrally closed in K and S is completely integrally closed in G.*

In [AA82, Theorem 6.2], Dan and David Anderson extended Gilmer's results to seminormality for semigroup rings over integral domains. Later in the same year, David took care of n-root closed and root closed semigroup rings (and seminormal ones), again over integral domains [An82, Corollary 2.5].

Theorem 5.3. ([AA82, Theorem 6.2] and [An82, Corollary 2.5]) *Let R be an integral domain with quotient field K and let S be a torsion-free commutative cancellative monoid with quotient group G.*

(a) *$R[S]$ is seminormal (equivalently, $(2,3)$-closed in $K(G)$) if and only if R is seminormal (equivalently $(2,3)$-closed in K) and S is $(2,3)$-closed in G.*
(b) *$R[S]$ is n-root closed in $K(G)$ if and only if R is n-root closed in K and S is n-root closed in G.*
(c) *$R[S]$ is root closed in $K(G)$ if and only if R is root closed in K and S is root closed in G.*

As with the polynomial ring $R[\mathrm{x}]$, Gilmer's and the Andersons' results can be extended to rings with nonzero zero divisors. The first step is to extend Gilmer's two theorems to $R[S]$ where R is von Neumann regular. The results in the next theorem are based on Theorem 18.8 of [Gi84] and Theorems 3.1 and 3.7 and Corollaries 3.2 and 3.3 of [Lu92].

Theorem 5.4. *Let R be a von Neumann regular ring and let S be a torsion-free commutative cancellative monoid with quotient group G.*

(a) *$R[G]$ is both integrally closed and completely integrally closed in $T(R[G])$.*
(b) *The integral closure of $R[S]$ in $T(R[S]) = T(R[G])$ is the ring $R[S']$ where S' is the integral closure of S in G.*
(c) *$R[S]$ is integrally closed in $T(R[S])$ if and only if S is integrally closed in G.*
(d) *$R[S]$ is completely integrally closed in $T(R[S])$ if and only if S is completely integrally closed in G.*

As with the proofs for integral domains, the key to proving the statement in (a) is to realize that having $f \in T(R[G])$ (almost) integral over $R[G]$ implies there is a finitely generated subgroup H of G such that f is in $T(R[H])$ and (almost) integral over $R[H]$. Then simply use the fact that H is isomorphic to $\mathbb{Z}^n$ for some n and therefore $R[H]$ is isomorphic to $R[\mathrm{x}_1,\ldots,\mathrm{x}_n][\mathrm{x}_1^{-1},\ldots,\mathrm{x}_n^{-1}]$. Since R is von Neumann regular, $R[\mathrm{x}_1,\ldots,\mathrm{x}_n][\mathrm{x}_1^{-1},\ldots,\mathrm{x}_n^{-1}]$ is (completely) integrally closed in its total quotient ring. Hence $f \in R[H] \subseteq R[G]$. For the other three statements, use statement (a), Gilmer's theorems and the fact that R_M is a field for each maximal/minimal ideal M. For example, if $f \in T(R[S])$ is integral over $R[S]$, then it is in $R[G]$. Next, since R_M is a field, $R_M[S']$ is the integral closure of $R_M[S]$. So the image of f is in $R_M[S']$ for each M. Since R is reduced, no nonzero coefficient of f is zero in R_M. Thus $f \in R[S']$.

We will end with one grand theorem that takes care of all but u-closed and t-closed semigroup rings. It is a combination of Corollaries 3.7 and 3.11 of [An99], Theorems 3.4 and 3.7 of [Lu92] and Theorem 5 of [Lu97].

Theorem 5.5. *Let R be a reduced ring and let S be a torsion-free commutative cancellative monoid with quotient group G.*

(a) *The integral closure of $R[S]$ is the ring $R^\dagger[S']$ where $R^\dagger$ is the integral closure of R in $Q_0(R)$ and S' is the integral closure of S in G.*
(b) *$R[S]$ is integrally closed if and only if R is integrally closed in $Q_0(R)$ and S is integrally closed in G.*
(c) *The complete integral closure of $R[S]$ in $T(R[S])$ is the ring $R^{\#}[S^*]$ where $R^{\#}$ is the complete integral closure of R in $Q_0(R)$ as a subring of $T(R[\mathrm{x}])$ and S^* is the complete integral closure of S in G.*
(d) *$R[S]$ is completely integrally closed in $T(R[S])$ if and only if R is completely integrally closed in $Q_0(R)$ as a subring of $T(R[\mathrm{x}])$ and S is completely integrally closed in G.*
(e) *$R[S]$ is n-root closed in $T(R[S])$ if and only if R is n-root closed in $Q_0(R)$ and S is n-root closed in G.*
(f) *$R[S]$ is root closed in $T(R[S])$ if and only if R is root closed in $Q_0(R)$ and S is root closed in G.*
(g) *$R[S]$ is $(2,3)$-closed in $T(R[S])$ if and only if R is $(2,3)$-closed in $Q_0(R)$ and S is $(2,3)$-closed in G.*

References

[Ak80] Akiba, T.: Integrally closedness of polynomial rings. Japan. J. Math., **6**, 67–75 (1980).

[AA82] Anderson, D.D., Anderson, D.F.: Divisorial ideals and invertible ideals in a graded integral domain. J. Algebra, **76**, 549–569 (1982).

[An82] Anderson, D.F.: Root closure in integral domains. J. Algebra, **79**, 51–59 (1982).

[An99] Anderson, D.F.: Root closure in commutative rings: a survey. In: Dobbs, D.E., Fontana, M., Kabbaj, S (eds) Advances in Commutative Ring Theory. Lecture Notes in Pure and Applied Mathematics, **205**, 55–71. Dekker, New York (1999).

[An06] Anderson, D.F.: Robert Gilmer's work on semigroup rings, this volume.

[BC79] Brewer, J., Costa, D.L.: Seminormality and projective modules of polynomial rings. J. Algebra, **58**, 208–216 (1979).

[BCM79] Brewer, J., Costa, D.L., McCrimmon, K.: Seminormality and root closure in polynomial rings and algebraic curves. J. Algebra, **58**, 217–226 (1979).

[BHM54] Butts, H.S., Hall, M., Mann, H.B.: On integral closure. Canad. J. Math., **6**, 471–473 (1954).

[E61] Endo, S.: On semi-hereditary rings. J. Math. Soc. Japan, **13**, 109–119 (1961).

[F91] Faith, C.: Annihilators, associated prime ideals and Kasch-McCoy commutative rings. Comm. Algebra, **119**, 1867–1892 (1991).

[Gi84] Gilmer, R.: Commutative Semigroup Rings. The University of Chicago Press, Chicago (1984).

[Gi92] Gilmer, R.: Multiplicative Ideal Theory. Queen's Papers in Pure and Applied Mathematics, **90**. Queen's University Press, Kingston (1992).

[GH80] Gilmer, R., Heitman, R.: On $Pic(R[X])$ for R seminormal. J. Pure Appl. Algebra, **16**, 251-267 (1980).

[Gra82] Gräter, J.: Integral closure and valuation rings with zero-divisors. Studia Sci. Math. Hungarica, **17**, 457–458 (1982).

[Gri70] Griffin, M.: Generalizing valuations to commutative rings. Queen's Math. Preprint No. 1970-40. Queen's Univ., Kingston (1970).

[H88] Huckaba, J.: Commutative Rings with Zero Divisors. Dekker, New York (1988).

[HK79] Huckaba, J., Keller, J.: Annihilation of ideals in commutative rings. Pac. J. Math., **83**, 375–379 (1979).

[La86] Lambek, J.: Lectures on Rings and Modules. Chelsea, New York (1986).

[Le47] Lesieur, L.: Sur les domaines d'intégrite intégralement fermes. C. R. Acad. Sci. Paris, Series A-B, **229**, 691–693 (1947).

[Lu89a] Lucas, T.: Characterization when $R[X]$ is integrally closed. Proc. Amer. Math. Soc., **105**, 861–867 (1989).

[Lu89b] Lucas, T.: Characterization when $R[X]$ is integrally closed, II. J. Pure Appl. Algebra, **61**, 49–52 (1989).

[Lu89c] Lucas, T.: Root closure and $R[X]$. Comm. Algebra, **17**, 2393–2414 (1989).

[Lu92] Lucas, T.: The complete integral closure of $R[X]$. Trans. Amer. Math. Soc., **330**, 757–768 (1992).

[Lu94a] Lucas, T.: Strong Prüfer rings and the ring of finite fractions. J. Pure Appl. Algebra, **84**, 59–71 (1994).

[Lu94b] Lucas, T.: The ring of finite fractions. In Cahen,P.-J., Costa, D.L., Fontana, M., Kabbaj, S. (eds) Commutative Ring Theory. Lecture Notes in Pure and Appl. Math., **153**, 181–191. Dekker, New York (1994).

[Lu97] Lucas, T.: The complete integral closure of $R(X)$. In Anderson, D.D. (ed) Factorization in Integral Domains. Lecture Notes in Pure and Appl. Math., **189**, 401–415. Dekker, New York (1997).

[O85] Onoda, N., Sugatani, T., Yoshida, K.-I.: Local quasinormality and closedness type criteria. Houston J. Math., **11**, 247–256 (1985).

[Pi98] Picavet, G.: Anodality, Comm. Algebra. **26**, 345–393 (1998).

[PP93] Picavet, G., Picavet-L'Hermite, M.: Morphisme t-clos. Comm. Algebra, **21**, 179–219 (1993).

[PP95] Picavet, G., Picavet-L'Hermite, M.: Anneaux t-clos. Comm. Algebra, **23**, 2643–2677 (1995).

[PP00] Picavet, G., Picavet-L'Hermite, M.: t-closedness. In Chapman, S., Glaz, S. (eds) Non-Noetherian Commutaive Ring Theory. 369–386. Kluwer, Dordrecht (2000).

[Pr32] Prüfer, H.: Untersuchungen über Teilbarkeitseigenschaften in Körpern. J. reine angew. Math., **168**, 1–36 (1932).

[Q71] Quentel, Y.: Sur la compacité du spectre minimal d'un anneau. Bull. Soc. Math. France, **99**, 265–272 (1971).

[R80] Rush, D.: Seminormality. J. Algebra, **67**, 377-384 (1980).

[Sa57] Samuel, P.: La notion de place dans un anneau. Bull. Soc. Math. France, **85**, 123–133 (1957).

[Sw80] Swan, R.: On seminormality. J. Algebra, **67**, 210–229 (1980).

Punctually free ideals

Jack Ohm

900 Fort Pickens Rd, 215, Pensacola Bch, FL 32561, veebc@earthlink.net

To Robert Gilmer, in celebration of his distinguished career as researcher and teacher, and in remembrance of the good times we had working together many years ago.

This is intended to be a brief introduction to ideals that are at every prime either 0 or principal generated by a regular element. Such ideals divide into two classes, those of mixed rank 0 or 1 and those of constant rank 1. In the f.g. case, the first class includes the projective ideals and the second the invertible ideals. We give some examples that delineate the boundaries of these subdivisions.

1 Projective ideals

1.1 The language

Abbreviations: f.g. = finitely generated, m.s. = multiplicative system, nhbd = neighborhood, rk = rank, PP = punctually principal, PT = punctually trivial, PF = punctually free. Notation: R denotes a commutative ring (with identity) and Q its total quotient ring. If A and B are nonempty subsets of R and C a nonempty subset of Q, then $(A\colon B)_C = \{\xi \in C \mid \xi B \subseteq A\}$. Also, $A^0 = \{r \in R \mid rA = 0\} = 0\colon A$.

Spec R denotes the set of prime ideals of R with the Zariski topology: given a prime p, for every $s \notin p$, we define a (basic) nhbd N_s of p by $N_s = \{\text{primes } q \mid s \notin q\}$. Note that the intersection of all nhbds of p is the set $\{\text{primes } q \mid q \subseteq p\}$.

For an ideal A of R, $V(A)$ denotes the closed subset of Spec R consisting of all prime ideals containing A, and $U(A)$ denotes the complementary open subset consisting of all primes not containing A.

If I is an ideal, p is a prime ideal, and $\mathcal{P}$ is a certain property, we say that I has $\mathcal{P}$

at p if $I_p = IR_p$ has $\mathcal{P}$,
on the nhbd N_s of p if I has $\mathcal{P}$ at every prime $q \in N_s$,
at N_s, or *at* s, if $I_s = IR_s$ has $\mathcal{P}$,
punctually if I has $\mathcal{P}$ at every prime of R, and
globally if I itself has $\mathcal{P}$.

Similarly, if S is a m.s. of R, we say that I has $\mathcal{P}$ at S if $I_S = IR_S$ has $\mathcal{P}$, etc. When the intent is clear, we shall often identify an element $a \in R$ with its canonical image $a/1 \in R_S$.

1.2 Finite generation of ideals

If A is an ideal and I is a f.g. ideal, then $A\colon I$ localizes well, i.e., for every m.s. S, $(A\colon I)_R R_S = (AR_S\colon I)_{R_S}$. By applying this observation to ideals of the form $A\colon b$, we see that

two ideals that are punctually equal are equal.

Moreover, if I is a f.g. ideal and p is a prime ideal, then $(I^0)_p = (I_p)^0$, so there is no ambiguity in speaking of the annihilator of I at p or the annihilator at p of I.

If I is f.g., only finitely many equations are needed to express that a subset of I generates I, so

if I is f.g. and a subset A of I generates I at a prime p, then A generates I at a nhbd of p;

moreover, by replacing R by the specified nhbd ring R_s, we see that

if A generates I <u>on</u> a nhbd N of p, then A generates I <u>at</u> N.

Since Spec R is quasicompact,

I is f.g. if (and only if) I is f.g. <u>at</u> a nhbd of every prime.

Note, however, that I may be f.g. at every prime, i.e., punctually f.g., but not be f.g. (see [Gil68, p. 584, Appendix 3] or [[HO72, p. 276, Example 2.2] for an example of an *almost Dedekind* domain which is not Dedekind).

1.3 Regular elements and annihilator ideals

An element $a \in R$ is said to be *regular* if $a^0 = 0$, i.e., if a is not a zerodivisor, and an ideal of R is said to be regular if it contains a regular element. An element is punctually regular iff it is regular.

If a^0 is f.g., then $a^0 = 0$ at a prime p iff $a^{00} = (1)$ at p; hence then the open set $U(a^{00})$ is the set of all primes at which a is regular. More generally, if A is an ideal such that a^0 is f.g. for every $a \in A$, then the set of primes at which A is regular is the open set $\cup \{U(a^{00}) \mid a \in A\}$.

If A is a regular ideal, then $A^0 = 0$; and the converse is true if R is noetherian, for then if A consists entirely of zerodivisors, it must be contained in one of the finitely many associated primes of zero and such a prime is of the form $0 \colon b$ for some nonzero $b \in R$.

1.4 Regular elements in a reduced ring R

If a is a zerodivisor of R, then $a^0 \neq 0$; so if R is reduced, then there exists a prime ideal q of R which does not contain a^0, and it follows that a is 0 at q. Thus,

if R is reduced and a is punctually nonzero, then a is a regular element.

(More generally, if I is an ideal, then this argument applied to the ideal $I \colon a$ shows that if R is reduced and punctually $a \notin I$, then for all $r \neq 0$ in R, $ra \notin I$.)

Applying this observation to the quasilocal ring obtained by localizing R at one of its primes p (note that if R is reduced, then so also is R_p), we see that

if R is reduced and a is nonzero at all primes $q \subseteq p$, then a is regular at p.

Now suppose R is reduced and I is an ideal of R which is principal at the prime p and is generated at p by the element $a \in I$. If I is nonzero at every prime $q \subseteq p$, then a is also nonzero at every such q, hence a is regular at p by the preceding remark. Thus,

if R is reduced and I is an ideal which is principal at the prime p and nonzero at every prime $q \subseteq p$, then I is rk 1 free at p.

Note that we have not assumed that I is f.g. here; instead, the reduced hypothesis allows us to use a condition on the primes contained in p rather than on the possibly larger sets that comprise the nhbds of p.

1.5 Punctually trivial (PT) ideals

An ideal A is said to be *trivial* if A equals 0 or (1). A punctually trivial ideal is an ideal which is trivial at every prime. Such a PT ideal is punctually idempotent, hence idempotent, i.e., $A^2 = A$. Moreover, a f.g. ideal is PT iff it is generated by an idempotent:

Proposition 1.1. *If A is an ideal, then the following are equivalent:*

(i) *A is f.g. and PT;*
(ii) *every prime ideal has a nhbd on which A is either constantly 0 or constantly (1);*

(iii) $A + A^0 = (1)$;
(iv) *there exists an ideal* B *such that* $A + B = (1)$ *and* $AB = 0$;
(v) A *is generated by an idempotent;*
(vi) A *is f.g. and* $A = A^2$.

Moreover, (iv) (clearly) implies

(vii) *there exists an ideal* B *such that Spec* R *is the disjoint union of the closed sets* $V(A)$ *and* $V(B)$,

and if R *is reduced then (vii) implies (iv).*

Proof. i) $\Longrightarrow$ ii): Since A is f.g., if 0 (resp., 1) generates A at a prime p, then 0 (resp., 1) generates A on a nhbd of p. ii) $\Longrightarrow$ i): If the ideal A is constantly 0 (resp., (1)) on a nhbd N of a prime p, then it is 0 (resp., (1)) <u>at</u> N; and if A is f.g. at a nhbd of every prime, then A is f.g. (see § 1.2). i) $\Longrightarrow$ iii): Since A is f.g., A^0 localizes well, hence A^0 is PT of opposite parity to A, which implies $A + A^0 = (1)$ punctually and therefore also globally. iii) $\Longrightarrow$ iv): Take $B = A^0$. iv) $\Longrightarrow$ v): Write $1 = a + b$ with $a \in A$ and $b \in B$, and then $A = (a)$ and $a^2 = a$. v) $\Longrightarrow$ vi): Trivial. vi) $\Longrightarrow$ i): It suffices to verify that if R, m is quasilocal and v) holds, then A is trivial. If $A \neq R$, then $A = mA$, and since A is f.g., this implies $A = 0$ by Nakayama's lemma.

Finally, suppose R is reduced. If there exists an ideal B such that $V(A) \cup V(B) = \text{Spec } R$ is a disjoint union, then $A+B = (1)$ and every prime contains AB; and since R is reduced, the latter condition implies $AB = 0$.

1.6 Punctually principal (PP) and punctually free (PF) ideals

If a f.g. ideal I of R is PF, i.e., if I is either 0 or regular principal at every prime, then I^0 is PT; and the converse holds if, say, R is noetherian (see § 1.3) or I is PP. Thus,

a f.g. ideal I is PF iff I is PP and I^0 is PT.

The next proposition shows that for such an ideal the annihilator I^0 acts like the kernel in a free presentation of I. Recall that an R-module M is said to be finitely presented if there exists an integer $n \geq 1$ and an exact sequence of R-modules $0 \to K \to R^n \to M \to 0$ with K f.g.; and when the kernel is f.g. in one such sequence, then it is f.g. in every such sequence (by the existence of pull-backs for modules).

Proposition 1.2. *Suppose the ideal* I *is f.g. and PP. Then* I *is finitely presented iff* I^0 *is f.g.*

Proof. If I is PP and f.g., then an element $a \in I$ which generates I at a prime p also generates I at a nhbd of p. Therefore at a nhbd of each prime we have a presentation of the form

$$0 \to I^0 \to R \to I \to 0.$$

If I is finitely presented, then it follows that I^0 is f.g. at a nhbd of each prime and is therefore f.g. Conversely, if I^0 is f.g., then there exists a nhbd of each prime at which the kernel K in a presentation $0 \to K \to R^n \to I \to 0$ is f.g.; and therefore K is f.g.

Corollary 1.3. *If I is a f.g. PF ideal, then the following are equivalent:*

(i) *rk I is constant on a nhbd of each prime.*
(ii) *rk I^0 is constant on a nhbd of each prime.*
(iii) *I^0 is f.g. (hence is generated by an idempotent).*
(iv) *I is finitely presented.*

Proof. Since I is f.g. PF, I^0 is PT, and rk I^0 is of opposite parity to that of I, hence i)$\Longleftrightarrow$ii). Moreover, ii)$\Longleftrightarrow$iii) by 1.1 and iii)$\Longleftrightarrow$iv) by 1.2.

1.7 Projective ideals

Recall that an ideal I (or, more generally, an R-module) is said to be *projective* if given a surjective homomorphism $L \to M$ of R-modules, the induced R-module homomorphism Hom $(I, L) \to$ Hom (I, M) is also surjective. Equivalently, I is projective iff I is a direct summand of a free R-module. (This notion is fundamentally module-theoretic rather than ideal-theoretic. The same comment applies to the notion of finitely presented of § 1.6.) The following lemma says that the R-module Hom (P, M) extends well through a flat extension of R when M is finitely presented.

Lemma 1.4. (See [Bou89, p. 23, Proposition 11] or [Eis95, p. 69, Proposition 2.10].) *Let R' be a flat R-algebra and P and M be two R-modules, and let $'$ denote tensor product w.r.t. R', e.g., $P' = P \otimes_R R'$. If P is finitely presented and R' is R-flat, then the canonical homomorphism $Hom_R(P, M)' \to Hom_{R'}(P', M')$ is an isomorphism of R'-modules.*

Theorem 1.5. *If I is a f.g. ideal, then I is projective iff I is PF and finitely presented.*

Proof. $\Longrightarrow$: Since I is f.g., there exists an exact sequence of R-modules

$$0 \to K \to R^n \to I \to 0, \qquad (1)$$

and since I is projective, the sequence splits, i.e., $R^n \cong K \oplus I$, and thus I is finitely presented. To prove that I is PF, it suffices to show that a f.g. projective I over a quasilocal R, m is free. If n is chosen minimal in a resolution (1), then $K \subseteq mR^n$. But it follows from the definition of projective that the homomorphism $K \to R^n$ has an inverse homomorphism from $R^n \to K$, and therefore $K = mK$, hence by Nakayama's lemma $K = 0$ and $R^n \cong I$.

$\Longleftarrow$: We must prove that given a surjective homomorphism $L \to M$ of R-modules, the induced homomorphism Hom $(I, L) \to$ Hom (I, M) is also surjective. Since I is finitely presented, Hom $(I,\ )$ localizes well by 1.4; and since I is PF, the induced homomorphism $\mathrm{Hom}(I, L) \to \mathrm{Hom}(I, M)$ is punctually surjective. But then the cokernel of this homomorphism is punctually 0 and therefore 0, so we have the desired surjectivity.

By combining 1.3 and 1.5, we obtain the following:

Corollary 1.6. *If I is a f.g. ideal, then I is projective. iff I is PF and of constant rk on a nhbd of each prime.*

If a f.g. ideal I is 0 at a prime p, then it is 0 at some nhbd of p; more generally, if I is principal at p, then an element of I which generates at p generates at some nhbd of p. But if I is principal and regular, i.e., rk 1 free, at p, then regularity does not necessarily spread out to a nhbd of p. What is needed to ensure that the regularity should spread out is that I^0 should be f.g.

Proposition 1.7. *Suppose I is a f.g. ideal such that I is rk 1 free at a prime p. If either*

(i) *I^0 is f.g., or*
(ii) *I is rk 1 free on a nhbd of p,*

then there exists a nhbd N of p such that I is rk 1 free at *N.*

Proof. Since I is f.g. and I is rk 1 free at p, I is principal at a nhbd N_1 of p and I^0 is 0 at p. If I^0 is f.g., then it is also 0 at a nhbd N_2 of p; and then I is principal and I^0 is 0 at $N = N_1 \cap N_2$, hence I is rk 1 free at N. Similarly, if I is rk 1 free on a nhbd of p, then there exists a nhbd N of p such that I is principal at N and rk 1 free on N. But this means I^0 is 0 on N and therefore 0 at N, hence I is principal and regular at N.

1.8 Two examples of a principal PF ideal which is not projective

These are both examples of a principal ideal I whose annihilator I^0 is PT but not f.g. By § 1.6 such an I is PF but not finitely presented and therefore also not projective.

The first example: Let R_0 be the product of infinitely many copies, say indexed by the natural numbers, of the field $Z/2Z$, and let S_0 be the direct sum ideal of R_0. Then S_0 is not f.g., and S_0 consists entirely of idempotents, hence is PT. Now let $R = R_0[X]$, let $\overline{R} = R/S_0X$, and let $(\overline{X})$ be the ideal of $\overline{R}$ generated by $\overline{X}$. Note first that $\overline{X}^0 = \overline{S_0}$. Since $\overline{S_0}$ is generated by idempotents, it is a PT ideal of $\overline{R}$. On the other hand, $\overline{S_0}$ is not principal and therefore cannot be f.g. by Proposition 1.1.

The second example (due to Roger Wiegand): Let

$$R_1 = \prod \{Z/2pZ \mid p = 3, 5, 7 \ldots\} = Z/6 \times Z/10 \times Z/14 \times \cdots,$$

and let R be the subring of R_1 generated by the direct sum ideal and 1. Note that R consists of those tuples which are eventually constant. Let I

be the principal ideal of R generated by $(2,2,2,\ldots)$. Then I^0 is the ideal of R generated by $f_3 := (3,0,0,0,\ldots), f_5 := (0,5,0,\ldots), f_7 = (0,0,7,\ldots),\ldots$. Since the f_i are idempotent, I^0 is PT. However I^0 is not a principal ideal of R since any single generator would have to be eventually 0, and therefore f_i, for i sufficiently large, could not be a multiple of it.

2 Invertible ideals

2.1 PF ideals of constant rk 1

Let I be an ideal of the ring R with total quotient ring Q. We define

$$I^{-1} = (R\colon I)_Q := \{\alpha \in Q \mid \alpha I \subseteq R\}.$$

Note that I^{-1} is an R-submodule of Q and

$$II^{-1} \subseteq R. \tag{2}$$

Moreover, for every regular element $b \in I$,

$$bI^{-1} = ((b)\colon I)_R, \tag{3}$$

and if I^{-1} is f.g., then there exists an ideal J of R and a regular element $b \in R$ such that $I^{-1} = J/b := \{a/b \mid a \in J\}$ and $IJ \subseteq (b)$.

The ideal I is said to be *invertible* if equality holds in (2). Equivalently, I is invertible iff there exist $a_1,\ldots,a_n \in I$ and $\alpha_1,\ldots,\alpha_n \in I^{-1}$ such that

$$a_1\alpha_1 + \cdots + a_n\alpha_n = 1. \tag{4}$$

By multiplying (4) by elements of I and I^{-1}, we see that the equation implies

$$I = (a_1,\ldots,a_n), \text{ and } I^{-1} = \alpha_1 R + \cdots + \alpha_n R. \tag{5}$$

Thus, if I is invertible, then both I and I^{-1} are f.g., and therefore there exists a regular element $b \in R$ and an ideal J of R such that $I^{-1} = J/b$ and

$$IJ = (b). \tag{6}$$

Moreover, this b is in I; hence an invertible ideal is both f.g. and regular.

If M is an R-submodule of Q such that $IM = R$, then $M = I^{-1}$ (hence $II^{-1} = R$ and I is invertible); for, $IM = R$ implies, on the one hand, $M \subseteq (R\colon I)_Q = I^{-1}$ and, on the other hand, $I^{-1} = I^{-1}(IM) = (II^{-1})M \subseteq M$.

If J is an ideal of R and b a regular element of R such that $IJ = (b)$, as in (6), then $J = bI^{-1} = ((b){:}I)_R$. In general, if I is invertible, then for every regular $b \in I$,

$$(b) = bI^{-1}I = ((b)\colon I)_R I. \tag{7}$$

In summary, the above remarks prove the following two propositions:

Proposition 2.1. *If I is an ideal of R, then the following are equivalent:*

(i) *I is invertible.*
(ii) *There exists an R-submodule M of Q such that $IM = R$.*
(iii) *There exist $a_1, \ldots, a_n \in I$ and $\alpha_1, \ldots, \alpha_n \in I^{-1}$ such that (4) holds.*

Moreover, if I is invertible, then I is f.g. regular and I^{-1} is the unique submodule M of Q such that $IM = R$.

The next proposition is the intrinsic (i.e., entirely inside R) criterion for invertibility.

Proposition 2.2. *The ideal I of R is invertible iff there exists an ideal J of R and a regular element $a \in R$ such that $IJ = (a)$. Moreover, when this is the case, then $J = ((a) \colon I)_R$, and for every regular element $b \in I$, $I((b) \colon I)_R = (b)$.*

We turn now to the punctual criterion for invertibility.

Theorem 2.3. *The ideal I of R is invertible iff*

$$\begin{cases} i)\ I \text{ is regular,} \\ ii)\ I \text{ is f.g.,} \\ iii)\ I \text{ is punctually rk 1 free.} \end{cases} \tag{8}$$

Condition (iii) means: for every prime ideal p of R, I_p is a principal ideal generated by a regular element of R_p.

Proof. $\Longrightarrow$: We have already seen that an invertible ideal is f.g. and regular. As for (iii), the invertibility criterion 2.2 shows that if p is a prime ideal of R and I is invertible, then I_p is an invertible ideal of R_p. Therefore it only remains to note that an invertible ideal of a quasilocal R is principal; and this is a consequence of the fact that at least one of the summands of the equation (4) must be a unit when R is quasilocal.

$\Longleftarrow$: Suppose (i), (ii), and (iii) hold for the ideal I, and let a be a regular element of I. By 2.2, it suffices to verify the equality $I((a) \colon I)_R = (a)$. Since I is f.g., the colon ideal localizes well, and since I is punctually invertible, the equality holds punctually and therefore also holds globally.

It is not true that (ii) and (iii) of (8) imply (i), for Bourbaki [Bou89, Exercise 12, p. 150] gives an example of a f.g. maximal ideal m, not regular and therefore also not invertible, such that its extension to R_m is both principal and regular. However, as noted in [BR72], ideals satisfying only (ii) and (iii) can be given the following intrinsic characterization:

Proposition 2.4. *An ideal I satisfies (ii) and (iii) of (8) iff it is a f.g. multiplication ideal with annihilator* 0.

(An ideal I is said to be a *multiplication ideal* if for every ideal $A \subseteq I$, there exists an ideal B such that $BI = A$, or equivalently, the equality $(A \colon I)I = A$ holds.)

Proof. A f.g. ideal I is a multiplication ideal with $I^0 = 0$ iff the same is true punctually; for the two relevant equalities $0\colon I = 0$ and $(A\colon I)I = A$ localize well because I is f.g., hence they hold iff they hold punctually. Therefore, it remains to prove that if I is an ideal of a quasilocal ring R, m, then I is principal regular iff I is a multiplication ideal.

If A is an ideal with $A \subseteq I = (b)$, b regular in R, then A/b is an ideal such that $A = (A/b)(b)$. Conversely, if I is a multiplication ideal with $I^0 = 0$, then for every $a \in I$, there exists an ideal B such that $BI = (a)$. If $B = R$, then $I = (a)$; if not, then $a \in mI$. Thus, either I is principal or $I = mI$. In the latter case, $I = 0$ by Nakayama's lemma since I is f.g., and this contradicts our assumption that $I^0 = 0$. Thus, $I = (a)$, and a is regular because $I^0 = 0$.

An intrinsic characterization of ideals satisfying only (iii) has been given in [AR97]: An ideal I satisfies (iii) iff it is a *cancellation ideal*, i.e., iff I has the property that $AI = BI$, for ideals A and B of R, implies $A = B$.

3 The Bourbaki example

This is an example (cf. [Bou89, Exercise 12, p. 150]) of a ring R' containing a f.g. maximal ideal m' such that m' is punctually rk 1 free but m' is not regular; hence m' satisfies (ii) and (iii) but not (i) of 2.3. Said differently, m' is a f.g. projective ideal of constant rk 1 which is not regular and therefore not invertible. Moreover, $R'_{m'}$ will be a discrete, rk 1 valuation *domain*.

3.1 The Principle of Idealization

Let R be a ring and M be an R-module. The principle of idealization turns the direct sum $R + M$ into a ring R', called the *idealization ring of R w.r.t. M*, by defining multiplication by $(r + e)(r' + e') = rr' + re' + r'e$ ($r, r' \in R$; $e, e' \in M$). If we identify R with its isomorphic image $R + 0$ in R', then the idealization ring R' may be characterized intrinsically as an overring of R having an ideal M such that $R' = R + M$ and $M^2 = 0 = R \cap M$.

3.2 Properties

1. An ideal p' of R' is prime iff $p = p' \cap R$ is a prime ideal of R and $p' = p + M$.
2. An element $r' = r + e$ ($r \in R$, $e \in M$) is a nonunit of R' iff r is a nonunit of R.
3. Let S be a m.s. of R, and let $S' = S + M$. Then S' is a m.s. of R' and $R'_{S'} = R'_S = R_S + M_S$ (idealization ring). In particular, if p is a prime ideal of R and $p' = p + M$, then $R'_{p'} = R_p + M_p$. (To be precise, the equalities are ring isomorphisms.)

3.3 The module M

Let $\mathcal{Q}$ be a nonempty set of prime ideals of R, and let M be the direct sum of the R-moduies R/p, $p \in \mathcal{Q}$:

$$M\colon = \bigoplus \{R/p \mid p \in \mathcal{Q}\}\,.$$

For $f \in M$, let $f(p)$ denote the pth coordinate of f; let Supp $f = \{q \in \mathcal{Q} \mid f(q) \neq 0\}$; and let e_p denote the element of M having 0 in all coordinates except the pth, and $\overline{1}$ in the pth. Note that the set $\{e_p \mid p \in \mathcal{Q}\}$ is a generating set for the R-module M. Let $Z(M)\colon = \{r \in R \mid$ there exists $f \neq 0 \in M$ such that $rf = 0\}$. Then

4. For $r \in R$ and $0 \neq f \in M$, $rf = 0$ iff $re_p = 0$ for all $p \in$ Supp f iff $r \in \cap \{p \mid p \in \text{Supp } f\}$. In particular, $re_p = 0$ iff $r \in p$, hence
$$Z(M) = \cup \{p \mid p \in \mathcal{Q}.\}$$
5. Suppose m is an ideal of R which is relatively prime to an ideal $p \in \mathcal{Q}$. Then there exist $r \in m$ and $t \in p, \notin m$ such that $1 = r + t$; and then $te_p = 0$ and $e_p = re_p$. Thus,
   if m is a maximal ideal of R such that $m \notin \mathcal{Q}$, then $M = mM$ and $M_m = 0$.
6. If S is a m.s. of R, then the kernel of the canonical homomorphism $M \to M_S$ is the submodule L of M generated by $\{e_p \mid p \in \mathcal{Q}$ and $p \cap S \neq \emptyset\}$, hence
$$M_S \cong \bigoplus \{R_S/pR_S \mid p \in \mathcal{Q} \text{ and } p \cap S = \emptyset\}\,.$$
   Proof: If $p \in \mathcal{Q}$ and $p \cap S \neq \emptyset$, then any element of $p \cap S$ annihilates e_p, so L is contained in the kernel of the canonical homomorphism $M \to M_S$. Conversely, any element f of the kernel is annihilated by an element s of S, so $s \in \bigcap \{p \in \mathcal{Q} \mid p \in \text{ Supp f}\}$ and f is therefore a linear combination of the generators e_p of L.
7. If $r \neq 0 \in R$, then the annihilator $(0\colon r)_M$ of r on M is generated by $\{e_p \mid r \in p$ and $p \in \mathcal{Q}\}$. In particular, if r is in only finitely many primes of $\mathcal{Q}$, then $(0\colon r)_M$ is finitely generated.

3.4 The idealization of R with respect to M

Let $R' = R + M$ be the idealization ring, with M as in 3.3 and $\mathcal{Q}$ a set of *maximal* ideals.

8. Every nonunit r' of R' is of the form $r' = r + e$ with r a nonunit of R and $e \in M$. If $r \in p$ for some $p \in \mathcal{Q}$, then $e_p r' = 0$, and since $e_p \neq 0$, r' is then a zerodivisor of R'. On the other hand, if r is regular in R and $r \notin Z(M) = \cup \{p \mid p \in \mathcal{Q}\}$, then r' is regular in R'.

9. If $m' = m + M$ is a maximal ideal of R' with $m \notin \mathcal{Q}$, then $R'_{m'} = R_m$ and $m'R'_{m'} = mR_m$.
   Proof: By 3, $R'_{m'} = R_m + M_m$, and by 5 $m \notin \mathcal{Q}$ implies $M_m = 0$.
10. If $m' = m + M$ is a maximal ideal of R' with $m \in \mathcal{Q}$, then by 6 $R'_{m'} = R_m + M_m = R_m + R_m/mR_m = R_m + R/m$.

3.5 The example

Choose R to be a domain containing a maximal ideal m which is not the radical of a principal ideal and such that mR_m is principal, e.g., a Dedekind domain with non-torsion class group, and let $\mathcal{Q}$ be the set of maximal ideals of R different from m. Since the ideal m is f.g., $m' = m + M$ is f.g. also; for by 5 $M = mM$, and therefore any generating set for m is also a generating set for m'. Moreover, $R'_{m'} = R_m$ and $m'R'_{m'} = mR_m$ by 9. It follows that m' is punctually rk 1 free, because at every prime of R' different from m', m' extends to the unit ideal. On the other hand, since m is not the radical of a principal ideal, the ideals of $\mathcal{Q}$ cover the nonunits of R; and therefore by 8 every nonunit of R' is a zerodivisor. Thus, the ideal m' is punctually rk 1 free but not regular.

3.6 A modification of Bourbaki's example.

If R is a *domain*, then there is a straightforward proof that a nonzero f.g. PF ideal is invertible, (the "if" direction of 2.3): Since the ideal I is punctually principal, for every prime ideal p of R there exist $s \notin p$ and $a_s \in I$ such that $(s/a_s)I \subseteq R$, i.e., such that $s/a_s \in I^{-1}$. (This uses the fact that a_s is regular in R.) Therefore 1 is in the ideal generated by all such s:

$$1 = r_1 s_1 + \cdots + r_n s_n, \quad r_i \in R,$$

hence

$$1 = (r_1 a_{s_1})(s_1/a_{s_1}) + \cdots + (r_n a_{s_n})(s_n/a_{s_n}),$$

which implies $II^{-1} = R$.

This proof depends on the fact that if I is an invertible ideal of a domain, then for every prime ideal p of R, there exists a regular element $a \in I$ such that $I_p = aR_p$.

> Question: Suppose I is an invertible ideal of a ring R and p is a prime ideal of R. Does there exist a *regular* (in R) element $a \in I$ such that $I_p = aR_p$?

We show here that the answer is *no* by giving an example of an invertible ideal I and a prime ideal p of R such that $I_p = R_p$ but every regular element of I is in p (and therefore cannot be a generator of I_p). If the set of zerodivisors of R is the union of finitely many prime ideals, e.g., if R is noetherian, then such an example cannot exist, for in that case I is covered by the union of the given prime p and the zerodivisor primes and therefore is contained in one of them.

We shall argue below that there exists a domain R with the following properties: R contains two distinct maximal ideals m_1 and m_2 such that m_1 is not the radical of a principal ideal and their product is principal, $m_1m_2 = (a)$. Now let M be the R-module

$$M: = \bigoplus \{R/m \mid m \text{ is a maximal ideal of } R \text{ different from } m_1 \text{ and } m_2\}$$

and form the ring $R' = R + M$ by the principle of idealization. Let $m_i' = m_i + M$, and note that $m_iR' = m_i(R+M) = m_i + M = m_i'$ because $m_iM = M$ by 5. Since the only maximal ideals of R containing a are m_1 and m_2, it follows from 8 that a is a regular element of R'; and therefore $m_1'm_2' = aR'$ implies m_1' is an invertible ideal of R' by 2.2. Since m_1 is not the radical of a principal ideal, every element of m_1 is in one of the other maximal ideals of R, hence by 8 every regular element of m_1' must be in the maximal ideal m_2' of R'. Thus, m_1' is an invertible ideal of R' which extends to the unit ideal at m_2', yet every regular element of m_1' is in m_2' and therefore cannot be a unit at m_2' , so we are done.

Description of the domain R

Let R be the affine ring of an elliptic curve $\mathbf{C}$ over an uncountable algebraically closed field, and let the 0 for the group structure on $\mathbf{C}$ be at a point of inflection. Any line through 0 intersects $\mathbf{C}$ in two other points A and B such that $A + B = 0$. Since there are uncountably many choices for A and only countably many points of finite order on $\mathbf{C}$, it follows that we may choose A to not have finite order and to be different from B. In terms of the corresponding prime ideals of the Dedekind domain R, we then have two distinct maximal ideals m_A and m_B such that m_Am_B is principal and m_A is not the radical of a principal ideal. (By Abel's theorem [Lan59, p. 36], addition on the curve translates to linear equivalence in the affine ring.)

Another choice for the domain R (due to W. Heinzer)

Let $R = k[x^2, x^3]$, where k is a field of characteristic 0 and x is an indeterminate, and let $m_1 = (x-1)k[x] \cap R$ and $m_2 = (x+1)k[x] \cap R$. Then m_1 and m_2 are maximal ideals of R because they have residue field k, and they are distinct since $x^3 - 1 \in m_1$ and $\notin m_2$.

Claim: m_i is not the radical of a principal ideal. For suppose there exists $f \in R$ such that, say, $\sqrt{(f)} = m_1$. Since $x^2 - 1$ is in m_1, then $x^2 - 1$ has a power which is a multiple in R, and a fortiori in $k[x]$, of f, and a similar assertion holds for $x^3 - 1$. Since $x - 1$ is the only common irreducible factor in $k[x]$ of these two polynomials, it follows that $x - 1$ is the only irreducible factor of f in $k[x]$, hence $f = a(x-1)^n$ for some nonzero $a \in k$ and integer $n \geq 1$. But char $k = 0$ implies such an f is not in R, a contradiction to our supposition.

It remains to note that $m_1 m_2 = (x^2-1)R$. One inclusion is clear, so suppose $g \in m_1 m_2$. Then $g \in (x^2-1)k[x]$, hence there exists $h \in k[x]$ such that $g = (x^2-1)h = (x^2-1)(h_0 + h_1 x + h_2 x^2 + \cdots)$. Since $g \in R$, the equality implies $h_1 = 0$; so $h \in R$, which is the desired conclusion.

4 Trace

Multiplication of an ideal I by an element α in the total quotient ring Q of R may be regarded as a homomorphism h_α of I to R which takes $a \in I$ to αa, i.e., $h_\alpha(a) = \alpha a$. Seen from this perspective, equation (4) becomes

$$1 = h_{\alpha_1}(a_1) + \cdots + h_{\alpha_n}(a_n). \tag{9}$$

If the ideal I contains a regular element a, then *every* homomorphism h of I to R is of the form $h = h_\alpha$, where $\alpha = h(a)/a$. For, if $a, b \in I$, then $ah(b) = h(ab) = bh(a)$, or $h(b) = (h(a)/a)b$. Thus,

if I is regular, we may identify the R-modules I^{-1} and $\mathrm{Hom}\,(I,R)$ by the isomorphism which takes α to h_α.

For $a \in I$, we define the trace ideal by

$$T(a) = \{h(a) \mid h \in \mathrm{Hom}\,(I,R)\},$$
$$T(I) = \text{ideal of } R \text{ generated by } \{T(a) \mid a \in I\}.$$

If I is regular, then by the identification of I^{-1} and $\mathrm{Hom}\,(I,R)$, we have $T(I) = II^{-1}$. Therefore if I is invertible, then $T(I) = R$. Conversely, if $T(I) = R$, then there exists $a \in I$ and $h \in \mathrm{Hom}\,(I,R)$ such that $h(a) = 1$, and such an element a must be regular, hence $II^{-1} = T(I) = R$.

This observation generalizes to a not necessarily regular ideal I as follows:

Proposition 4.1. *If I is an ideal of R, then the following are equivalent:*

(i) *There exists an idempotent element e of R such that $T(I) = eR$.*
(ii) *I is a f.g. projective module.*
(iii) *I is a f.g.* trace module.

(An ideal I is called a *trace module* if for every $a \in I$, $a \in T(a)I$.)

Proof. (i) $\Longrightarrow$ (iii): By (i) there exist $a_1, \ldots, a_n \in I$ and $h_1, \ldots, h_n \in \mathrm{Hom}\,(I,R)$ such that

$$e = h_1(a_1) + \cdots + h_n(a_n). \tag{10}$$

Since the identity map is a homomorphism of I into R, $I \subseteq T(I) = eR$. Therefore, for all $a \in I$,

$$a = ae = ah_1(a_1) + \cdots + ah_n(a_n) = a_1h_1(a) + \cdots + a_nh_n(a) \in T(a)I, \quad (11)$$

and $I = (a_1, \ldots, a_n)$.

(i) $\Longrightarrow$ (ii): Let $\phi\colon R^n \to I$ be the surjective homomorphism defined by $e_i \mapsto a_i$, where e_i denotes the ith canonical generator of R^n, and let $\phi'\colon I \to R^n$ be the homomorphism defined by $a \mapsto h_1(a)e_1 + \cdots + h_n(a)e_n$. Then

$$\begin{aligned} \phi\phi'(a) &= \phi(h_1(a)e_1 + \cdots + h_n(a)e_n) \\ &= h_1(a)a_1 + \cdots + h_n(a)a_n \\ &= a[h_1(a_1) + \cdots + h_n(a_n)] \\ &= ae \\ &= a. \end{aligned} \quad (12)$$

Thus $\phi\phi'\colon I \to I$ is the identity, which is what is needed to prove that I is projective.

(ii) $\Longrightarrow$ (iii): This time we start with a surjective homomorphism $\phi\colon R^n \to I$ and are given a homomorphism $\phi'\colon I \to R^n$ such that $\phi\phi' =$ identity; and we define a homomorphisms $h_i\colon I \to R$ by following ϕ' with projection on the ith factor of R^n. Then for all $a \in I$, we have

$$\begin{aligned} a &= \phi\phi'(a) \\ &= \phi(h_1(a)e_1 + \cdots + h_n(a)e_n) \\ &= h_1(a)\phi(e_1) + \cdots + h_n(a)\phi(e_n) \in T(a)I \end{aligned} \quad (13)$$

(iii) $\Longrightarrow$ (i): Consider first the case that R, m is quasilocal. If $T(I) \neq R$, then $T(I) \subseteq m$, hence by Nakayama's lemma, (iii) implies $I = 0$, and therefore also $T(I) = 0$.

Now for the case of arbitrary R, since $T(I)$ localizes well (see [OR72]), the equality $I = T(I)I$ holds punctually, and therefore by the quasilocal case $T(I)$ is PT. Thus, to prove that $T(I)$ is generated by an idempotent, it only remains to see that $T(I)$ is f.g. Since I is f.g. and $I = T(I)I$, there exist $b_1, \ldots, b_n \in T(I)$ such that $I = T_0I$, where T_0 is the ideal generated by the b_i. But then for $h \in \mathrm{Hom}\,(I, R)$ and $a \in I$, there exist $a_1, \ldots, a_n \in I$ such that $a = b_1a_1 + \cdots + b_na_n$, hence $h(a) = b_1h(a_1) + \cdots + b_nh(a_n) \in T_0$, so $T_0 = T(I)$.

A related notion to that of trace ideal, one which does not appeal directly to homomorphisms for its definition, is that of content ideal; see [OR72].

5 Spreading out freeness

It goes without saying that most of the preceding generalizes to modules, sometimes routinely and sometimes not so routinely. In this section and the next we give a couple of illustrations of this remark. The present section examines the question of when freeness at a prime ideal spreads out to a nhbd of the prime. To begin with, note that we have given examples in 1.8 of a f.g. (in fact, principal) ideal which is free at a prime p but is not free at a nhbd of p, the obstacle being that regularity at p may not spread out to a nhbd.

Usually some kind of finiteness condition is needed to insure that freeness does spread out. For instance, if M is an R-module, then a set of elements $a_1, \ldots, a_n \in M$ is linearly independent iff the ideals $A_1 = (0\colon a_1)_R$, $A_2 = (a_1R\colon a_2)_R$, $A_3 = (a_1R + a_2R\colon a_3)_R$, $\ldots$, $A_n = (a_1R + \cdots + a_{n-1}R\colon a_n)_R$ are all zero. Note that these ideals localize well because their denominators are f.g. (see [Bou89, p. 23, Proposition 12]).

Proposition 5.1. *Let M be a f.g. R-module which is free at the prime ideal p, and let $a_1, \ldots, a_n \in M$ be a basis of M at p. If the ideals $A_1, \ldots, A_n$ are f.g., then M is free at a nhbd of p.*

Proof. Since the ideals A_i are zero at p and are f.g., they are zero at a nhbd of p.

For the remainder of this section we fix a domain D. Moreover, we let $S = D \setminus 0$.

Proposition 5.2. *Let M be a D-module, and suppose M is f.g. at a nhbd of the prime ideal p of D. If M is free at p, then there exists a nhbd of p at which M is free.*

Proof. By replacing D by the nhbd ring D_s at which M is f.g., we may assume to begin with that M is f.g.; and therefore a set of elements of M which generates at p also generates at some nhbd of p. Moreover, since D is a domain, the canonical map $D \to D_p$ is injective, hence a set of elements of M which is linearly independent at p is globally linearly independent and therefore linearly independent at every nhbd of p.

Often a (global) f.g. hypothesis can be replaced by the weaker f.g. at a nhbd of p hypothesis, as was done in 5.2.

By an *overring* R of the domain D we mean a ring R together with a ring isomorphism $\phi\colon D \to R$. We then identify D with $\phi(D)$ and regard R as a ring extension of D.

Corollary 5.3. *Suppose R is an overring of D and M is an R-module. If there exist s_1 and s_2 in S such that M_{s_1} is a f.g. R_{s_1}-module and R_{s_2} is a f.g. D_{s_2}-module, then $M_{s_1 s_2}$ is a f.g. $D_{s_1 s_2}$-module and there exists $s \in S$ such that M_s is a free D_s-module.*

Proof. Since the D-module M is f.g. at a nhbd of the prime ideal 0 of D and also free at 0, it follows from 5.2 that M is free at a nhbd of 0.

Definitions 1 *Let M be a D-module. If $\phi\colon D \to R$ is an overring of D, then M will be called a* compatible R-module *if M has an R-module structure such that for all $d \in D$ and $a \in M$, $da = \phi(d)a$.*

We say that M is generically free *if there exists $s \in S$ such that M_s is a free D_s-module, i.e., if M is free at a nhbd of the prime ideal 0 of D.*

We call an overring R of D a finiteness ring *for M if R is a f.g. ring extension of D, M is a compatible R-module, and there exists $s \in S$ such that M_s is a f.g. R_s-module. Moreover, we say that M is* quasi-finite *if there exists a finiteness ring for M.*

If R is an overring of the domain D, we let $\mathbf{d(R)}$ = max *n such that there exists a chain $p_0 < p_1 < \cdots < p_n$ of prime ideals of R with $p_i \cap D = 0$. Equivalently, $d(R) =$ Krull dim R_S.*

If the D-module M is quasi-finite, we define $\mathbf{d(M)}$ = min *of all $d(R)$ such that R is a finiteness ring for M.*

Theorem 5.4. *If D is a domain and M is a quasi-finite D-module, then M is generically free.*

In other words, if there exists a f.g. ring extension R of D and an element $s \in S$ such that M is an R-module and M_s is a f.g. R_s-module, then there exists $s' \in S$ such that $M_{s'}$ is a free $D_{s'}$-module. Thus, under this weak finiteness condition on M, the punctual freeness of M at the prime 0 can be spread out to a nhbd of 0. This generalizes the non-graded part of Grothendieck's *generic flatness (or freeness) lemma*; see [Eis95, p. 312], [Mat70, p. 156]. (We do not assume that D is noetherian.) Theorem 5.4 will be proved in a series of lemmas.

Lemma 5.5. [Normalization lemma] (See [Bou89, p. 347].) *If R is a f.g. overring of the domain D, then there exists $s \in S$ and $x_1, \ldots, x_t \in R$ $(t \geq 0)$ such that $x_1, \ldots, x_t$ are algebraically independent over D and R_s is a f.g. $D_s[x_1, \ldots, x_t]$-module.*

If $t = 0$, then the intent of the lemma is that R_s is a f.g. D_s-module. Note also that, in the notation of the lemma, $d(R) = t$ and $D_s[x_1, \ldots, x_t] = D[x_1, \ldots, x_t]_s$.

Corollary 5.6. *If M is a quasi-finite D-module and $d(M) = t$, then there exist elements $x_1, \ldots, x_t$ algebraically independent over D such that $D[x_1, \ldots, x_t]$ is a finiteness ring for M.*

Proof. Since M is quasi-finite, there exists a finiteness ring R for M with $d(R) = d(M) = t$, hence by definition there exists $s_2 \in S$ such that M_{s_2} is a f.g. R_{s_2}-module. By 5.5 there exist $x_1, \ldots, x_t \in R$ algebraically independent over D and $s_1 \in S$ such that R_{s_1} is a f.g. $D[x_1, \ldots, x_t]_{s_1}$-module. Therefore if $s = s_1 s_2$, then M_s is a f.g. R_s-module and R_s is a f.g. $D[x_1, \ldots, x_t]_s$-module, so M_s is a f.g. $D[x_1, \ldots, x_t]_s$-module.

Lemma 5.7. *Suppose R is an overring of D and L is a f.g. R-module, say $L = Ra_1 \cdots + Ra_n$ $(a_i \in L)$. Let $L_1 = Ra_1$, $L_2 = (Ra_1 + Ra_2)/L_1$, $\ldots$, $L_n = L/L_{n-1}$. If L_i is a generically free D-module for $i = 1, \ldots, n$, then L is a generically free D-module.*

Proof. If $0 \to A \to B \to C \to 0$ is an exact sequence of D-modules and there exist $s_1, s_2 \in S$ such that A_{s_1} is D_{s_1}-free and C_{s_2} is D_{s_2}-free, then $0 \to A_s \to B_s \to C_s \to 0$, where $s = s_1 s_2$, is also exact and A_s and C_s are D_s-free, hence B_s is D_s-free.

Lemma 5.8. *Suppose* $R = D[x_1, \ldots, x_t]_s$, *with* $x_1, \ldots, x_t$ $(t \geq 1)$ *algebraically independent over* D *and* $s \in S$. *If* L *is a cyclic* R*-module, i.e., if* $L = R/I$, I *an ideal of* R, *then either*

(i) $I = 0$, *and then* $L = L_s$ *is a free* D_s*-module; or*
(ii) $I \cap D \neq 0$, *and then* $L_{s_1} = 0$ *for any nonzero* $s_1 \in I \cap D$; *or*
(iii) $I \neq 0$, $I \cap D = 0$, *and then* $d(L) < t$.

Proof. The assertions (i) and (ii) are immediate, so suppose $I \neq 0$ and $I \cap D = 0$. Since $I \cap D = 0$, R/I is a f.g. overring of D; moreover L is a cyclic R/I-module. Thus, R/I is a finiteness ring for L. But every chain of prime ideals of R/I lying over 0 in D lifts to a corresponding chain of prime ideals of R lying over 0 in D and containing I. Since $I \neq 0$, such a chain can be properly extended by adjoining the prime 0. Therefore $d(L) \leq d(R/I) < d(R) = t$.

Proof (Proof of 5.4). The proof is by induction on $d(M) = t$. By 5.6 there exists a finiteness ring for M of the form $D[x_1, \ldots, x_t]$, with $x_1, \ldots, x_t$ algebraically independent over D, and therefore there exists $s \in S$ such that M_s is a f.g. $D[x_1, \ldots, x_t]_s$-module. If $t = 0$, then M_s is a f.g. D_s-module and the theorem follows from 5.2.

Suppose then $t > 0$ and every quasi-finite D-module L with $d(L) < t$ is generically free. Let $L = M_s$ and $R = D[x_1, \ldots, x_t]_s$. Claim: There exists $s_1 \in S$ such that L_{s_1} is a free D_{s_1}-module. If we prove this, we are done, since then $L_{s_1} = M_{ss_1}$ is also a free D_{ss_1}-module. By 5.7 it suffices to prove the Claim for a cyclic R-module L, and this case follows from 5.8.

6 Fitting ideals

Let M be a f.g. R-module. Given a set of generators $a_1, \ldots a_n$ for M, we let

$$K = \{(b_1, \ldots, b_n) \in R^n \mid b_1 a_1 + \cdots + b_n a_n = 0\},$$

and thereby obtain the short exact sequence

$$0 \to K \to R^n \to M. \to 0 \tag{14}$$

of R-modules. We define an infinite sequence of ideals

$$0 = F_0(M) \subseteq F_1(M) \subseteq F_2(M) \subseteq \cdots \subseteq F_n(M) \subseteq F_{n+1}(M) = R \subseteq \cdots$$

by

F_1 is the ideal of R generated by the set of all $\det A$ such that A is an $n \times n$ matrix whose rows are n-tuples from K,

F_2 is the ideal of R generated by the set of all $\det A$ such that A is an $(n-1) \times (n-1)$ submatrix of an $(n-1) \times n$ matrix whose rows are n-tuples from K,

$\vdots$

F_n is the ideal of R generated by the set of all $\det A$ such that A is a 1×1 submatrix of a $1 \times n$ matrix whose row is an n-tuple from K,

$F_{n+1} = R$,

$\vdots$

Properties that follow from the definition:

1) In forming the determinants $\det A$, one need only consider matrices whose rows lie in a given generating set for K.

2) The ideals $F_i = :F_i(M)$ localize well, i.e., for every m.s. S of R, $F_i(M)_S = F_i(M_S)$. (Here the intended generating set for M_S is $a_1/1, \ldots, a_n/1$.)

3) F_1 is contained in the annihilator ideal M^0 of M, by Cramer's rule. On the other end, if $a \in M^0$, then aI_n, where I_n denotes the $n \times n$ identity matrix, is one of the matrices A in the definition of F_1, hence

$$(M^0)^n \subseteq F_1 \subseteq M^0. \tag{15}$$

4) M is free on the generators $a_1, \ldots, a_n$ iff $F_n = 0$.

The result that makes these (Fitting) ideals $F_i(M)$ useful is

Theorem 6.1. (Fitting) *If M is a f.g. R-module, then the ideals $F_i(M)$ depend only on M and not on the choice of generating set for M.*

See [Kap74, p. 146] for a quick outline of the proof, and [Bro93, p. 149] for a detailed treatment. (The numbering of our Fitting ideals agrees with that of [Kap74].)

Corollary 6.2. *If M is punctually generated by n elements, then $(M^0)^n \subseteq F_1 \subseteq M^0$.*

Proof. We have already observed that these inclusions hold when M itself is generated by n elements, and inclusions that hold punctually hold globally.

In particular, if M is punctually generated by one element (e.g., if M is a punctually free ideal), then $F_1(M) = M^0$.

Proposition 6.3. *If M is free of rk $r \geq 0$, then $F_r(M) = 0$ and $F_{r+1}(M) = R$; and the converse holds if R, m is quasilocal.*

Proof. $\Longrightarrow$: We can choose a resolution $0 \to K \to R^r \to M \to 0$ with $K = 0$. $\Longleftarrow$: Suppose $F_r = 0$ and $F_{r+1} = R$. Take a resolution (14) with n minimal. It follows then that $K \subseteq mR^n$, hence $F_n \subseteq m$ and $F_{n+1} = R$. Therefore we must have $n = r$ and $F_n = 0$, which implies $K = 0$.

Corollary 6.4. *M is punctually free iff the ideals $F_i(M)$ are punctually trivial.*

Corollary 6.5. *The f.g. module M is punctually free of constant rk r (hence projective) iff $F_r(M) = 0$ and $F_{r+1}(M) = R$.*

For example, if M is a nonzero ideal of a Dedekind domain, then $F_1(M) = 0$ and $F_2(M) = R$.

Proposition 6.6. [CLSG79] *A f.g. module M is punctually free and is of constant rk on a nhbd of each prime (i.e., M is projective) iff each ideal $F_i(M)$ is generated by an idempotent.*

Proof. $\Longrightarrow$: By 6.4 M is punctually free implies the ideals $F_i(M)$ are punctually trivial, hence idempotent; so it remains to check that the ideals F_i are f.g. If M is of constant rk on a nhbd N of the prime p, then by 6.3 for all i F_i is either 0 on N or (1) on N, and therefore is either 0 <u>at</u> N or (1) <u>at</u> N. But an ideal which is f.g. <u>at</u> a nhbd of every prime is f.g.

$\Longleftarrow$: If each $F_i(M)$ is generated by an idempotent e_i, then by 6.4 M is punctually free. Moreover, by 6.3, M is of rk r at a prime p iff $e_r \in p$ and $e_{r+1} \notin p$; and then M is of rk r on the clopen set $V(e_r) \cap V(1 - e_{r+1})$.

7 Concluding remarks

Let I be a nonzero f.g. PF ideal. We have seen that I is invertible iff it is regular, and I is projective iff rk I is constant on a nhbd of each prime; thus, such an I is invertible iff it is a regular projective ideal. Moreover, the Bourbaki example shows that such an I may be projective of constant rk 1 and still not be invertible. From a local perspective the two notions seem very close, yet their global characterizations appear to be quite different, with invertibility being an ideal-theoretic concept and projectivity a module-theoretic concept.

A primary reference for the ideal-theoretic topics presented here is Gilmer's influential book Multiplicative Ideal Theory (now available in three editions [Gil68], [Gil72], [Gil92]); for example, one finds there subject headings for invertible ideals, cancellation ideals, almost Dedekind domains, etc. On the other hand, the notion of projective and its offshoots are best pursued in Bourbaki.

My thanks to W. Heinzer for his comments and encouragement.

References

[AR97] D.D. Anderson and M. Roitman. A characterization of cancellation ideals. *Proc. Amer. Math. Soc.*, 125:2853–2854, 1997.

[Bou89] N. Bourbaki. *Commutative Algebra.* Chapters 1–7. Springer-Verlag, 1989.

[BR72] J. Brewer and E. Rutter. A note on finitely generated ideals which are locally principal. *Proc. Amer. Math. Soc.*, 31:429–432, 1972.

[Bro93] W.C. Brown. *Matrices Over Commutative Rings.* Marcel Dekker Inc., New York, 1993. Pure and Applied Mathematics, No. 169.

[CLSG79] A. Campillo-López and T. Sánchez-Giralda. Finitely generated projective modules and Fitting ideals. *Collectanea Mathematica*, 30, no. 2:3–8, 1979.

[Eis95] D. Eisenbud. *Commutative Agebra with a View Toward Algebraic Geometry.* Springer-Verlag, New York, 1995.

[Gil68] R. Gilmer. *Multiplicative Ideal Theory*, volume 12 of *Queen's Papers in Pure and Applied Mathematics.* Queen's University, Kingston, ON, 1968.

[Gil72] R. Gilmer. *Multiplicative Ideal Theory.* Marcel Dekker Inc., New York, 1972. Pure and Applied Mathematics, No. 12.

[Gil92] R. Gilmer. *Multiplicative Ideal Theory*, volume 90 of *Queen's Papers in Pure and Applied Mathematics.* Queen's University, Kingston, ON, 1992. Corrected reprint of the 1972 edition.

[HO72] W. Heinzer and J. Ohm. Noetherian intersections of integral domains. *Trans. Amer. Math. Soc.*, 167:291–308, 1972.

[Kap74] I. Kaplansky. *Commutative Rings.* University of Chicago Press, 1974.

[Lan59] S. Lang. *Abelian Varieties.* Interscience, 1959.

[Mat70] H. Matsumura. *Commutative Algebra.* Benjamin, 1970.

[OR72] J. Ohm and D.E. Rush. Content modules and algebras. *Math. Scand.*, 31:49–68, 1972.

Holomorphy rings of function fields

Bruce Olberding*

Department of Mathematical Sciences, New Mexico State University, Las Cruces, New Mexico 88003-8001 olberdin@nmsu.edu

Dedicated to Robert Gilmer

1 Introduction

In his 1974 text, *Commutative Ring Theory*, Kaplansky states that among the examples of non-Dedekind Prüfer domains, the main ones are valuation domains, the ring of entire functions and the integral closure of a Prüfer domain in an algebraic extension of its quotient field [Kap74, p.72]. A similar list today would likely include Kronecker function rings, the ring of integer-valued polynomials and real holomorphy rings. All of these examples of Prüfer domains have been fundamental to the development of multiplicative ideal theory, as is evidenced in the work of Robert Gilmer over the past 40 years. These rings have been intensely studied from various points of views and with different motivations and tools. In this article we make some observations regarding the ideal theory of holomorphy rings of function fields.

A *holomorphy ring* is an intersection of valuation rings having a common quotient field F. The terminology arises from viewing elements of F as functions on collections of valuation rings having quotient field F. To formulate this more precisely, let F be a field and D be a subring of F. The *Zariski-Riemann space* of F is the collection $\Sigma(F|D)$ of all valuation rings V containing D and having quotient field F. If D is the prime subring of F, then we write $\Sigma(F)$ for $\Sigma(F|D)$. One can introduce a topology on $\Sigma(F|D)$ in a natural way [ZS75, p. 110]. In Section 2 we will consider the Zariski patch topology on $\Sigma(F|D)$.

If $S \subseteq \Sigma(F|D)$, then $x \in F$ is *holomorphic on* S if x has no pole on S. More precisely, for each $V \in S$, let $\phi_V{:}F \rightarrow F_V \cup \{\infty\}$ be the place corresponding to V, where F_V is the residue field of V. Then x assigns to V the value $\phi_V(x)$. Thus x is holomorphic on S if and only if x is finite on each $V \in S$; if and

* The author was supported in part by a grant from the National Security Agency.

only if $x \in V$ for all $V \in S$. The holomorphic functions form a subring H of F called the *holomorphy ring of* S. Evidently, $H = \bigcap_{V \in S} V$.

The holomorphy rings we consider originate with a 1965 theorem of Dress that states that if F is a field in which -1 is not a square, then the subring of F generated by $\{1/(1+t^2): t \in F\}$ is a Prüfer domain such that the set of valuation overrings is the set of valuation rings in $\Sigma(F)$ for which -1 is not a square in the residue field [Dre65]. In Section 2 of the 1969 paper [Gil69] Gilmer provides a very general foundation for such examples by showing that if F is a field and $f(X)$ is a monic nonconstant polynomial in $F[X]$ having no root in F, then the integral closure H of the subring of F generated by $\{1/f(t): t \in F\}$ is a Prüfer domain having quotient field F. Moreover, the valuation overrings of H are precisely the valuation rings in $\Sigma(F)$ such that f does not have a root in the residue field. Thus by considering the polynomial $f(X) = 1 + X^2$, the result of Dress is an immediate consequence of Gilmer's generalization. Unaware of Gilmer's work in [Gil69], P. Roquette in 1973 proved a similar result in Theorem 1 of [Roq73]. Later, in Corollary 2.6 of his 1994 article [Lop94] K. A. Loper also rediscovered this result. More recently, D. Rush has extended this theorem in several interesting ways [Rus01]. We give here Roquette's formulation, since it will be the most convenient for our purpose.

Theorem 1.1. (Gilmer [Gil69]; Roquette [Roq73]; Loper [Lop94]) *Let F be a field, let $S \subseteq \Sigma(F)$, and let $H = \bigcap_{V \in S} V$. Suppose that there exists a nonconstant monic polynomial $f(X) \in H[X]$ such that for every $V \in S$, there is no root of $f(X)$ in the residue field of V. Then H is a Prüfer domain with quotient field F and for every finitely generated ideal I of H, I^k is a principal ideal, where k is a power of deg $(f(X))$.*

The following special case is of interest. Let K be a field that is not algebraically closed, and let F be a field extension of K. Since K is not algebraically closed, there exists a monic nonconstant polynomial $f(X) \in K[X]$ such that $f(X)$ does not have a root in K. For any valuation ring V having quotient field F, we may identify K with its image in the residue field F_V of V. If we take the intersection H of all valuation rings V with quotient field F such that f does not have a root in the residue field F_V (assuming such a valuation ring exists!), then by the theorem H is a Prüfer domain.

Taking this a step farther, we may restrict our intersection to valuation rings V in $\Sigma(F|K)$ such that the residue field F_V has the property that a polynomial in $K[X]$ has a root in K if and only if it has a root in F_V. We in fact assume a stronger property: a field K is *existentially closed* in an extension field L if for any $m, k > 0$ and every choice of polynomials $f_1, \ldots, f_k, g$ in $K[X_1, \ldots, X_m]$ such that $f_1, \ldots, f_k$ have a common zero in L^m that is not a zero of g, then they also have a common zero in K^m that is not a zero of g; equivalently, every finitely generated K-subalgebra A of L admits a K-homomorphism $\phi: A \to K$ [BJ85, Theorem 1.1]. We define the *absolute K-holomorphy ring* H of $F|K$ to be the intersection of all valuation rings V in

$\Sigma(F|K)$ such that K is existentially closed in the residue field of V. If there does not exist a valuation ring V in $\Sigma(F|K)$ such that K is existentially closed in F, then we set $H = F$. We will consider only holomorphy rings defined in this way via the notion of existential closure. We refer to [BS86] and [KP84] for some closely related constructions of holomorphy rings using the notion of existential closure. These approaches all encompass the important example of the *real holomorphy ring* of $F|K$; that is, the case where K is real closed, F is formally real and the holomorphy ring of interest is the intersection of all the formally real valuation rings in $\Sigma(F|K)$. That this case falls into our setting follows from the fact that if K is a real closed field and F is an extension field of K, then F is formally real if and only if K is existentially closed in F [BCR98, Proposition 4.1.1]. Because of its importance in real algebraic geometry this case motivates much of the research on holomorphy rings. See for example [Sch82b],[Bec82], [BK89] and [Ber95]. In addition real holomorphy rings have provided the only known examples of Prüfer domains having finitely generated ideals that cannot be generated by two elements; see [OR], this volume, and its references for more on this aspect of real holomorphy rings.

Another interesting example, and one that is central to our approach later, is the case where K is a field of characteristic 0 that is not algebraically closed and F is a finitely generated subfield of a purely transcendental extension of K. Then by the following basic proposition K is existentially closed in F, and we may consider the absolute K-holomorphy ring H of $F|K$. By Proposition 3.1 H is a Prüfer domain having Krull dimension the same as the transcendence degree of $F|K$.

Proposition 1.2. (Ribenboim [Rib84, Proposition 1]) *Let $K \subseteq L$ be infinite fields such that K is existentially closed in L. If F is a subfield of a purely transcendental extension of L, then K is existentially closed in F.*

We restrict throughout this article to the case that $F|K$ is a function field. For in this case if K is existentially closed in F, then there are infinitely many valuation rings in $\Sigma(F|K)$ having residue field K (Theorem 2.2). Hence the absolute K-holomorphy ring of $F|K$ is distinct from F. Moreover, as we show in Corollary 3.5 H is simply the intersection of the DVRs in $\Sigma(F|K)$ with residue field K. As we see in Section 4, the ideal theory of the K-holomorphy ring is quite complicated. This is due to the richness of the valuation theory of function fields, as is discussed in Section 2, where we rely heavily on recent work of F. Kuhlmann.

2 Discrete valuations of function fields

In this section we survey several strong results regarding the existence of "good" valuations on function fields $F|K$. All the main results of the two subsequent sections are consequences of these existence theorems. The basic idea, and one which we will exploit in the next section, is to trade arbitrary valuations for discrete valuations with nice residue fields that behave the same way with respect to a given finite set of data. There are many other interesting existence and local uniformization results for such valuations given in [Kuh04a], but we cite here only those needed for our purposes in the next sections. For more background on valuations, see [ZS75].

Let K be a field, and let F be an extension field of K. Then F is a *function field in n variables over K* (or, *$F|K$ is a function field in n variables*) if F is a finitely generated field extension of K of transcendence degree n. A *valuation on $F|K$* is a valuation on the field F such that $v(x) = 0$ for all $x \in K$. Let V be the valuation subring of F corresponding to v. We will use the following notation.

- F_v = residue field of v. We view K as a subfield of F_v.
- G_v = value group of v.
- $m(v)$ = maximal ideal of V.
- $\phi_v: F \to F_v \cup \{\infty\}$ is the *place* corresponding to v defined by $\phi_v(x) = x + m(v)$ if $x \in V$ and $\phi_v(x) = \infty$ otherwise.
- $\dim_K(v)$ = transcendence degree of F_v over K.
- $\mathrm{rr}(v)$ = dimension of $\mathbb{Q} \otimes_{\mathbb{Z}} G$ as a vector space over $\mathbb{Q}$. Thus $\mathrm{rr}(v)$ is the rational rank of the abelian group G_v.
- $\mathrm{rank}(v)$ = the rank of the totally ordered abelian group G_v. This number coincides with the Krull dimension of V. Moreover, $\mathrm{rank}(v) \leq \mathrm{rr}(v)$

When we wish to emphasize the valuation ring V rather than the valuation v we write F_V, G_V, $m(V)$, ϕ_V, $\mathrm{rr}(V)$, $\dim_K(V)$ and $\mathrm{rank}(V)$ analogously. The rational rank and dimension of a valuation on $F|K$ are governed by Abhyankar's inequality [Abh56, Theorem 1]:

$$\dim_K(v) + \mathrm{rr}(v) \leq n, \tag{1}$$

where n is (as above) the transcendence degree of F over K.

The valuation v on $F|K$ is *discrete* if G_v is isomorphic as a totally ordered abelian group to the lexicographic sum $\bigoplus_{i=1}^{k} \mathbb{Z}$ for some $k > 0$. The valuation ring V corresponding to v is *discrete* if v is discrete. We follow the usual convention and reserve the abbreviation *DVR* for a discrete valuation ring V of rank 1; equivalently, G_V is isomorphic to $\mathbb{Z}$. Thus a discrete valuation ring need not by a DVR. In this way our terminology follows [Gil92] but differs from [Kuh04a].

We define a valuation v of $F|K$ to be *good* if v is discrete, $F_v|K$ is a function field and K is existentially closed in F_v. A valuation ring V in $\Sigma(F|K)$ is *good* if its corresponding valuation is good. We denote by $\mathrm{Good}_{r,d}(F|K)$ the set of good valuation rings V in $\Sigma(F|K)$ such that $\mathrm{rr}(V) = r$ and $\dim_K(V) = d$. Since V is discrete the rational rank is also the Krull dimension of V. If $F|K$

has more than 1 variable, then there exist bad (i. e. not good) valuations on $F|K$; for example, see [MLS39] or [Kuh04b].

A valuation v on $F|K$ is *K-rational* if $K = F_v$. Similarly, a valuation ring V in $\Sigma(F|K)$ is *K-rational* if $K = F_V$. If V is in $\mathrm{Good}_{r,0}(F|K)$, then V is K-rational since K is existentially closed in F_V.

We state now a powerful theorem regarding the existence of valuations on $F|K$ with residue field finitely generated over K. The version we give was proved in 1984 by Kuhlmann and Prestel using the Ax-Kochen-Ershov Theorem. Bröcker and Schülting proved a similar result in characteristic 0 using Hironka's resolution of singularities [BS86, Theorem 2.12]. A more direct proof using a curve selection lemma is given in the case $K = \mathbb{R}$ by Andras in [And85, Theorem 5.1]. Recently Kuhlmann has obtained the most general version, which is valid in arbitrary characteristic [Kuh04a, Theorem 1]. If G is a totally ordered abelian group and $\mathbb{Z}$ is the group of integers, then any direct sum of the form $\mathbb{Z} \oplus \cdots \oplus \mathbb{Z} \oplus G \oplus \mathbb{Z} \oplus \cdots \oplus \mathbb{Z}$ is a *discrete lexicographic r-extension of G* with respect to the lexicographic ordering on the sum, where r is the number of copies of $\mathbb{Z}$.

Theorem 2.1. (Kuhlmann–Prestel [KP84, Main Theorem]) *Let $F|K$ be a function field in n variables such that K has characteristic 0. Let v be a valuation on $F|K$, and let $x_1, \ldots, x_m \in F$. Let r_1 and d_1 be integers such that*

$$\dim_K(v) \leq d_1 \leq n-1 \text{ and } \mathrm{rr}(v) \leq r_1 \leq n - d_1.$$

Then there exists a valuation w on $F|K$ such that:

(i) *$\dim_K(w) = d_1$ and F_w is a subfield, finitely generated over K, of a purely transcendental extension of F_v;*
(ii) *$\mathrm{rr}(w) = r_1$ and G_w is a finitely generated subgroup of an arbitrarily chosen discrete lexicographic $(r_1 - \mathrm{rr}(v))$-extension of G_v; and*
(iii) *$\phi_v(x_i) = \phi_w(x_i)$ and $v(x_i) = w(x_i)$ for all $i = 1, 2, \ldots, m$.*

We will need in Lemma 3.3 to replace a valuation v that is positive on a given finite set of elements by a K-rational rank one discrete valuation that is also positive on these same elements. The first step in doing so is to replace the valuation v with a K-rational discrete valuation that is nonnegative on these same elements. This we do via Theorem 2.2:

Theorem 2.2. (Kuhlmann [Kuh04a, Theorem 23]) *Let $F|K$ be a function field in n variables such that K is existentially closed in F. Let D be a finitely generated K-subalgebra of F. Then there are infinitely many discrete K-rational valuation rings in $\Sigma(F|D)$ of Krull dimension n.*

The second step is to replace this K-rational discrete valuation with a K-rational rank one discrete valuation that is nonnegative on the initially given set of finitely many elements. This step is enabled by the following theorem, which is formulated topologically. The *Zariski patch topology* on $\Sigma(F|K)$

is defined by basic open sets of the form $U(x_1,\ldots,x_k;y_1,\ldots,y_m) = \{V \in \Sigma(F|K): x_1,\ldots,x_k \in V; y_1,\ldots,y_m \in m(V)\}$, where $x_1,\ldots,x_k,y_1,\ldots,y_m \in F$.

Theorem 2.3. (Kuhlmann [Kuh04a, Corollary 5]) *If K is a perfect field, then the K-rational DVRs lie dense with respect to the Zariski patch topology in the space of all K-rational valuation rings in $\Sigma(F|K)$.*

3 Intersection representations of holomorphy rings

Let $F|K$ be a function field, and let D be K-subalgebra of F. We define the *relative K-holomorphy ring* of $F|D$ to be the intersection of all valuation rings V in $\Sigma(F|D)$ such that K is existentially closed in F_V. The *absolute K-holomorphy ring* of F is the relative K-holomorphy ring of $F|K$. Throughout the rest of this article we work under the following standing hypotheses:

- K is a field of characteristic 0 that is not algebraically closed,
- $F|K$ is a function field in $n > 0$ variables,
- K is existentially closed in F,
- D is a finitely generated K-subalgebra of F, and
- H is the relative K-holomorphy ring of $F|D$.

Recall from the introduction that these standing hypotheses are satisfied in the case (a) where $F|K$ is a formally real function field with K a real closed field, or (b) K is a non-algebraically closed field of characteristic 0 and F is a finitely generated subfield of a purely transcendental extension of K.

Proposition 3.1. *H is a Prüfer domain having Krull dimension n and quotient field F. Moreover, there exists $k > 0$ such that for all finitely generated ideals I of H, I^k is a principal ideal of H.*

Proof. By Theorem 1.1 H is a Prüfer domain. Let M be a maximal ideal of H. Then H_M is a valuation domain, so by Abhyankar's inequality (1), M has height at most n. Moreover, by Theorem 2.2 there is a valuation overring of H of Krull dimension n. Thus H has Krull dimension n. Finally, since K is not algebraically closed, there exists a monic nonconstant polynomial $f(X) \in K[X]$ such that f does not have a root in K. Thus the final assertion follows from Theorem 1.1. □

Remark 3.2. The integer k in Proposition 3.1 can be chosen to be a power of the greatest common divisor of the degrees of all monic nonconstant polynomials $f \in K[X]$ such that f has no root in K [Gil69, Theorem 2.2]. Hence if this gcd is 1, then H is a Bézout domain.

We prove below a representation theorem for H in terms of good valuation rings, as defined in Section 2, in $\Sigma(F|D)$. It is a consequence of the following key lemma, which is a basic application of the existence results in Section 2.

Lemma 3.3. *Let d and r be integers such that $0 \leq d < n$ and $1 \leq r \leq n - d$. The set* $\text{Good}_{r,d}(F|K)$ *is dense with respect to the patch topology in the subspace S of valuation rings V in $\Sigma(F|K)$ such that K is existentially closed in F_V.*

Proof. Let $x_1, \ldots, x_k, y_1, \ldots, y_m \in F$, let $U = U(x_1, \ldots, x_k; y_1, \ldots, y_m)$, and suppose that $U \cap S$ is not empty. We prove first that there exists a K-rational DVR in $U \cap S$. To do this, it is enough by Theorem 2.3 to show that there exists a K-rational valuation ring in $U \cap S$. Let V be a valuation ring in $S \cap U$ with corresponding valuation v. Then $v(x_i) \geq 0$ for all $i = 1, \ldots, k$ and $v(y_j) > 0$ for all $j = 1, \ldots, m$. Thus by Theorem 2.1 there exists a valuation w on $F|K$ such that $w(x_i) = v(x_i)$ for all $i = 1, 2, \ldots, k$; $w(y_j) = w(y_j)$ for all $j = 1, 2, \ldots, m$; and F_w is a finitely generated subfield of a purely transcendental extension of F_v. Since K is by assumption existentially closed in F_v we have by Proposition 1.2 that K is existentially closed in F_w.

Let W be the valuation ring corresponding to w. Since K is existentially closed in F_w, we have by Theorem 2.2 that there exists a K-rational valuation ring A in $\Sigma(F_w|K)$ such that $\phi_w(x_i) \in A$ for all $i = 1, 2, \ldots, k$. Now there is a valuation ring $B \subseteq W$ such that $B/m(W) = A$; in particular, $F_B = K$. For each $i = 1, 2, \ldots, k$ we have $\phi_w(x_i) \in A$, so that $x_i + m(W) \in B/m(W)$. Hence $x_1, \ldots, x_k \in B$. Moreover, $y_1, \ldots, y_m \in m(W) \subseteq m(B)$. Consequently, B is a K-rational valuation ring in U, so by Theorem 2.3 there is a K-rational DVR W in $U \cap S$.

We use W now to show that $U \cap \text{Good}_{r,d}(F|D)$ is nonempty. Let w be the valuation ring corresponding to W. Since w is K-rational, $\dim_K(w) = 0$, and since w is discrete of rank one, $\text{rr}(w) = 1$. Hence by Theorem 2.1 there exists a valuation w' on $F|K$ such that $\dim_K(w') = d$; $\text{rr}(w') = r$; $F_{w'}$ is a finitely generated subfield of a purely transcendental extension of K; $w'(x_i) = w(x_i)$ for $i = 1, \ldots, k$; and $w(y_j') = w(y_j)$ for $j = 1, \ldots, k$. By Proposition 1.2 K is existentially closed in $F_{w'}$, so $W \in \text{Good}_{r,d}(F|D)$. □

Schülting, using resolution of singularities, proves in Lemma 2.5 of [Sch82a] that if K is a real closed field, then H is the intersection of the K-rational valuation rings in $\Sigma(F|D)$. Becker in Theorem 1.14 of [Bec82] gives a different proof for the same result using a trace formula for quadratic forms. In Theorem 3.13 of [BS86] Bröcker and Schülting characterize in terms of direct limits of rings arising from blow-ups which collections of valuation overrings of H can be intersected to obtain H. However they assume as part of their standing hypotheses that K is existentially closed in $K((t))$. The fields with this latter property are shown in Theorem 15 of [Kuh04a] to be precisely the *large fields*, that is, the fields K such that for every smooth curve over K, the set of K-rational points is infinite if it is nonempty. On the other hand, in Theorem 2 of [KP84], Kuhlmann and Prestel prove in a context similar to the present one that the *absolute* K-holomorphy ring is an intersection of valuation rings in $\text{Good}(F|K)_{1,d}$, where d is some fixed number $< n$. Thus the generality of the following theorem appears to be new, even for real closed fields, but is not surprising in light of these other results.

Theorem 3.4. *Let d and r be integers such that $0 \leq d < n$ and $1 \leq r \leq n-d$. Then H is the intersection of the valuation rings in* $\mathrm{Good}_{r,d}(F|D)$.

Proof. The inclusion $H \subseteq \bigcap_{V \in \mathrm{Good}_{r,d}(F|K)} V$ is clear. To prove the reverse inclusion, let $z \in F \setminus H$. We show that there exists $W \in \mathrm{Good}_{r,d}(F|K)$ such that $z \notin W$. Since K is not algebraically closed, there exists a monic nonconstant polynomial $f \in K[X]$ such that $f(x) \neq 0$ for all $x \in F$. Define $y = (f(z))^{-1}$. We claim that $y \in H$. Indeed, let $V \in \Sigma(F|D)$ such that K is existentially closed in F_V. Suppose that $y \notin V$. Then $f(z) = y^{-1} \in m(V)$ since V is a valuation ring. Write $f(X) = \sum_{i=1}^{t} a_i X^i$. Then since $f(z) \in m(V)$, we have $0 = \phi_V(f(z)) = \sum_{i=1}^{t} a_i \phi_V(z)^i$. Thus $\phi_V(z)$ is a root of $f(X)$ in F_V. Now K is existentially closed in F_V, so this forces f to have a root in K, contrary to the choice of f. Hence $y \in V$. Therefore, $y \in H$.

Next observe that y is not a unit in H. For if y is a unit in H, then $f(z) \in H$, and since H is integrally closed, $z \in H$, a contradiction. Write $D = K[x_1, \ldots, x_k]$, where $x_1, \ldots, x_k \in F$. Since y is not a unit in H the set $U(x_1, \ldots, x_k; y)$ contains a valuation ring such that K is existentially closed in its residue field. Therefore, by Lemma 3.3 there is a valuation ring W in $U(x_1, \ldots, x_k; y) \cap \mathrm{Good}_{r,d}(F|K)$. Hence $y \in m(W)$, so $f(z) = y^{-1} \notin W$. Consequently, $z \notin W$. Since $W \in \mathrm{Good}_{r,d}(F|K)$ this shows that $\bigcap_{V \in \mathrm{Good}_{r,d}(F|D)} V \subseteq H$. □

Corollary 3.5. *H is the intersection of the K-rational DVRs in $\Sigma(F|D)$.*

Proof. By Theorem 3.4 H is the intersection of all the valuation rings in the set $\mathrm{Good}_{1,0}(F|K)$. Each valuation ring V in $\mathrm{Good}_{1,0}(F|K)$ is a DVR such that F_V is algebraic over K and K is existentially closed in F_V. It follows that $F_V = K$. □

Corollary 3.6. *H is a completely integrally closed Prüfer domain.*

Proof. By Corollary 3.5 H is an intersection of completely integrally closed domains. Hence H is completely integrally closed. □

Corollary 3.7. *A valuation ring $V \in \Sigma(F|D)$ is an overring of H if and only if K is existentially closed in F_V.*

Proof. If $V \in \Sigma(F|D)$ and K is existentially closed in F_V, then clearly $H \subseteq V$. Conversely, suppose that $V \in \Sigma(F|D)$ and $H \subseteq V$. We claim that K is existentially closed in F_V. It is enough to show that for every $m, k > 0$ and every choice of polynomials $f_1, \ldots, f_k \in K[X_1, \ldots, X_m]$, whenever $f_1, \ldots, f_k$ have a common zero in $(F_V)^m$, they have a common zero in K^m [BJ85, Theorem 1.1]. Let $m, k > 0$, and let $f_1, \ldots, f_k \in K[X_1, \ldots, X_m]$. Suppose that for some $\mathbf{a} = (a_1, \ldots, a_m) \in V^m$, $f_1(\mathbf{a}), \ldots, f_k(\mathbf{a}) \in m(V)$. Let

$I = (f_1(\mathbf{a}), \ldots, f_k(\mathbf{a}))H$. By Proposition 3.1 some power of I is a principal ideal of H, say xH. We claim that x is contained in the maximal ideal of some K-rational DVR in $\Sigma(F|D)$. Indeed, by Corollary 3.5 H is the intersection of the K-rational DVRs in $\Sigma(F|D)$. Hence if x is not contained in the maximal ideal of any of these DVRs, then $x^{-1} \in H \subseteq V$, a contradiction to the fact that $x \in m(V)$. Therefore there is a K-rational DVR W in $\Sigma(F|D)$ such that $x \in m(W)$. Thus $xH \subseteq m(W)$ since $H \subseteq W$, and since xH is a power of I, $I \subseteq m(W)$. Hence $f_1(\mathbf{a}), \ldots, f_k(\mathbf{a}) \in m(W)$. Since $K = W/m(W)$, we may write for each $i = 1, 2, \ldots, m$, $a_i = u_i + x_i$ for some $u_i \in K$ and $x_i \in m(W)$. Let $\mathbf{u} = (u_1, \ldots, u_m)$. Then for each $i = 1, \ldots, k$, $f_i(\mathbf{u}) \in m(W) \cap K = \{0\}$. Hence $f_1, \ldots, f_k$ have a common zero in K. □

Let R be a commutative ring. For $x_1, \ldots, x_k, y \in R$, let $U(x_1, \ldots, x_k, y) = \{P \in \mathrm{Spec}(R): x_1, \ldots, x_k \in P \text{ and } y \notin P\}$. The *patch topology* on $\mathrm{Spec}(R)$ is given by the basic open sets $U(x_1, \ldots, x_k, y)$, where $x_1, \ldots, x_k, y \in R$. We say that a prime ideal P of H is *good* if H_P is a good valuation ring in $\Sigma(F|D)$. As usual, the *dimension* of a prime ideal P of a commutative ring R is the Krull dimension of the ring R/P. The *height* of P is the Krull dimension of R_P.

Corollary 3.8. *Let d and h be integers such that $0 \leq d < n$, $0 < h \leq n$ and $d + h \leq n$. Then the set of good prime ideals of H of height h and dimension d is dense in* $\mathrm{Spec}(H)$ *with respect to the patch topology.*

Proof. By Corollary 3.7 $\Sigma(F|H)$ is precisely the set of valuation rings V in $\Sigma(F|D)$ such that K is existentially closed in F_V. Thus by Lemma 3.3 $\mathrm{Good}_{h,d}(F|D)$ is dense in $\Sigma(F|H)$. Moreover, since H is a Prüfer domain, the mapping $\Sigma(F|H) \to \mathrm{Spec}(H)$ that sends a valuation to its center on H is bijective. It is easily checked that this mapping is in fact a homeomorphism with respect to patch topologies. Hence the corollary follows. □

The next corollary was proved for the case where K is a real closed field in Lemma 2.5 of [Sch82a].

Corollary 3.9. *Let H_0 be the absolute K-holomorphy ring of F. Then $H = H_0[D]$.*

Proof. Clearly $H_0[D] \subseteq H$. Conversely, since $H_0[D]$ is an overring of the Prüfer domain H_0, then $H_0[D]$ is a Prüfer domain, and hence an intersection of its valuation overrings. By Corollary 3.7 each valuation overring V of $H_0[D]$ has the property that $V \in \Sigma(F|D)$ and K is existentially closed in F_V. Hence $H \subseteq V$. It follows that $H \subseteq H_0[D]$. □

To conclude this section we show that if $F|K$ has ≤ 2 variables, then H is a Hilbert ring. This is noted for real closed fields on p. 1256 of [Sch82a].

Lemma 3.10. *If P is a prime ideal of H, then H/P is the relative K-holomorphy ring of its quotient field with respect to the image of D in H/P.*

Proof. Let L denote the quotient field of H/P, and let $\alpha: H_P \to L$ denote the canonical map. Let $B = \alpha(D)$. A subring W of L is a valuation ring in $\Sigma(L|B)$ if and only if $\alpha^{-1}(W)$ is a valuation ring in $\Sigma(F|B)$ [FHP97, Proposition 1.1.8]. Moreover, if W is a valuation ring, then $F_{\alpha(W)} = L_W$. Thus the valuation rings W in $\Sigma(L|B)$ such that K is existentially closed in L_W are precisely the images of the valuation rings V in $\Sigma(F|D)$ such that $V \subseteq H_P$ and K is existentially closed in F_V. The claim now follows. □

Proposition 3.11. *If the function field $F|K$ has ≤ 2 variables, then H is a Hilbert ring.*

Proof. If P is a nonmaximal prime ideal of H, then from Proposition 3.1 it follows that H/P is a Prüfer domain of Krull dimension 1. The quotient field of H/P is isomorphic to H_P/P_P. Since $F|K$ has ≤ 2 variables, H_P/P_P is a finitely generated field extension of K [ZS75, Corollary 1.5.10]. Thus since H/P is an integrally closed domain of Krull dimension 1, H/P is a Dedekind domain. Moreover, by Corollary 3.7 K is existentially closed in H_P/P_P. Thus by Lemma 3.10 and Theorem 2.2 H/P is the intersection of infinitely many valuation rings. Hence H/P has infinitely many maximal ideals, and the intersection of these maximal ideals is the zero ideal of H/P. Hence P is the intersection of the maximal ideals in H containing it. □

Remark 3.12. H need not be a Hilbert ring if $F|K$ has > 2 variables. Schülting in [Sch82a, p. 1257] constructs such an example in the case where K is a real closed field. The obstacle in degree > 2 is that in this case the residue field of a valuation need not be finitely generated over K.

4 Radical ideals of holomorphy rings

We maintain the standing hypotheses of Section 3. We examine in this section the representation of radical ideals in holomorphy rings as an intersection of prime ideals. With the exception of a version of Corollary 4.9, which we discuss below, the results in this section are evidently new, even for real holomorphy rings. We prove first some preliminary lemmas for Prüfer domains.

Lemma 4.1. *Let I be a nonzero ideal of a Prüfer domain R. Then $I^{-1} = R$ if and only if I is not contained in any proper finitely generated ideal of R.*

Proof. Observe that $I^{-1} = R$ if and only if every finite intersection J of principal fractional ideals of R containing I contains R. Since R is a Prüfer (hence coherent) domain, any such ideal J is finitely generated. On the other hand, every finitely generated ideal of R is invertible, hence a finite intersection of principal fractional ideals. Thus the lemma follows. □

If I is an ideal of a ring R and P is a prime ideal of R, then P is a *Zariski-Samuel associated prime ideal* of I if there exists $x \in R$ such that $\sqrt{I :_R x} = P$.

If R is a Noetherian ring, then every ideal has a Zariski-Samuel associated prime ideal, but for non-Noetherian commutative rings, this need not be the case; see for example [FHO05] and its references.

Lemma 4.2. *Let R be a Prüfer domain, let $\mathcal{C}$ be a collection of height 1 maximal ideals of R, and let $\mathcal{D} = \mathrm{Max}(R) \setminus \mathcal{C}$. Suppose that $\mathcal{C}$ and $\mathcal{D}$ are nonempty and $R = \bigcap_{M\in\mathcal{C}} R_M = \bigcap_{M\in\mathcal{D}} R_M$. Then the following statements hold for R.*

(i) *No nonzero prime ideal of R is the radical of a finitely generated ideal.*
(ii) *No nonzero finitely generated ideal of R has a Zariski-Samuel associated prime ideal.*
(iii) *Every nonzero finitely generated proper ideal of R has infinitely many prime ideals minimal over it.*

Proof. (i) Let P be a nonzero nonmaximal prime ideal of R. Since R is a Prüfer domain and P is a nonmaximal prime ideal of R, then $P^{-1} = \mathrm{End}(P)$ [FHP97, Corollary 3.1.8]. Since R is an intersection of DVRs, R is a completely integrally closed domain. Hence $P^{-1} = \mathrm{End}(P) = R$. Also, since R is a Prüfer domain and P is a nonmaximal prime ideal, $P^{-1} = R_P \cap (\bigcap_\alpha R_{M_\alpha})$, where $\{M_\alpha\}$ is the collection of prime ideals of R not containing P [FHP97, Theorem 3.26]. Thus since $P^{-1} = R$, we have for any maximal ideal M containing P, $R_M = R_P \cap (\bigcap_\alpha R_{M_\alpha})R_M$. Since R_M is a valuation domain, this forces $\bigcap_\alpha R_{M_\alpha} \subseteq R_P$. If P is the radical of a finitely generated ideal I, then since I is invertible, $I = q_1 R \cap \cdots \cap q_k R$ for some $q_1, \ldots, q_k$ in the quotient field of R. Thus there exists i such that $I \subseteq R \cap Rq_i \subseteq P$. Since the radical of I is P, it follows that $q_i^{-1} \in \bigcap_\alpha R_{M_\alpha} \setminus R_P$, a contradiction. Therefore, P is not the radical of a finitely generated ideal of R.

We consider next the case of maximal ideals. If M is a maximal ideal of R and M is the radical of a finitely generated ideal I of R, then using an argument such as the one above, we have $\bigcap_{N\neq M} R_N \not\subseteq R_M$, where N ranges over the maximal ideals of R distinct from M. However, if $M \in \mathcal{C}$, then $\bigcap_{N\in\mathcal{D}} R_N = R \subseteq R_M$, a contradiction. On the other hand, if $M \notin \mathcal{C}$, then $\bigcap_{N\in\mathcal{C}} R_N = R \subseteq R_M$, which is also a contradiction. Hence M is not the radical of a finitely generated ideal of R.

(ii) If I is a finitely generated ideal with a Zariski-Samuel associated prime ideal P, then $P = \sqrt{I{:}_R b}$ for some $b \in R$. However, since R is a Prüfer domain, $I{:}_R b$ is a finitely generated ideal, so P is the radical of a finitely generated ideal. Thus by (i), $P = 0$. Hence $I = 0$.

(iii) If an ideal I of a Prüfer domain has only finitely many prime ideals minimal over it, then each prime ideal minimal over I is a Zariski-Samuel associated prime of I [FHO05, Lemma 5.9]. If I is finitely generated, then by (ii) this forces $I = 0$. □

Lemma 4.3. *If J is a nonzero finitely generated radical ideal of a Prüfer domain R, then every prime ideal containing J is a maximal ideal of R.*

Proof. Suppose that P is a prime ideal of R that is minimal over J, and let M be a maximal ideal of R containing P. Since J is a radical ideal of R and R_M is a valuation domain, $J_M = P_M$. Hence P_M is a nonzero prime ideal of R_M that is finitely generated, which since R_M is a valuation domain is possible only if $P_M = M_M$. Hence we conclude that P is a maximal ideal of R. □

Lemma 4.4. *Let R be a domain, and let $\mathcal{C}$ be a collection of prime ideals of R such that $R = \bigcap_{P \in \mathcal{C}} R_P$. If I is a nonzero divisorial ideal of R, then $I = \bigcap_{P \in \mathcal{C}} I_P$.*

Proof. We may assume that I is a proper ideal of R. Clearly, $I \subseteq \bigcap_{P \in \mathcal{C}} I_P \subseteq R$. To prove that the first inclusion reverses, let $x \in R \setminus I$. Then $x^{-1}I \cap R$ is a proper divisorial ideal of R, so there exists q in the quotient field of R such that $x^{-1}I \cap R \subseteq qR \cap R \neq R$. Now $qR \cap R = \bigcap_{P \in \mathcal{C}}(qR_P \cap R_P)$, so there exists $P \in \mathcal{C}$ such that $x^{-1}I \cap R \subseteq qR \cap R \subseteq P$. Consequently, $x \notin I_P$. This proves the lemma. □

Lemma 4.5. *Let R be a Prüfer domain, and let $\mathcal{C}$ be a collection of height-one maximal ideals of R. Suppose that $\mathcal{C}$ is nonempty and $R = \bigcap_{M \in \mathcal{C}} R_M$. If J is a nonzero radical ideal contained in a proper finitely generated ideal I of R, then I is a radical ideal and any prime ideal containing I is a maximal ideal of R.*

Proof. It suffices by Lemma 4.3 to show that I is a radical ideal of R. We claim in fact that I is the intersection of the ideals in $\mathcal{C}$ that contain it. If $M \in \mathcal{C}$ and $I \subseteq M$, then $J_M \subseteq I_M$. Since J is a nonzero radical ideal of R and R_M has Krull dimension 1, it must be that $J_M = M_M$. Hence $I_M = M_M$. Thus by Lemma 4.4, $I = (\bigcap_{M \in \mathcal{C}, M \supseteq I} I_M) \cap R = (\bigcap_{M \in \mathcal{C}, M \supseteq I} M_M) \cap R = \bigcap_{M \in \mathcal{C}, M \supseteq I} M$. □

Recall the standing hypotheses of this section and the last.

Lemma 4.6. *Let d and h be nonnegative integers. Then P is a good prime ideal of height h and dimension d if and only if H_P is a discrete valuation ring of Krull dimension h, $(H_P/P_P)|K$ is a function field in d variables and K is existentially closed in H_P/P_P.*

Proof. Suppose that P is a good prime ideal of height h and dimension d. By assumption H/P has Krull dimension d. Also, by Lemma 3.10 H/P is the relative holomorphy ring of its quotient field with respect to the image of D in H/P. Thus by Proposition 3.1 d is the transcendence degree of the quotient field of H/P over K. Since the quotient field of H/P is isomorphic to H_P/P_P, the assertion follows.

Conversely, it suffices to show P has dimension d. Let B denote the image of D in H_P/P_P. By Theorem 2.2 there exists a K-rational discrete valuation ring V in $\Sigma(H_P/P_P|B)$ such that V has Krull dimension d. Let W be the subring of H_P such that $W/P_P = V$. Then W is a K-rational discrete valuation ring of Krull dimension $d + h$ and P_P is a prime ideal of W of height h. Since $P = P_P \cap H$, it follows that P has dimension d. □

Theorem 4.7. *If $F|K$ has transcendence degree > 1, then the following statements hold for H.*

(i) *No nonzero prime ideal of H is the radical of a finitely generated ideal.*
(ii) *No nonzero finitely generated ideal of H has a Zariski-Samuel associated prime ideal.*
(iii) *If I is a proper nonzero finitely generated ideal of H and d and h are integers such that $0 \leq d < n$, $0 < h \leq n$ and $d + h \leq n$, then there exist infinitely many good prime ideals of H of dimension d and height h that are minimal over I.*
(iv) *$J^{-1} = H$ for all nonzero radical ideals J of H.*

Proof. To prove (i) and (ii) it is enough to show that H satisfies the hypotheses of Lemma 4.2. By Corollary 3.5 H is an intersection of K-rational DVRs. For each such DVR V, $V = H_M$ for some maximal ideal M of H. Thus $H = \bigcap_{M \in \mathcal{C}} H_M$, where $\mathcal{C}$ is the collection of height-one maximal ideals in M.

We claim next that H is the intersection of all H_M, where $M \in \mathrm{Max}(H) \backslash \mathcal{C}$. Indeed, since $F|K$ has transcendence degree > 1, we have by Theorem 3.4 that H is the intersection of the valuation rings in $\mathrm{Good}(F|D)_{1,1}$. If V is such a valuation overring, then $V = H_P$ for some prime ideal P of H, and by Lemma 4.6 P is a nonmaximal prime ideal of H. It follows that $H = \bigcap_{M \in \mathrm{Max}(H) \backslash \mathcal{C}} H_M$. Thus H satisfies the hypotheses of Lemma 4.2, and (i) and (ii) follow.

(iii) For each proper nonzero finitely generated ideal I of H, let $\mathcal{G}_{d,h}(I)$ denote the collection of dimension d, height h good prime ideals of H that are minimal over I. To prove (iii) it is enough to show that for each nonzero proper finitely generated ideal I of H, there are at least 2 prime ideals in $\mathcal{G}_{d,h}(I)$ that are minimal over I. For suppose there exists a proper nonzero finitely generated ideal I of H such that $\mathcal{G}_{d,h}(I)$ is finite. Then by prime avoidance there exists $x \in H$ such that $\mathcal{G}_{d,h}(I + xH)$ has one element, a contradiction.

Let I be a proper nonzero finitely generated ideal of H. By Proposition 3.1 there exists $e > 0$ such that $I^e = xH$ for some $x \in I$. Since $\sqrt{I} = \sqrt{xH}$ we may assume without loss of generality that $I = xH$. By Theorem 3.4 H is the intersection of the K-rational DVRs of H. Thus since x is a non-unit in H, there exists a K-rational rank one discrete valuation v on $F|D$ such that $v(x) > 0$.

By Theorem 2.1 there exists a discrete valuation ring W in $\Sigma(F|D)$ such that W has Krull dimension h; $F_W|K$ is a function field in d variables; K is existentially closed in F_W (apply Proposition 1.2); and $IW = m(W)^e$, where $e = v(I)$ (for by the theorem we may select the position of G_v arbitrarily in the value group of w). Therefore, by Lemma 4.6 $P := m(W) \cap H$ is a member of $\mathcal{G}_{d,h}(I)$ since $H_{m(W) \cap H} = W$.

We claim next that P is not the only member of $\mathcal{G}_{d,h}(I)$. Let $y \in H \setminus P$ such that $I + yH \neq H$. (If no such y exists, then P is necessarily maximal with $P = \sqrt{I}$, contrary to (i).) As above we may find a K-rational discrete rank one valuation v' such that $v'(I + yH) > 0$. Thus $v'(I) > 0$ and $v'(y) > 0$.

Now, using again Theorem 2.1, there exists a valuation overring W' of H such that W' has Krull dimension h; $F_{W'}|K$ is a function field in d variables; K is existentially closed in $F_{W'}$; and IW' and yW' are $m(W')$-primary ideals of W'. We have used in this last assertion that $v'(I)$ and $v'(y)$ are linearly dependent elements of $G_{v'}$ and the position of $G_{v'}$ can be chosen arbitrarily in the value group of W'. Thus by Lemma 4.6 $m(W') \cap H$ is a member of $\mathcal{G}_{d,h}(I)$. Since $y \in (m(W') \cap H) \setminus m(W)$, it follows that $\mathcal{G}_{d,h}(I)$ contains at least 2 elements.

(iv) Let J be a nonzero radical ideal of R, and suppose that $J^{-1} \neq H$. Then by Lemma 4.1 J is contained in a proper finitely generated ideal I of H. Now by Theorem 3.4 H is the intersection of the valuation rings in $\mathrm{Good}_{1,1}(F|D)$, so by Lemma 4.6 H is an intersection of the localizations H_P, where P is a dimension 1 prime ideal. Thus by Lemma 4.4 I is contained in a nonmaximal prime ideal of H. This implies that J is contained in a nonmaximal prime ideal of H, which contradicts Lemma 4.5. □

Remark 4.8. In [Nak53] Nakano gives an example of a Prüfer domain of Krull dimension 1 such that no nonzero finitely generated ideal has a Zariski-Samuel associated prime ideal. By Theorem 4.7 there exists for any $n > 1$ a Prüfer domain of Krull dimension n also having this property.

In [Sch82a] Schülting proves a Nullstellensatz for finitely generated ideals of real holomorphy rings. For an ideal J of H, let $\mathcal{V}(J)$ to be the collection of all K-rational places ϕ in $F|K$ such that $\phi(x) = 0$ for all $x \in I$. For a collection $\mathcal{V}$ of K-rational places ϕ in $F|K$, define $\mathcal{I}(\mathcal{V}) = \{x \in F{:}\phi(x) = 0$ for all $\phi \in \mathcal{V}\}$. Schülting proves in Theorem 2.6 of [Sch82a] that if I is a proper finitely generated ideal of H, where K is real closed, then $\sqrt{I} = \mathcal{I}(\mathcal{V}(I))$. In terms of ideals, this is equivalent to the assertion that $\sqrt{I}$ is the intersection of the maximal ideals M containing I such that H_M is a K-rational valuation ring. Thus $\sqrt{I}$ is an intersection of prime ideals of H of dimension 0.

In the following corollary we extend Schülting's theorem in several ways by generalizing it from invertible ideals of absolute real holomorphy rings to divisorial ideals of relative K-holomorphy rings, where K need not be real closed. We also may restrict to good prime ideals and allow constraints on the dimension and height of the prime ideals involved in the intersection representation.

Corollary 4.9. *Assume that $n > 1$, and let I be a proper nonzero divisorial ideal of H. Let d and h be integers such that $0 \leq d < n$ and $1 \leq h \leq n-d$. Then $\sqrt{I}$ is the intersection of infinitely many good prime ideals P of dimension d and height h.*

Proof. We prove that $\sqrt{I}$ is the intersection of good prime ideals P of dimension d and height h, for then by Theorem 4.7(iii) this intersection is infinite.

Since every nonzero divisorial ideal of a Prüfer domain is an intersection of finitely generated ideals, it is enough to prove the theorem in the case where I is a proper finitely generated ideal. Let J be the intersection of all good prime ideals P of dimension d and height h that contain I, and let $a \in J$. Since by Theorem 3.1 some power of I is a principal ideal, we may assume without loss of generality that $I = bH$ for some $b \in I$.

We claim that $1/b \in H[1/a]$. By Theorem 3.4 and Corollary 3.9, $H[1/a]$ is the intersection of all $V \in \text{Good}_{h,d}(F|K)$ such that $a^{-1} \in V$. Let $V \in \text{Good}_{h,d}(F|K)$ and suppose that $a^{-1} \in V$. If $1/b \notin V$, then $b \in m(V)$. Hence $I = bH \subseteq m(V)$. Since $V \in \text{Good}_{h,d}(F|K)$ we have by the choice of J that $a \in J \subseteq m(V)$, a contradiction to the assumption that $a^{-1} \in m(V)$. Therefore, $1/b \in V$, and we conclude that $1/b \in H[1/a]$. Hence there exists $m > 0$ such that $a^m \in bH \subseteq I$. This proves that $J = \sqrt{I}$. □

Corollary 4.10. *Assume that $n > 1$. Let I be a proper nonzero divisorial ideal of H, and let d be an integer such that $0 \leq d < n$. Let $\mathcal{X}$ be the set of good prime ideals of height 1 and dimension d containing I. There is a sequence $\{e_P\}_{P\in\mathcal{X}}$ of positive integers such that for any $k \geq 0$, if $\mathcal{Y}_k = \{P \in \mathcal{X}{:}e_P \geq k\}$, then $I = \bigcap_{P\in\mathcal{Y}_k} P^{e_P}$.*

Proof. Let $\mathcal{X}$ denote the set of good prime ideals of height 1 and dimension d that contain I. By Theorem 3.4 and Lemma 4.6 $H = \bigcap_{P\in\mathcal{X}} H_P$. Thus since I is divisorial, we have by Lemma 4.4 that $I = \bigcap_{P\in\mathcal{X}} I_P \cap H$. Since H_P is a DVR, we have for each $P \in \mathcal{X}$, that there exist $e_P > 0$ such that $I_P = P_P^{e_P}$. Since H is a Prüfer domain, $P_P^{e_P} \cap H = P^{e_P}$, and we conclude that $I = \bigcap_{P\in\mathcal{X}} P^{e_P}$.

Now let $k > 0$, and let $\mathcal{Y}_k$ be as above. Define $\mathcal{Z}_k = \mathcal{X}\backslash\mathcal{Y}_k$. We may assume that $\mathcal{Z}_k$ is not empty. Write $J = \bigcap_{P\in\mathcal{Z}_k} P^{e_P}$ and $B = \bigcap_{P\in\mathcal{Y}_k} P^{e_P}$. We claim that $J^{-1} = H$. Indeed, if $A = \bigcap_{P\in\mathcal{Z}_k} P$, then $A^k \subseteq J$, so $J^{-1} \subseteq (A^k)^{-1}$. Since by Theorem 4.7 $A^{-1} = H$, an easy induction shows that $(A^k)^{-1} = H$. Hence $J^{-1} = H$, as claimed. Now $JB \subseteq J \cap B = I = (I^{-1})^{-1}$, so $BJI^{-1} \subseteq H$. Hence $BI^{-1} \subseteq J^{-1} = H$, so $B \subseteq (I^{-1})^{-1} = I$. Thus $B = I$. □

The previous corollaries are framed in terms of divisorial ideals. However, I do not know of an example in the present context of a non-invertible divisorial ideal:

Question 4.11. For which (if any) choices of K, F and D does there exist a nonzero divisorial ideal I of H such that I is not invertible?

An integral domain for which every nonzero divisorial ideal is invertible is termed a "generalized Dedekind domain" in [Zaf86] and a "pseudo-Dedekind domain" in [AK89]. The ring of entire functions is such a Prüfer domain [Zaf86], while the ring of integer-valued polynomials on $\mathbb{Z}$ is not [Lop97], though like the holomorphy ring H and the ring of entire functions it is completely integrally closed.

References

[Abh56] S. Abhyankar. On the valuations centered in a local domain. *Amer. J. Math.*, 78:321–348, 1956.

[AK89] D. D. Anderson and B. G. Kang. Pseudo-Dedekind domains and divisorial ideals in $R[X]_T$. *J. Algebra*, 122(2):323–336, 1989.

[And85] C. Andradas. Real places in function fields. *Comm. Algebra*, 13(5):1151–1169, 1985.

[BCR98] J. Bochnak, M. Coste, and M.-F. Roy. *Real algebraic geometry*, volume 36 of *Ergebnisse der Mathematik und ihrer Grenzgebiete (3) [Results in Mathematics and Related Areas (3)]*. Springer-Verlag, Berlin, 1998.

[Bec82] E. Becker. The real holomorphy ring and sums of $2n$th powers. In *Real algebraic geometry and quadratic forms (Rennes, 1981)*, volume 959 of *Lecture Notes in Math.*, pages 139–181. Springer, Berlin, 1982.

[Ber95] R. Berr. On real holomorphy rings. In *Real analytic and algebraic geometry (Trento, 1992)*, pages 47–66. de Gruyter, Berlin, 1995.

[BJ85] E. Becker and B. Jacob. Rational points on algebraic varieties over a generalized real closed field: a model theoretic approach. *J. Reine Angew. Math.*, 357:77–95, 1985.

[BK89] M. A. Buchner and W. Kucharz. On relative real holomorphy rings. *manuscripta math.*, 63(3):303–316, 1989.

[BS86] L. Bröcker and H.-W. Schülting. Valuations of function fields from the geometrical point of view. *J. Reine Angew. Math.*, 365:12–32, 1986.

[Dre65] A. Dress. Lotschnittebenen mit halbierbarem rechtem Winkel. *Arch. Math.*, 16:388–392, 1965.

[FHO05] L. Fuchs, W. Heinzer, and B. Olberding. Commutative ideal theory without finiteness conditions: primal ideals. *Trans. Amer. Math. Soc.*, 357(7):2771–2798, 2005.

[FHP97] M. Fontana, J. A. Huckaba, and I. J. Papick. *Prüfer domains*, volume 203 of *Monographs and Textbooks in Pure and Applied Mathematics.* Marcel Dekker Inc., New York, 1997.

[Gil69] R. Gilmer. Two constructions of Prüfer domains. *J. Reine Angew. Math.*, 239/240:153–162, 1969.

[Gil92] R. Gilmer. *Multiplicative ideal theory*, volume 90 of *Queen's Papers in Pure and Applied Mathematics.* Queen's University, Kingston, ON, 1992.

[Kap74] I. Kaplansky. *Commutative rings.* The University of Chicago Press, Chicago, Ill.-London, 1974.

[KP84] F.-V. Kuhlmann and A. Prestel. On places of algebraic function fields. *J. Reine Angew. Math.*, 353:181–195, 1984.

[Kuh04a] F.-V. Kuhlmann. Places of algebraic function fields in arbitrary characteristic. *Adv. Math.*, 188(2):399–424, 2004.

[Kuh04b] F.-V. Kuhlmann. Value groups, residue fields, and bad places of rational function fields. *Trans. Amer. Math. Soc.*, 356(11):4559–4600 (electronic), 2004.

[Lop94] K. A. Loper. On Prüfer non-D-rings. *J. Pure Appl. Algebra*, 96(3):271–278, 1994.

[Lop97] K. A. Loper. Ideals of integer-valued polynomial rings. *Comm. Algebra*, 25(3):833–845, 1997.

[MLS39] S. MacLane and O. F. G. Schilling. Zero-dimensional branches of rank one on algebraic varieties. *Ann. of Math. (2)*, 40:507–520, 1939.

[Nak53] N. Nakano. Idealtheorie in einem speziellen unendlichen algebraischen Zahlkörper. *J. Sci. Hiroshima Univ. Ser. A.*, 16:425–439, 1953.

[OR] B. Olberding and M. Roitman. The minimal number of generators of an invertible ideal, this volume.

[Rib84] P. Ribenboim. Remarks on existentially closed fields and Diophantine equations. *Rend. Sem. Mat. Univ. Padova*, 71:229–237, 1984.

[Roq73] P. Roquette. Principal ideal theorems for holomorphy rings in fields. *J. Reine Angew. Math.*, 262/263:361–374, 1973.

[Rus01] D. Rush. Bezout domains with stable range 1. *J. Pure Appl. Algebra*, 158:309–324, 2001.

[Sch82a] H.-W. Schülting. On real places of a field and their holomorphy ring. *Comm. Algebra*, 10(12):1239–1284, 1982.

[Sch82b] H.-W. Schülting. Real holomorphy rings in real algebraic geometry. In *Real algebraic geometry and quadratic forms (Rennes, 1981)*, volume 959 of *Lecture Notes in Math.*, pages 433–442. Springer, Berlin, 1982.

[Zaf86] M. Zafrullah. On generalized Dedekind domains. *Mathematika*, 33(2):285–295 (1987), 1986.

[ZS75] O. Zariski and P. Samuel. *Commutative algebra. Vol. II.* Springer-Verlag, New York, 1975.

The minimal number of generators of an invertible ideal

Bruce Olberding[1] and Moshe Roitman*[2]

[1] Department of Mathematical Sciences, New Mexico State University, Las Cruces, New Mexico 88003-8001 olberdin@nmsu.edu
[2] Department of Mathematics, University of Haifa 31905, Israel mroitman@math.haifa.ac.il

Dedicated to Robert Gilmer

1 Introduction

All rings in this paper are commutative with unity; we will deal mainly with integral domains. Let R be a ring with total quotient ring K. A fractional ideal I of R is *invertible* if $II^{-1} = R$; equivalently, I is a projective module of rank 1 (see, e.g., [Eis95, Section 11.3]). Here, $I^{-1} = (R{:}I) = \{x \in K \mid xI \subseteq R\}$. Moreover, a projective R-module of rank 1 is isomorphic to an invertible ideal. (We use the term "ideal" in the sense of an integral ideal.)

We denote the minimal number of generators of an ideal I of R by $\nu_R(I)$. If R is a Dedekind domain, equivalently, a domain in which each nonzero ideal is invertible, then $\nu_R(I) \leq 2$ for each nonzero ideal I; moreover, I is strongly 2-generated, in the sense that one of the generators can be an arbitrary nonzero element of I. A Dedekind domain is characterized as an integrally closed Noetherian domain of Krull dimension 1. It turns out that of these three properties, Krull dimension 1 always implies that an invertible ideal is 2-generated, as was shown by Sally and Vasconcelos in [SV74]. R. Heitmann generalized this fact to arbitrary finite Krull dimension: an invertible ideal of an n-dimensional domain R is strongly $(n+1)$-generated (see Section 3). Moreover, this result is sharp, in the sense that for each $n \geq 1$ there exists an n-dimensional domain R, even Prüfer, with an invertible ideal requiring $n+1$ generators: see the examples in Sections 4, 5 and 6.

The general problem of determining the minimal number of generators for an invertible ideal of a domain was first studied by Gilmer and Heinzer in [GH70]. Among other fundamental results, they provide sufficient conditions

* M. Roitman thanks the Department of Mathematical Sciences of New Mexico State University for its hospitality.

for an invertible ideal to have the property that it can be generated by two elements. The question whether a finitely generated ideal of a Prüfer domain can be always be generated by 2 elements was first raised by Gilmer around 1964 [Swa84]. Recall that a Prüfer domain is an integral domain in which each nonzero finitely generated ideal is invertible. In a 1979 paper Schülting gave an example of a Prüfer domain with an invertible ideal that cannot be generated by 2 elements. Schülting's result was generalized by Swan and Kucharz. We discuss in Sections 4 and 5 the different approaches used by these authors. In Section 4 we give a new, more direct proof of Schülting's example, and in Section 5, we elaborate on some steps in the proof of Kucharz's theorem on holomorphy rings of finitely generated ideals. In doing so we highlight his method for constructing in a formally real function field of degree $n > 0$ over a real closed field an invertible ideal in the holomorphy ring that cannot be generated by n elements.

These approaches to the n-generator property in holomorphy rings are related to an unpublished example of Chase that dates back some 45 years of an affine $\mathbb{R}$-domain of Krull dimension n that has an invertible ideal that cannot be generated by n elements (the n-generator property means that every finitely generated ideal is n-generated). Swan provided a proof for Chase's example in [Swa62]. Later, in [Gil69a], Gilmer provided a more direct proof for Chase's example. We review Chase's example in Section 3, and we observe that Gilmer's argument extends to affine domains over formally real fields by using an algebraic version of the Borsuk-Ulam theorem for polynomial mappings.

In Section 6, we use reductions of ideals to obtain sufficient conditions for a Prüfer overring of a Noetherian domain to have the n-generator property.

The construction of many examples is over a formally real field. We recall that a field K is *formally real* if -1 is not a sum of squares of elements in K; K is formally real if and only if it admits a total order. A field K is *real closed* if it is formally real, and it is not properly contained in a formally real field that is algebraic over K. Thus $\mathbb{R}$ is real closed.

Of course, this survey is not exhaustive.

2 Two-generated invertible ideals

In this section we discuss briefly the special case of invertible ideals that can be generated by two elements. We define for invertible ideals I and J of a domain A:

$$\beta(I, J) = [IJ] - [I] - [J] + [A] \in K_0(A).$$

In [MS76] Murthy and Swan give the following characterization of two-generated invertible ideals.

Theorem 2.1. (Murthy-Swan [MS76, Lemma 6.3]) *Let A be a Noetherian ring with maximal spectrum of dimension ≤ 2. Let I be an invertible ideal of A. The the following are equivalent:*

(i) *I is 2-generated.*
(ii) *$I \oplus I^{-1}$ is free (that is, $I \oplus I^{-1} \cong A^2$).*
(iii) $\beta(I, I) = 0$.

Since the image of the map β in the preceding theorem is contained is $SK_0(A)$, it follows that if $SK_0(A) = 0$, then every invertible ideal in A is 2-generated [MS76, Corollary 6.4]. As shown in [MS76], the converse is false in general, but one of the partial converses presented in the cited paper is used to provide examples of invertible ideals of Noetherian rings of Krull dimension 2 that require more than 2 generators [MS76, Corollary 6.7].

Gilmer and Heinzer note in Lemma 1 of [GH70] that if I is a finitely generated ideal of a domain R, then $\nu_R(I/I^2) \leq \nu_R(I) \leq \nu_R(I/I^2) + 1$. In particular, if I/I^2 is a principal R-module, then I is 2-generated. The converse is false in general, even for invertible ideals in a Prüfer domain, as shown in [GH70, Example 1].

In a recent paper McAdam and Swan introduce a special class of invertible 2-generated ideals. An ideal I of a domain R is an *S-ideal* if there exists an ideal J of R such that $I + J = R$ and IJ is a principal ideal. By [MS05, Proposition 1.5], an ideal I in an arbitrary domain is finitely generated and I/I^2 is principal if and only if I is an S-ideal. Thus S-ideals are invertible and 2-generated. By [MS05, Corollary 1.7], a nonzero fractional ideal is isomorphic to a an S-ideal if and only if it is invertible and 2-generated. However, an integral ideal that is isomorphic to an S-ideal is not necessarily an S-ideal. Indeed, let R be an integral domain with a nonzero Jacobson radical; in particular, we may consider a DVR. Then, if a is a nonzero element in $\mathrm{Jac}(R)$, the ideal aR is principal, so is isomorphic to the S-ideal R, but there is no ideal J so that $aR + J = R$ (cf. [MS05, Corollary 1.7]). Moreover, if every 2-generated ideal in an integral domain R is an S-ideal, that is, if I/I^2 is principal for each 2-generated ideal I, then each finitely generated ideal is an S-ideal, thus it is a 2-generated invertible ideal; hence R is a Prüfer domain (cf. [Sch79, Proposition 3.4]).

If I is a nonprincipal 2-generated invertible ideal, then there exists an infinite sequence $x_1, x_2, x_3, \ldots$ of distinct elements of I so that I is generated by x_i and x_j for all $i \neq j$ (see [MS05, Proposition 1.8] and its proof). Moreover, assuming that I is an S-ideal, the previous statement can be extended to all powers of I as follows:

Theorem 2.2. (McAdam-Swan [MS05, Proposition 1.8]) *Let I be a nonprincipal 2-generated invertible ideal of a domain R. Then there is an infinite list $x_1, x_2, x_3, \ldots$ of distinct elements of I such that if $n \geq 1$ and if y and z are monomials of degree n consisting of products of powers of some of these x_i such that no x_i appears in both y and z, then $I^n = (y, z)R$.*

3 n-generated invertible ideals

In [Bas61, p. 541] Bass attributes to Chase an example for each $n > 0$ of an affine $\mathbb{R}$-domain D_n of Krull dimension n such that D_n has an invertible ideal I_n with $\nu(I_n) = n+1$. The first published proof for the example is given by Swan in [Swa62, p. 270]. We first describe briefly Swan's approach, then present Gilmer's proof of the example. This example plays an important role in the work on holomorphy rings surveyed in the next section.

By [Swa62, Theorem 2, page 268], if X is a compact Hausdorff space, then the finitely generated projective modules over the ring $C(X)$ of continuous $\mathbb{R}$-valued functions on X are, up to isomorphism, the modules of global sections of the vector bundles over X. We will use this correspondence between projective modules and vector bundles in Chase's example. Let γ_1 be the standard real line bundle over the real projective space $X = \mathbb{P}^n$. Recall that $\mathbb{P}^n$ is the set of lines through the origin of $\mathbb{R}^{n+1}$, and that for each $l \in \mathbb{P}^n$ the fibre of γ_1 at l is the line l itself, more precisely, is the set of all pairs (l, z), where $z \in l$. We may identify $\mathbb{P}^n$ with the sphere S^n having each pair of antipodal points z and $-z$ identified. Hence, the ring of continuous functions $C(\mathbb{P}^n)$ can be identified with the ring of continuous functions $f{:}S^n \to \mathbb{R}$ satisfying $f(z) = f(-z)$ for all z in S^n; that is, with the ring of even continuous functions on S^n. We consider the module of global sections of γ_1 as the set of all continuous functions $f{:}\mathbb{P}^n \to R^{n+1}$ so that $f(l) \in l$ for each line $l \in \mathbb{P}^n$. We may view such a function f as defined on S^n and satisfying $f(z) = f(-z) = \varphi(z)z$ for some continuous real function $\varphi(z){:}S^n \to \mathbb{R}$. Thus we may identify f with φ, which is an odd function, since f is even. Hence the module of global sections of γ_1 may be identified with the set of odd continuous functions. As proved in [Swa62], the module of global sections of γ_1 requires $n+1$ generators. Moreover, the module of global sections of $\gamma_1 \oplus \sigma^k$, where σ is the trivial line bundle and $k \geq 0$ requires $n+k+1$ generators. Restricting everything to polynomials, we have the following

Theorem 3.1. (Chase) *Let D_n be the subring of*

$$\mathbb{R}[X_0, \ldots, X_n]/(1 - (X_0^2 + \cdots + X_n^2)) = \mathbb{R}[x_0, \ldots, x_n]$$

consisting of the cosets of the even polynomials. Let I_n be the D_n-module $(x_0, \ldots, x_n)D_n$. Then D_n has Krull dimension n, I_n is an invertible fractional D_n-ideal, and $\nu_{D_n}(I_n) = n+1$. More generally, for each $k \geq 0$, the minimal number of generators of the projective D_n-module $I_n \oplus D_n^k$ equals $n+k+1$.

That D_n has Krull dimension n follows from the fact that the Krull dimension of an affine domain equals the transcendence degree of its fraction field.

Note that the even polynomials mentioned in the theorem are the polynomials in which all monomials are of even degree, thus $D_n = \mathbb{R}[X_iX_j, 0 \leq i, j \leq n]/(1 - (X_0^2 + \cdots + X_n^2)) = \mathbb{R}[x_ix_j, 0 \leq i, j \leq n]$. Also the set of cosets of odd polynomials is I_n.

In [Gil69a] Gilmer gives a brief and direct justification for a similar example over $\mathbb{R}$. We will present Gilmer's proof in the case $k = 0$ and for an arbitrary formally real field, using an algebraic version of the Borsuk-Ulam Theorem for polynomials: each n odd polynomials have a common zero on the sphere S^n. This version of the Borsuk-Ulam theorem was proved by Gilmer for $K = \mathbb{R}$ in [Gil69a, Lemma 1], while an algebraic proof for real closed fields was given in 1982 by Knebusch [Kne82, Theorem 1]. Later Arason and Pfister published an elegant proof for an arbitrary real closed field, which we reproduce here [AP82, Satz 1]:

Proof. Let $f_1, \ldots, f_n$ be odd polynomials in $K[X_1, \ldots, X_n, X_{n+1}]$, where K is a real closed field, and let $F_1, \ldots, F_n$ be their homogenizations in $K[X_0, X_1, \ldots, X_n, X_{n+1}]$. Since each monomial occurring in f_i for $1 \leq i \leq n$ is of odd degree, we see that X_0 occurs in F_i just in even degrees (these degrees are the differences between $\deg f_i$ and the degrees of the monomials occurring in f_i). For each $1 \leq i \leq n$, plug in $X_0 \to \sqrt{\sum_{i=1}^{n+1} X_i^2}$ in F_i to obtain a *homogeneous polynomial* $G_i = (X_1, \ldots, X_{n+1})$ of odd degree. By the real Bézout's Theorem (see [Sha94, Section 2.2, Chapter IV] and [Lan53]) the n polynomials G_i (which are of odd degrees) have a common zero $(a_1, \ldots, a_{n+1})$ so that not all a_i are zero. Since K is real closed, we may replace each a_i by $\frac{a_i}{\sqrt{\sum_{i=1}^{n+1} a_i^2}}$, thus we may assume that $\sum_{i=1}^{n+1} a_i^2 = 1$. We have for all i, $f_i(a_1, \ldots, a_{n+1}) = F_i(1, a_1, \ldots, a_{n+1}) = G_i(a_1, \ldots, a_{n+1}) = 0$. □

With this general version of the Borsuk-Ulam theorem, we can extend Gilmer's proof of Chase's example to formally real fields:

Proof. We may replace K with its real closure, since if the image of $I_n = (x_0, \ldots, x_n)$ in the resulting ring cannot be generated by n elements, neither can I_n. Thus we assume that K is a real closed field. The fractional ideal I_n of D_n is clearly invertible. It consists of all the cosets of odd polynomials in $K[X_0, \ldots, X_n]$. If I_n is generated as an D_n-module by n elements, then these elements (viewed as functions $S^n \to K$) have a common zero by the Borsuk-Ulam Theorem as quoted above, although $x_0, \ldots, x_n$ have no common zero on S^n, a contradiction. □

Gilmer's original example in [Gil69a] is

$$\mathbb{R}[X_i X_j : 1 \leq i, j \leq n][\frac{1}{X_1^2 + \cdots + X_n^2}].$$

Since D_{n-1} is a homomorphic image of this ring, it follows from Theorem 3.1 that the invertible ideal $I = X_1(X_1, \ldots, X_n)$ cannot be generated by less than n elements. Note that $I^2 = (X_1^2)$ is a principal ideal.

By Theorem 3.1, for each $n > 0$, there is an integral domain of Krull dimension n with an invertible ideal I so that $\nu(I) = n + 1$. The bound $n + 1$ is sharp. The case $n = 1$ was proved by Sally and Vasconcelos in [SV74]. The general case is due to Heitmann:

Theorem 3.2. (Heitmann [Hei76, Theorem 3.1]) *Let R be a domain of Krull dimension n. Then each invertible ideal of R is strongly $(n+1)$-generated (that is, one of the nonzero generators can be arbitrarily chosen).*

For a converse, a strongly 2-generated ideal in an arbitrary domain is invertible [LM88, Theorem 1].

We sketch the proof of Theorem 3.2 just for Prüfer domains. See also Section II.2.3 of [FHP98]. Let R be a Prüfer domain, let $\mathcal{A}$ be a finite set of nonzero finitely generated ideals of R, and let P be a prime ideal of R. We define $m(P,\mathcal{A})$ as the maximal k for which there is a chain $\sqrt{I_0} \subsetneq \sqrt{I_1} \subsetneq \cdots \subsetneq \sqrt{I_k} \subseteq R_P$, where $I_0, dots, I_k \in \mathcal{A}$. Since R is Prüfer, a radical ideal of PR_P is prime, hence $m(P,\mathcal{A}) \leq \text{ ht}P \leq n$. We say that an element $a \in R$ *generates the ideal I at a prime P* if $I_P = R_P a$.

Lemma 3.3. (Heitmann) *Let R be a Prüfer domain, and let $I = (x_1, \ldots, x_m)R$ be invertible, where the generators $x_1, \ldots, x_m$ are nonzero.*

Let $\mathcal{A} = \{x_1 I^{-1}, \ldots, x_m I^{-1}\}$, let A be a closed subspace of Spec$R$ *so that $m(P,\mathcal{A}) \leq k$ for all primes $P \in A$, and set $C = \{P \in A \,|\, m(P,\mathcal{A}) = k\}$. Then:*

(i) [Hei76, Lemma 2.3] *There exists an element in I that generates I at every prime in C.*
(ii) [Hei76, Proposition 2.2] *C is closed in* Spec$R$ *(this holds for an arbitrary finite set $\mathcal{A}$ of invertible ideals so that $m(P,\mathcal{A}) \leq k$ for all primes $P \in A$).*

We now complete the proof of Theorem 3.2:

Let r be an arbitrary nonzero element of R, and let $\{x_1, \ldots, x_m\}$ be a finite set of nonzero generators of I that contains r. We proceed by descending induction on n starting with $n = \dim R$. For the sake of clarity we assume that $n = 2$. Let C_2 be the set of primes P in $\mathcal{A}_2$ with $m(P,\mathcal{A}) = 2$. The set C_2 is closed by Lemma 3.3 (ii), hence by Lemma 3.3 (i), there exists an element $y_2 \in R$ that generates I at every prime in C_2. Let A_1 be the set of primes at which y_2 does not generate I, thus $A_1 =$ support $I/(y_2)$ is a closed set of primes. Clearly, $m(P,\mathcal{A}) \leq 1$ for all $P \in A_1$ (if P is a prime so that $m(P,\mathcal{A}) > 1$, then, by construction, y_2 generates I at P, thus $P \notin A_1$). Let C_1 be the set of primes P in A_1 with $m(P,\mathcal{A}) = 1$. Again, C_1 is a closed set of primes, and there exists an element y_1 of I that generates I at every prime of C_1. Finally, let $A_0 =$ support $I/(y_2, y_1)$. As before, $m(P,\mathcal{A}) = 0$ for all $P \in A_0$; since the ideals $\sqrt{x_i I^{-1} R_P}$ are comparable in R_P, and since one of them is equal to R_P, we see that all of them are equal to R_P. Hence each x_i (in particular, the given element r) generates I at every prime in A_0. We conclude that the set $\{y_n, \ldots, y_1, r\}$ generates I at every prime of R, thus it is a set of generators for I. □

If R is a Prüfer domain and M is a projective R-module of finite rank d, then M is isomorphic to a direct sum of d ideals [Proposition 6.1, CE56]. Thus if R has Krull dimension n, then by Theorem 3.2, M can be generated by $d(n+1)$ elements. A stronger result, due to Heitmann, is true:

Theorem 3.4. (Heitmann [Theorem 3.2, Hei76]) *If R is a Prüfer domain of dimension n and M is a rank d projective R-module, then M may be generated by $n+d$ elements.*

4 Schülting's example

In [Sch79] Schülting proved using techniques from valuation theory and the geometry of curves, in particular Bézout's Theorem, that if $F = \mathbb{R}(X, Y)$, then the fractional ideal $(1, X, Y)$ of the real holomorphy ring of F cannot be generated by 2 elements. See also [FHP97]. In this section we give a relatively simple proof of Schülting's example.

4.1 The real Bézout Theorem

Before discussing Schülting's example, we review some material related to the real Bézout Theorem, although we will not use the theorem itself. Let K be a field. Let $F(X_0, X_1, X_2)$ be a homogeneous form of positive degree in $K[X_0, X_1, X_2]$. A *parametrization* (a branch) of F is a zero $\mathbf{x}(t) = (x_0(t), x_1(t), x_2(t))$ of F in the power series ring $K[[t]]$ so that there is no nonzero element $z(t) \in K[[t]]$ satisfying $z(t)x_i(t) \in K$ for all i. A parametrization as above is *irreducible* if there is no integer $r > 1$ so that $x_i(t) \in K[[t^r]]$ for all i. Two parametrizations $\mathbf{x}(t)$ and $\mathbf{x}'(t)$ are *equivalent* if there is a power series $\varphi(t) \in tK[[t]]$ so that $\mathbf{x}'(t) = \mathbf{x}(\varphi(t))$. A *place* of the curve F in the sense of [Wal62, §2.2,Chap. IV] is an equivalence class of irreducible parametrizations. If F is irreducible, then such a place induces a place in the usual sense (and so, a valuation) on the quotient field of $K[X_0, X_1, X_2]/(F)$. If G is a homogeneous form, and $\mathbf{x}(t)$ is a parametrization of F, then the *order* of G at $\mathbf{x}(t)$, is defined as $\mathrm{ord}_t G(\mathbf{x}(t))$.

We will use in Schülting's example the following real version of [Wal62, Theorem 5.5], which is related to the real Bézout Theorem:

Proposition 4.1. *Let K be a real closed field, and let $F(X_0, X_1, X_2)$ and $G(X_0, X_1, X_2)$ be two nonzero coprime homogeneous forms over K of degrees m and n. Then the sum of the orders of G at all the places of F over K is congruent modulo 2 to mn.*

Proof. Since K is real closed, the algebraic closure of K equals $K(i)$, where $i = \sqrt{-1}$. The sum of the orders of G at all places of F over $K(i)$ is equal to mn by [Wal62, Theorem 5.5]. On the other hand, parametrizations of F over $K(i)[[t]]$, but not over $K[[t]]$, come in conjugate pairs over $K[[t]]$; moreover, the order of G at two conjugate parametrizations is the same. The Proposition follows. □

Proposition 4.2. *Let K be a real closed field. Then a polynomial of odd degree in $K[X_1, \ldots, X_n]$, where $n \geq 2$, has infinitely many roots in K^n.*

Proof. By changing indeterminates, it is enough to consider a polynomial f of odd degree that is monic in X_n, and so that its total degree equals its degree in X_n. Hence for each sequence $(a_1, \ldots, a_{n-1})$ in K^{n-1}, the polynomial $f(a_1, \ldots, a_{n-1}, X_n) \in K[X_n]$ has a root in K since it is of odd degree and since K is real closed. □

For the next Proposition, recall that if f is a nonzero polynomial in $K[X, Y]$, then $[0, x, y] \in \mathbb{P}^2(K)$ is a zero of the homogenization F of f if and only if $(x, y) \in \mathbb{P}^1(K)$ is a zero of the leading form of f; thus there is a one-to-one correspondence between the zeros of f (that is, of F) at infinity and the zeros of the leading form of f in $\mathbb{P}^2(K)$.

Proposition 4.3. *Let K be a real closed field, and let f be a nonzero polynomial in $K[X, Y]$. Then the number of zeros of f at infinity, counting multiplicities, is finite and it is congruent modulo 2 to its degree.*

Proof. Let H be the form of highest degree in f. Since K is real closed, we may decompose H over K as follows

$$H = H_0 \prod_{i=1}^{m} (\alpha_i X - \beta_i Y),$$

where H_0 is a product of irreducible polynomials over K of degree 2, and α_i, β_i are scalars in K, not both equal to zero for $1 \leq i \leq m$. Thus $\deg H_0$ is even, and so $m \equiv \deg H \pmod 2$. □

4.2 Real holomorphy rings

Let F be a formally real field, and let D be a subring of F. By a *valuation ring V of $F|D$* we mean a valuation ring V such that $D \subseteq V \subseteq F$ and V has quotient field F. The *real holomorphy ring* $H(F|D)$ of $F|D$ is the intersection of all formally real valuation rings V of $F|D$ (that is, the valuation rings containing D and having formally real residue field and field of fractions F). The real holomorphy ring is a Prüfer domain; see for example [Gil69b]. In case K is a real closed field and $F|K$ is a formally real function field (i. e. a formally real finite field extension of K), then the real holomorphy ring of $F|K$ has Krull dimension equal to the transcendence degree of $F|K$; see for example [Olb], where the ideal theory of holomorphy rings of function fields is discussed.

Proposition 4.4. *Let $K \subseteq L$ be formally real fields, and let D be a subring of K. Then $H(K|D) \subseteq H(L|D)$.*

Proof. Let (V, M) be a formally real valuation ring of $L|D$. Hence $V \cap K$ is a valuation ring for $K|D$, and the field $(V \cap K)/(M \cap K)$ is formally real since it is a subfield of V/M. Thus $H(K|D) \subseteq V$, hence $H(K|D)$ is contained in the intersection of all formally real valuation rings of $L|D$, that is, $H(K|D) \subseteq H(L|D)$. □

We will use the following simple criterion in the proof of Schülting's example.

Proposition 4.5. *Let F be a formally real field, and let D be a subring of F. Let H be the real holomorphy ring of $F|D$. Let $f_1, \ldots, f_n$ be elements of F. Then the minimal number of generators of the fractional ideal $(f_1, \ldots, f_n)H$ is n if and only if there is no unit $u \in H$ so that $u(f_1^2 + \cdots + f_n^2)$ is a sum of $n-1$ squares in F.*

Proof. This is a consequence of the fact that in a real holomorphy ring H, if $f_1, \ldots, f_n \in F$, then $(f_1, \ldots, f_n)^2 H = (f_1^2 + \cdots + f_n^2)H$; see for example [Sch79, Lemma 2.4]. □

If K is a formally real field, then K admits a total order. Hence one may formulate in the usual way the notion of a bounded set. Similarly, we can use limits, absolute values, etc. We recall that if ϕ is a nonzero rational function in $K(X_1, \ldots, X_n)$, then $\deg \phi$ (the *degree* of ϕ) is unambiguously defined as $\deg f - \deg g$, where f and g are nonzero polynomials, and $\phi = \frac{f}{g}$.

Proposition 4.6. *Let K be a formally real field, and let H be the real holomorphy ring of $K(X_1, \ldots, X_n)$ with respect to K. Then:*

(i) *H is the set of all bounded rational functions, thus the degree of every nonzero rational function in H is ≤ 0. Hence the degree of a unit in H is 0.*
(ii) *Let $\frac{f}{g}$ be a unit in H, where f and g are nonzero polynomials. Then the homogenizations of f and g have the same zeros over K.*

Proof. Statement (i) is a particular case of [Sch82b, Theorem, p. 435].

As for statement (ii), let F and G be the homogenizations of f and g, respectively. Thus the function $\frac{F}{G}(x_0, \ldots, x_n)$ is bounded on the set of points $[x_0, \ldots, x_n] \in \mathbb{P}^n(K)$ so that $x_0 \neq 0$, and $G(x_0, \ldots, x_n) \neq 0$. It follows that each zero of G over K is also a zero of F. By symmetry, we see that F and G have the same zeros over K. □

4.3 Proof of Schülting's example

In this section we give a proof for Schülting's example:

Theorem 4.7. (Schülting [Sch79]) *Let K be a formally real field, and let H be the real holomorphy ring of $K(X, Y)$. Then the fractional ideal $(1, X, Y)H$ of H requires 3 generators.*

Proof. By Lemma 4.4 we may assume that K is a real closed field. Assume by contradiction that the fractional ideal $(1, X, Y)$ is 2-generated. By Proposition 4.5 there exist rational functions ϕ, ψ in $K(X, Y)$ so that $u := \frac{1+X^2+Y^2}{\phi^2+\psi^2}$ is a unit in H. In particular, by Proposition 4.6 $\deg u = 0$.

We may write $\phi = s\frac{f}{h}, \psi = s\frac{g}{h}$, where s, f, g, h are polynomials in $K(X, Y)$, s and h are nonzero, and the following two pairs are coprime: $\langle f, g\rangle$ and $\langle s, h\rangle$ (if, e.g., $\phi = 0$, we may choose $f = 0$ and $g = 1$).

Assume that the polynomial h has infinitely many zeros in K^2. Since h and s are coprime, they have just finitely many zeros in K^2. Similarly, also f and g have just finitely many zeros in K^2. Thus h has a zero $\mathbf{p}$ in K^2, so that $s(f^2 + g^2)$ does not vanish at $\mathbf{p}$. This contradicts the assumption that u is a unit. It follows that h has just finitely many zeros. Similarly, s has just finitely many zeros. Since a polynomial of odd degree has infinitely many roots in K by Lemma 4.2, it follows that s and h have no factors of odd degree.

We may assume that $\deg\phi \geq \deg\psi$ (or $\psi = 0$), thus by Lemma 4.6, $\deg\phi^2 = \deg(1 + X^2 + Y^2) = 2$. Hence $\deg\phi = 1$. Since $\phi = \frac{sf}{h}$ and since $\deg s$ and $\deg h$ are even, it follows that $\deg f$ is odd. First assume that $\deg\phi > \deg\psi$. The number of zeros at infinity in $\mathbb{P}^2(K)$, counting multiplicities, of a nonzero polynomial in $K[X, Y]$ is finite and it is congruent modulo 2 to its degree (Proposition 4.3). Applying this to the polynomials sf and h, we obtain that u or u^{-1} is not bounded, contradicting our assumption that u is a unit in H.

Now assume that $\deg\phi = \deg\psi = 1$, thus both f and g have odd degrees. The polynomial f has a prime factor q of odd degree. By Proposition 4.1, the sum of the orders of G (the homogenization of g) at all the places of Q over $K[[t]]$ is odd; and for H and S these sums are even. Thus summing $\text{ord}_t(G(\mathbf{x}(t)) + \text{ord}_t(S(\mathbf{x}(t)) + \text{ord}_t H(\mathbf{x}(t))$ for all places of Q over $K[[t]$ we obtain an odd integer. It follows that there exists a parametrization $\mathbf{x}(t)$ of $Q(X_0, X_1, X_2)$ in $K[[t]]$ so that $\text{ord}_t G(\mathbf{x}(t)) + \text{ord}_t S(\mathbf{x}(t)) + \text{ord}_t H(\mathbf{x}(t))$ is odd. There exist polynomials $\tilde{x}_0(t), \tilde{x}_1(t), \tilde{x}_2(0) \in K[t]$ so that $\tilde{x}_i(t) \equiv x_i(t) \pmod{t^m}$ for each i, where

$$m = \max(\text{ord}_t G(\mathbf{x}(t)), \text{ord}_t S(\mathbf{x}(t)), \text{ord}_t H(\mathbf{x}(t))),$$

and $S(\mathbf{x}(t))$ and $G(\mathbf{x}(t))$ do not vanish (if, e.g., $x_0(t) = \sum_{i=1}^{\infty} a_i t^i$, let $\tilde{x}_0(t) = \sum_{i=1}^{r} a_i t^i$, where $r > m$. Moreover, e.g., if $G(\tilde{\mathbf{x}}(t)) = 0$ for all choices of $\tilde{\mathbf{x}}(t)$ for $r \gg 0$, then $G(\mathbf{x}(t)) = 0$, a contradiction). Clearly, $\text{ord}_t(x_i(t)) = \text{ord}_t(\tilde{x}_i(t)$ for all i.

Recall that $u = \frac{h^2(1+X^2+Y^2)}{s^2(f^2+g^2)}$. Let

$$z(t) = \frac{H(\tilde{\mathbf{x}}(t))^2}{S(\tilde{\mathbf{x}}(t))^2(F(\tilde{\mathbf{x}}(t))^2 + G(\tilde{\mathbf{x}}(t))^2}.$$

Then $\text{ord}_t z(t)$ is congruent to 2 modulo 4, hence is not zero. Thus we may write $z(t) = t^m z_0(t)$, where m is nonzero, and $z_0(0) \neq 0$. If $m > 0$, then u^{-1} is not bounded. If $m < 0$, then u is not bounded. This contradicts the fact that u is a unit in H. □

The preceding proof has a natural meaning. We use the notation of the proof. Since F and G are of odd degrees, by the real Bézout Theorem, the number of the K-points of intersection of the curves F and G is odd, counting multiplicities. Similarly, the number of the common K-zeros of F and S, and also of F and H, is even. Assume that all these intersection points are simple and that F, G and S have no common K-zeros. It then easily follows that either H has a K-zero that is not a zero of $S(F^2 + G^2)$, or conversely. In either case u is not a unit in the holomorphy ring. In order to get rid of the additional assumptions of simplicity, we count points over $K[[t]]$ rather than over K; more precisely, we use places as in Proposition 4.1.

In [Swa84] Schülting's result is extended as follows [Swa84, Theorems 1 and 2]:

1. For any integer $n \geq 1$, there is a Prüfer domain R of Krull dimension n and an ideal I_n of R with $\nu_R(I_n) = n + 1$.
2. There is a Prüfer domain R such that for every integer $n \geq 0$ there is an ideal I_n of R with $\nu_R(I_n) = n + 1$.

To prove these theorems, Swan uses Chase's example.

5 Kucharz's Theorem

Kucharz has obtained the most general result regarding the n-generator problem and holomorphy rings:

Theorem 5.1. (Kucharz [Kuc91, Theorem 1]) *Let K be a real closed field, and let F be a formally real function field over K of transcendence degree n. Then the real holomorphy ring of $F|K$ has an invertible ideal I that cannot be generated by n elements.*

We describe briefly the proof of Kucharz's Theorem. We highlight mainly the construction of the ideal I in the statement of the theorem by combining results from [BS86], [Kuc89] and [Kuc91]. Let K be a real closed field, and let $F|K$ be a formally real function field of transcendence degree $n > 0$. By a *K-variety* we mean an integral separated scheme of finite type over K. For a projective K-variety X with function field F, we denote by $\mathcal{O}_X$ the structure sheaf on X. Since F is formally real, the set $X(K)$ of K-rational points is nonempty [BJ85, Theorem 1.1], and since K is real closed, $X(K)$ is precisely the set of points x of X such that x is the center of a K-rational valuation on $F|K$ [BS86, Remark 3.4]. Finally we define $\mathcal{R}(X) = \bigcap_{x \in X(K)} \mathcal{O}_{X,x}$.

As we shall see in the course of the proof, for each projective K-variety X with function field F of transcendence degree n over K, the maximal ideals of the Noetherian regular domain $\mathcal{R}(X)$ cannot be generated by n elements, and these maximal ideals extend to finitely generated ideals of the real holomorphy ring of $F|K$ that cannot be generated by n elements. Moreover, each maximal ideal of $\mathcal{R}(X)$ can be extended to an invertible prime ideal of a Noetherian overring of $\mathcal{R}(X)$ (defined in terms of the chosen maximal ideal) that cannot be generated by n elements.

(1) *There exists a nonsingular projective K-variety X with function field F.* Since K has characteristic 0 this is a consequence of Hironaka's resolution of singularities [Hir64].

(2) *There is a regular affine K-domain D such that $X(K) \subseteq \mathrm{Spec}(D) \subseteq X$ and $\mathcal{R}(X) = D_S$, where $S = D \setminus \bigcup_{x \in X(K)} \mathfrak{m}_x$ and $\mathfrak{m}_x$ is the ideal of functions of D vanishing at x.* This is proved in Proposition 3.3 of [BS86]. Here the assumption that K is not algebraically closed is essential. For in this case there is a form $f \in K[X_0, X_1, \ldots, X_n]$ that has no nontrivial zero over K. Thus $X(K)$ lies in the complement U of the zero-set of f, and this complement is affine. Let $D = K[U]$, the coordinate ring of U. Thus D is a regular affine K-domain. Then, as in [BS86, Proposition 3.3], $\mathcal{R}(X) = D_S$.

(3) *The ring $\mathcal{R}(X)$ is a Noetherian regular domain whose maximal ideals are precisely the ideals of the form $\mathfrak{m}_x\mathcal{R}(X)$, $x \in X(K)$.* This follows from (2); see [BS86, Proposition 3.3].

(4) *Fix $x \in X(K)$, and let $\pi: Y \to X$ be the blow-up of X at the point x. Then $\mathfrak{m}_x\mathcal{R}(Y)$ is the ideal of functions of $\mathcal{R}(Y)$ that vanish on the K-rational points of $\pi^{-1}(x)$.* This is proved in Lemma 4.5 of [Kuc89] for the case $K = \mathbb{R}$, but the proof extends without modification to the case where K is a real closed field.

(5) *The ideal $\mathfrak{m}_x\mathcal{R}(Y)$ is an invertible prime ideal of $\mathcal{R}(Y)$.* By (4), the ideal $\mathfrak{m}_x\mathcal{R}(Y)$ is the ideal of functions of $\mathcal{R}(Y)$ that vanish on the K-rational points of $\pi^{-1}(x)$. Hence $\mathfrak{m}_x\mathcal{R}(Y)$ is a prime ideal. Since Y is nonsingular, $\mathcal{R}(Y)$ is a regular Noetherian domain (apply (3) to the variety Y). Thus the height one prime ideal $\mathfrak{m}_x\mathcal{R}(Y)$ is invertible.

(6) *If $\phi: Z \to Y$ is a composition of finitely many blowing-ups along non-singular centers, then $\mathfrak{m}_x\mathcal{R}(Z)$ cannot be generated by n elements.* This is the crux of the proof. It involves calculations in the Chow group of the rational equivalence classes of cycles on Z of codimension n; see Lemma 4 and pp. 5-6 of [Kuc91].

(7) *If $x \in X(K)$, then $\mathfrak{m}_xH$ cannot be generated by n elements.* By way of contradiction, suppose that $\mathfrak{m}_xH$ is generated by n elements. The ring H is isomorphic to a direct limit of the rings $\mathcal{R}(Z)$, where Z ranges over the projective K-varieties that are compositions of finitely many blowing-ups along non-singular centers; see for example [BS86, Proposition 3.5]. From this it follows that there is such a projective K-variety Z such that $\mathfrak{m}_x\mathcal{R}(Z)$ can be generated by n elements. But this contradicts (6), so $\mathfrak{m}_xH$ cannot be generated by n elements.

To exemplify Kucharz's construction we start with the $\mathbb{R}$-projective variety $X = \mathbb{P}^n(\mathbb{R})$, thus $K = \mathbb{R}$ and F is the field of rational functions in $K(X_0, \ldots, X_n)$ of degree 0, that is, $F = K(\frac{X_1}{X_0}, \ldots, \frac{X_n}{X_0})$. The ring $\mathcal{R}(X)$ is the ring of rational functions of degree 0 having in the denominator a form with no real nontrivial zeros (thus necessarily of even degree). Let $f = X_0^2 + X_1^2 + \cdots + X_n^2$, hence D is the set of the rational functions in

F having as denominator a power of f. Let $x = [1, 0, \ldots, 0]$, the origin of the affine space $\mathbb{R}^n$. The ideal $\mathfrak{m}_x$ of D is generated by the elements $\frac{X_i X_j}{X_0^2+\cdots+X_n^2}$ for all $0 \leq i, j \leq n$ except $i = j = 0$. We have $\mathcal{R}(X) = D_S$, where S is the multiplicative subset of D consisting of rational functions of the form $\frac{g}{f^m}$, for $m \geq 0$, and g a $\mathbb{R}$-form of degree $2m$ that has no real zeros.

The blow-up of $X = \mathbb{P}^n(\mathbb{R})$ at $\mathbf{x} = [1, 0, \ldots, 0]$ is given by $\pi: Y \to X$, where Y is the closed subvariety of $\mathbb{P}^n(\mathbb{R}) \times \mathbb{P}^{n-1}(\mathbb{R})$ defined by the equations $x_i y_j = x_j y_i$ for $i, j = 1, \ldots, n$ (here a point in $\mathbb{P}^n(\mathbb{R})$ is denoted by $[x_0, \ldots, x_n]$, and a point in $\mathbb{P}^{n-1}(\mathbb{R})$ is denoted by $[y_1, \ldots, y_n]$). The ring of rational functions on Y is the field of rational functions that are homogeneous of degree 0 both in $X_0, \ldots, X_n$ and in $Y_1, \ldots, Y_n$ in the fraction field of the integral domain $K[X_0 \ldots, X_n; Y_1, \ldots, Y_n]/(X_i Y_j - X_j Y_i)_{1 \leq i, \leq j \leq n} = K[x_0 \ldots, x_n; y_1, \ldots, y_n]$. Since $\frac{y_i}{y_j} = \frac{x_i}{x_j}$, we see that this field equals $F = K(\frac{X_1}{X_0}, \ldots, \frac{X_n}{X_0})$ and it contains D. The ideal $\mathfrak{m}_x R(Y)$ is generated by the n elements $\frac{x_i}{x_0}$ for $1 \leq i \leq n$.

The extension of the ideal $\mathfrak{m}_x$ to $H(F|\mathbb{R})$ cannot be generated by less than n elements by Kucharz's Theorem.

Using Kucharz's theorem and an argument similar to Proposition 4.5, one obtains:

Corollary 5.2. (Kucharz [Corollary 2, Kuc91]) *Let K be a real closed field, and let F be a formally real function field over K of transcendence degree n. Then there exist elements $f_1, \ldots, f_{n+1}$ of F such that for all $k > 0$, the sum $f_1^{2k} + \cdots + f_{n+1}^{2k}$ cannot be written in the form $g_1^{2k} + \cdots + g_n^{2k}$ for any $g_1, \ldots, g_n \in K$.*

Remark 5.3. If F is a function field over $\mathbb{R}$, then a positive element h in the real holomorphy ring of F is a unit if and only for every $m > 0$, h can be written a sum of $2m$-powers of elements of F [Bec82, Theorem 1.21].

This criterion is significant for the theory of real holomorphy rings to the classical problem of sums of squares in function fields, as in Hilbert's 17th problem.

In [BK89] Buchner and Kucharz consider in the case $K = \mathbb{R}$ the question of when (in the above notation) a finitely generated $H(F|\mathbb{R})$-subalgebra of F has the property that every finitely generated ideal can be generated by n elements. Let D be a finitely generated K-subalgebra of F such that D has quotient field F. Then the compositum $H[D]$ of H and D is precisely the real holomorphy ring $H(F|D)$, i.e. the intersection of all formally real valuation overrings of H containing D as defined above [Sch82a, Lemma 2.5]. Define $V(D)$ to be the set of all $\mathbb{R}$-algebra homomorphisms from D to $\mathbb{R}$. Thus the elements in $V(D)$ are in one-to-one correspondence with the maximal ideals M of D such that $D/M = \mathbb{R}$. If $f \in D$, then we may define $\hat{f}: V(D) \to \mathbb{R}$ by $\hat{f}(g) = g(f)$ for all $g \in V(D)$. We give $V(D)$ the coarsest topology such that each $\hat{f}$, $f \in D$, is continuous. A point $g \in V(D)$ is *nonsingular* if the localization of D at the maximal ideal $\{f \in D: \hat{f}(g) = 0\}$ is a regular local ring.

Theorem 5.4. (Buchner-Kucharz [BK89, Theorem 1.1]) *Let $F|\mathbb{R}$ be a function field of transcendence degree n, and let D be a finitely generated $\mathbb{R}$-subalgebra of F with quotient field F. If each point of $V(D)$ is nonsingular, then the following conditions are equivalent.*

(i) *Every finitely generated ideal of the real holomorphy ring of $F|D$ can be generated by n elements.*
(ii) *$V(D)$ has no compact connected component.*

6 Prüfer domains and the n-generator property

In this section we first make some observations regarding reductions of ideals of a domain D and the n-generator property for Prüfer domains containing D. If I is a finitely generated ideal of D, then an ideal $J \subseteq I$ is a *reduction* of I if there exists $k > 0$ such that $I^{k+1} = JI^k$; equivalently, the integral closure of I is the integral closure of J [BH93, Exercise 10.2.10]. Katz proves in [Kat94, p. 80] that if D is a Noetherian domain of Krull dimension n having infinite residue fields, then every ideal of D has a reduction that can be generated by $n+1$ elements.

Lemma 6.1. *Let D be a domain, and let R be an overring of D. If I is a fractional ideal of D that is isomorphic to an ideal with an n-generated reduction and IR is an invertible ideal of R, then IR can be generated by n elements.*

Proof. By assumption there exists a nonzero element q in the quotient field of D and an n-generated ideal J of D such that $(qI)^{k+1} = J(qI)^k$ for some $k > 0$. Then $(qI)^{k+1}R = J(qI)^kR$, and since qIR is invertible, $IR = q^{-1}JR$. Hence IR can be generated by n elements. □

Proposition 6.2. *Let D be a domain, let $n > 0$ and let R be a Prüfer overring of D. If every finitely generated ideal of D is isomorphic to an ideal of D that has an n-generated reduction, then every finitely generated ideal of R can be generated by n elements.*

Proof. Since every finitely generated nonzero ideal of R is invertible, this follows from Lemma 6.1 and the fact that every finitely generated ideal of R is extended from a finitely generated fractional ideal of D. □

Applying results from the literature on n-generated reductions, we obtain:

Corollary 6.3. *Let D be a Noetherian domain of Krull dimension n, and let R be a Prüfer overring of D. Then in each of the following cases R has the property that every finitely generated ideal can be generated by n elements.*

(i) *D has a nonzero Jacobson radical and infinite residue fields.*

(ii) $D = K[X_1, \ldots, X_n]$, *where K is an infinite field.*

Proof. Statement (i) follows from a theorem of Katz which states that under the assumptions in (i) any ideal contained in the Jacobson radical of D has a n-generated reduction [Kat94, p. 80]. Thus if I is an ideal of D and x is a nonzero element of the Jacobson radical of D, then xI has an n-generated reduction. Hence by Proposition 6.2 every finitely generated ideal of R can be generated by n elements. Statement (ii) follows from Proposition 6.2 and the fact that in case (ii), ideals of D have n-generated reductions; see [Lyu86]. □

Using Theorem 5.4 one can generate examples where Corollary 6.3 fails:

Corollary 6.4. *Let D be affine $\mathbb{R}$-domain of Krull dimension n. If $V(D)$ has a compact connected component, then there exists an ideal of D that does not have an n-generated reduction.*

Proof. By Theorem 5.4 the relative real holomorphy ring H of $F|D$ has an ideal that cannot be n-generated. Hence by Proposition 6.2 D has an ideal that does not have an n-generated reduction. □

As noted above, Katz has shown that if D is a Noetherian domain of Krull dimension n having infinite residue fields, then every ideal of D has an $(n+1)$-generated reduction. Using the proof of Kucharz's theorem (Section 5), we can give an example that illustrates the sharpness of this bound in a strong way. Our notation is that of Section 5.

Proposition 6.5. *Let K be a real closed field, and let $F|K$ be a formally real function field of transcendence degree $n > 0$. Let X be a nonsingular projective K-variety with function field isomorphic to F. Then $\mathcal{R}(X)$ is a regular Noetherian domain such that every maximal ideal has height n but no maximal ideal has an n-generated reduction.*

Proof. Let $\mathfrak{m}$ be a maximal ideal of $\mathcal{R}(X)$. Then by observation (3), Section 5, $\mathfrak{m} = \mathfrak{m}_x\mathcal{R}(X)$ for some $x \in X(K)$. By (7), $\mathfrak{m}H$ cannot be generated by n elements. Hence by Lemma 6.1, $\mathfrak{m}$ cannot have an n-generated reduction. □

In order to obtain interesting invertible ideals in a Prüfer domain, a Prüfer domain containing a field needs to be an intersection of sufficiently many valuation overrings. The following observation, which does not seem to have been noted before, makes this last statement more precise.

Theorem 6.6. *Let λ be a cardinal and let K be a field of cardinality $> \lambda$. Then, an intersection of λ valuation domains containing K and that are contained in a common field is a Bézout domain.*

Proof. Let $\mathcal{V}$ be a set of λ valuation domains as in the Theorem, and let T be the intersection of the rings in $\mathcal{V}$. Let f and g be nonzero elements of T. Let $V \in \mathcal{V}$. Suppose that there exists a scalar c in K such that $f + cg$

is not a generator of the ideal $(f,g)V$. Since for each scalar $a \neq c$, we have $(f+ag, f+cg)V = (f,g)V$, and since V is a valuation domain, we obtain that $\frac{f+cg}{f+ag} \in V$. Hence $f+ag$ is a generator of the ideal $(f,g)V$. It follows that for each $V \in \mathcal{V}$ there is at most one scalar c_V so that the element $f+c_V g$ does not generate the ideal $(f,g)V$. Thus for all scalars a, except for at most λ scalars, the element $f+ag$ generates the ideal $(f,g)V$ for all $V \in \mathcal{V}$. Since $|K| > \lambda$, there exists such a scalar a. Hence $\frac{f}{f+ag}, \frac{g}{f+ag} \in T$. Thus $(f,g)T = (f+ag)T$; so T is a Bézout domain. □

In [GH00] Gabelli and Houston pose the problem of whether the construction of rings via pullbacks could be used to produce examples of Prüfer domains having invertible ideals requiring more than 2 generators. Recently, Houston has answered this in the negative. Consider the diagram

$$\begin{array}{ccc} R & \longrightarrow & D \\ \downarrow & & \downarrow \\ T & \xrightarrow{\varphi} & T/M, \end{array}$$

where T is a domain, M is a maximal ideal of T, $\varphi{:}T \to T/M$ is the canonical homomorphism, D is a subring of K and $R = \varphi^{-1}(D)$.

Theorem 6.7. (Houston [Hou]) *Let I be an ideal of R not contained in M such that $\varphi(I)$ is an n-generated ideal of D and IT is an m-generated ideal of T. Then I can be generated by $max\{2,n,m\}$ elements of R.*

An example given in [Hou] shows that if $I \subseteq M$, then the number of generators of IT is not enough to determine the number of generators of I.

7 Conjectures

We end with some generalities. Let R be a domain with a finitely generated invertible ideal $I = (x_1,\ldots,x_n)R$. Hence there exist elements $y_1,\ldots,y_n$ in the field of fractions of R so that $x_iy_j \in R$ for all i,j and $\sum_{i=1}^n x_iy_i = 1$. Thus we conjecture

Conjecture 7.1. Let k be an integral domain. Let

$$R = k[X_1,\ldots,X_n,\{X_iY_j{:}1 \le i,j \le n\}, \frac{1}{X_1Y_1+\cdots+X_nY_n}].$$

Then the the minimal number of generators up to radical of the prime invertible ideal $(X_1,\ldots,X_n)$ of R is n.

We obtain an equivalent conjecture, if we replace the ring R by

$$k[X_1,\dots,X_n,\{X_iY_j{:}1\le i,j\le n\}]/\left(1-\frac{1}{X_1Y_1+\cdots X_nY_n}\right).$$

Clearly, it is enough to prove this conjecture for k a field. If we require in Conjecture 7.1 that the minimal number of generators of I is n, then, in this formulation the conjecture holds at least if k is a formally real field.

Here is an equivalent formulation of Conjecture 7.1:

Let k be a field. Let $f_1,\dots,f_m$ $(m<n)$ be polynomials in the ring $k[\mathbf{X},\mathbf{Y}]=k[X_1,\dots,X_n;Y_1,\dots,Y_n]$ so that $\deg_{\mathbf{X}}M>\deg_{\mathbf{Y}}M$ for each monomial M occurring in one of the polynomials f_i. Then $\sum_{i=1}^n X_iY_i\notin\sqrt{(f_1,\dots,f_m)k[\mathbf{X},\mathbf{Y}]}$.

We now present the background for Conjecture 7.2 below. Let T be a Prüfer domain with a finitely generated ideal $I=(x_1,\dots,x_n)T$ with $\nu_T(I)=n$. Thus there exist elements $y_1=\frac{a_1}{b_1},\dots,y_n=\frac{a_n}{b_n}$ in the field of fractions of T so that $a_i,b_i\in T$ for all i; $x_iy_j\in T$ for all i,j and $\sum_{i=1}^n x_iy_i=1$. Let R be the subring of T generated by the elements x_i,a_i,b_i. Then R is a Noetherian domain, and the ideal $(x_1,\dots,x_n)R$ requires n generators. If T contains a field k, we may choose for R an affine subring of T over k. Swan's example [Swa84, Theorem 1] is a Prüfer overring of Chase's example, and it is obtained by using the fact that a sufficient condition for a domain R of characteristic $\neq 2$ to be Prüfer is that $\frac{1}{1+x^2}\in R$ for all $x\in$ Qf (R) (for other variants see [Swa62]). If we take the smallest overring containing Chase's example D_n (Theorem 3.1) that contains for $t_1,\dots,t_m$ in $\mathbb{R}(x_0,\dots,x_n)$ the element $1+\frac{1}{\sum_{i=1}^m t_i^2}$, then we obtain the real holomorphy ring of $\mathbb{R}(x_0,\dots,x_n)$.

Motivated by these examples, we conjecture that we can start with a "generic" affine example and extend it to a Prüfer domain with a finitely generated ideal which requires n generators, as follows.

A necessary and sufficient condition for a domain R to be Prüfer is that for each nonzero element $x\in$ Qf (R) there exists an element $t\in R$ so that $tx,\frac{1-t}{x}\in R$ (see [Gil68, Part II., Theorem (21.2)(f)]). The ring $\mathcal{P}^\infty(R)$ is constructed using this property ([FHPR93] and [FHP97, Section 8.3]). Note that $\mathcal{P}^\infty(R)$ is not an overring of R.

Conjecture 7.2. Let k be a field, and let $R=k[X_1,\dots,X_n,\{X_iY_j\},\frac{1}{X_iY_i}]$. Then, the ideal $(X_1,\dots,X_n)$ of $\mathcal{P}^\infty(R)$ requires n generators up to radical.

References

[AP82] J. K. Arason and A. Pfister. Quadratische Formen über affinen Algebren und ein algebraischer Beweis des Satzes von Borsuk-Ulam. *J. Reine Angew. Math.*, 331:181–184, 1982.

[Bas61] Hyman Bass. Projective modules over algebras. *Ann. of Math. (2)*, 73:532–542, 1961.

[Bec82] Eberhard Becker. The real holomorphy ring and sums of $2n$th powers. In *Real algebraic geometry and quadratic forms (Rennes, 1981)*, volume 959 of *Lecture Notes in Math.*, pages 139–181. Springer, Berlin, 1982.

[BH93] Winfried Bruns and Jürgen Herzog. *Cohen-Macaulay rings*, volume 39 of *Cambridge Studies in Advanced Mathematics*. Cambridge University Press, Cambridge, 1993.

[BJ85] Eberhard Becker and Bill Jacob. Rational points on algebraic varieties over a generalized real closed field: a model theoretic approach. *J. Reine Angew. Math.*, 357:77–95, 1985.

[BK89] M. A. Buchner and W. Kucharz. On relative real holomorphy rings. *Manuscripta Math.*, 63(3):303–316, 1989.

[BS86] Ludwig Bröcker and Heinz-Werner Schülting. Valuations of function fields from the geometrical point of view. *J. Reine Angew. Math.*, 365:12–32, 1986.

[Eis95] David Eisenbud. *Commutative algebra*, volume 150 of *Graduate Texts in Mathematics*. Springer-Verlag, New York, 1995.

[FHP97] Marco Fontana, James A. Huckaba, and Ira J. Papick. *Prüfer domains*, volume 203 of *Monographs and Textbooks in Pure and Applied Mathematics*. Marcel Dekker Inc., New York, 1997.

[FHPR93] Marco Fontana, James A. Huckaba, Ira J. Papick, and Moshe Roitman. Prüfer domains and endomorphism rings of their ideals. *J. Algebra*, 157(2):489–516, 1993.

[GH70] Robert Gilmer and William Heinzer. On the number of generators of an invertible ideal. *J. Algebra*, 14:139–151, 1970.

[GH00] Stefania Gabelli and Evan Houston. Ideal theory in pullbacks. In *Non-Noetherian commutative ring theory*, volume 520 of *Math. Appl.*, pages 199–227. Kluwer Acad. Publ., Dordrecht, 2000.

[Gil68] Robert W. Gilmer. *Multiplicative ideal theory*. Queen's Papers in Pure and Applied Mathematics, No. 12. Queen's University, Kingston, Ont., 1968.

[Gil69a] Robert Gilmer. A note on generating sets for invertible ideals. *Proc. Amer. Math. Soc.*, 22:426–427, 1969.

[Gil69b] Robert Gilmer. Two constructions of Prüfer domains. *J. Reine Angew. Math.*, 239/240:153–162, 1969.

[Hei76] Raymond C. Heitmann. Generating ideals in Prüfer domains. *Pacific J. Math.*, 62(1):117–126, 1976.

[Hir64] Heisuke Hironaka. Resolution of singularities of an algebraic variety over a field of characteristic zero. I, II. *Ann. of Math. (2) 79 (1964), 109–203; ibid. (2)*, 79:205–326, 1964.

[Hou] Evan Houston. Generating ideals in pullbacks. to appear.

[Kat94] D. Katz. Generating ideals up to projective equivalence. *Proc. Amer. Math. Soc.*, 120(1):79–83, 1994.

[Kne82] Manfred Knebusch. An algebraic proof of the Borsuk-Ulam theorem for polynomial mappings. *Proc. Amer. Math. Soc.*, 84(1):29–32, 1982.

[Kuc89] Wojciech Kucharz. Invertible ideals in real holomorphy rings. *J. Reine Angew. Math.*, 395:171–185, 1989.

[Kuc91] Wojciech Kucharz. Generating ideals in real holomorphy rings. *J. Algebra*, 144(1):1–7, 1991.

[Lan53] Serge Lang. The theory of real places. *Ann. of Math. (2)*, 57:378–391, 1953.

[LM88] David C. Lantz and Mary B. Martin. Strongly two-generated ideals. *Comm. Algebra*, 16(9):1759–1777, 1988.

[Lyu86] Gennady Lyubeznik. A property of ideals in polynomial rings. *Proc. Amer. Math. Soc.*, 98(3):399–400, 1986.

[MS76] M. Pavaman Murthy and Richard G. Swan. Vector bundles over affine surfaces. *Invent. Math.*, 36:125–165, 1976.

[MS05] Stephen McAdam and Richard G. Swan. A special type of invertible ideal. In *Arithmetical properties of commutative rings and monoids*, volume 241 of *Lect. Notes Pure Appl. Math.*, pages 356–362. Chapman & Hall/CRC, Boca Raton, FL, 2005.

[Olb] B. Olberding. Ideals in holomorphy rings of function fields. this volume.

[Sch79] Heinz-Werner Schülting. Über die Erzeugendenanzahl invertierbarer Ideale in Prüferringen. *Comm. Algebra*, 7(13):1331–1349, 1979.

[Sch82a] Heinz-Werner Schülting. On real places of a field and their holomorphy ring. *Comm. Algebra*, 10(12):1239–1284, 1982.

[Sch82b] Heinz-Werner Schülting. Real holomorphy rings in real algebraic geometry. In *Real algebraic geometry and quadratic forms (Rennes, 1981)*, volume 959 of *Lecture Notes in Math.*, pages 433–442. Springer, Berlin, 1982.

[Sha94] Igor R. Shafarevich. *Basic algebraic geometry. 1.* Springer-Verlag, Berlin, 1994.

[SV74] Judith D. Sally and Wolmer V. Vasconcelos. Stable rings. *J. Pure Appl. Algebra*, 4:319–336, 1974.

[Swa62] Richard G. Swan. Vector bundles and projective modules. *Trans. Amer. Math. Soc.*, 105:264–277, 1962.

[Swa84] Richard G. Swan. n-generator ideals in Prüfer domains. *Pacific J. Math.*, 111(2):433–446, 1984.

[Wal62] Robert J. Walker. *Algebraic curves.* Dover Publications Inc., New York, 1962.

About minimal morphisms

Gabriel Picavet[1] and Martine Picavet-L'Hermitte[2]

[1] Laboratoire de Mathématiques, Université Blaise Pascal, 63177 Aubière Cedex, France, Gabriel.Picavet@math.univ-bpclermont.fr
[2] Laboratoire de Mathématiques, Université Blaise Pascal, 63177 Aubière Cedex, France, Martine.Picavet@math.univ-bpclermont.fr

La lecture de l'œuvre de R. Gilmer, nos rencontres avec lui, avec ses élèves et d'autres collègues nous ont fait découvrir des aspects de l'Algèbre Commutative, peu connus en France. Nous sommes fiers et heureux d'apporter notre contribution à l'hommage rendu à ce mathématicien chaleureux.

1 Introduction and notation

As usual, an overring of a commutative ring R, with total quotient ring $\mathrm{Tot}(R)$, is an R-subalgebra of $\mathrm{Tot}(R)$. For a multiplicative subset S of R, the ring R_S is called the localization of R at S. Then for $S = R\setminus P$, P a prime ideal of R, the ring R_P is called the localization of R at P. We say that a ring is local if it has only one maximal ideal. This terminology does not completely agree with the definitions used in a paper published in 1967 by Gilmer and Heinzer [GH67]. It can be considered as the seminal work on minimal morphisms, although it is not easy to determine whether further papers rely on it. Gilmer-Heinzer's paper introduces *the minimal overring* of an integral domain D. Namely, the overring D_1 of D is called the unique minimal overring of D in case D_1 is the least element in the set of all overrings C of D such that $D \neq C$, ordered under inclusion. But this notion should not be confused with the usual meaning of minimal. Anyway, consider a minimal overring C of D in the usual sense, then there are no intermediate rings between D and C, except D and C. Later on in 1970, Ferrand and Olivier were inspired by a paper of Levelt, who considers finite injective ring morphisms $A \to B$, such that the length of the A-module B/A is ≤ 1, calling them *simple monomorphisms* [Lev69]. Before giving the Ferrand-Olivier's definition of *minimal ring morphisms*, we do say that some results of Gilmer and Heinzer are valid in the general context, but Ferrand and Olivier were unaware of the Gilmer-Heinzer work. By the way, we are not sure that our list of references is complete since minimal morphisms appear sporadically in the literature, under various forms and names.

We now give some notation and recall some results. The spectrum (resp. minimal spectrum, maximal spectrum, zero-divisor set, integral closure) of a ring R are respectively denoted by Spec(R), Min(R), Max(R), Z(R), R'. The residual field k(P) at $P \in \mathrm{Spec}(R)$ is R_P/PR_P.

A ring epimorphism is an epimorphism of the category of commutative rings, for instance a localization. The dominion D(A, B) of a ring morphism $f{:}A \to B$ is the set of all $b \in B$, such that $b \otimes 1 = 1 \otimes b$ in $B \bigotimes_A B$. Then D(A, B) is an intermediate ring between $f(A)$ and B. Among ring morphisms $f{:}A \to B$, an epimorphism is characterized by $B = \mathrm{D}(A, B)$, while $A = \mathrm{D}(A, B)$ characterizes strict monomorphisms. An epimorphism, which is either finite or has a zero-dimensional domain, is surjective. Flat epimorphisms $A \to B$ are characterized by Spec(B) $\to$ Spec(A) is injective and $A_{f^{-1}(Q)} \to B_Q$ is an isomorphism for each $Q \in \mathrm{Spec}(B)$. A faithfully flat epimorphism is bijective. These facts may be found in Lazard's paper [Laz69].

Definition 1.1. [FO70] *A ring morphism $f{:}A \to B$ is called* minimal *in case f is injective, nonsurjective, and for each decomposition $f = g \circ h$ into injective morphisms, then either g or h is bijective.*

Actually, identifying A with its image $f(A)$, which will be implicitely done in the following, a minimal ring morphism is nothing but a *minimal extension* of rings $A \subset B$; that is, there are no intermediate rings strictly between A and B. In this case, A is also called a *maximal subring* of B. Such an extension is also dubbed *immediate* or *adjacent.*

Clearly, if $A \to B$ is an injective non surjective ring morphism, then f is minimal if and only if $B = A[x]$ for each $x \in B \setminus A$.

We now introduce some key results, whose proofs are easy.

Lemma 1.2. [FO70, Lemme 1.3] *If $A \to B$ is a minimal ring morphism and S is a multiplicative subset of A, then $A_S \to B_S$ is either minimal or an isomorphism.*

Lemma 1.3. [FO70, Lemme 1.4] *Consider the commutative diagram of ring morphisms defined by $v \circ f = f' \circ u$, where $f{:}A \to B$, $v{:}B \to B'$, $u{:}A \to A'$, and $f'{:}A' \to B'$.*

(1) *If f is minimal, f' injective and v surjective, then f' is either minimal or an isomorphism.*
(2) *If the diagram is a pullback ($A \simeq A' \times_{B'} B$), v is surjective and f' is minimal, then f is minimal.*
(3) *If the diagram is a pushout ($B' \simeq A' \bigotimes_A B$), f is minimal and f' is injective and nonsurjective, then the diagram is a pullback. Hence, if I' is the conductor of f', then $u^{-1}(I')$ is the conductor of f.*

Corollary 1.4. *An injective ring morphism $A \to B$, with I as a common ideal of A and B, is minimal if and only if $A/I \to B/I$ is minimal.*

The next two examples of minimal ring morphisms have different behaviors. If A is a one-dimensional valuation ring, with quotient field K, then $A \to K$ is minimal, and a flat epimorphism because it is a localization. Now if $L|K$ is a field extension, whose degree is 2, then $K \to L$ is minimal and a finite ring morphism. We will see that a minimal ring morphism is either finite or a flat epimorphism. These two conditions are mutually exclusive, because a finite epimorphism is surjective.

The following section explores minimal ring morphisms, whose domains are fields.

2 Minimal ring morphisms whose domains are fields

The following result is crucial, because proofs can often be reduced to the case of a field base ring.

Lemma 2.1. [FO70, Lemme 1.2] *Let K be a field and $f{:}K \to A$ an injective ring morphism (equivalently, $A \neq 0$). Then f is minimal if and only if one of the following conditions is verified:*

(1) *A is a field and $K \to A$ is a minimal ring morphism.*
(2) *f identifies with the diagonal morphism $K \to K \times K$.*
(3) *f identifies with the natural map $K \to D_K(K){:} = K[X]/(X^2)$.*

Hence $K \to A$ is finite and $A = K[x]$ for any $x \in A \setminus K$.

Thus the minimal morphisms, whose domains are fields are known, except when condition (1) holds. We examine below this condition. A field extension $L|K$ is minimal if and only if $K \to L$ is a minimal ring morphism. A minimal field extension is clearly either separable or purely inseparable. Let n be a positive integer. Recall that a transitive subgroup G of the symmetric group Σ_n is called primitive if any subgroup H of G, fixing some element of the set of all integers $\{1, \dots, n\}$, is maximal in G (see for instance [DM96]). Note that Σ_n is primitive. Note also that the alternate group A_n is primitive for $n \geq 4$. [Bou70, Exercices 13 and 14, p.I-131]. We summarize unpublished results of A. Philippe [Phi69].

Proposition 2.2. *Let $K \subset L$ be a field extension.*

(1) *If $L|K$ is purely inseparable, with characteristic p, then $L|K$ is minimal if and only if $[L{:}K] = p$.*
(2) *If $[L{:}K] = p$, a prime integer, then $L|K$ is minimal. The converse holds if $L|K$ is Galois.*
(3) *If $L|K$ is separable and $L = K[x]$, let $p(X)$ be the minimal polynomial of x over K, with splitting field N. Then $L|K$ is minimal if and only if the Galois group of $N|K$ is primitive.*

This proposition may suggest that $L|K$ is minimal amounts to saying that $[L{:}K]$ is a prime integer. This is false, for it is enough to consider a Galois extension $L|K$ with primitive Galois group Σ_n where $n > 1$. Then Σ_{n-1} is the subgroup fixing n. Denote by K_1 its fixed field, then $K_1|K$ is minimal and $[K_1{:}K] = n$. Moreover, the Galois group of $L|K_1$ is Σ_{n-1}. We thus get a decomposition into minimal extensions $K \subset K_1 \subset \cdots \subset K_{n-2} \subset K_{n-1} = L$ with $[K_{i+1}{:}K_i] = n - i$.

A. Philippe proved that a field K has a minimal extension, with degree n for each integer $n > 1$, in case either K is a nonalgebraic finite type extension of the Galois field F_p (p a prime integer) or K is a finite type extension of the field of rational numbers Q. Actually, such fields have finite Galois extensions, with Σ_n as Galois group.

A field extension, with degree ≤ 3, is minimal. The following facts may be found in [AP03] and corroborate an above result. Let x be any root of $x^4 - nx - 1 = 0$, where n is an integer. Then the quartic extension $Q(x)|Q$ is minimal if and only if $n \notin \{-4, 0, 4\}$. More generally, let $p(X) \in K[X]$ be a separable irreducible polynomial, with degree 4, splitting field E, and cubic resolvent $r(X) \in K[X]$. If x is any root of $p(x) = 0$, then $K(x)|K$ is minimal if and only if $r(X)$ is irreducible or also, if and only if the Galois group of $E|K$ is either Σ_4 or A_4.

3 Characterizations of minimal morphisms

Ferrand and Olivier proved the following fundamental result.

Theorem 3.1. [FO70, Théorème 2.2] *Let $f{:}A \to B$ be a minimal ring morphism.*

(1) *There exists $M \in \mathrm{Max}(A)$ such that $A_P \to B_P$ is an isomorphism for each $P \in \mathrm{Spec}(A) \setminus \{M\}$.*
(2) *The following conditions are equivalent:*
   (a) *There is a prime ideal in B lying over M.*
   (b) $MB = M$.
   (c) *$A \to B$ is a strict monomorphism.*
   (d) *f is finite.*
   (e) *$\mathrm{Spec}(B) \to \mathrm{Spec}(A)$ is surjective.*
(3) *If the conditions of* (2) *are not verified, f is a flat epimorphism.*

Therefore, a minimal ring morphism is either integral (equivalently, finite) or a flat epimorphism and these conditions are mutually exclusive, as already remarked. The maximal ideal M exhibited in (1) is unique and is called the *crucial maximal ideal* of $A \to B$ by Dechéne [Dec78]. We thank her for her kind authorization of using results of her unpublished thesis. The next result clarifies the role of this ideal.

Proposition 3.2. [Dec78, Proposition 2.11] *Let $f{:}A \to B$ be a minimal ring morphism with crucial ideal M. The following statements hold:*

(1) $A{:}B \in \mathrm{Spec}(A)$ *and* $\{M\} = \mathrm{Max}(A) \cap \mathrm{V}(A{:}B)$.
(2) *If f is finite, then* $A{:}B = M$.
(3) *If f is a flat epimorphism, $A/(A{:}B)$ is a one-dimensional local domain,* $A{:}B \in \mathrm{Max}(B)$ *and* $A_{A:B} = B_M$.

M. Picavet exhibited a classification of minimal integral morphisms, used in the theory of algebraic orders [PL86] and [PL96]. But as observed in a recent paper [DMP05], this classification works in the general case. The terminology was suggested by P. Samuel.

Theorem 3.3. [DMP05, Corollary II.2] *Let $A \to B$ be an injective ring morphism, with conductor $A{:}B$. Then $A \to B$ is minimal and finite if and only if $A{:}B \in \mathrm{Max}(A)$ and one of the following three conditions holds:*

(1) **inert case**: *$A{:}B \in \mathrm{Max}(B)$ and $A/(A{:}B) \to B/(A{:}B)$ is a minimal field extension.*
(2) **decomposed case**: *There exist $N_1, N_2 \in \mathrm{Max}(B)$ such that $A{:}B = N_1 \cap N_2$ and the natural maps $A/(A{:}B) \to B/N_1$ and $A/(A{:}B) \to B/N_2$ are each isomorphisms.*
(3) **ramified case**: *There exists $N \in \mathrm{Max}(B)$ such that $N^2 \subseteq A{:}B \subset N$, $[B/(A{:}B){:}A/(A{:}B)] = 2$ and the natural map $A/(A{:}B) \to B/N$ is an isomorphism.*

In this case, there are at most two prime ideals in B lying over $A{:}B$.

Some conclusions of this theorem are reminiscent of the Gilmer-Heinzer's paper [GH67].

As observed by Dechéne, we can elucidate the case when $A{:}B = 0$.

Lemma 3.4. [Dec78, Corollary 2.14] *Let $A \to B$ be a minimal morphism, where B is an integral domain. Then $A{:}B = 0$ if and only if either $A \to B$ is a minimal field extension or A is a one-dimensional valuation domain, and B is the quotient field of A.*

Combining previous results, we get the following proposition.

Proposition 3.5. *Let $A \to B$ be an injective ring morphism. Then $A \to B$ is minimal and a flat epimorphism if and only if $A/(A{:}B)$ is a one-dimensional valuation ring and $B/(A{:}B)$ is its quotient field.*

There are minimal morphisms $A \to B$ between integral domains whose conductors are nonzero.

Example 3.6. Let R be a two-dimensional local regular domain and P a prime ideal of R, such that R/P is a one-dimensional valuation ring, with quotient field $\mathrm{k}(P)$; so that P has height one. Consider the CPI extension $A{:} = R +$

$PR_P = R_P \times_{k(P)} R/P$ of R and $B := R_P$, then $A \to B$ is minimal by Lemma 1.3. As $k(P) \simeq B \bigotimes_A (R/P)$, $A \to B$ cannot be finite, and $A \to B$ is a flat epimorphism with conductor PR_P.

Proposition 3.7. *Let $A \to B$ be a minimal integral ring morphism, with crucial ideal M. Then $A \to B$ is flat if and only if A_M is a field. Hence, if A is either an integral domain or local, and $A \to B$ is flat, then A is a field.*

Proof. If $A \to B$ is flat, $A_M \to B_M$ verifies the same conditions. By [FO70, Lemme 4.3.1], A_M is a field. The converse is easy, because $A_N \to B_N$ is an isomorphism for each maximal ideal $N \neq M$. □

We derive that a Prüfer domain A is a field if there is a minimal integral and torsion-free morphism $A \to B$, for instance if B is an overring of A.

If $A \to B$ is a minimal morphism and A is zero-dimensional, then $A \to B$ is integral for an epimorphism, whose domain is zero-dimensional, is surjective. Therefore, B is zero-dimensional.

As a minimal integral morphism $A \to B$ is strict, an element $a \in A$, which is invertible in B, is invertible in A. It follows that if B is local, so is A and $A \to B$ is local.

Example 3.8. If K is a field and n an even integer, then [FGH88] shows that R is a maximal element in the class of all local K-algebras, with integral closure $K[[X]]$ and fixed conductor $X^n K[[X]]$, if and only if R is Gorenstein. Next consider a minimal integral morphism $R \to K[[X]]$ where $K \subseteq R$. As $K[[X]]$ is local, its conductor is $X^2K[[X]]$ and the above remark shows that $R = K + X^2K[[X]]$ is a local ring. By [FGH88], R is a Gorenstein ring. Actually, $R = K + X^2K[[X]] \to K[[X]]$ is minimal because so is $K \to K[X]/X^2K[X] \simeq K[[X]]/X^2K[[X]]$ by Lemma 2.1.

Let $f{:}R \to S$ be an injective ring morphism, such that $f(R \setminus \mathrm{Z}(R)) \subseteq S \setminus \mathrm{Z}(S)$ (Condition (R)). Then there is an injective ring morphism $f'{:}\mathrm{Tot}(R) \to \mathrm{Tot}(S)$, inducing f. The Condition (R) is verified when either f is flat or essential; that is, for each $s \neq 0$ in S, there is some $t \in S$ such that $st \in R \setminus \{0\}$. This last condition holds for an injective integral ring morphism between integral domains. In case $R{:}S$ contains a regular element of R, we see that f' is an isomorphism and S is an overring of R. The same result holds if $\mathrm{Tot}(R)$ is zero-dimensional and f is a flat injective epimorphism. Moreover, if $\mathrm{Tot}(R)$ is zero-dimensional and $P \in \mathrm{Spec(R)}$, then $P \subseteq \mathrm{Z}(R)$ is equivalent to P is a minimal prime ideal. Hence, we get the following result, which generalizes the case of a minimal morphism between integral domains [SSY92, Theorem 4].

Proposition 3.9. *Let $A \to B$ be a minimal morphism, such that* $\mathrm{Tot}(A)$ *is zero-dimensional, then B is an overring of A in case one of the following conditions holds.*

(1) *$A \to B$ is finite, verifies Condition* (R) *and $A{:}B \notin$* $\mathrm{Min}(A)$.

(2) *$A \to B$ is a flat epimorphism.*

It follows that if A and B are integral domains and A is not a field, then B is an overring of A.

4 The case of a flat epimorphism

The following results come from Ferrand-Olivier's paper [FO70]. A minimal morphism $A \to B$, which is a flat epimorphism, is integrally closed and verifies Samuel's condition P_2; that is, for $x, y \in B$, then $xy \in A$ implies $x \in A$ or $y \in A$. We have the following characterization:

Theorem 4.1. [FO70, Proposition 3.3] *Let $f{:}A \to B$ be a minimal morphism, with crucial ideal M. Then $A \to B$ is a flat epimorphism if and only if $(A/(A{:}B))_M$ is a one-dimensional valuation ring. In this case, $(A{:}B)_M = 0$ if $A{:}B$ is finitely generated over A.*

In view of Lazard's paper [Laz69, Proposition 2.5], a flat epimorphism $A \to B$, with spectral image $X \subseteq \mathrm{Spec}(A)$, can be identified to $A \to \widetilde{A}(X)$, where $\widetilde{A}$ is the affine scheme associated to A and $\widetilde{A}(X) = \widetilde{A}_{|X}(X)$ is the ring of sections of the affine scheme $(X, \widetilde{A}_{|X})$. If $A \to B$ is minimal and a flat epimorphism, with crucial ideal M, then $X = \mathrm{D}(M) = \mathrm{Spec}(A) \setminus \{M\}$ is an affine open subset and hence is quasi-compact; so that $M = \sqrt{(a_1, \ldots, a_n)}$. It follows that B is the dominion of the flat morphism $A \to A[X_1, \ldots, X_n]/(a_1X_1 + \cdots + a_nX_n - 1)$, with spectral image $\mathrm{D}(M)$ [Pic03, 2.7, 2.8]. If $M = \sqrt{(a)}$, we see that $B = A_a$, where $a \in M$.

If A is an integral domain, then $\widetilde{A}(\mathrm{D}(M)) = \cap[A_P \mid P \in \mathrm{D}(M)]$, the so-called *Kaplansky transform* $S(M)$.

Corollary 4.2. [FO70, Proposition 3.3] *Let A be a ring and M a maximal ideal of A, such that A_M is a one-dimensional valuation ring. Then $\mathrm{D}(M) = \mathrm{Spec}(A) \setminus \{M\}$ is a quasi-compact (open) subset if and only if $A \to \widetilde{A}(\mathrm{D}(M))$ is a minimal morphism and a flat epimorphism.*

In case A is Noetherian, the correspondence suggested by the previous results establishes a bijection between the set of all minimal morphisms, which are flat epimorphisms, and the set of all maximal ideals M of A, such that A_M is a discrete valuation ring.

5 Unique minimal overrings

To begin with, we observe that Dobbs proved recently the following result [Dob]. For a commutative ring R and an R-module E, the canonical ring morphism $f{:}R \to R(+)E$ (the idealization) is a minimal morphism if and only if E is simple. Hence, each nonzero ring R has at least $|\mathrm{Max}(R)|$ minimal (finite) ring morphisms. It may be asked whether every ring has also a minimal morphism, which is a flat epimorphism. The answer is negative because a flat epimorphism, whose domain is absolutely flat, is surjective. In view of Proposition 3.9, it seems more reasonable to restrict the question to overrings.

Gilmer-Heinzer's paper [GH67] deals with unique minimal overrings introduced in Section 1. As observed in [GH67, p.138], an integral domain A, with a unique minimal overring B, is local. Such an integral domain is a valuation ring if A is integrally closed, and $B \subseteq A'$ if A is not. Their primary interest lies in the case where A' is the unique minimal overring of A. A satisfactory answer is given for QQR-domains A. They are integral domains A such that each of their overrings is an intersection of localizations (resp. at prime ideals) of A.

Using Theorem 3.3, we can complete a result of [GH67].

Theorem 5.1. [GH67, Theorem 3.3] *Let (A, M) be a local integral domain, which is not a valuation ring. The following statements are equivalent:*

(1) *A is a QQR-domain.*
(2) *A' is the unique minimal overring of A and the maximal ideals of A' are unbranched.*
(3) *$A \to A'$ is a minimal morphism, M is unbranched in A, and A' is a Prüfer domain.*

If these conditions are verified, $A \to A'$ is a minimal morphism, which is either decomposed or inert.

Note that $D \to D'$ may be a minimal morphism, while D' is not the unique minimal overring (see [GH67, Example, p.140]), where a lacking condition is D' is Prüfer. The condition "each maximal ideal of D' is unbranched" implies that $D \to D'$ is not ramified. Actually, [GH67, Example 4.3] exhibits a decomposed case and A' has two maximal ideals, while an inert case is given by [GH67, Example 4.1].

But the characterization of non-integrally closed domains, with a unique minimal overring, seems to be incomplete. In this direction, we can quote the following result.

Proposition 5.2. *Let A be a non-integrally closed local domain.*

(1) [GH67, Theorem 2.4] *If A' is Prüfer and $A \to A'$ is minimal, then A' is the unique minimal overring of A.*
(2) [GH67, Proposition 2.5] *If A has a unique minimal overring B and $A \to B$ is decomposed (equivalently, $\mathrm{Max}(B)$ has two elements), then $A' = V_1 \cap V_2$, where the (V_i, N_i)'s are valuation rings, and $N_1 \cap N_2 = M$ is the maximal ideal of A.*
(3) [GH67, Proposition 2.6] *If A has a unique minimal overring B and $A \to B$ is inert (equivalently, (B, N) is local and N is the maximal ideal M of A), then A' is a valuation ring, with maximal ideal M.*

The case where $A \to B$ is ramified remains unsolved [GH67, p.141].

To our best knowledge, concrete characterizations of domains having at least a minimal overring are known for integrally closed domains and Noetherian domains. We first develop the integrally closed case, studied by Papick [Pap76], and revisited by Ayache [Aya03]. Let A be an integrally closed domain and B a minimal overring. In this case, $A \to B$ is a flat epimorphism. Assume that A is a local domain and B a minimal overring. Then by [Aya03, Section 1], B is unique and local. Actually, we have $B = A_P$, where $P \in \mathrm{Spec}(A)$ and $P = PA_P$ [Pap76, Lemma 2.7]. In fact, the last condition means that P is a divided prime ideal [Dob76].

Proposition 5.3. [Aya03, Theorem 1.2] *Let (A, M) be an integrally closed local domain. Then A has a minimal overring (necessarily unique) B if and only if there is a divided prime ideal $P \subset M$ of A, such that A/P is a one-dimensional valuation ring. In this case, $B = A_P$.*

In case A is a valuation ring, a minimal overring is a unique minimal overring. This is a necessary condition when A is coherent [Pap76, Theorem 2.8]. Ayache provided a characterization for an arbitrary integrally closed domain, in the style of Theorem 4.1.

Theorem 5.4. [Aya03, Theorem 2.4] *Let A be an integrally closed domain. Then A has a minimal overring B if and only if the two following conditions are verified:*

(1) *There is some maximal ideal M in A, such that* $\mathrm{D}(M) = \mathrm{Spec}(A) \setminus \{M\}$ *is quasi-compact, and hence $M = \sqrt{(a_1, \ldots, a_n)}$.*
(2) *There is some prime ideal $P \subset M$, such that $PA_P = PA_M$ and $(A/P)_M$ is a one-dimensional valuation ring.*

In this case, $B = \widetilde{A}(\mathrm{D}(M)) = \mathrm{S}(M) = A_{a_1} \cap \cdots \cap A_{a_n}$, *the Kaplansky transform of M, and $M{:}M = A$* [Aya03, Proposition 2.6 and Lemma 2.8].

It follows that for a divided integrally closed domain (A, M), then $M{:}M$ cannot be a minimal overring of A. In the same way, a PVD (A, M), whose associated valuation ring is $M{:}M$, cannot have a minimal overring, except if A is a valuation ring. Note that there exist local orders A such that $A \to A'$ is ramified and A' is a valuation ring (see Section 9).

If A is a Prüfer domain and $M \in \mathrm{Max}(A)$, then $\mathrm{S}(M)$ is a minimal overring of A if and only if $\mathrm{D}(M)$ is quasi-compact [Aya03, Theorem 5.2]. But A may have more than one minimal overring (see [Pap76, Section 3] and [Aya03, Section 5]). In particular, a principal ideal domain, with n maximal ideals, has $n + 1$ minimal overrings (see [Aya03, Corollary 5.8] for a more precise result).

We now examine the Noetherian case. A first approach appeared in [SSY92]. Later on, Matsuda eliminated superfluous conditions [Mat92]. Before giving Matsuda's results, we state a crucial lemma, which has its own interest.

Lemma 5.5. [SSY92, Lemma 6] *Let A be a Noetherian integral domain containing an A-sequence a, b. Then $A[b/a]$ is never a minimal overring of A and is strictly contained in $A[1/a]$.*

Theorem 5.6. *Let A be a Noetherian integral domain.*

(1) [SSY92, Theorem 3] *If A is one-dimensional, then A has a minimal overring.*
(2) [Mat92, Theorem A'] *If* $\mathrm{Dim}(A) \geq 2$ *and A has a depth-one maximal ideal, then A has a minimal overring.*
(3) [Mat92, Theorem B'] *If every maximal ideal of A has depth ≥ 2, and hence* $\mathrm{Dim}(A) \geq 2$, *then A has no minimal overring.*

We can add that the integral domain hypothesis is necessary. We present here two new results without proofs, due to the lack of place. Clearly, a one-dimensional non-integrally closed Noetherian ring has a minimal overring.

Proposition 5.7. *Let A be a local integrally closed Noetherian ring. If A is not an integral domain, then A does not have a minimal overring.*

Theorem 5.8. *Let A be a one-dimensional reduced Noetherian ring, which is not an integral domain. Then A has no minimal overring if and only if A is integrally closed and local.*

Stable domains have been recently investigated by Olberding. A survey on these rings may be found in [Olb01c]. An integral domain A, with quotient field K, is called *stable* if each of its non-zero ideals I is projective over the ring $I{:}_K I$. The following results are deduced from Olberding's papers on the subject [Olb01a], [Olb01b] and [Olb02]. If (A, M) is a local domain with quotient field K, we set $A_1 = M{:}_K M$ and A_1 is an overring of A.

As Olberding observed, stable domains are related to *divisorial domains*. An integral domain is *divisorial* in case each of its nonzero ideals is divisorial and *totally divisorial* if each of its overrings is divisorial. Actually, an integral domain R is totally divisorial if and only if R is a stable divisorial domain [Olb01b, Theorem 3.12]. We use below the terminology introduced in Theorem 3.3.

Theorem 5.9. *Let (A, M) be a local stable domain, with quotient field K. Assume that $A \neq A_1$. The following statements hold:*

(1) *A_1 is local, with maximal ideal M_1, $A_1 = M_1{:}_K M_1$, and $[A_1/M_1{:}A/M] = 2$. In this case $A \to A_1$ is minimal and either ramified or inert, with $M = M_1$.*
(2) *A_1 is not local. In this case, $A \to A_1$ is minimal and decomposed.*

In both cases, the maximal ideals of A_1 are principal.

Proof. In view of [Olb02, Lemma 4.1], $M:_K M$ is an integral extension of A; so that $M:_K M \subseteq A'$. As A' has at most two maximal ideals [Olb01a, Corollary 2.4], $M:_K M$ has at most two maximal ideals. Next consider [Olb02, Proposition 4.2]. If A_1 is local with maximal ideal M_1, then A_1 is a minimal overring of A if $A_1 = M_1:_K M_1$. If A_1 is not local, its maximal spectrum has less than 2 elements and hence, exactly two maximal ideals. Moreover, $A/M \to A_1/M$ has rank two; so that $A \to A_1$ is minimal. The statement about maximal ideals may be found in [Olb02, Proposition 4.2]. □

Remark 5.10. Note that in case $A \subset A_1 \subset M_1:_K M_1$ and A_1 is local with maximal ideal M_1, we have $M_1^2 \subseteq M \subset M_1$ and $A/M \to A_1/M_1$ is an isomorphism [Olb02, Proposition 4.2]. In this case, by Theorem 3.3, $A \to A_1$ is minimal (ramified) if and only if $A/M \to A_1/M$ has rank 2, or equivalently, the rank of M_1/M over A/M is 1. This case occurs for a totally divisorial domain, whence stable (see an above quoted result).

Bazzoni and Salce proved the following results about a local divisorial domain. On one hand, we have that a local divisorial domain, whose maximal ideal is principal, is a valuation ring [BS96, Lemma 5.5]. In this case, we can use Ayache's results to look at minimal overrings of A. On the other hand, we have the following.

Theorem 5.11. [BS96, Theorem 5.7 and Remark p.855-856] *Let (A, M) be a local divisorial domain, such that M is not principal. Then A_1 is a unique minimal overring of A. Moreover, one and only one of the following statements holds:*

(1) *A_1 is local, with maximal ideal $M_1 \supset M$ and $A \to A_1$ is ramified.*
(2) *A_1 is a valuation ring, with maximal ideal M, and $A \to A_1$ is inert.*
(3) *A_1 has two maximal ideals and $A \to A_1$ is decomposed. Moreover, A_1 is a Prüfer domain, which is the intersection of two valuation rings (V_1, N_1) and (V_2, N_2), such that $M = N_1 \cap N_2$.*

Notice that we recover Bazzoni-Salce's result for totally divisorial domains in Olberding's result by using the above remark. Indeed, the missing case for $A \to A_1$ being minimal is when A_1 is local and $A_1 \neq M_1:_K M_1$. But M_1/M is simple both as an A-module and as an A_1-module; so that the rank of M_1/M over A/M is 1.

As already remarked for QQR-domains, the ramified case is difficult to handle and Bazzoni and Salce considered totally divisorial domains. They built a denumerable chain of minimal morphisms from A to A'.

6 Maximal subrings

Let A be a ring, with characteristic morphism $c: Z \to A$, where Z is the ring of all integers. The ring A is called finitely generated if $Z/\ker(c) \to A$ is finitely generated. In case A is finitely generated and has a subring $C \neq A$, an application of Zorn's lemma shows that there exists a maximal subring B of A; so that $B \to A$ is minimal.

Visweswaran defines a *maximal non-Noetherian subring* B of a Noetherian ring as a subring B, which is maximal for the non-Noetherian property [Vis90, Section 2]. He considers a domain $R = K[y_1, \cdots, y_t]$, where K is a field, with a finite Krull dimension $m > 0$. Let I be a nonzero proper ideal of R and D a subring of K, then $S := D + I$ is a subring of R. He completely determines when S is a maximal non-Noetherian subring of R [Vis90, Proposition 2.1]. Surprisingly, such a subring is a maximal subring if and only if D is not a field.

Proposition 6.1. [Vis90, Proposition 2.2(1)] *If D is not a field, then S is a maximal subring of R if and only if D is a rank one valuation ring, with quotient field K, and $I = (y_1 - a_1, \dots, y_t - a_t)$ for some $a_1, \dots, a_t \in K$.*

When D is a field, there is still a (complicated) characterization when S is a maximal subring [Vis90, Proposition 2.2(2)]. Using this material, he proves that $T := K[X] + (1 + XY)K[X, Y]$ is a maximal subring of $A := K[X, Y]$, which is integrally closed in A and not completely integrally closed. It follows that $T \to A$ is a flat epimorphism. Moreover, T_M is 2-dimensional and either Noetherian or a discrete valuation ring for each $M \in \mathrm{Max}(T)$, each maximal ideal of T is finitely generated, and T is not Laskerian. The reader may also consult [AJ02] for a more general setting.

Consider the class of one-dimensional local Noetherian analytically unramified rings A, such that $\widehat{A}$ has two minimal prime ideals. A ring A of this class is called *maximal with fixed conductor* if $A{:}A' \subset T{:}A'$ for each strict overring $T \subseteq A'$ of A [BDF00]. A ring R in the above class is maximal with fixed conductor if and only if R has a unique minimal overring [BDF00, Proposition 4.10].

To conclude this section, consider as did Gilmer and Huckaba, an integral domain D, with quotient field K, and $\alpha \in K \setminus D$ [GH74]. They define *rings being maximal without α*.

Proposition 6.2. [GH74, Proposition 1] *An integral domain D is maximal without α for some $\alpha \in K$ if and only if D has a unique minimal overring. In this case, D is local and $D[\alpha]$ is the unique minimal overring.*

In case D is maximal without α and if α is not integral over D, then D is a valuation ring and $D \to D[\alpha]$ is a flat epimorphism.

7 Constructions of minimal morphisms and transfer of the minimal morphism property

Next we examine the transfer of the minimal overring property to polynomial rings, which is linked to Goldman domains. An integral domain A is called a *G-domain* if there is some $M \in \mathrm{Max}(A[X])$ such that $M \cap A = 0$.

Proposition 7.1. [Aya03, Corollary 6.3] *Let A be an integrally closed domain, then $A[X]$ has a minimal overring if and only if A is a G-domain. It follows that $A[X_1, \ldots, X_n]$ has no minimal overrings for $n \geq 2$.*

We proved that if $A \to B$ is a minimal morphism, so is $A(X) \to B(X)$, where $A(X)$ and $B(X)$ are the Nagata rings of A and B [DMP05]. This was also proved by Dechéne in a more elementary way. A similar statement for polynomial rings fails miserably.

On the level of algebraic extensions, we can offer the following result.

Proposition 7.2. [Dec78, Remark 5.13] *Let $A \to B$ be a minimal integral morphism and let α be algebraic over A. If $B = A + Ay + \cdots + Ay^k$ and $A[\alpha] = A[\alpha, y\alpha, \ldots, y\alpha^k]$, then $A[\alpha] \to B[\alpha]$ is minimal.*

In the language of Dechéne, a 2-adjacent extension $A \to B$ is nothing but an integral minimal morphism $A \to B$, where A is a local domain and B an integral domain, with two maximal ideals. Her interest in these extensions relies on the fact that the Catenary Chain Conjecture (CCC) can be expressed with the help of 2-adjacent extensions. We recall that the CCC is: "the integral closure of a catenary local domain satisfies the chain condition cc for prime ideals". Note that a 2-adjacent extension appears in a famous Nagata's example of a catenary local Noetherian domain, which is not universally catenary [Nag62, p.p.203-205].

Theorem 7.3. [Dec78, Theorem 6.20] *The CCC holds if and only if any 2-adjacent extension of a catenary local Noetherian domain is catenary.*

By using the conductor, it is easily seen that for a field K and its polynomial ring $K[X]$, the extension $K[X^2, X^3] \to K[X]$ is minimal. More generally, if $R \to B$ is a finitely generated extension, there exists some subring A of B such that $A \to B$ is minimal [Dec78, Example 3.4].

We have also the following example [Dec78, Example 3.7]. Let $R \to B$ be an integral extension, where R is local and $\mathrm{Max}(B) = \{M_1, \ldots, M_n\}$ with $n > 1$. Pick x_i in each M_i, and let J be the Jacobson radical of $R[x_1, \ldots, x_n]$. Setting $A{:} = R + J$, we get that $A \to A[x_1]$ is a 2-adjacent extension and $A[x_1, \ldots, x_k] \to A[x_1, \ldots, x_{k+1}]$ is minimal for $k = 1, \ldots, n-1$. Actually, n is the maximum length of any chain of minimal morphisms between A and $A[x_1, \ldots, x_n]$ [Dec78, Example 6.31].

Proposition 7.4. [Dec78, Example 3.17] *Let A be a ring and M a maximal ideal of A. Then setting $B{:} = A[X]/(MX, X^2 - X)$, we get a minimal morphism $A \to B$ with conductor M. Moreover, $A_M \to B_M$ is 2-adjacent.*

8 Extensions of rings with finitely many subextensions

In a recent paper [Gil03], Gilmer studies two conditions on the set of overrings of an integral domain A, labelled as (FO) and (FC):

(FO) A has only finitely many overrings.

(FC) Each chain of distincts overrings of A is finite.

We add here another condition introduced by Anderson, Dobbs and Mullins for an extension of rings [ADM99]:

(FIP) An extension of rings $A \to B$ has FIP if there are only finitely many intermediary A-algebras between A and B.

It is obvious that these conditions generate chains of minimal morphisms from A to the quotient field K of A. Gilmer (re)establishes some properties of minimal (integral) morphisms in order to get characterizations of FO-domains and FC-domains. A key lemma is the following:

Lemma 8.1. [Gil03, Lemma 2.11] *Let $R \to S$ be a minimal integral extension of domains. If S/M^2 is Artinian for each maximal ideal M of S, then R/P^2 is Artinian for each maximal ideal P of R.*

Among many results, we quote the next theorem.

Theorem 8.2. [Gil03, Theorem 2.14] *Let A be an integral domain with conductor C. Then A is an FC-domain if and only if A' is a Prüfer domain with finite spectrum, $A \to A'$ is finite and A/C is an Artinian ring.*

The characterization of FO-condition is less achieved: A is an FO-domain if and only if A' is a Prüfer domain with finite spectrum and the extension $A/C \to A'/C$ has FIP. As Gilmer observed, a more definitive form of the condition FIP for an extension $R \to S$, where R is Artinian and S is a PIR is needed. The case of an Artinian reduced based ring is completely solved in [DMP05]. The interested reader may look at this paper for more information on FIP and for references. Due to the lack of place, we cannot give results of this paper. Gilmer states that the two conditions (FO) and (FC) are equivalent when R is integrally closed. Their class coincides with the class of Prüfer domains with finite spectra. In view of [DMP05, Corollary V.7], R is an integrally closed FO-domain if and only if R is a strong G-domain, with only finitely many prime ideals.

Recall that an extension of rings $A \subseteq B$ is called a *normal pair* if each intermediate ring C between A and B is integrally closed in B. Then Jaballah proved the following.

Proposition 8.3. [Jab99, Theorem 3.3] *Let $A \subseteq B$ be a normal pair defining $f{:}A \to B$, where B is an overring of A. Assume that $n = |\mathrm{Spec}(A) \setminus {}^a f(\mathrm{Spec}(B))|$ is finite. Then each maximal chain of intermediary rings has length n.*

Hence, for a Prüfer domain A, with only finitely many prime ideals, each maximal chain of overrings of A has a length equal to $|\mathrm{Spec}(A)| - 1$.

9 One-dimensional Noetherian rings

We first give some new results for one-dimensional Noetherian rings.

Proposition 9.1. *Let A be a one-dimensional Noetherian integral domain, such that $A \to A'$ is finite. Then A is stable if and only if A is totally divisorial.*

Proof. A' is h-local (each nonzero ideal of A' is contained in at most finitely many maximal ideals of A', and each nonzero prime ideal of A' is contained in a unique maximal ideal of A'), since A' is a Dedekind domain. Then the equivalence follows from [Olb01b, Proposition 3.8]. □

When (A, M) is a one-dimensional Noetherian local domain, we get a converse to Theorem 5.9. We use the notation of Section 5.

Theorem 9.2. *Let (A, M) be a one-dimensional Noetherian local domain, with quotient field K. Assume that $A \neq A_1$ and suppose in addition that $A_1 = M_1{:}_K M_1$ if (A_1, M_1) is local. Then A is stable if and only if $A \to A'$ is minimal and $[A'/M{:}A/M] = 2$.*

Proof. One implication is Theorem 5.9(1) in the inert case and Theorem 3.3 in the ramified or decomposed cases.

Conversely, assume that $A \to A'$ is minimal and $[A'/M{:}A/M] = 2$. First, A' is a Dedekind domain which has at most two maximal ideals. Let B be an A-submodule of A' containing A. Then B contains M as an A-submodule; whence B/M is an A/M-subspace of A'/M containing A/M. But $[A'/M{:}A/M] = 2$ gives either $B = A$ or $B = A'$. Hence, B is a ring and A is stable, in view of [Olb01c, Proposition 4.2]. □

Corollary 9.3. *Let (A, M) be a one-dimensional Noetherian local domain such that $A \to A'$ is minimal finite. Then A is stable if and only if A is divisorial. In this case, $[A'/M{:}A/M] = 2$.*

Proof. One implication is Proposition 9.1. Conversely, assume that A is divisorial. Then $[A'/M{:}A/M] = 2$ when $A \to A'$ is ramified or decomposed by Theorem 3.3. Assume that $A \to A'$ is inert, with conductor M. Then $A' = A_1$ with the previous notation. Since A is divisorial, A'/A is a simple A-module [BS96, Remark (3) p.856]. Then, any A-submodule of A' is a ring (see the proof of Theorem 9.2). Hence, A is stable in all cases. □

Remark 9.4. A one-dimensional Noetherian integral domain, such that $A \to A'$ is finite, may be divisorial without being stable. Consider the following example [PL87, Example, p.28-29]. Let t be a root of $t^3 - t - 1 = 0$. Then $A = Z[5t]$, where Z is the ring of all integers, is a one-dimensional Noetherian divisorial domain (Gorenstein) and $M = (5, 5t)$ is a maximal ideal of A. But $A_1 = M{:}M$ is not divisorial. Therefore, A_M is not stable, because not totally divisorial.

The reader is referred to [Swa80] and [PPL93] for the definitions of the seminormalization $\overset{+}{_B}A$ and of the t-closure $\overset{t}{_B}A$ of A in B.

Theorem 9.5. [PL96, Theorem 3.4] *Let B be a finite overring of a one-dimensional Noetherian ring A, with conductor I, whose radical is R in A and S in B. Let us consider the canonical decomposition $A \to \overset{+}{_B}A \to \overset{t}{_B}A \to B$. Then*

(1) *$A \to \overset{+}{_B}A$ admits only decompositions into ramified morphisms, with lengths $\mathrm{L}_A(S/R)$.*
(2) *$\overset{+}{_B}A \to \overset{t}{_B}A$ admits only decompositions into decomposed morphisms, with lengths $\mathrm{Card}(\mathrm{V}_B(I)) - \mathrm{Card}(\mathrm{V}_A(I))$.*
(3) *$\overset{t}{_B}A \to B$ admits only decompositions into inert morphisms. The length of such a decomposition $\overset{t}{_B}A \to \cdots \to A_i \to A_{i+1} \to \cdots \to B$ is equal to $\sum n_i$, where n_i is the length of the chain $A/P \to \ldots \to A_i/P_i \to A_{i+1}/P_{i+1} \to \ldots \to B/P'$, with $P' \in \mathrm{V}_B(I)$ and $P = P' \cap A$, $P_i = P' \cap A_i$.*

We end by two criteria related to orders, whose definition follows.

Let D be a Dedekind domain with quotient field Q and let K be a finite separable field extension of Q. An *order* A, related to the extension $K|Q$, is a D-subalgebra A of K, such that $D \to A$ is finite and such that A contains a basis of the Q-vector space K. An order A is divisorial if and only if A is Gorenstein. A finite overring of an order is an order.

Proposition 9.6. [PL87, Proposition 9, Section 3] *Let A be a divisorial order and B a finite overring of A, with conductor I. Then $A \to B$ is minimal if and only if I is a maximal ideal of A.*

Let D be a PID and consider an order A related to a finite separable field extension K of the quotient field Q of D. Let $\delta(A)$ be the discriminant of any basis of the D-module A, up to a unit of D. For a finite overring B of A, the *relative discriminant* of B over A is $\delta(A, B) = \delta(A)\delta(B)^{-1}$ and is a squared element of D.

Proposition 9.7. [PL86, Proposition 13 and Remarques p. 55-57] *Let A be an order related to a PID D and B a finite overring of A.*

(1) *If $\delta(A, B)$ is the square of an irreducible element of D, then $A \to B$ is minimal.*
(2) *If $A \to B$ is minimal, then $\delta(A, B) = p^{2k}$, where p is an irreducible element of D and k is a positive integer.*
(3) *Let $[K{:}Q] = 3$ and $K = Q(t)$. For $A = D[t]$, the morphism $A \to B$ is minimal if and only if $\delta(A, B)$ is the square of an irreducible element of D.*

References

[ADM99] D.D. Anderson, D. E. Dobbs, and B. Mullins. The primitive element theorem for commutative algebras. *Houston J. Math.*, 25:603–623, 1999.

[AJ02] A. Ayache and N. Jarboui. Maximal non-Noetherian subrings of a domain. *J. Algebra*, 248:806–823, 2002.

[AP03] T. Albu and L. Panaitopol. Quartic field extensions with no proper intermediate field. *Rev. Roumaine Math. Pures Appl.*, 48:1–11, 2003.

[Aya03] A. Ayache. Minimal overrings of an integrally closed domain. *Comm. Algebra*, 31:5693–5714, 2003.

[BDF00] V. Barucci, M. D'Anna, and R. Fröberg. The semigroup of values of a one-dimensional local ring with two minimal primes. *Comm. Algebra*, 28:3607–3663, 2000.

[Bou70] N. Bourbaki. *Algèbre, Chapitres I-III.* Hermann, Paris, 1970.

[BS96] S. Bazzoni and L. Salce. Warfield domains. *J. Algebra*, 185:836–868, 1996.

[Dec78] L. I. Dechéne. Adjacent extensions of rings. Ph.D. thesis, University of California at Riverside, 1978.

[DM96] J. D. Dixon and B. Mortimer. *Permutation Groups.* Springer-Verlag, New York, 1996.

[DMP05] D. E. Dobbs, B. Mullins, G. Picavet, and M. Picavet-L'Hermitte. On the FIP property for extensions of commutative rings. *Comm. Algebra*, 33:3091–3119, 2005.

[Dob76] D. E. Dobbs. Divided rings and going-down. *Pacific J. Math*, 67:353–363, 1976.

[Dob] D. E. Dobbs. Every commutative ring has a minimal ring extension. *Comm. Algebra*, to appear.

[FGH88] R. Fröberg, C. Gottlieb, and R. Häggkvist. Gorenstein rings as maximal subrings of $k[[X]]$ with fixed conductor. *Comm. Algebra*, 16:1621–1625, 1988.

[FO70] D. Ferrand and J. P. Olivier. Homomorphismes minimaux d'anneaux. *J. Algebra*, 16:461–471, 1970.

[GH67] R. Gilmer and W. J. Heinzer. Intersections of quotient rings of an integral domain. *J. Math Kyoto Univ.*, 7-2:133–150, 1967.

[GH74] R. Gilmer and J. A. Huckaba. Maximal overrings of an integral domain not containing a given element. *Comm. Algebra*, 2:377–401, 1974.

[Gil03] R. Gilmer. Some finiteness conditions on the set of overrings of an integral domain. *Proc. Amer. Math. Soc.*, 131:2337–2346, 2003.

[Jab99] A. Jaballah. A lower bound for the number of intermediary rings. *Comm. Algebra*, 27:1307–1311, 1999.

[Laz69] D. Lazard. Autour de la platitude. *Bull. Soc. Math. France*, 97:81–128, 1969.

[Lev69] A. H. M. Levelt. Foncteurs exacts à gauche. *Invent. Math.*, 9:114–140, 1969.

[Mat92] R. Matsuda. Remarks on Sato-Sugatani-Yoshida's results concerning minimal overrings. *Math. J. Toyama Univ.*, 15:39–42, 1992.

[Nag62] M. Nagata. *Local Rings.* Interscience Publishers, New York, 1962.

[Olb01a] B. Olberding. On the classification of stable domains. *J. Algebra*, 243:177–197, 2001.

[Olb01b] B. Olberding. Stability, duality, 2-generated ideals and a canonical decomposition of modules. *Rend. Sem. Mat. Univ. Padova*, 106:261–290, 2001.

[Olb01c] B. Olberding. Stability of ideals and its applications. In D. D. Anderson and I. J. Papick, editors, *Ideal Theoretic Methods in Commutative Algebra*, pages 319–341. Dekker, New York, 2001.

[Olb02] B. Olberding. On the structure of stable domains. *Comm. Algebra*, 30:877–895, 2002.

[Pap76] I. J. Papick. Local minimal overrings. *Can. J. Math.*, 28:788–792, 1976.

[Phi69] A. Philippe. Extension minimale de corps. Mémoire de D.E.A., Paris, 1969.

[Pic03] G. Picavet. Geometric subsets of a spectrum. In M. Fontana, S. Kabbaj, and S. Wiegand, editors, *Commutative Ring Theory and Applications*, pages 387–417. Dekker, New York, 2003.

[PL86] M. Picavet-L'Hermitte. Ordres d'entiers et morphismes minimaux. Thèse de doctorat d'Etat, Université de Clermont II, 1986.

[PL87] M. Picavet-L'Hermitte. Ordres de Gorenstein. *Ann. Sci. Univ. Clermont II*, 24:1–32, 1987.

[PL96] M. Picavet-L'Hermitte. Decomposition of order morphisms into minimal morphisms. *Math. J. Toyama Univ.*, 19:17–45, 1996.

[PPL93] G. Picavet and M. Picavet-L'Hermitte. Morphismes t-clos. *Comm. Algebra*, 21:179–219, 1993.

[SSY92] J. Sato, T. Sugatani, and K. Yoshida. On minimal overrings of a Noetherian domain. *Comm. Algebra*, 20:1735–1746, 1992.

[Swa80] R. G. Swan. On seminormality. *J. Algebra*, 67:210–229, 1980.

[Vis90] S. Visweswaran. Intermediate rings between $D + I$ and $K[y_1, \ldots, y_t]$. *Comm. Algebra*, 18:309–345, 1990.

What v-coprimality can do for you

Muhammad Zafrullah

57 Colgate Street, Pocatello, ID 83201, zafrullah@lohar.com

1 Introduction

Let D be an integral domain with quotient field K. Two elements $x, y \in D\backslash\{0\}$ are said to be v-coprime if $xD \cap yD = xyD$. A saturated multiplicative set $S \subseteq D\backslash\{0\}$ is a splitting set of D if every $x \in D\backslash\{0\}$ can be written as $x = ds$ where $s \in S$ and d is v-coprime to every member of S. The notions of v-coprimality and splitting sets can be traced back to the work of Gilmer and Parker [31] and Mott and Schexnayder [34]. These authors worked on generalizing the following theorem due to Nagata [35]: Let D be a Noetherian domain and let S be a multiplicative set generated by principal nonzero primes of D. If D_S is a UFD then so is D . The purpose of this article is to present a brief survey of the notion of v-coprimality, its applications, its morphs and its generalizations; as a bouquet of flowers from the garden that sprang up from the seeds planted by Gilmer, Mott, Parker and Schexnayder. The space constraints make it hard to present the full view of the garden, but I will do my best to provide a sizeable bouquet. Before I get down to describing what I aim to do, it seems expedient to give a brief description of the tools that I will be using throughout this survey.

Let $F(D)$ be the set of nonzero fractional ideals of D, $A^{-1} = \{x \in K : xA \subseteq D\}$, $A_v = (A^{-1})^{-1} = \bigcap_{A \subseteq cD} cD$, $c \in K\backslash\{0\}$.

A function $*$ on $F(D)$ is called a star operation, if for all $a \in K\backslash\{0\}$ and $A, B \in F(D)$, the following hold.

(1*) $(a)^* = (a)$, $(aA)^* = aA^*$, (2*) $A \subseteq A^*$ and $A \subseteq B \Rightarrow A^* \subseteq B^*$ (3*) $(A^*)^* = A^*$.

Given that $*$ is a star operation on $F(D)$ and $A, B \in F(D)$, we have $(AB)^* = (A^*B)^* = (A^*B^*)^*$. These equations are said to define the "$*$-multiplication". The function on $F(D)$ defined by $A \to A_v$ is a star operation such that for any star operation $*$ and for any $A \in F(D)$ we have $A^* \subseteq A_v$. To each star operation $*$ we can associate $*_f$ defined by $A^{*_f} = \cup\{F^* : 0 \neq F$ is a finitely generated D-submodule of $A\}$ for $A \in F(D)$. Call $*$ of finite type if

$A^* = A^{*_f}$ for all $A \in F(D)$. Indeed for any star operation $*$ the operation $*_f$ is of finite type. The well known t-operation is given by $t = v_f$. (So, if A is finitely generated then obviously $A_t = A_v$.) The identity function $A \mapsto A$ on $F(D)$ is the d-operation. If $\{D_\alpha\}$ is a family of overrings of D such that $D = \cap D_\alpha$ then the function $A \mapsto A^* = \cap AD_\alpha$ is also a star operation. An integral ideal P of D is called a prime $*$-ideal if P is a $*$-ideal and a prime ideal. If $*$ is of finite type, a proper integral ideal M that is maximal with respect to being a $*$-ideal is called a maximal $*$-ideal and is necessarily prime. Moreover, every proper $*$-ideal is contained in at least one maximal $*$-ideal. For $*$ of finite type the set of maximal $*$-ideals is usually denoted by $*$-$Max(D)$. It can be shown that $D = \bigcap_{M \in *-Max(D)} D_M$. The star operation induced by $\{D_M\}_{M \in t\text{-}Max(D)}$ is usually denoted by w. Obviously every prime ideal contained in a maximal t-ideal is a w-ideal.

An ideal $A \in F(D)$ is a $*$-ideal of finite type if $A = B^*$ for some f.g. $B \in F(D)$, and A is $*$-invertible if $(AB)^* = D$ for some $B \in F(D)$. A t-invertible t-ideal is of finite type, and every invertible ideal is a v-ideal. D is a Prüfer v-multiplication domain (PVMD) if each two generated nonzero ideal of D is t-invertible. Clearly a GCD domain is a PVMD. For details on star operations the reader may consult sections 32 and 34 of Gilmer's book [30] or, especially for the w-operation, the survey [44].

The paper is split into six sections. In section 2, I briefly treat the notion of v-coprimality. I indicate ways in which v-coprimality is different from coprimality. I also introduce the more general notion of $*$-coprimality by saying that two elements $x, y \in D$ are $*$-coprime if $(x, y)^* = D$ and characterize v-coprimality. Then I show that v-coprimality is similar to disjointness in partially ordered groups. In section 3, I show how v-coprimality has been used in circumstances, to do with divisibility, where coprimality has no effect. In section 4, I describe the splitting sets, indicating some properties. I also provide a brief historical background on them. In section 5, I give some examples and applications of splitting sets giving various forms and generalizations of Nagata-type theorems. In the 6th and last section, I indicate the kind of generalizations of splitting sets that have interested me and my co-workers.

2 v-coprimality

In this section I briefly treat the notion of v-coprimality. I show the ways in which v-coprimality is different from coprimality. I also introduce the more general notion of $*$-coprimality by saying that two elements $x, y \in D$ are $*$-coprime if $(x, y)^* = D$ and characterize v-coprimality. Then I show that v-coprimality is similar to disjointness in partially ordered groups.

Definition 2.1. *Two nonzero elements $x, y \in D$ are called v-coprime if $(x, y)_v = D$ (i.e. $xD \cap yD = xyD$ or equivalently $(x, y)^{-1} = D$).*

It is easy to establish that $(a,b)_v = D \Leftrightarrow ((a,b) \subseteq (c/d) \Rightarrow c|d)$. So $(x,y)_v \neq D \Leftrightarrow$ (there exist $c,d \in D$ such that $c \nmid d$ but$(a,b) \subseteq c/d)$. Now, ordinarily $x,y \in D$ are said to be coprime if x and y have no nonunit common factor in D. Note that x,y being v-coprime implies x,y coprime but not conversely; as is apparent from the following discussion. Next, the negation of coprimality is much cleaner than the negation of v-coprimality. Indeed it is useful to note that $\text{GCD}(a,b) = 1 \Leftrightarrow \forall_{x\in D}((x \mid a,b) \Rightarrow x \mid 1)$ and so the negation of "a,b are coprime" would be $\exists_{x\in D}((x \mid a,b) \wedge x \nmid 1)$. Yet the negation of $(a,b)_v = D$ gives only that $(a,b)_v \neq D$, and $(a,b)_v \neq D$ does not imply that a,b have a nonunit common factor. For example in the ring $D = F[[X^2,X^3]]$ where F is a field, $(X^2,X^3)_v \neq D$ but $\text{GCD}(X^2,X^3) = 1$. On the other hand in some integral domains, such as GCD domains the notions of coprime and v-coprime coincide. I must note that v-coprimality is the ring theoretic equivalent of orthogonality (disjointness) in directed p.o. groups. Recall that a,b in a p.o. group $(G,\leq,\cdot)$ are disjoint if, $\inf(a,b) = a \wedge b$ exists and is e the identity, or equivalently $\sup(a,b) = a \vee b$ exists and is $a \cdot b$ [30, page 156] (We use orthogonal as a synonym of disjoint as we deal only with positive elements.) Indeed given $G(D) = \{kD{:}k \in K\backslash\{0\}\}$ partially ordered by $aD \leq bD$ if and only if $bD \subseteq aD$ for $a,b \in K\backslash\{0\}$, $G(D)$ represents the group of divisibility of D. Let $G(D)^+$ represent the positive cone of $G(D)$; then for $rD, sD \in G(D)^+$, $rD \vee sD \in G(D)$ translates to $rD \cap sD$ being principal, and if $rD \cap sD = rsD$ then in $G(D)$ we have $rD \vee sD = rDsD$ which forces rD and sD to be disjoint in $G(D)$.

Let $*$ be a general star operation and call $x,y \in D$ $*$-coprime if $(x,y)^* = D$. Since $A^* \subseteq A_v$ for every star operation $*$, we know that x,y being $*$-coprime implies x,y v-coprime but not conversely. For example in $F[X,Y]$ where F is a field, $(X,Y)_v = D$ but $(X,Y) = (X,Y)_d \neq D$. Finally no proper $*$-ideal contains a pair of $*$-coprime elements.

Proposition 2.2. *For a general star operation $*$ on $F(D)$ the following hold: (1) r, $s \in D$ are $*$-coprime to $x \in D$ if and only if $(rs,x)^* = D$. (2) For $r_1,r_2,...,r_n \in D$, $(r_1r_2...r_n,x)^* = D$ if and only if $(r_i,x)^* = D$. (3) $(r,x)^* = D$ if and only if every factor of r is $*$-coprime to x.*

Proof. (1). Suppose r and s are $*$-coprime to x and consider $(x,rs)^* = (x,rx,rs)^* = (x,(rx,rs)^*)^* = (x,r(x,s)^*)^* = (x,r)^* = D$. Conversely suppose $(rs,x)^* = D$ and consider $(x,r)^* = (x,rs,r)^* = ((x,rs)^*,r)^* = (D,r)^* = D$. Next, (2) and (3) follow from (1).

The following proposition lists some properties of $*$-coprime (and hence v-coprime) elements.

Proposition 2.3. *For a general star operation $*$ on $F(D)$ and for $r,s \in D$ the following hold. (i) $(r,s)^* = D \Leftrightarrow (r^n,s)^* = D \Leftrightarrow (r^n,s^m)^* = D$ for any natural m,n. (ii) $(r,s)^* = D$ and $r|sy \Rightarrow r|y$. (iii) $(r,s)^* = D \Rightarrow D = D_r \cap D_s$, here $D_r = D_S$ where $S = \{r^n{:}n$ ranges over nonnegative integers$\}$.*

(iv) $(r,s)_v = D \Leftrightarrow D = D_r \cap D_s$. *(v)* $(r,s)_v = D$ *if and only if* r *and* s *do not share any prime t-ideals. (vi) Let* $x = \frac{r}{s} \in K \backslash D$, *if* s *has a nonunit factor that is* $*$*-coprime with* r *then* x *cannot be integral over* D.

Proof. (i) is direct. For (ii) note: $(r) = (r,sy)^* = (r,ry,sy)^* = ((r,y(r,s)^*)^* = (r,y)^*$. For (iii) let $(r,s)^* = D$ and consider $h \in D_r \cap D_s$. Then for some natural numbers m,n $hr^m, hs^n \in D$. So $D \supseteq (hr^m, hs^n)^* = h(r^m,s^n)^* = hD$ (by (i)). For (iv) use (iii) to establish that $(r,s)_v = D \Rightarrow D = D_r \cap D_s$. For the converse assume $D = D_r \cap D_s$ and note that $(r,s)^* = (r,s)D_r \cap (r,s)D_s = D$. But $(r,s)^* = D \Rightarrow (r,s)_v = D$. For (v) note that if $(r,s)_v = D$ then r,s cannot be in a proper integral t-ideal and hence cannot be in a prime t-ideal. Conversely if r,s do not share a prime t-ideal then r,s do not share a maximal t-ideal. But then $(r,s)_w = \bigcap_{M \in t-Max(D)} (r,s)D_M = \bigcap_{M \in t-Max(D)} D_M = D$. But as $(r,s)_w \subseteq (r,s)_v$ we have the conclusion. For (vi) use the fact that if $\frac{r}{s}$ is integral over D then $s|r^n$ for some n; then use (i).

Using the aforementioned similarity between disjoint (or orthogonal) elements in (directed) partially ordered groups and v-coprime elements in integral domains we can associate with each nonempty $S \subseteq D\backslash\{0\}$ the m-complement $S^\perp = \{t \in D{:}(t,s)_v = D \text{ for all } s \in S\}$ and state the following proposition which comes from a recent paper by David Anderson and Chang [10].

Proposition 2.4. *Let* D *be an integral domain,* S, S_1 *and* S_2 *nonempty subsets of* $D\backslash\{0\}$ *and let* $\{S_\alpha\}$ *be a nonempty family of nonempty subsets of* D. *(1)* $S^\perp = (< S >)^\perp = (\overline{< S >})^\perp$, *where* $< S >$ *denotes the set multiplicatively generated by* S *and* $\overline{< S >}$ *denotes the saturation of* $< S >$. *(We do not entertain empty sets nor empty products.) (2)* $S^\perp$ *is a saturated multiplicative set. (3) If* $S_1 \subseteq S_2$ *then* $(S_1)^\perp \supseteq (S_2)^\perp$. *(4)* $S \cap S^\perp \subseteq U(D)$ *where* $U(D)$ *denotes the set of units of* D. *(Equality if* $S \supseteq U(D)$.*) (5)* $S \subseteq (S^\perp)^\perp = S^{\perp\perp}$ *(notation). (6)* $S^\perp = (S^{\perp\perp})^\perp = S^{\perp\perp\perp}$ *(notation). (7)* $(\cup_\alpha S_\alpha)^\perp = (< \cup_\alpha S_\alpha >)^\perp = \cap (S_\alpha)^\perp$. *(8)* $(S_1S_2)^\perp = (S_1)^\perp \cap (S_2)^\perp$. *(9) If* $S_1 \cap S_2 \neq \phi$ *then* $(S_1)^\perp (S_2)^\perp \subseteq (S_1 \cap S_2)^\perp$. *(10)* $D = D_{<S>} \cap D_{S^\perp}$.

The proofs are simple. For instance (1) and (2) can be proved using Proposition 2.2. Next, (3), (4) and (5) were treated in ([5, Proposition 2.4]). For (6), applying (3) to (5) we get $S^\perp \supseteq [(S^\perp)^\perp]^\perp$ and also applying (5) to $S^\perp$we get $S^\perp \subseteq [(S^\perp)]^\perp]^\perp$. In case of (7), for each α, $S_\alpha \subseteq \cup S_\alpha$ implies by (3) that $(S_\alpha)^\perp \supseteq (\cup S_\alpha)^\perp$ which in turn means that $\cap_\alpha (S_\alpha)^\perp \supseteq (\cup S_\alpha)^\perp$. For the reverse inclusion note that $x \in \cap_\alpha (S_\alpha)^\perp$ implies that x is v-coprime to each member of S_α for each α and so x is v-coprime to each member of $\cup S_\alpha$. The equation in (8) can be established using Proposition 2.2 For (9) note that $\phi \neq S_1 \cap S_2 \subset S_i$ $(i = 1,2)$ and so $(S_i)^\perp \subseteq (S_1 \cap S_2)^\perp$, but by (2), $(S_1 \cap S_2)^\perp$ is multiplicative and saturated.

As pointed out in [10] the inclusions in Proposition 2.4 can be proper. Our notation is different from that of [10]. This is partly in solidarity with [5] where it was noted that in the study of partially ordered groups $S^{\perp}$ is used to denote the set of elements orthogonal to elements of S, and partly because I want to see if some of the multiplicative ideal theory of domains can be used in studying partially ordered groups and monoids.

3 Applications of v-coprimality I (Divisibility)

The notion of v-coprimality is useful when we need to sift through factors in the presence of properties weaker than the GCD property. Recall that D is an almost GCD(AGCD) domain (monoid) if for each pair of nonzero elements x, y there is a natural number n such that $(x^n, y^n)_v$ is principal. The notion of AGCD domains was introduced in [41]. It was studied further in [8] and in [25]. It was shown in [8, Lemma 3.3] that D is an AGCD domain if and only if for each set $x_1, x_2, ..., x_n$ of nonzero elements of D there is a natural number m such that $(x_1^m, x_2^m, ..., x_n^m)_v$ is principal.

Here is a brief demonstration of the use of v-coprime elements. For this let us agree to call a quasi-local domain (D, M) t-local if the maximal ideal M is a t-ideal.

Proposition 3.1. *Let (D, M) be a t-local AGCD domain. Then for each pair $x, y \in D\backslash\{0\}$ there is a natural number n such that $x^n | y^n$ or $y^n | x^n$.*

Proof. Let $x, y \in D\backslash\{0\}$. If either of x, y is a unit we have nothing to prove. So, let $x, y \in M\backslash\{0\}$. Since D is AGCD, $(x^n, y^n)_v = dD$ for some natural n and $d \in D$. Or $(\frac{x^n}{d}, \frac{y^n}{d})_v = D$. Because M is a t-ideal, $\frac{x^n}{d}, \frac{y^n}{d}$ cannot both be in M, forcing one of $\frac{x^n}{d}, \frac{y^n}{d}$ to be a unit and making d an associate of x^n or of y^n.

The extent to which v-coprimality can be of use in bringing about uniqueness and order where there appears to be none is apparent in the study of factorization in integral domains that do not have the unique factorization property. Recall for instance that an integral domain in which every nonzero nonunit is expressible as a product of primary elements is called a weakly factorial domain (WFD). Now, every nonzero nonunit of a WFD can be written uniquely as a product of mutually v-coprime primary elements. Weakly factorial domains were introduced by Anderson and Mahaney [6] and further studied by Anderson and Zafrullah [7]. In a recent survey of "alternate" factorization in integral domains Anderson [1] treats WFD's in greater detail. In a GCD domain, as mentioned earlier, the notions of coprime and v-coprime coincide, making the study of alternate factorizations much easier. It was in GCD domains that I started studying my kind of unique factorization. An element $r \in D$ is said to be a rigid element if r is a nonzero nonunit such that for all pairs $x, y \mid r$ we have $x \mid y$ or $y \mid x$. In [39] the following result was proved.

Proposition 3.2. *If in a GCD domain D, an element x is a product of finitely many rigid elements then x can be uniquely expressed as a product of finitely many coprime rigid elements.*

It is well known that D is a GCD domain if and only if its group of divisibility $G(D)$ is a lattice ordered group. Noting that the rigid element in a GCD domain is the same as the basic element ($b \in G^+, [0, b]$ is totally ordered) in a lattice ordered group. I decided to translate Conrad's condition F from [22] to the GCD domain setup as: A GCD domain D satisfies Conrad's condition F if every nonzero nonunit of D is divisible by at most a finite number of mutually coprime nonunits. Clearly, in a GCD domain, a nonzero nonunit that has no coprime factors is a rigid element. Using this I was able in [40] to prove the following result.

Proposition 3.3. *A GCD domain D is a ring of Krull type if and only if D satisfies Conrad's condition F.*

Recall from [32] that an integral domain D is a ring of Krull type if D has a family $\mathcal{F}$ of prime ideals such that for each $P \in \mathcal{F}$, D_P is a valuation domain and $D = \bigcap_{P \in \mathcal{F}} D_P$ is a locally finite intersection. The work on GCD domains satisfying Conrad's condition F came in handy when I became involved in a similar study of almost GCD domains in [25]. But in AGCD domains coprime and v-coprime do not coincide. For example a Dedekind domain with nonzero torsion class group is an AGCD domain but since such a domain is not a PID it must have a prime ideal P that is not principal and this forces P to have at least two non-associated irreducible elements x, y. Now being non-associated, x and y are coprime and being in P, x and y are not v-coprime, because being invertible P is a v-ideal. To cut the long story short we brought in new definitions. We called for a nonzero nonunit x the set $S(x) = \overline{< x >}$ the span of x and we called r an almost rigid element if for each m and for all $x, y \mid r^m$ there exists $n = n(x, y)$ such that $x^n \mid y^n$ or $y^n \mid x^n$. Thus in an AGCD domain an element r is almost rigid if and only if $S(x)$ contains no pair of v-coprime nonunits. We also showed that if r is almost rigid then the set $P(r) = \{x \in D : (r, x)_v \neq D\}$ is a maximal t-ideal. Calling 'of finite t-character' a domain in which every nonzero nonunit belongs to at most a finite number of maximal t-ideals we proved the following theorem.

Theorem 3.4. *An almost GCD domain D is of finite t-character if and only if for no nonzero nonunit $x \in D$, $S(x)$ contains an infinite set of nonzero nonunit mutually v-coprime elements.*

Indeed a ring of Krull type of Griffin [32] is a ring of finite t-character. In the AGCD (and hence GCD) situation an upper bound on the number of mutually v-coprime elements delivers some interesting results. The following result will facilitate the appreciation of those results.

Proposition 3.5. *If D is a domain with only a finite number of maximal t-ideals then D is a semi-quasi-local domain with each maximal ideal a t-ideal.*

Proof. Suppose that D has finitely many maximal t-ideals say $\mathcal{P}_1, \mathcal{P}_2, ..., \mathcal{P}_r$; we can assume all of them to be distinct. Now since for every nonunit d in D, dD is contained in some maximal t-ideal we conclude that $\mathcal{P}_1 \cup \mathcal{P}_2 \cup ... \cup \mathcal{P}_r$ consists of all nonunits of D. Next since every element of a nonzero prime ideal M is a nonunit we conclude that for each maximal ideal M we have $M \subseteq \mathcal{P}_1 \cup \mathcal{P}_2 \cup ... \cup \mathcal{P}_r$. But then it is well known that M must be contained in one of the maximal t-ideals say $\mathcal{P}_i$ (see e.g. [33, Theorem 83]). But since M is maximal we have $M = \mathcal{P}_i$. From this it is easy to show that $\mathcal{P}_1, \mathcal{P}_2, ..., \mathcal{P}_r$ are precisely the maximal ideals of D.

Recall that D is an almost Bezout domain if for each pair $a, b \in D \backslash \{0\}$ there is a natural number $n = n(a, b)$ such that the ideal (a^n, b^n) is principal.

Corollary 3.6. *An AGCD (a GCD) domain D having only a finite maximal set $S = \{x_1, x_2, ..., x_n\}$ of mutually v-coprime (resp. coprime) nonunits is a semilocal almost Bezout (resp. Bezout) domain.*

Proof. The idea is that if there is a maximal set S of mutually v-coprime nonunits then these nonunits are each almost rigid. To each almost rigid element x_i we have a unique maximal t-ideal $P(x_i)$. Now these are the only maximal t-ideals of D. For if not and there is a maximal t-ideal P such that $P \neq P(x_i)$ for each i then $x_1 x_2 ... x_n \notin P$. Then $(x_1 x_2 ... x_n, P)_t = D$. That is there are $y_1, y_2, ..., y_r \in P$ such that $(x_1 x_2 ... x_n, y_1, y_2, ..., y_r)_v = D$. This means that $x_1 x_2 ... x_n, y_1, y_2, ..., y_r$ do not share any prime t-ideals, which in turn means that $x_1 x_2 ... x_n, y_1^m, y_2^m, ..., y_r^m$ do not share any prime t-ideal where m is such that $(y_1^m, y_2^m, ..., y_r^m)_v = pD$. But then $p \in P$ and so is a nonunit and we end up with $D = (x_1 x_2 ... x_n, y_1^m, y_2^m, ..., y_r^m)_v = (x_1 x_2 ... x_n, (y_1^m, y_2^m, ..., y_r^m)_v)_v = (x_1 x_2 ... x_n, p)_v$. Consequently, $x_1, x_2, ..., x_n, p$ are mutually v-coprime contradicting the maximality of S. Now we end up with finitely many maximal t-ideals and Proposition 3.5 applies. For the almost Bezout part let a, b be two nonzero elements of D; then since D is AGCD there is n such that $(a^n, b^n)_v = dD$. Or $(\frac{a^n}{d}, \frac{b^n}{d})_v = D$. This means that $\frac{a^n}{d}, \frac{b^n}{d}$ do not share any maximal t -ideal. But since all the maximal ideals are maximal t-ideals we conclude that $(\frac{a^n}{d}, \frac{b^n}{d}) = D$. But then $(a^n, b^n) = dD$.

Now here is an anecdotal proof of the fact that v-coprimality helps.

Proposition 3.7. *If an integral domain D contains two nonunits a, b such that $(a, b)_v = D$ but $(a, b) \subsetneq D$ then Spec(D) is infinite.*

Proof. Exercise.

4 Applications of v-coprimality II (Splitting sets)

In this section we define and briefly describe the splitting and lcm-splitting sets and provide brief historical background on them.

Definition 4.1. *A saturated multiplicative set S of D is a splitting multiplicative set if each $x \in D \backslash \{0\}$ can be written as $x = ds$ where $s \in S$ and d is v-coprime to every member of S.*

It follows that if S is a splitting set then the m-complement $S^{\perp} = \{t{:}(t,s)_v = D,\ s \in S\}$ is also a splitting set. Note also that if S is a splitting set then $S \cap S^{\perp} = U(D)$. Here are a few characterizations of splitting sets whose proofs can be found in [2].

Theorem 4.2. *The following are equivalent for a saturated multiplicative set S: (1) S is a splitting set. (2) $< SD >$, the p.o. subgroup of $G(D)$ generated by $\{sD{:}s \in S\}$, is a cardinal summand of $G(D)$, the group of divisibility of D, i.e., there is a p.o. subgroup H of $G(D)$ such that $< SD > \oplus_c H = G(D)$. (3) If A is a principal integral ideal of D_S then $A \cap D$ is a principal ideal of D. (That is, principal integral ideals of D_S contract to principal ideals of D).(4) There is a multiplicative subset T of D such that (a) each element d of $D \backslash \{0\}$ can be written as $d = st$ where $s \in S$ and $t \in T$ and (b) any of the following equivalent conditions holds: (i) If $s_1t_1 = s_2t_2$, where $s_i \in S$ and $t_i \in T$ then $s_2 = s_1u$ and $t_2 = t_1u^{-1}$, where $u, u^{-1} \in D$. That is $d = st$ is unique up to associates. (ii) If $d = st$ $(s \in S, t \in T)$, then $dD_S \cap D = tD$. (iii) For each $s \in S$ and $t \in T$, $(s,t)_v = D$. (iv) For each $t \in T$, $tD_S \cap D = tD$.*

Some forms of the statements in Theorem 4.2 can be found in [31] and [34].

Definition 4.3. *A splitting set S is an lcm-splitting set if in addition every element of S has an lcm with every element of D.*

Proposition 4.4. *The following are equivalent for a saturated multiplicative set S: (i) S is lcm-splitting. (ii) $s_1D \cap s_2D$ is principal for $s_i \in S$. (iii) $s_1D \cap s_2D = sD$ for s, $s_i \in S$. (iv) $D_{S^{\perp}}$ is a GCD domain.*

Remark 4.5. It may be noted that if S in Proposition 4.4 is generated by primes then $D_{S^{\perp}}$ is a UFD [2, Proposition 2.6].

Since [2, Proposition 2.6] covers a lot of ground it seems best to quote it as a theorem.

Theorem 4.6. *The following conditions are equivalent for a saturated multiplicative set S of D: (1) S is generated by a set of prime elements $\{p_\alpha\}$ satisfying (a) for each α, $\bigcap_{n=1}^{\infty} p_\alpha^n D = 0$, and (b) for any sequence $\{p_{\alpha_n}\}$ of* *nonassociate members of* $\{p_\alpha\}$, $\bigcap_{n=1}^{\infty} p_{\alpha_n} D = 0$. *(2) S is generated by a splitting set of principal primes. (3) S is generated by a set of principal prime elements and S is a splitting set. (4) S is a splitting set and D_T is a UFD, where $T = S^\perp$ is the m-complement for S.*

(Note that a set $\{p_\alpha\}$ of principal prime elements is a splitting set of principal primes if the saturation of the set multiplicatively generated by $\{p_\alpha\}$ is a splitting set.) The set described in Theorem 4.6 is called a UF set in [34].

Using the statement of Definition 4.1 and using part (10) of Proposition 2.4 we can state the following result.

Proposition 4.7. *Let S be a splitting multiplicative set in D and let $T = S^\perp$ then $S = T^\perp = S^{\perp\perp}$ and $D = D_S \cap D_T$.*

Proposition 4.8. *Let S be a splitting set of D. (1) If P is a prime t-ideal, then P intersects S or P intersects $S^\perp$ but not both. (Any prime t-ideal that intersects both would be forced to contain a v-coprime pair, which is impossible.) (2) ([2]) If A is a nonzero ideal of D then $A_t D_S = (AD_S)_t$. (So, P is a prime t-ideal of D if and only if PD_S or PD_T is a prime t-ideal of the respective quotient ring.) (3) ([2]) Let $T = S^\perp$ and let $s_1, s_2, ..., s_m \in S$; $t_1, t_2, ..., t_m \in T$. Then $(s_1t_1, s_2t_2, ..., s_mt_m)_v = ((s_1, s_2, ..., s_m)(t_1, t_2, ..., t_m))_v$.*

The above proposition provides the sort of insight that gives you tools. For instance, from (1) we gather that if S is a splitting set it partitions the set of nonzero w-prime ideals into ones that intersect S and those that intersect $S^\perp$. Also look up [9]. From (2) we infer that if S is a splitting set then t-ideals of D extend to t-ideals of D_S, and v-ideals of finite type extend to v-ideals of finite type, something that usually does not hold. (For this see the discussion on pages 2522 and 2523 of [45].) Finally (3) is a gem. It shows that if S is a splitting set every v-ideal of finite type can be written as the v-image of a product of two ideals; one generated completely by elements from S and the other generated completely by elements from the m-complement T. Now assume that we are dealing with a t-invertible t-ideal A then we know that A is a v-ideal of finite type. So $A = (s_1t_1, s_2t_2, ..., s_mt_m)_v = ((s_1, s_2, ..., s_m)(t_1, t_2, ..., t_m))_v$ as a t-product of two t-invertible t-ideals. Now recall that under t-multiplication the group of t-invertible t-ideals modulo the group of nonzero principal ideals is called the t-class group of D, $Cl_t(D)$; if D is a GCD domain t-invertible t-ideals are principal and so $Cl_t(D) = 0$.These observations led the authors of [2] to prove the following result.

Theorem 4.9. *If S is a splitting set of D then $Cl_t(D) \simeq Cl_t(D_S) \times Cl_t(D_{S^\perp})$.*

In case S is an lcm-splitting set
$(s_1t_1, s_2t_2, ..., s_mt_m)_v = ((s_1, s_2, ..., s_m)(t_1, t_2, ..., t_m))_v$ becomes
$(s_1t_1, s_2t_2, ..., s_mt_m)_v = s(t_1, t_2, ..., t_m)_v$ where $s = (s_1, s_2, ..., s_m)_v$. Consequently we have the following result.

Theorem 4.10. [2, Theorem 4.1] *Let D be an integral domain, S an lcm-splitting set, and T the m-complement for S. Then $D = D_S \cap D_T$, where D_T is a GCD domain. Every finite type integral t-ideal A of D has the form $A = s(AD_S \cap D) = s(t_1, t_2, ..., t_m)_v$ where $s \in S$, $t_1, t_2, ..., t_m \in T$. Moreover the map $Cl_t(D) \to Cl_t(D_S)$ given by $[A] \mapsto [AD_S]$ is an isomorphism.*

Splitting sets originated in efforts to produce generalizations of Nagata's theorem mentioned earlier. They first appeared in Gilmer and Parker [31] as Δ-sets, and in [34] as UF sets. Both the Δ and UF sets can now be described as splitting sets generated by prime elements. In [31] we also see [31, Proposition 2.2] stating some conditions that are equivalent to "x, y being LCM-prime (i.e. v-coprime)" and a statement ([31, Proposition 3.1]) which in the language of this and many recent papers can be rephrased as: If S is an lcm-splitting set of D and if D_S is a GCD domain then so is D. Then Mott and Schexnayder [34] gave the splitting sets the proper setting, they showed that a splitting set splits the group of divisibility of the domain into a cardinal product of two subgroups. Apparently a wish to give a more general form of Nagata's theorem had existed prior to [31], [34] and [38]. For example Samuel in [37] had restated Nagata's theorem for Krull domains and Cohn [20] had published his "Nagata's Theorem for Schreier domains". Indeed as pointed out in [31], Cohn had something like the following result in [21]: Suppose D is atomic (i.e. every nonzero nonunit of D is a finite product of irreducible elements) and S is a multiplicative set generated by some primes of D. If D_S is a UFD then so is D. Theorem 177 in [33] can also be cited as an example. In the next section I present, briefly, the current state of the art as far as applications and examples of splitting sets are concerned.

5 Splitting sets: Examples and Applications

In this section I plan to give examples of splitting sets along with their extreme cases and the various forms and generalizations of Nagata's theorem for UFD's.

Examples lend insight which is so very important for understanding. Understanding on the other hand causes further appetite for understanding, which would come from insight, and insight depends on examples and often reasoning. So, I bring in below some examples and some results that I hope will enhance the readers' understanding and at the same time whet the readers' appetite for more.

In a Noetherian (Krull, or atomic) domain D the saturation of every multiplicative set S generated by nonzero principal primes is an lcm-splitting set (look up [2, Corollary 2.7]). For general situations a good example of a splitting set is what Gilmer and Parker [31] (page 69) describe as a multiplicative set generated by a family of primes $\{p_\alpha\}$ in D such that no nonzero element of D is divisible by infinitely many members of $\{p_\alpha\}$ or by infinitely many powers of any member of $\{p_\alpha\}$. More generally in any domain the set $U(D)$ of units of D is a splitting set and so is $D\backslash\{0\}$. (Call these trivial splitting sets.) Here is a good example of how splitting sets can have an effect.

Theorem 5.1. [7, Theorem] *An integral domain D is a weakly factorial domain if and only if every saturated multiplicative subset of D is a splitting set.*

Now here is an "appetite" question: All we have seen is an lcm-splitting set and trivial splitting sets. Is there a clear nontrivial example of a splitting set that is not an lcm-splitting set? For the answer, suppose that every nontrivial splitting set in the whole world (universe?) is an lcm-splitting set and let p be a prime in a Noetherian domain D. Also suppose that D is not a UFD. Now $S = \overline{<p>}$ is a splitting set and so is $S^\perp$ and by our assumption $S^\perp$ is lcm-splitting. But then by Theorem 4.6 D_S is a UFD. Since S is generated by a prime, by Nagata's theorem, we have that D is a UFD, a contradiction. On the basis of these arguments we can say the following.

Remark 5.2. There are sufficiently many splitting sets that are not lcm-splitting sets.

We have seen that there are plenty of splitting sets that are not lcm-splitting; however there are situations in which a splitting set has to be lcm-splitting. Of these one comes from [9] and it would look best stated as a theorem.

Theorem 5.3. [9, Theorem 2.2] *Let S be a multiplicatively closed subset of D. If S is a splitting set of $D[X]$ then S is an lcm-splitting set in D and hence in $D[X]$. Conversely if S is an lcm-splitting set of D with m-complement T then S is an lcm-splitting set of $D[X]$ with m-complement $T' = \{f \in D[X]$: $(A_f D_T)_v = D_T\}$, where A_f denotes the ideal generated by the coefficients of f.*

Of immediate everyday interest is the following corollary:

Corollary 5.4. *D is a GCD domain if and only if $S = D\backslash\{0\}$ is a splitting set in $D[X]$.*

Ordinary splitting sets can come in handy in some other decision-making processes. For the next one we need to prepare a little. If $x \in D$ such that $x = a_1 a_2 ... a_n$ where a_i are atoms, we say that x has an atomic factorization of length n. An integral domain D is a half factorial domain (HFD) if D is atomic and for each nonzero nonunit $x \in D$ the length $n = n(x)$ of atomic factorizations of x is fixed. The following results are to appear in [24].

Proposition 5.5. *Let $K \subseteq B$ be an extension of integral domains such that K is a field, $D = K + XB[X]$ and let $S = \{f \in D$:$f(0) \neq 0\}$. Then D is an HFD if and only if S is a splitting set.*

If, in the above proposition, the field K is replaced by a domain A that is not necessarily a field, we have a more interesting situation.

Proposition 5.6. *Let $A \subseteq B$ be an extension of integral domains, $D = A + XB[X]$ and let $S = \{f \in D{:}f(0) \neq 0\}$. Suppose that S is a splitting set of D and that each element of S has all factorizations of fixed length in D. Then D is an HFD.*

To appreciate something, we need to know about what it looks like and we need to know about what it does. We have already seen some of "what it does", and here we shall concentrate on Nagata-type theorems.

Theorem 5.7. *Let S be an lcm-splitting set in D generated by principal primes. Then D is a UFD (satisfies ACC on principal ideals, is atomic, is integrally closed, is completely integrally closed) if and only if the same holds for D_S.*

The proof(s) of the above Theorem schema can be picked from [3]. In [3] we essentially establish a link between the factorization properties of D_S with those of D when S is a splitting set generated by primes. The results on integral closure and complete integral closure came off the particular thought processes we had at that time.

Theorem 5.8. *Let S be a splitting set in D generated by principal primes. If D_S is a Mori domain, (Krull domain, PVMD, GCD domain, almost GCD domain), then so is D.*

The results stated for PVMD's and GCD domains and AGCD domains will follow from more general results (Theorem 5.9). For the Mori domains I cannot recall any references. So, I will just give a proof. First recall that D is a Mori domain if D has ACC on integral v-ideals. An integral domain D is a Krull domain if and only if D is completely integrally closed and Mori ([27]). (A UFD is well known to be a Krull domain.) It is known that a locally finite intersection of Mori domains is Mori [36]; for a quick reference see [42, Corollary 4]. Now let D_S be Mori, then by Remark 4.5, $D_{S^\perp}$ is a UFD which is Mori and $D = D_S \cap D_{S^\perp}$ an intersection of two Mori domains.

Theorem 5.9. *(Nagata type theorem for general lcm-splitting sets): If S is lcm-splitting, then D_S is a PVMD (GCD domain, AGCD domain) if and only if D is.*

For PVMD's, and GCD domains see [2, Theorem 4.3]. Now suppose that D_S is an AGCD domain. Then $D = D_S \cap D_{S^\perp}$ where D_S is AGCD and $D_{S^\perp}$ is GCD. Let a, b be two nonzero elements of D. Then $a, b \in D_S$ and so there exists n such that $((a^n, b^n)D_S)_v = hD_S$. Since S is lcm-splitting $hD_S = ((a^n, b^n)D_S)_v = (a^n, b^n)_v D_S$. But then by Theorem 4.10 $(a^n, b^n)_v = s((a^n, b^n)_v D_S \cap D)$ which is principal because S is a splitting set and $(a^n, b^n)_v D_S$ is principal. Conversely, it is easy to verify that if D is almost GCD and if S is a multiplicative set then D_S is AGCD.

Theorem 5.10. (cf. [43, Corollary 1.5]) *If D is a GCD domain and S a saturated multiplicative set of D then the $D+XD_S[X]$ construction is a GCD domain if and only if S is a splitting set of D .*

Let me note that [43, Corollary 1.5] has a problem. I do not mention that S must be a saturated set (as I do in Theorem 5.10). The oversight could have been caused by the fact that if S is a multiplicative set and if $\overline{S}$ the saturation of S then $D+XD_S[X] = D+XD_{\overline{S}}[X]$. Theorem 1 of [43] too can be restated as: Let D be a GCD domain and let S be a saturated multiplicative set in D. Then $D+XD_S[X]$ is a GCD domain if and only if for every PF-prime P of D with $P \cap S = \phi$ there exists $d \in P$ such that d is not divisible by any nonunit of S. (A PF prime in a GCD domain is just a prime t-ideal.) In [43, Corollary 1.5] too, replacing multiplicative S, saturated multiplicative S will do the trick. (For [43, Theorem 1], Evan Houston has suggested the following statement: Let D be a GCD domain and let S be a multiplicative set in D. Then $D+XD_S[X]$, is a GCD domain if and only if for every PF-prime P of D with $P \cap S = \phi$ there exists $d \in P$ such that d is not divisible by any nonunit of $\overline{S}$. I am thankful to Evan.) .

I end this section with an odd sort of result. The space constraints prevent me from making any statements about it, but the result is pretty interesting without any introductions.

Theorem 5.11. [26] *Let S be an lcm-splitting set in a coherent domain D. If the integral closure of D_S is a GCD domain then so is the integral closure of D. Consequently, if D is Noetherian and if S is a splitting set generated by principal prime elements of D and if the integral closure of D_S is a UFD then so is the integral closure of D.*

Indeed it would be interesting to see if (a) this theorem can be put to some interesting use and (b) if some of the restrictions can be relaxed.

6 Generalizations of splitting sets

When a certain approach appears to be successful in one area, some researchers are tempted to see if it can be mimicked in another area or in the same area but in a different form. This is how notions get generalized. Generalization does not always have to be bad or trivial. Sometimes you generalize because you want to get a feel of what you are studying. Remaining in the box would only let you follow the beaten tracks, but a stroll outside the box has the potential of opening new doors. I am happy to announce that all the generalizations of v-coprimality and of splitting sets have proved to be useful in enhancing our understanding of divisibility.

The first generalization that I would like to present arose in connection with AGCD domains. The idea germinated in [25] but took real shape in [5], and has been further studied by Chang [17]. A saturated multiplicative set S of D is called an almost splitting set if for each $d \in D \backslash \{0\}$ we can find a natural number $n = n(d)$ such that $d^n = st$ where $s \in S$ and t is v-coprime to every element of S. Like the splitting sets almost splitting sets can be characterized as saturated multiplicative sets S such that for each $d \in D \backslash \{0\}$ $d^n D_S \cap D$ is principal for some natural number n. We can also define almost lcm-splitting sets as almost splitting sets S such that for $s \in S$ and for each (nonzero) $d \in D$ we have $s^n D \cap d^n D$ principal for some $n = n(s, d)$. It may be noted that in an AGCD domain an almost splitting set is automatically an almost lcm-splitting set. Like splitting sets the almost splitting sets are in abundance. For instance every saturated multiplicative subset of a Dedekind domain with torsion class group is an almost splitting set; see [5, Theorem 2.11] for a more general result. Here are a couple of results that can be stated within the setup of this survey.

Theorem 6.1. [5, Theorem 3.12] *Let $S \subseteq D \backslash \{0\}$ be a saturated multiplicative set. Then $D + XD_S[X]$ is an AGCD domain if and only if (i) D and $D_S[X]$ are AGCD domains, and (ii) S is an almost splitting set.*

A quick corollary to this result is that if D is integrally closed AGCD domain and if S is a saturated multiplicative set, then $D + XD_S[X]$ is an AGCD domain if and only if S is an almost splitting set. This gives us a lot of examples of AGCD domains. Just to mention a general one, let D be a Dedekind domain with torsion class group and let S be a multiplicative subset of D. Then $D + XD_S[X]$ is an integrally closed AGCD domain that is also coherent. For the "coherent" part see [23, Theorem 4.32].

Theorem 6.2. [17, Proposition 2.6] *Let D be integrally closed. Then $D \backslash \{0\}$ is an almost splitting set of $D[X]$ if and only if D is an AGCD domain .*

Of course there are points of difference; for instance if S is almost splitting $Cl_t(D)$ is no longer isomorphic to $Cl_t(D_S) \times Cl_t(D_{S^\perp})$ [17, Example 2.9].

The next generalization of splitting sets is an example of an essential generalization. As I mention in [45], I kept looking for a result that would allow me to construct a PVMD $D + XD_S[X]$ domain from D a PVMD. It did not happen until we hit upon the notion of a t-splitting set. A (saturated) multiplicative set S of D is a t-splitting set if for each nonzero nonunit $d \in D$ we have $(d) = (AB)_t$ where A and B are ideals with $A \cap S \neq \phi$ and B is such that $(B, s)_t = D$ for each $s \in S$. The t-splitting sets S are characterized by: If A is a principal ideal of D_S then $AD_S \cap D$ is t-invertible. In [4] we proved that for D a PVMD and S a multiplicative set of D the construction $D + XD_S[X]$ is a PVMD if and only if S is a t-splitting set of D. (GCD domains are a special case of PVMD's.) In [18] the t-splitting sets are further explored, and there we bring forth Nagata-type Theorems that do not seem to have anything to do with the GCD property or the UFD property. Here is a quick example:

Proposition 6.3. [18, Corollary 3.8] *Let X be an indeterminate over D, $G = \{f \in D[X]{:}(A_f)_v = D\}$ and let S be a nonempty multiplicative subset of G. Then $D[X]$ is a Krull (resp. Mori, integrally closed, completely integrally closed, essential, UMT, Prüfer v-multiplication) domain if $D[X]_S$ is.*

I usually tend to think of the set $G = \{f \in D[X]{:}(A_f)_v = D\}$ as the Gilmer set, because I first saw Gilmer use it in [29] and since then I have often made good use of this set.

From t-splitting sets of elements we graduated to t-splitting sets of ideals in [19]. Let D be an integral domain, S a multiplicative set of ideals of D and $D_S = \{x \in K{:}\ xA \subseteq D$ for some $A \in S\}$ the S-transform of D in the sense of Arnold and Brewer[16]. If I is an ideal of D, then $I_S = \{x \in K{:}\ xA \subseteq I$ for some $A \in S\}$ is an ideal of D_S containing I. Denote by $S^{\perp}$ the set of ideals B of D with $(A+B)_t = D$ for all $A \in S$. Call $S^{\perp}$ the t-complement of S. Denote by $sp(S)$ the "saturation" of S (set of all ideals C of D such that $C_t \supseteq A$ for some $A \in S$). Call S a t-splitting set of ideals if every nonzero principal ideal dD can be written as $dD = (AB)_t$ where $A \in sp(S)$ and $B \in S^{\perp}$.

It turns out that S being t-splitting is equivalent to $sp(S)$ being t-splitting and that if S is generated by principal ideals and t-splitting then it is the usual t-splitting set defined above. Moreover if S is t-splitting then (i) so is $S^{\perp}$, and (ii) for each $C \in S$, C_t contains a t-invertible t-ideal of $sp(S)$. (So, a splitting set of ideals S is v-finite in Gabelli's terminology [28]) In fact if S_i is the set of all t-invertible t-ideals in $sp(S)$ then S_i is a t-splitting set with t-complement $S^{\perp}$. It turns out that a lot of results proved for (t-)splitting sets carry through to this more general setting albeit with some new interpretations. Here's a sampling of some of the results proved in [19].

Proposition 6.4. *Let S be a t-splitting set of ideals of D. Then for every nonzero ideal I of D we have $I_t = (AB)_t$ with $A \in sp(S)$ and $B \in S^{\perp}$, and this "splitting" of I is unique up to t-closures.*

Proposition 6.5. *Let S be a multiplicative set of ideals of D. Then S is t-splitting if and only if S is v-finite and every nonzero principal ideal of D_S contracts to a t-invertible t-ideal.*

Proposition 6.6. *A t-splitting set of ideals induces a natural cardinal product decomposition of the ordered monoid of fractional t-ideals of D under the t-product and ordered by the usual reverse inclusion.*

Finally here's something to remind you of the earlier "Nagata-type Theorems".

Proposition 6.7. *Let F be a family of height one t-invertible prime t-ideals of D such that every nonzero nonunit of D belongs to at most a finite number of members of F. Let S be a multiplicative set generated by members of F. Then the following hold: (i) D is a PVMD if and only if so is D_S. (ii) D is a Krull domain if and only if so is D_S. (iii) D is of finite t-character if and only if so is D_S.*

Now, a word about a gap that needs to be filled. In jumping from splitting sets to t-splitting sets we overlooked the possibility of studying, say, d-$*$-splitting sets, d for divisibility. It appears to me that there is a whole world of results parallel to those we know about splitting sets. Let me give an example. Call a saturated multiplicative set S a d-d-splitting set if every element $x \in D\backslash\{0\}$ can be written as $x = st$ where $s \in S$ and t is d-coprime to every member of S. Recall that d-coprime $\equiv$ comaximal. Example: A saturated multiplicative set S generated by height one principal maximal ideals such that no nonzero member of D is divisible by an infinite set of nonassociated primes from S.

Proposition 6.8. *If S is a d-d-splitting set generated by height one principal maximal ideals. Then D is a PID (Noetherian, Prüfer) if and only if D_S is.*

The proofs are straightforward and so are left to the reader, but I hope the point is made.

Here are some papers that I could not include because given the space I could not do justice to them (different jargon) or I came to know about them so late in the day that finding a suitable section was too hard. For the use of splitting sets in the direct sum decomposition of the ideal class group of a Dedekind domain see [14]. For splitting sets in the locally half factorial (D_x is half factorial for each $x \in D\backslash\{0\}$) setup see [13]. The splitting sets also show up in the study of elasticity of factorization in [15]. Recently, in [11] the notion of homogeneous splitting sets has been introduced, in the graded domain environ. Finally, on seeing an earlier version of this paper, Chang has sent me a preprint of a recent paper of his with David Anderson and Jeanam Park [12] that contains a general theory of splitting sets. For a finite character star operation $*$ they call a saturated multiplicative set S a g^*-splitting set if each $d \in D\backslash\{0\}$ can be written as $d = st$ where $s \in S$ and t is $*$-coprime with every element of S. It turns out that every g^*-splitting set is a splitting set but not conversely. They also study the $*$-complement of a subset $\phi \neq S \subseteq D\backslash\{0\}$ and redo quite a few results on splitting sets. Interesting reading.

Acknowledgement. This survey is an expanded version of the two talks I gave recently, one at the 994th AMS, March 2004, meeting at Tallahassee, Fla., and the other at the Workshop on Commutative Rings and their Modules held at Cortona, Italy in June 2004. I am thankful to the organizers of those meetings and to the editors of the Gilmer Volume for letting me write this survey. I would not be writing it if Robert Gilmer had not written [30] and he had not been a constant support in the earlier part of my career. The presentation of this article would have more bad punctuations and typos if Dan Anderson, Evan Houston Gyu Whan Chang and the referee had not helped me. I am grateful to them all.

References

1. D.D. Anderson, Non-Atomic Unique Factorization in Integral domains, *in Arithmetical properties of commutative rings and monoids,* 1–21, Lect. Notes Pure Appl. Math., 241, Chapman & Hall/CRC, Boca Raton, FL, 2005.
2. D.D. Anderson, D.F. Anderson and M. Zafrullah, Splitting the t-class group, J. Pure Appl. Algebra 74(1991), 17- 37.
3. D.D. Anderson, D.F. Anderson and M. Zafrullah, Factorization in integral domains II, J.Alg. 152(1992), 78-93.
4. D.D. Anderson, D.F. Anderson and M. Zafrullah, The ring $R + XR_S[X]$ and t-splitting sets, Arab. J. Sci. Eng. Sect. C 26(2001), 3–16.
5. D.D. Anderson, T. Dumitrescu, M. Zafrullah, Almost splitting sets and AGCD domains, Comm. Algebra 32(2004), 147–158
6. D.D. Anderson and L. Mahaney, On primary factorization, J. Pure Appl. Algebra 54(1988), 141- 154.
7. D.D. Anderson and M. Zafrullah, Weakly factorial domains and groups of divisibility, Proc. Amer. Math. Soc. 109(1990), 907–913.
8. D.D. Anderson and M. Zafrullah, Almost Bézout domains, J. Algebra 142(1991), 285–309.
9. D.D. Anderson and M. Zafrullah, Splitting sets in integral domains, Proc. Amer. Math. Soc. 129(2001), 2209–2217.
10. D.F. Anderson and G. Chang, The m-complement of a multiplicative set, *in Arithmetical properties of commutative rings and monoids, 180–187,* Lect. Notes Pure Appl. Math., 241, Chapman & Hall/CRC, Boca Raton, FL, 2005.
11. D. F. Anderson and G. Chang, Homogeneous splitting sets of a graded integral domain, J. Algebra, 288(2005), 527-544.
12. D. F. Anderson, G. Chang and J. Park, A general theory of splitting sets, Comm. Algebra (to appear).
13. D. F. Anderson, and J. Park, Locally half-factorial domains. Houston J. Math. 23(1997), 617–630.
14. D.F. Anderson, Jeanam Park, Splitting multiplicative sets in Dedekind domains, Bull. Korean Math. Soc. 38(2001), 389–398.
15. D. Anderson, J. Park, Gi-Ik Kim, Heung-Joon Oh, Splitting multiplicative sets and elasticity, Comm. Algebra 26(1998), 1257–1276.
16. J. Arnold and J. Brewer, On flat overrings, ideal transforms and generalized transforms of a commutative ring, J. Algebra, 18(1971), 254-263
17. G. Chang, Almost splitting sets in integral domains, J. Pure Appl. Algebra 197(2005), 279-292.
18. G. Chang, T. Dumitrescu and M.Zafrullah, t-splitting sets in integral domains, J. Pure Appl. Algebra, 187(2004),71-86
19. G. Chang, T. Dumitrescu and M.Zafrullah, t-splitting multiplicative sets of ideals in integral domains, J. Pure Appl. Algebra 197(2005), 239–248.
20. P. Cohn, Bezout rings and their subrings, Proc. Cambridge Philos. Soc.64(1968), 251-264.
21. P. Cohn, *Free rings and their relations*, Academic Press, New York, 1971.
22. P. Conrad, Some structure theorems for lattice ordered groups, Trans. Amer. Math. Soc. 99(1961), 212-240.
23. D. Costa, J. Mott and M.Zafrullah, The construction $D + XD_S[X]$, J. Algebra 53(1978), 423–439.

24. J. Coykendall, T. Dumitrescu and M. Zafrullah, The half factorial property and domains of the form $A + XB[X]$, Houston J. Math. (to appear)
25. T. Dumitrescu, Y. Lequain, Mott, J. and M. Zafrullah, Almost GCD domains of finite t-character, J. Alg 245(2001), 161-181.
26. T. Dumitrescu and M. Zafrullah, LCM-splitting sets in some ring extensions, Proc. Amer. Math. Soc. 130 (2002), 1639–1644.
27. R. Fossum, *The divisor class group of a Krull domain*, Ergebnisse der Mathematik und ihrer grenzgebiete B. 74, Springer-Verlag, Berlin, Heidelberg, New York, 1973.
28. S. Gabelli, On Nagata's Theorem for the class group II*, in Commutative algebra and algebraic geometry (Ferrara), 117-142,* Lecture Notes in Pure and Appl. Math., 206, Dekker, New York, 1999.
29. R. Gilmer, An embedding theorem for HCF-rings, Proc. Cambridge Philos. Soc. 68 1970 583–587.
30. R. Gilmer, *Multiplicative Ideal Theory*, Marcel Dekker, New York, 1972
31. R. Gilmer and T. Parker, Divisibility properties in semigroup rings, Michigan Math. J. 21 (1974), 65–86.
32. M. Griffin, Rings of Krull type, J. Reine Angew. Math. 229(1968) 1-27.
33. I. Kaplansky, *Commutative rings*, Allyn and Bacon, Inc., Boston, Mass. 1970.
34. J.Mott and M. Schexnayder, Exact sequences of semi-value groups, J. Reine Angew. Math. 283/284(1976), 388–401.
35. M. Nagata, A remark on the unique factorization theorem, J. Math. Soc. Japan 9(1957), 143-145.
36. N. Raillard, Sur le anneaux de Mori, Thesis, Paris, VI (1976)
37. P. Samuel, *Lectures on unique factorization domains*, Notes by M. Pavman Murthy, Tata Institute of Fundamental Research Lectures on Mathematics, No. 30 Tata Institute of Fundamental Research, Bombay 1964.
38. M. Schexnayder, Groups of divisibility, Dissertation, Florida State University, 1973.
39. M. Zafrullah, Semirigid GCD domains, Manuscripta Math. 17(1975), no. 1, 55–66.
40. M. Zafrullah, Rigid elements in GCD domains, J. Natur. Sci. Math. 17(1977), 7–14.
41. M. Zafrullah, A general theory of almost factoriality. Manuscripta Math. 51(1985), 29–62.
42. M. Zafrullah,Two characterizations of Mori domains, Math. Japon. 33(1988), 645–652.
43. M. Zafrullah, The $D + XD_S[X]$ construction from GCD-domains, J. Pure Appl. Algebra 50 (1988), 93–107.
44. Zafrullah, M. Putting t-invertibility to use, *Non-Noetherian Commutative Ring Theory*, 429-457 (Editors Chapman, S. and Glaz, S.) Kluwer Academic Publishers, 2000.
45. M. Zafrullah, Various facets of rings between $D[X]$ and $K[X]$, Comm. Algebra 31(2003), 2497–2540.

Some questions for further research

Robert Gilmer

Department of Mathematics, Florida State University, Tallahassee, FL 32306-4510
gilmer@math.fsu.edu

Mathematics can be compared to a living organism with many different systems, each affecting and interacting with the others. For this reason it is impossible to anticipate what new topics commutative algebraists may address in the future, but it is possible to identify certain gaps in areas of past research. That is what this paper attempts to do. It identifies eight areas, with two or three questions in each. The selection of questions has been strongly influenced by preparations made for a survey talk, "Forty years of commutative ring theory", that I gave at the conference Venezia 2002 (see [0G]). While the purpose of that talk was not to discuss open questions per se, it was a personal goal of mine to mention a few such questions that I considered to be significant. The talk included six open questions. One, concerning the content formula for polynomials, has since been resolved, and four of the other five are repeated here. As one would expect, most of the questions in this paper have appeared elsewhere. To be sure, in compiling the current list I consulted the following lists of questions on commutative ring theory that appear in the literature.

1. Zero-Dimensional Commutative Rings. In: Anderson, D.F. and Dobbs, D.E. (ed) Lect. Notes Pure Appl. Math. **171**. Dekker, New York 363-372 (1995)

2. Notices Amer. Math. Soc., Amer. Math. Soc., Providence, RI 175–176 (1973)

3. Non-Noetherian Commutative Ring Theory. Chapman, S.T. and Glaz, S. (ed) Kluwer, Dordrecht/Boston/New York 459–476 (2000)

Some of the questions in this article appear on one of these three lists. I need to repeat here a caveat from the talk I gave in Venice: I claim no superiority for the questions chosen. It seemed advisable to write about subjects on which I am reasonably knowledgeable. There is much in the area of commutative ring theory that is worthwhile, but where my own acquaintance is quite limited.

In order to limit the number of references in this paper, where possible I have made citations of the form "see [X] and its references", rather that cite both X and its relevant references. One risk for an author in writing an article of this type is that of omission of significant contributions to questions. I've tried to be thorough in researching such contributions, but oversights are still quite possible. I apologize in advance to any author whose relevant work has been slighted.

1 Prime spectra of rings of integer-valued polynomials

Suppose D is an integral domain with quotient field K, n is a positive integer, and H is a subset of K^n. The ring of integer-valued polynomials on H, denoted $\text{Int}(H, D)$ is the subring of $K[X_1, \ldots, X_n]$ consisting of those polynomials f such that $f(h) \in D$ for each $h \in H$. $\text{Int}(D^n, D)$ is denoted $\text{Int}(D^n)$ and $\text{Int}(D^1)$ written as $\text{Int}(D)$. The rings of integer-valued polynomials originally studied by Polya and Ostrowski (1919) were of the form $\text{Int}(\mathcal{O}_K)$, where $\mathcal{O}_K$ is the ring of algebraic integers of the finite algebraic number field K. The more recent extensive body of work on integer-valued polynomials stems from the work of P.-J. Cahen and J.-L. Chabert on the subject during the period 1971–73. Prime spectra of rings of integer-valued polynomials have been the object of much study during the past 35 years, and yet basic questions concerning $\text{Spec}(\text{Int}(D))$, for example, remain open. Here are two; though closely related, we list them separately for the sake of discussion.

IV1) If D is a finite-dimensional domain, is $\text{Int}(D)$ also finite-dimensional?

IV2) With D as in IV1, is $\dim(\text{Int}(D)) \leq \dim(D[X])$?

The paper [1C] of Cahen is the first in the literature that was devoted explicitly to the study of the Krull dimension of $\text{Int}(D)$; for later developments see the papers [1FIKT] and [1FK] and their references. Cahen developed a sound basis for the further study of $\dim(\text{Int}(D))$, proving several important results. In particular he showed that $\dim(\text{Int}(D)) \geq \dim(D[X]) - 1$ and that $\dim(\text{Int}(D)) = \dim(D[X]) = \dim(D) + 1$ if D is a Jaffard domain (hence if D is Noetherian or a Prüfer domain). He notes that no example is known where $\dim(\text{Int}(D)) > \dim(D[X])$, and in [1FK] Fontana and Kabbaj conjecture that this inequality fails for all domains — that is, they conjecture that IV2 has an affirmative answer. If this is correct it obviously settles IV1 as well. Cahen showed that the crucial issue in determining $\dim(\text{Int}(D))$ is the dimension of the fiber in $\text{Int}(D)$ of a maximal ideal M of D with finite associated residue field D/M. In the case of polynomial rings the fiber in $D[X]$ of each prime ideal of D has dimension 1. In the case of $\mathrm{Int}(D)$ this remains true for primes P of D such that D/P is infinite, for $\mathrm{Int}(D_P) = D_P[X]$ in that case. However, both [1C] and [1FIKT] contain examples showing that for a maximal ideal M of finite index in D, the dimension of the fiber in $\mathrm{Int}(D)$ of M can be any positive integer.

A long-term problem in this area is that of determining all sequences of non-negative integers that can be realized in the form $\{\dim(\mathrm{Int}(D^i))\}_{i=0}^{\infty}$, where $\dim(\mathrm{Int}(D^0))$ means $\dim(D)$. A solution to this problem requires, of course, resolution of IV1 and IV2. Related to this problem, it is known that for an infinite domain D, $\mathrm{Int}(D^{n+1}) = \mathrm{Int}(\mathrm{Int}(D^n)) = \mathrm{Int}(D, D^n)$ [1CC, Prop. XI.1.1].

2 Prime spectra of power series

There is formal parallelism between this section and the preceding one, but one would not expect more than a tangential connection between results of these two areas. Also, the dimension theory of power series rings is at a more advanced stage of development than is the corresponding theory for rings of integer-valued polynomials. This is reflected in the questions posed for this section.

PS1) If R is a ring such that $R[[X]]$ is finite-dimensional, is $R[[X,Y]]$ also finite-dimensional?

PS2) If $R[[X]]$ is finite-dimensional, is its dimension at most $2(\dim R)+1$?

Question PS1 is implicit from work of Condo and Coykendall in [2CC] and of Coykendall in [2C]. As noted in [2C], all known examples of rings R with R and $R[[X]]$ finite-dimensioinal are such that $R[[X_1, \ldots, X_n]]$ is finite-dimensional for all n. If PS1 has an affirmative answer this is true in general.

Much of the dimension theory of power series rings was developed by Arnold in a series of papers he authored or coauthored during the period 1973–82 (see references to Section 3 of [0G]). Results of Arnold and others bring PS2 into sharper focus. First, Arnold showed that finite-dimensionality of $R[[X]]$ implies that R is an SFT-ring, and in an SFT-ring $P[[X]] = \sqrt{P \cdot R[[X]]}$ for each $P \in \mathrm{Spec}(R)$. This means that primes of $R[[X]]$ lying over P in R correspond to proper prime ideals of the domain $(R/P)[[X]]_{(R/P)^*}$, and in particular this domain and the fiber in $R[[X]]$ of P have the same dimension. If D is a Prüfer domain, Arnold showed that $D[[X]]_{D^*}$ is a Dedekind domain, so the dimension of the fiber in $D[[X]]$ of each $P \in \mathrm{Spec}(D)$ is 2, and this was a key factor in his proof that $\dim D[[X]] = \dim(D)+1$ in this case.

At any rate, the results just described mean that in considering PS2, domains of the form $D[[X]]_{D^*}$, where D is SFT and $D[[X]]$ is finite-dimensional are of great interest. Domains of this type (not necessarily with $D[[X]]$ finite-dimensional) have received attention in several papers in the literature, especially for the case where D is a valuation domain, See, for example, [2KP] where, among numerous interesting results, it is shown that if (V, M) is a rank-one nondiscrete valuation domain, then $M[[X]]$ contains an infinite strictly decreasing sequence of prime ideals lying over (0) in V. Coykendall [2C] used this result of Kang and Park in his construction of a one-dimensional SFT-domain V_1 such that $V_1[[X]]$ is neither SFT nor finite-dimensional.

Again a long-term problem in the dimension theory of power series rings is that of determining all sequences of nonnegative integers realizable in the form $\{\dim R, \dim R[[X_1]], \dim R[[X_1, X_2]], \ldots\}$. A solution seems inaccessible under the current state of knowledge.

A general reference for this section is Chapitre 11 of [2B].

3 Unique factorization in power series rings

While unique element factorization in integral domains has played an important role in commutative algebra since its inception, it is a topic that has been marked by periods of great activity followed by long periods of relative dormancy. Since the appearance of Zak's paper [3Z] in 1980 there has been a lot of work on weaker forms of unique factorization, notably on the half-factorial property, but little activity on unique factorization itself during the same period. This was true in spite of the fact that several basic questions concerning unique factorization remain open. Before listing two such questions, we review a bit of history of the subject. The last spurt of activity on factorial domains (or UFDs) was motivated in large part by work of P. Samuel on the subject in the early 1960's. In particular Samuel published in 1960 a well-known paper [3S] in which he addressed a question of Krull: If D is factorial, is $D[[X]]$ also factorial? While [3S] contained several results providing a positive answer to this question when additional conditions are imposed on D, Samuel established a negative answer to Krull's question by giving an example of a two-dimensional local factorial domain A such that $A[[X]]$ is not factorial. The domain A is not complete, and Samuel implicitly raises the question of whether the power series ring over a two-dimensional complete factorial domain is again factorial. P. Salmon provided a negative answer in 1966. (See [3L] for a more detailed account of the developments just sketched.) In the early 1970's work on unique factorization entered another lull. Now for the two questions.

UF1) If $D[[X]]$ is factorial, is $D[[X, Y]]$ also factorial?

UF2) If K is a field and $\{X_i\}_{i=1}^{\infty}$ is a set of indeterminates over K, is $K[\{X_i\}_1^{\infty}][[Y]]$ factorial?

I've found little in the literature related to UF1[2]. Bayart's paper [3B] treats it briefly; if D is a regular UFD, results of [3S] show that $D[[X_1, \ldots, X_n]]$ is a regular UFD for each positive integer n, so UF1 has an affirmative answer in this case. A domain D is factorial if and only if it is a GCD-domain satisfying the ascending chain conditions for principal ideals (a.c.c.p.). The a.c.c.p. is inherited by $D[[X]]$ from D, so UF1 has an affirmative answer if $D[[X, Y]]$ is a GCD-domain. See [3AKP] for results concerning transfer of the GCD-property from D to $D[[X]]$.

UF2 is closely related to several developments connected to appearance of Samuel's 1960 paper. For one thing $K[X_1, \ldots, X_n]$ is regular for each n so $K[X_1, \ldots, X_n][[Y]]$ is factorial by results of [3S]. In 1959 Y. Mori showed that if the completion $\widehat{D}$ of a Zariski ring (D, I) is factorial, then D is also factorial. Although $E = K[\{X_i\}_1^\infty][[Y]]$ is not Noetherian, one is nonetheless tempted to emulate Mori's approach by considering the completion of E in its $(\{X_i\}_1^\infty, Y)$-adic topology. This yields the infinite power series ring denoted $K[[\{X_i\}_1^\infty, Y]]_2$ consisting of all series $\sum_{i=1}^\infty f_i$, where $f_i \in K[\{X_i\}_1^\infty, Y]$ is 0 or a form of degree i for each i. While a 1963 paper of Cashwell and Everett (see [3DD] and its references) implies that the full power series ring $K[[\{T_a\}]]_3$ in an arbitrary set $\{T_a\}$ of indeterminates over K is factorial, the question of factoriality of $K[[\{T_i\}_1^\infty]]_2$ remains open.

4 Splitting polynomials

Consider the following questions.

SP1) Suppose R is a subring of the commutative unitary ring S. If each monic polynomial over R has a root in S, is each monic over R a product of monic linear polynomials in $S[X]$?

SP2) Suppose F is an infinite field and $F - \{f(a) | a \in F\}$ is finite for each nonconstant polynomial f over F. Is F algebraically closed?

[2] The general paradigm for UF1 is the following. Let E be a ring-theoretic property and let $A_n(R)$ be an extension ring of the ring R defined in terms of n indeterminates. Call R *n-stably E* if property E is inherited by $A_n(R)$ from R and call R *stably E* if E is inherited by $A_k(R)$ from R for all k. One asks in general if the property of being $(k+1)$-stably E or stably E follows from k-stability of E. For properties E that are not, in general, inherited by $A_1(R)$ from R, these questions tend to be difficult. UF1 is this type of question, where E is the factorial property and $A_n(R)$ is the power series ring over R in n indeterminates.

SP1 arose from one of the standard proofs that each field admits an algebraic closure. To wit, let F be a field and let $\{f_i\}_{i\in I}$ be the set of nonconstant monic polynomials over F. Let $\{X_i\}_{i\in I}$ be a set of indeterminates over F, let $D = F[\{X_i\}]$, and let $I = (f_i(X_i))$. The ideal I is proper, so I is contained in a maximal ideal M of D. Thus $L = D/M$ is (up to isomorphism) an algebraic extension field of F and each nonconstant polynomial over F has a root in L. Using normality and the primitive element theorem, one can show that L is, in fact, an algebraic closure of F. This is the point (normality/primitive element theorem) at which the proof for fields breaks down for a more general base ring.

Primary interest in SP1 is in the case where R is an integral domain. There the question has the following equivalent form. Let K be the quotient field of R and let K^* be an algebraic closure of K. For each nonconstant monic f_i over R, let θ_i be a root of f_i in K^*. SP1 is equivalent to asking whether $R[\{\theta_i\}]$ is the integral closure of R in K^*. Even for $R = \mathbb{Z}$ the answer to this form of SP1 was not known until P. Roquette (unpublished) proved in 2002 that the question has an affirmative answer. Roquette's proof extends to the case of a Dedekind domain R, all of whose residue fields are perfect.

A condition reminiscent of the primitive element theorem is of interest in the domain case of SP1. The condition can be described as follows. Let R, K, and K^* be as in the previous paragraph. Assume that R is integrally closed and let R^* be integral closure of R in K^*. Consider the condition (*) that each finitely generated extension of R in R^* is contained in a simple extension of R in R^* — that is, for all $u, v \in R^*$ there exists $w \in R^*$ such that $R[u, v] \subseteq R[w]$. If R satisfies (*) and S is an integral domain, it can be shown that SP1 has an affirmative answer. Thus the class of integral domains satisfying (*) is of interest. In particular, does $\mathbb{Z}$ satisfy (*)? If so, then each ring of algebraic integers in a finite algebraic number field satisfies (*). It is well known that for algebraic integers u, v, the domain $\mathbb{Z}[u, v]$ need not be a simple ring extension of $\mathbb{Z}$.

SP2 may seem dissimilar to SP1, but having worked on each question I have found myself confronted by closely related issues in the two. SP2 actually arose in a paper [4P] in model theory. There Podewski defines a field F to be almost algebraically closed (a.a.c.) if F satisfies the hypothesis of SP2. According to Wagner [4W], Reineke has conjectured (unpublished) than an a.a.c. field is algebraically closed. The notion of an a.a.c. field is related to that of a minimal field (a concept in logic). In fact, minimal $\Rightarrow$ a.a.c. $\Rightarrow$ root-closed (so an a.a.c. field is perfect). Wagner showed that a minimal field of nonzero characteristic is algebraically closed, but SP2 is apparently open for all characteristics.

5 Ideal class groups

One of the signal achievements in multiplicative ideal theory during the last forty years was L. Claborn's proof that each abelian group can be realized as the ideal class group of a Dedekind domain (see Section 1 of [0G] and its references). Each of the questions below has its historical roots in this theorem of Claborn. In [5C] Claborn expanded his original existence theorem to allow one to impose some restrictions on a realizing Dedekind domain with a specified ideal class group; this is in the spirit of CG1.

CG1) Can every abelian group be realized as the ideal class group of a Dedekind domain that is also a half-factorial domain (HFD)?

CG2) Can every finite abelian group be realized as the ideal class group of the ring of algebraic integers of some finite algebraic number field?

Zaks posed CG1 in his 1980 paper [3Z, p. 301] in which he introduced the concept of an HFD and developed a basic theory of these domains. For references to work on CG1 see [5GG], where it is shown that each Warfield group is the class group of a Dedekind HFD.

CG2 is Problem 12 in the list of 35 unsolved problems at the end of W. Narkiewicz's book [5N]; Narkiewicz chronicles progress on CG2 in the historical notes on Section III.2 of [5N]. One naturally expects a solution to CG2 to entail tools from algebraic number theory.

6 Intersections of valuation domains

For a ring-theoretic property E, the general problem area to be considered here is to determine conditions (sufficient or equivalent) on a family $\{V_\lambda\}$ of valuation domains on a field K in order that $\cap_\lambda V_\lambda = D$ should have property E. In this generality much is known in regard to this problem. Krull domains and domains of finite character, for example, are defined as intersections of such families, suitably restricted. And characterizations of Dedekind and factorial domains in terms of such families are known.

The case in which E is the condition of being a Prüfer domain is probably the most prominent current case of this problem. In this vein there is Nagata's proof that a finite intersection of valuation domains is a Prüfer domain. There is also work of A. Dress, Roquette, Becker, Schülting and Kucharz on real holomorphy rings (see [6S], [0G, Sect. 5] and [6O] and their references). Papers [6G] and [6L1] give sufficient conditions for an intersection $\cap V_\lambda$ to be a Prüfer domain (the constructions in these papers are related to that of A. Dress). Loper has written a nice survey article [6L2] on this work, including some extensions by himself and others; see also [6E, Sect. 2.2].

Many questions have been raised in the literature where results on the problem just discussed would reasonably be expected to be useful. The following questions are examples.

VI1) Can each integrally closed domain D be expressed in the form $D_1 \cap D_2$, where each D_i is a Bezout domain and an overring of D?

VI2) If D is a Dedekind domain, is each overring of D a finite intersection of quotient rings of D?

VI3) Is each Krull domain an intersection of a finite number of factorial or PID-overrings?

VI1 was raised in [6AA, p. 97]; Cahen (unpublished) weakened VI1 to ask whether each integrally closed domain is an intersection of two overrings, each of which is a Prüfer domain.

VI2 is considered in [6GG], where it is shown that the answer is affirmative if the maximal spectrum of D is countable.

VI3 is Question 2 of [6AA], where an affirmative answer is established for a Krull domain with countable divisor class group (in which case two overrings suffice).

By replacing "finite" by "two" in VI2 and VI3, one obtains more restrictive questions, and the reverse replacement in VI1 results in a less restrictive question. In the general case each of the (six) resulting questions is open.

7 Prüfer domains

Prüfer domains play a central role in multiplicative ideal theory. One reason for their importance is the wide range of contexts in which they arise. Here are three open questions concerning Prüfer domains.

PD1) What are the Prüfer overrings of a two-dimensional regular local ring? of $\mathbb{Z}[X]$? of $\mathbb{Z}[[X]]$?

PD2) Does each Prüfer overring arising in PD1 have the two-generator property?

PD3) If subset E of the domain D is such that $\mathrm{Int}(E, D)$ is a Prüfer domain, does $\mathrm{Int}(E, D)$ have the two-generator property?

PD1 may stem from an old question (unpublished) of M.P. Murthy, who asked for a determination of all Prüfer domains with quotient field $\mathbb{Q}(X)$. This is essentially the problem of determining all Prüfer overrings of $\mathbb{Z}[X]$. In a 1936 paper [7M], S. MacLane determined, using a concept he called key polynomials, all extensions of a rank-one discrete valuation on a field K to the simple transcendental extension field $K(X)$. This means that in a sense all valuations on $\mathbb{Q}(X)$ are known and the problem of determining all Prüfer overrings of $\mathbb{Z}[X]$ can be viewed as a case of the general problem described in Section 6.[3]

The two-generator problem was the engine that drove much work on Prüfer domains for almost twenty years — until the appearance of Swan's paper [7S], in which he proved existence of Prüfer domains with finitely generated ideals requiring arbitrarily large numbers of generators (see also [0G, Sect. 5]). Because of its historical context, the two-generator property remains an important property to consider for any new class of Prüfer domains arising in the literature.

A. Loper in [7L1] completed the solution of the problem of determining equivalent conditions on D in order that $\text{Int}(D)$ should be Prüfer; there are partial results in the literature dealing with the corresponding problem for $\text{Int}(E, D)$ [7CCL], [7CLS]. For D Noetherian it is known that if $\text{Int}(D)$ is a Prüfer domain, then it has the two-generator property [7Mc], [7C]. (For a Noetherian domain D, $\text{Int}(D)$ is a Prüfer domain if and only if D is a Dedekind domain with finite residue fields.)

8 One-dimensional quasilocal domains

For many questions concerning integral domains D, the case where D is one-dimensional attracts special attention. The hypothesis of one-dimensionality may allow one to answer a question for that case, though not in general. Viewed from a different perspective, one-dimensionality may limit the existence of counterexamples for that case, though counterexamples exist in higher dimensions. Here are two specific open questions concerning one-dimensional quasilocal domains.

D1) If D and E are one-dimensional quasilocal domains with common quotient field K, does $D \cap E$ have quotient field K?

D2) Suppose (D, M) is a one-dimensional quasilocal domain. If D is not Noetherian, must M be expressible as the union of a strictly ascending sequence of ideals of D?

[3] A. Loper and F. Tartarone have a paper in preparation in which they characterize Prüfer overrings of $\mathbb{Z}[X]$ that are contained in $\mathbb{Q}[X]$; their work makes strong use of key polynomials.

D1 is raised in Question 1.6 of [8H]; Section 1 of that paper contains results that are related to D1, as does Section 2 of [8GH1]. In particular [8GH1, p. 246] contains the observation that if D1 has a negative answer, then there is an example where $D \cap E = F$ is a field and $K = F(X, Y)$, where $\{X, Y\}$ is algebraically independent over F.

D2 is a special case of a more general question raised in Section 2 of [8AGH]. A sufficient condition for an affirmative answer to D2 is that there exists an ideal $I < M$ such that M/I is countably generated, but not finitely generated, as a D-module. This condition need not hold in general, but E. Enochs has shown (see [8GH2, Th. 3.1]) that a non-finitely generated module over a Noetherian ring has a homomorphic image that is countable generated but not finitely generated.

References

[0G] Gilmer, R., Forty years of commutative ring theory. In: Facchini, A., Houston, E., and Salce, L. (ed) Rings, Modules, Algebras, and Abelian Groups. Lect. Notes Pure Appl. Math., **236**. Dekker, New York 229–256 (2004)

[1C] Cahen, P.-J., Dimension de l'anneau des polynomes a valeurs entieres. manuscr. math., **67**, 333–343 (1990)

[1CC] Cahen, P.-J. and Chabert, J.-L.: Integer-Valued Polynomials. Amer. Math. Soc. Surveys Monogr. No. 48. Providence, RI (1997)

[1FIKT] Fontana, M., Izelgue, L., Kabbaj, S., and Tartarone, F., Polynomial closure in essential domains and pullbacks. Lect. Notes Pure Appl. Math. **205**. Dekker, New York 307–321 (1999)

[1FK] Fontana, M. and Kabbaj, S., Essential domains and two conjectures in dimension theory. Proc. Amer. Math. Soc., **132**, 2529–2535 (2004)

[2B] Benhissi, A., Les Anneaux de Séries Formelles, Queen's Papers Pure Appl. Math., **124**, Kingston, Ont., Canada (2003)

[2CC] Condo, J.T. and Coykendall, J., Strong convergence properties in SFT-rings. Commun. Alg., **27**, 2073–2085 (1999)

[2C] Coykendall, J., The SFT-property does not imply finite dimension of power series rings. J. Algebra, **256**, 85–96 (2002)

[2KP] Kang, B.G. and Park, M.H., A localization of a power series ring over a valuation domain. J. Pure Appl. Alg., **140**, 107–142 (1999)

[3AKP] Anderson, D.D., Kang, B.G., and Park, M.H., GCD-domains and power series rings. Commun. Alg., **30**, 5955–5960 (2002)

[3B] Bayart, M., Series formelles sur un anneau factorial. C.R. Acad. Sci. Paris Ser. A, **277**, 449–450 (1973)

[3DD] Deckart, D. and Durst, L.K., Unique factorization in power series rings and semigroups. Pac. J. Math., **16**, 239–242 (1966)

[3L] Lipman, J., Unique factorization in complete local rings. Proc. Symp. Pure Math., **29**, 531–546 (1975)

[3S] Samuel, P., On unique factorization domains. Ill. J. Math., **5**, 1–17 (1961)

[3Z] Zaks, A., Half-factorial domains. Israel J. Math., **37**, 281–302 (1980)

[4P] Podewski, K.-P., Minimale Ringe. Math.-Phys. Semesterber., **22**, 193–197 (1973)

[4W] Wagner, F.O., Minimal fields. J. Symbol. Logic, **65**, 1833–1835 (2000)

[5C] Claborn, L., Specified relations in the ideal class group. Mich. Math. J., **15**, 249–255 (1968)

[5GG] Geroldinger, A. and Göbel, R., Half-factorial subsets in infinite abelian groups. Houston J. Math., **29**, 841–858 (2003)

[5N] Narkiewicz, W., Elementary and Analytic Theory of Algebraic Numbers. Monografie Matematyczne No. 57, Polish Scientific Publishers (PWN), Warsaw (1974)

[6AA] Anderson, D.D. and Anderson, D.F., Finite intersections of PID or factorial overrings. Canad. Math. Bull., **28**, 91–97 (1985)

[6E] Ershov, Y.I.: Multi-Valued Fields. Kluwer, New York (2001)

[6G] Gilmer, R., Two constructions of Prüfer domains. J. Reine Angew. Math., **239**, 153–162 (1970)

[6GG] Gilmer, R. and Grams, A., Finite intersections of quotient rings of a Dedekind domain. J. London Math. Soc., **12**, 257–261 (1976)

[6L1] Loper, K.A., On Prüfer non-D-rings. J. Pure Appl. Algebra, **96**, 271–278 (1994)

[6L2] Loper, K.A., Constructing examples of integral domains by intersecting valuation domains. In Non-Noetherian Commutative Ring Theory. Kluwer, Dordrecht 325–340 (2000)

[6O] Olberding, B., Holomorphy rings in function fields, this volume.

[6S] Schülting, H.-W., Über die Erzeugendenanzahl invertierbarer Ideale in Prüferringen, Comm. Alg., **7**, 1331–1349 (1979)

[7CCL] Cahen, P.-J., Chabert, J.-L., and Loper, K.A., High dimension Prüfer domains of integer-valued polynomials, J. Korean Math. Soc., **38**, 915–935 (2001)

[7C] Chabert, J.-L., Un anneau de Prüfer, J. Algebra, **107**, 1–16 (1987)

[7CLS] Chapman, S.T., Loper, K.A., and Smith, W.W., The strong two-generator property in rings of integer-valued polynomials determined by finite sets, Arch. Math. **78**, 372–375 (2002)

[7L1] Loper, A., A classification of all D such that $\operatorname{Int}(D)$ is a Prüfer domain, Proc. Amer. Math. Soc., **126**, 657–660 (1998)

[7M] MacLane, S., A construction for absolute values, Trans. Amer. Math. Soc., **40**, 363–395 (1936)

[7Mc] McQuillan, D., On Prüfer domains of polynomials, J. Reine Angew. Math., **358**, 162–178 (1985)

[7S] Swan, R., n-Generator ideals in Prüfer domains, Pac. J. Math., **111**, 433–446 (1984)

[8AGH] Arnold, J.T., Gilmer, R., and Heinzer, W., Some countability conditions in a commutative ring, Ill. J. Math., **21**, 648–665 (1977)

[8GH1] Gilmer, R. and Heinzer, W., The quotient field of an intersection of integral domains, J. Algebra, **70**, 238–249 (1981)

[8GH2] Gilmer, R. and Heinzer, W., On the cardinality of subrings of a commutative ring, Canad. Math. Bull., **29**, 102–108 (1986)

[8H] Heinzer, W., Noetherian intersections of integral domains II, Lecture Notes Math., **311**, 107–119. Springer-Verlag, New York/Berlin (1973)

Robert Gilmer's published works

In the editors' correspondence with Robert Gilmer concerning this volume Robert expressed the opinion that the greatest contribution he has made to the topic of multiplicative ideal theory consists of the four books (three of which are titled Multiplicative Ideal Theory*) that he has written in the area. At our request, he has written out the following statement concerning how these four books came to be.*

The original version of *Multiplicative Ideal Theory*, published in 1968 in the series Queens Papers in Pure and Applied Mathematics, was written while I was at the University of Tübingen in 1967. I wrote it at the invitation of Paulo Ribenboim, whom I had not met at the time of the invitation. While in Germany I met Krull, my mathematical idol at the time; he gave me a number of his reprints, including those of most of his papers in the Beiträge series. Krull died in 1971, and I've always been happy to have met him. The second and third versions of *Multiplicative Ideal Theory* are dedicated to Krull. Earl Taft was editor of Marcel-Dekker's monograph series (Pure and Applied Mathematics) about this time and in late 1968 he contacted me at the January 1969 AMS meeting in Biloxi to discuss the possibility of my writing a book for that series. I suggested that I could write a revised version of *Multiplicative Ideal Theory* that included exercises. My suggestion was accepted and the red Marcel-Dekker book was the result; I usually refer to this as RMIT. It appeared in 1972; it took longer to write because, whereas I had been on leave when writing MIT, I was teaching fulltime while writing RMIT. I remember having read how Nathan Jacobson said he wrote his books: after-dinner until bedtime. I probably did some of the work on RMIT during that time, but I think most of its work was relegated to summers. I learned from David Anderson at the AMS meeting in Louisville, KY in January, 1984 that RMIT was out of print. Then, while attending a conference on integer-valued polynomials at CIRM in Luminy in December of 1990, Paulo Ribenboim said he thought Queens would be interested in publishing a version of RMIT that included corrections of errors I was aware of in the 1972 version. Permissions/authorizations had to be obtained, but they were and this corrected version was published in 1992. I refer to it as RRMIT. It still contains errors, one of which was introduced by Queens, where Multiplicative is misspelled on the title page (first t is omitted).

Part of how *Commutative Semigroup Rings* came about is recounted in a footnote to my paper in the Venezia 2002 Conference Proceedings, but I'll tell the full story here. Tom Parker was a student at Florida State University 1968-73. He was in a cluster of 6 students who started dissertations under me; the other five were Anne Grams, Eduardo Bastida, Joe Hoffmann, Roger Taylor, and Bruce Glastad (who finished with Joe Mott). I gave these students a choice of either having me suggest a thesis topic for them or of selecting their own topic. Initially Parker said he wanted me to suggest a topic in field theory; he worked on one for 6-8 weeks, then said he wanted to switch and work on commutative semigroup rings. There was little on the subject, per se, in the literature at the time (approximately January of 1972), and I believe that Parker's exposure to the subject consisted of one exercise in D. G. Northcott's book *Rings, Modules, and Multiplicities.* I told him it would be okay for him to work in commutative semigroup rings, but that was much too broad for a dissertation. He wound up working on three aspects, with his main work being on the nil and Jacobson radicals of a commutative semigroup ring. Parker's thesis stimulated my own interest in the subject, and *Commutative Semigroup Rings* is dedicated to him. I wound up writing several papers on the topic (most of which were joint), and in 1981 I agreed to write a monograph on CSR for the University of Chicago series Lecture Notes in Mathematics. I did this mainly during a Fall 1982 sabbatical spent at the University of Connecticut; in fact, 22 of the 25 chapters were written at UConn, then it took me six months to write the remaining 3 chapters while teaching at Florida State.

References

1. Rings in which semi-primary ideals are primary, Pacific J. Math. 12(1962), 1273-1276.
2. Commutative rings containing at most two prime ideals, Michigan J. Math. 10(1963), 263-268.
3. Finite rings having a cyclic multiplicative group of units, Amer. J. Math. 85(1963), 247-252.
4. On a classical theorem of Noether in ideal theory, Pacific J. Math. 13(1963), 579-583.
5. Rings in which the unique primary decomposition theorem holds, Proc. Amer. Math. Soc. 14(1963), 777-781. (Dissertation Publication)
6. Integral domains with quotient overrings (with Jack Ohm), Mathematische Annalen 153(1964), 97-103.
7. Extension of results concerning rings in which semi-primary ideals are primary, Duke Math. J. 31(1964), 73-78.

8. Integral domains which are almost Dedekind, Proc. Amer. Math. Soc. 15(1964), 813-818.
9. Multiplication rings as rings in which ideals with prime radical are primary (with Joe L. Mott), Trans. Amer. Math. Soc. 114(1965), 40-52.
10. Primary ideals and valuation ideals (with Jack Ohm), Trans. Amer. Math. Soc. 117(1965), 237-250.
11. The cancellation law for ideals of a commutative ring, Canadian J. Math. 17(1965), 281-287.
12. A class of domains in which primary ideals are valuation ideals, Mathematische Annalen 161(1965), 247-254.
13. Some containment relations between classes of ideals of a commutative ring, Pacific J. Math. 15(1965), 497-502.
14. Domains in which valuation ideals are prime powers, Archiv der Mathematik 17(1966), 210-215.
15. Eleven nonequivalent conditions on a commutative ring, Nagoya Math. J. 26(1966),183-194.
16. On the uniqueness of the -representation of an ideal (with Lawrence Husch), American Math. Monthly 73(1966), 876-877.
17. Overrings of Prüfer domains, Journal of Algebra 4(1966), 331-340.
18. The pseudo-radical of a commutative ring, Pacific J. Math. 19(1966), 275-284.
19. On the complete integral closure of an integral domain (with William Heinzer), J. Australian Math. Soc. 6(1966), 351-361.
20. Primary ideals and prime power ideals (with H.S. Butts), Canadian J. Math 18(1966), 1183-1195.
21. A counterexample to two conjectures in ideal theory, American Math. Monthly 74(1967), 195-197.
22. If $R[X]$ is Noetherian, R contains an identity, American Math. Monthly 74(1967), 700.
23. A class of domains in which primary ideals are valuation ideals. II. Math. Ann. 171(1967), 93-96.
24. Contracted ideals with respect to integral extensions, Duke Math. J. 34(1967), 551-572.
25. A note on the quotient field of $D[[X]]$, Proc. Amer. Math. Soc. 18(1967), 1138-1140.
26. Overrings of Prüfer domains. II. (with William Heinzer), J. of Algebra 7(1967), 281-302.
27. Intersection of quotient rings of an integral domain (with William Heinzer), J. Math. Kyoto University 7(1967), 133-150.
28. Some applications of the Hilfssatz von Dedekind-Mertens, Math. Scand. 20(1967), 240-244.
29. A note on two criteria for Dedekind domains, L'Enseignement Mathematique 13(1968), 253-257.
30. *Multiplicative ideal theory*, Queen's Papers in Pure and Applied Mathematics, No. 12 Queen's University, Kingston, Ontario, 1968.
31. On proper overrings of integral domains (with Joe L. Mott), Monatshefte für Mathematik 72(1968), 61-71.
32. Irredundant intersections of valuation rings (with William Heinzer), Math. Zeit., 103(1968), 306-317.

33. On the existence of exceptional field extensions (with William Heinzer), Bull. Amer. Math. Soc. 74(1968), 545-547.
34. Idempotent ideals and union of nets of Prüfer domains (with Jimmy T. Arnold), J. Sci. Hiroshima Univ. Ser. A-I 32(1967), 131-145.
35. Primary ideals and valuation ideals. II. (with William Heinzer), Trans. Amer. Math. Soc. 131(1968), 149-162.
36. R-automorphisms of $R[X]$, Proc. London Math. Soc. 43(1968), 328-336.
37. On a condition of J. Ohm for integral domains, Canadian J. Math. 29(1968), 970-983.
38. Rings of formal power series over a Krull domain (with William Heinzer), Math. Zeit. 106(1968), 379-387.
39. A note on the algebraic closure of a field, Amer. Math. Monthly 75(1968), 1101-1102.
40. A note on semigroup rings, Amer. Math. Monthly 76(1969), 36-37.
41. Integral dependence in power series rings, J. of Algebra 11(1969), 488-502.
42. A note on generating sets of invertible ideals, Proc. Amer. Math. Soc. 22(1969),426-427.
43. Power series rings over a Krull domain, Pacific J. Math. 29(1969), 543-550.
44. Commutative rings in which each prime ideal is principal, Math. Ann. 184(1969), 151-158.
45. The unique primary decomposition theorem in commutative rings without identity, Duke Math. J. 36(1969), 737-747.
46. Two constructions of Prüfer domains, J. Reine Angew. Math. 239(1970), 153 -162.
47. R-automorphisms of $R[[X]]$, Mich. J. Math. 17(1970), 15-21.
48. On the contents of polynomials (with Jimmy T. Arnold), Proc. Amer. Math. Soc. 24(1970), 556-562.
49. On the number of generators of an invertible ideal (with William Heinzer), J. of Algebra 14(1970), 139-151.
50. An existence theorem for non-Noetherian rings, Amer. Math. Monthly 77(1970), 621-623.
51. Integral domains with Noetherian subrings, Comment. Math. Helv. 45(1970), 129-134.
52. An embedding theorem for HCF-rings, Proc. Camb. Philos. Soc. 68(1970), 583-587.
53. Some results on contracted ideals (with Joe Mott), Duke Math J. 37(1970), 751-761.
54. Contracted ideals in Krull domains, Duke Math. J. 37(1970), 769-774.
55. An example in the theory of ordered semigroups, Bol Soc. Mat. Mexicana 15(1970), 42-43.
56. A theorem on Noetherian rings, Delta 2(1971), 12-14.
57. Integrally closed subrings of an integral domain (with Joe Mott), Trans. Amer.Math. Soc. 154(1971), 239-250.
58. On polynomial rings over a Hilbert ring, Michigan Math. J. 18(1971), 205-212.
59. An algebraic proof of a theorem of A. Robinson (with Joe Mott), Proc. Amer. Math. Soc. 29(1971), 461-466.
60. Domains with integrally closed subrings, Math. Japon. 16(1971), 9-12.
61. *Multiplicative ideal theory,* Pure and Applied Mathematics, No. 12. Marcel Dekker, Inc., New York, 1972.

62. Generating sets for a field as a ring extension of a subfield, Rocky Mountain J. Math. 2(1972), 111-118.
63. On Prüfer rings, Bull Amer. Math. Soc. 78(1972), 223-224.
64. Integral domains whose overrings are ideal transforms (with Jim Brewer), Math. Nachrichten 51(1971), 255-267.
65. The transform formula for ideals (with James A. Huckaba), J. Algebra 21(1972), 191-215.
66. On factorization into prime ideals, Comment Math. Helv. 47(1972), 70-74.
67. On commutative rings of finite rank, Duke Math. J. 39(1972), 381-384.
68. Rings whose proper subrings have property P (with Robert Lea and Matthew O'Malley), Acta. Math. Sci. (Szeged) 33(1972), 69-76.
69. The equality $(A \cap B)^n = A^n \cap B^n$ for ideals (with Anne Grams), Canad. J. Math. 24(1972), 792-798.
70. On solvability by radicals of field extensions, Math. Ann. 199(1972), 263-277.
71. On solvability by radicals of finite fields (with Robert Fray), Math. Ann. 199(1972), 279-291.
72. Prüfer-like conditions on the set of overrings of an integral domain, Proc. Kansas Conf. Comm. Algebra, Springer-Verlag, New York(1973), 90-102.
73. Overrings and divisorial ideals of rings of the form $D + M$ (with Eduardo R. Bastida), Mich, J. Math. 20(1973), 79-95.
74. Non-Noetherian rings for which each proper subring is Noetherian (with Matthew O'Malley), Math. Scand. 31(1972), 118-122.
75. On ideal-adic topologies for a commutative ring, L'Enseignement Math. 18(1972), 201-204.
76. Lagrange resolvents, Aequationes Math. 8(1972), 216-220.
77. Dimension sequences for commutative rings (with Jimmy T. Arnold), Bull. Amer. Math. Soc. 79(1973), 407-409.
78. The n-generator property for commutative rings, Proc. Amer. Math. Soc. 38(1973), 477-482.
79. A note on rings with only finitely many subrings, Scripta Math. 29(1973), 37-38.
80. Associative rings of order p^3 (with Joe Mott), Proc. Japan. Acad. 49(1973), 795-799.
81. On factorization of polynomials modulo n, Canadian Math. Bull. 16(1973), 521-523.
82. A note on unqiue factorization domains, Delta 3(1973), 7-8.
83. On a purely inseparable extension of a normal extension of a field (with Brian Wesselink), Abh. Math. Sem. Univ. Hamburg 40(1974), 94-99.
84. Dimension sequences of commutative rings, Ring Theory, Proc. Conf. Univ. Oklahoma, Lecture Notes in Pure and Applied Math., Vol. 7, Marcel Dekker, New York (1974), 31-46.
85. Finite element factorization in group rings, Ring Theory Proc. Conf. Univ. Oklahoma, Lecture Notes in Pure and Applied Math., Vol. 7, Marcel Dekker, New York (1974), 47-61.
86. Semigroup rings as Prüfer rings (with Tom Parker), Duke Math. J. 41(1974), 219-230.
87. A two-dimensional non-Noetherian factorial ring, Proc. Amer. Math. Soc. 44(1974), 25-30.
88. D-rings (with James A. Huckaba), J. Algebra 28(1974), 414-432.

89. Divisibility properties in semigroup rings (with Tom Parker), Michigan J. Math. 21(1974), 65-86.
90. The integral closure need not be a Prüfer domain (with Joseph F. Hoffmann), Mathematika 21(1974), 233-238.
91. Maximal overrings of an integral domain not containing a given element (with James A. Huckaba), Comm. in Algebra 2(1974), 377-401.
92. The dimension sequence of a commutative ring (with Jimmy T. Arnold), Amer. J. Math. 96(1974), 385-408.
93. Dimension theory of commutative rings without identity (with Jimmy T. Arnold), J. Pure Applied Algebra 5(1974), 209-231.
94. Polynomial rings over a commutative von Neumann regular ring, Proc. Amer. Math. Soc. 49(1975), 294-296.
95. On polynomial and power series rings over a commutative ring, Rocky Mt. J. Math. 5(1975), 157-175.
96. Dimension theory of commutative polynomial rings, Algebra and Logic, Lecture Notes in Mathematics #450, Springer-Verlag (1975), 140-154.
97. Dimension theory of commutativer power series rings, Algebra and Logic, Lecture Notes in Mathematics #450, Springer-Verlag (1975), 155-162.
98. Nilpotent elements of commutative semigroup rings (with Tom Parker), Michigan J. Math. 22(1975), 97-108.
99. A characterization of Prüfer domains in terms of polynomials (with Joseph F. Hoffmann), Pacific J. Math. 60(1975), 81-85.
100. Zero divisors in power series rings (with Anne Grams and Tom Parker), J. Reine Angew. Math. 278/279(1975), 141-164.
101. Finite intersections of quotient rings of a Dedekind domain (with Anne Grams), J. London Math. Soc. 12(1976), 257-261.
102. The dimension theory of commutative semigroup rings (with Jimmy T. Arnold), Houston J. Math. 2(1976), 299-313.
103. A non-Noetherian two-dimensional Hilbert domain with principal maximal ideals (with William Heinzer), Michigan J. Math. 23(1976), 353-362.
104. Cardinality of generating sets of ideals of a commutative ring (with William Heinzer), Indiana Univ. Math. J. 26(1977), 791-798.
105. R-endomorphisms of $R[[X_1, ..., X_n]]$ (with Matthew O'Malley), J. Algebra 48(1977), 30-45.
106. Units of semigroup rings (with Mark L. Teply), Communications in Algebra 5(1977), 1275-1303.
107. Idempotents of commutative semigroup rings (with Mark L. Teply), Houston J. Math. 3(1977), 369-385.
108. Modules that are finite sums of simple submodules, Publ. Math. Debrecen 24(1977), 5-8.
109. Some countability conditions on a commutative ring (with Jimmy T. Arnold and William Heinzer), Illinois J. Math. 21(1977), 648-665.
110. Two questions concerning dimension sequences (with Jimmy T. Arnold), Arch. der Math. 29(1977), 497-503.
111. A note on the integral closure of a Dedekind domain, Bull. Calcutta Math. Soc. 68(1978), 37-38.
112. On the divisors of monic polynomials over a commutative ring (with William Heinzer), Pacific J. Math. 78(1978), 121-131.
113. Ideals contracted from each extension ring (with Stephen McAdam), Communications in Algebra 7(1979), 287-311.

114. The Noetherian property for quotient rings of infinite polynomial rings (with William Heinzer), Proc. Amer. Math. Soc. 76(1979), 1-7.
115. The group of units of a commutative semigroup ring (with Raymond C. Heitmann), Pacific J. Math. 85(1979), 49-64.
116. On Pic($R[X]$) for R seminormal (with Raymond C. Heitmann), J. Pure Appl. Algebra 16(1980), 251-257.
117. The Laskerian property, power series rings, and Noetherian spectra (with William Heinzer), Proc. Amer. Math. Soc. 17(1980), 13-16.
118. Some countability conditions in commutative ring extensions (with William Heinzer), Trans. Amer. Math. Soc. 264(1981), 217-234.
119. The quotient field of an intersection of integral domains (with William Heinzer), J. Algebra 70(1981), 238-249.
120. On the ideal class group of a flat overring, Archiv der Math. 37(1981), 48-51.
121. Commutative Ring Theory, *Emmy Noether A Tribute to Her Life and Work*, Marcel Dekker, Chapter 8, pp. 131-143, New York, 1981.
122. On R-homomorphisms of power series rings (with Matthew O'Malley), Advances in Math. 42(1981), 154-168.
123. Extension of an order to a simple transcendental extension, Contemp. Math. 8(1982), 113-118.
124. Ideals contracted from a Noetherian extension ring (with William Heinzer), J. Pure Appl. Algebra 24(1982), 123-144.
125. Finitely generated ideals of the ring of integer-valued polynomials (with William W. Smith), J. Algebra 81(1983), 150-164.
126. The coefficient field of a semigroup algebra (with Eugene Spiegel), Math. Rep. Acad. Sci. Canada 5(1983), 111-116.
127. Cardinality of generating sets for modules over a commutative ring (with William Heinzer), Math. Scand. 52(1983), 41-57.
128. Principal ideal rings and a condition of Kummer (with William Heinzer), J. Algebra 83(1983), 285-292.
129. Noetherian pairs and hereditarily Noetherian rings (with William Heinzer), Arch. Math. 41(1983), 131-138.
130. On Jönsson modules over a commutative ring (with William Heinzer), Acta Sci. Math. 46(1983), 3-15.
131. *Commutative semigroup rings.* Chicago Lectures in Mathematics. University of Chicago Press, Chicago, 1984.
132. Integer-valued polynomials and the strong two-generator property (with William W. Smith), Houston J. Math. 11(1985), 65-74.
133. Hilbert subalgebras generated by monomials, Commun. in Algebra 13(1985), 1187-1192.
134. Commutative monoid rings as Hilbert rings, Proc. Amer. Math. Soc. (1985), 15-18.
135. Coefficient rings in isomorphic semigroups rings (with Eugene Spiegel), Commun. in Algebra 13(1985), 1780-1810.
136. Finitely generated intermediate rings (with William Heinzer), J. Pure Appl. Algebra 37(1985), 237-264.
137. Cardinality of subrings of a commutative ring (with William Heinzer), Canad. Math. Bull. 29(1986), 102-108.
138. Zero divisors in commutative rings, Amer. Math. Monthly 93(1986), 382-387.
139. Property E in commutative monoid rings, Proc. Internat. Conf. Group and Semigroup Rings, 13-18.

140. Conditions concerning prime and maximal ideals of commutative monoid rings, Proc. Internat. Conf. Group and Semigroup Rings, 19-26.
141. Commutative monoid rings with finite maximal or prime spectrum, Proc. Internat. Conf. Group and Semigroup Rings, 27-34.
142. Chain conditions in commutative semigroup rings, J. Algebra 103(1986), 592-599.
143. Jönsson ω_0-generated algebraic field extensions (with William Heinzer), Pacific J. Math. 128(1987), 81-116.
144. Generators of ideals containing monics (with B. Nashier and W. Nichols), Archiv der Math. 49(1987), 407-413.
145. On Jönsson algebras over a commutative ring (with William Heinzer), J. Pure Appl. Algebra 49(1987), 133-159.
146. The prime spectra of subalgebras of affine algebras and their localizations (with B. Nashier and W. Nichols), J. Pure Appl. Algebra 57(1989), 47-65.
147. On the heights of prime ideals under integral extensions (with B. Nashier and W. Nichols), Arch. der Math. 52(1989), 47-52.
148. Fields that admit a nonlinear permutation polynomial (with J.V. Brawley), J. Algebra 123 (1989), 111-119.
149. On Pic($D[a]$) for a principal ideal domain D (with W. Heinzer), Canad. Math. Bull. 32 (1989), 114-116.
150. On the imbedding of a direct product into a zero-dimensional commutative ring (with W. Heinzer), Proc. Amer. Math. Soc. 106 (1989), 631-637.
151. Sets that determine integer-valued polynomials, J. Number Theory 33(1989), 95-100.
152. Prüfer domains and rings of integer-valued polynomials, J. Algebra 129(1990), 502-517.
153. The ring of integer-valued polynomials of a Dedekind domain (with W. Heinzer, D. Lantz and W. Smith), Proc. Amer. Math. Soc 108(1990), 673-681.
154. On the Picard groups of a class of non-seminormal domains (with M. Martin), Commun. in Algebra 18(1990), 3263-3293.
155. Noetherian rings of integer-valued polynomials, Proc. of the 1990 Marseille Recontre autour des Polynômes à valeurs entières, CNRS Publication No. 225, 1991, pp. 13-16.
156. On the irreducibility of pure difference binomials, Semigroup Forum, 43(1991), 146-150.
157. *Multiplicative ideal theory,* Corrected reprint of the 1972 edition. Queen's Papers in Pure and Applied Mathematics, 90. Queen's University, Kingston, Ontario, 1992.
158. Products of commutative rings and zero-dimensionality (with W. Heinzer), Trans. Amer. Math. Soc. 331(1992), 663-680
159. Zero-dimensionality in commutative rings (with W. Heinzer), Proc. Amer. Math. Soc. 115(1992), 881-893.
160. The family of residue fields of a zero-dimensional ring (with W. Heinzer), J. Pure Appl. Algebra 82(1992), 131-153.
161. An application of Jönsson modules to some questions concerning proper subrings (with W. Heinzer), Math. Scand. 70(1992), 34-42.
162. Artinian subrings of a commutative ring (with W. Heinzer), Trans. Amer. Math. Soc. 336(1993), 295-310.

163. The Noetherian property in rings of integer-valued polynomials (with W. Heinzer and D. Lantz), Trans. Amer. Math. Soc. 338(1993), 187-199.
164. Commutative rings with periodic multiplicative semigroup, Commun. Algebra 161(1993), 4025-4028.
165. Primary ideals with finitely generated radical in a commutative ring (with W. Heinzer), manuscripta math. 78(1993), 201-221.
166. On t-invertibility and comparability (with J. Mott and M. Zafrullah), Proc. Fez Conference, Marcel Dekker Lecture Notes Pure Appl. Math. Vol. 153,1994, pp.141-150
167. Imbeddability of a commutative ring in a finite-dimensional ring (with W. Heinzer), manuscripta math. 84(1994), 401-414.
168. Homomorphic images of an infinite product of zero-dimensional rings (with W. Heinzer), Commun. in Algebra 23(1995), 1953-1965.
169. Zero-dimensionality and products of commutative rings, Zero-Dimensional Commutative Rings, Marcel-Dekker, New York, 1995, pp. 15-25.
170. Zero-dimensional extension rings and subrings, Zero-Dimensional Commutative Rings, Marcel-Dekker, New York, 1995, pp. 27-39.
171. Residue fields of zero-dimensional rings subrings, Zero-Dimensional Commutative Rings, Marcel-Dekker, New York, 1995, pp. 41-55.
172. Infinite products of zero-dimensional commutative rings (with W. Heinzer), Houston J. Math. 21(1995), 247-259.
173. Zero-dimensional subrings of commutative rings, Abelian Groups and Modules, Proc. 1994 Padova Conf., Kluwer, Dordrecht, pp. 209-219.
174. On the distribution of prime ideals within the ideal class group (with W. Heinzer and W.W. Smith), Houston J. Math. 22(1996), 51-59.
175. On complete integral closure and Archimedean valuation domains, J. Austral. Math. Soc. 61(1996), 377-380.
176. Every local ring is dominated by a one-dimensional local ring (with W. Heinzer), Proc. Amer. Math. Soc. 125(1997), 2513-2520.
177. An intersection condition for prime ideals, Factorization in Integral Domains, Pure and Appl. Math., vol. 189, Dekker, New York, 1997, pp. 327-331.
178. Prime ideals of finite height in polynomial rings (with W. Heinzer), Houston J. Math. 24(1998), 9-20.
179. On the embedding of a commutative ring in a local ring (with W. Heinzer), Ill. J. Math., 43(1999), 192-210.
180. The ideal of polynomials vanishing on a commutative ring, Proc. Amer. Math. Soc., 127(1999), 1265-1267.
181. Primary decomposition of ideals in polynomial rings (with G. Fu), Advances in Commutative Ring Theory, Pure and Applied Math., vol. 205, Dekker, New York, 1999, 369-390.
182. On the polynomial equivalence of subsets E and f(E) of Z, Archiv der Math., 73(1999), 355-365.
183. Finite generation of powers of ideals (with W. Heinzer and M. Roitman), Proc. Amer. Math. Soc., 127(1999), 3141-3151.
184. Commutative rings of dimension 0, Non-Noetherian Commutative Ring Theory: Recent Advances, Mathematics and its Appl., Kluwer, 2000, pp. 229-249.
185. A new criterion for embeddability in a zero-dimensional ring, Ideal-Theoretic Methods in Commutative Algebra, Lecture Notes Pure Appl. Math., Vol.220, Marcel-Dekker, New York, 2001, pp. 223-229.

186. A note on finitely generated ideals of an arithmetical ring, Internat. J. Commut. Rings, 1(2002), 27-30.
187. Ideal decompositions in commutative rings, The Concise Handbook of Algebra, Kluwer, Dordrecht, 2002, pp. 153-156.
188. Krull dimension of commutative rings, The Concise Handbook of Algebra, Kluwer, Dordrecht, 2002, pp. 156-159.
189. Some finiteness conditions on the set of overrings of an integral domain, Proc. Amer. Math. Soc. 131(2002), 2337-2346.
190. Forty years of commutative ring theory, Rings, Modules, and Abelian Groups, Marcel-Dekker, New York, 2004, pp. 229-256.

Commutative Algebra at Florida State 1963-1968

Jim Brewer and Bill Heinzer

We were both graduate students at Florida State in the middle 1960's. The research environment there at this time in commutative algebra was very stimulating. We would like to try to describe what we believe made this so.

Robert Gilmer arrived in Tallahassee in time for the start of the 1963-1964 academic year. The Herstein-level course in those days was a year's course in linear algebra using the text by Hoffman and Kunze. Robert taught this course to an audience consisting of around twenty beginning graduate students and a couple of undergraduates. His goal was to finish the book and he succeeded.

Robert had an immediate impact on the culture of the mathematics department. He organized Saturday morning touch football games for graduate students and junior faculty. As a consequence, the relationship between the students and the faculty was very cordial. This also enhanced Robert's appeal to those graduate students with an athletic background. Robert later said that he could learn something about the character of a student by seeing him play sports that he couldn't learn in a classroom.[1] So successful was this that he subsequently organized Thursday evening volleyball games for the same group.

In the fall of 1964 - spring 1965, Robert taught the beginning graduate algebra course using the English translations of the two volumes of van der Waerden's *Modern Algebra.* Quoting from Saunders Mac Lane in the 1997 *Notices of the AMS* Volume 44, Number 3, page 322 "It is, in my view, the most influential text in algebra of the 20th century". Robert taught this in a standard lecture format, but the essence of the course was the homework. This consisted of ten problems per week assigned on one Monday and taken

[1] At that time graduate students were usually male. However, Elizabeth Magarian, a Ph.D. student of Joe Mott, was an active participant in the graduate courses taught by Robert that are described below. Also two of Robert's Ph.D. students, Anne Grams and Mary Dowlen, are women.

up the next.[2] Each problem was graded all right or all wrong. Robert returned these on Wednesday with dittoed solutions. Often the solution of a problem was attributed to a student whose solution Robert found attractive. Robert was very demanding, but it was evident to his students that he was working at least as hard as they were. This quality also enhanced Robert's appeal.

Nick Heerema and Fred Kreimer also made valuable contributions to our education and to the commutative algebra environment at Florida State. For example, in the fall of 1964, Heerema gave a course using Nagata's *Local Rings*, and Kreimer gave a course using Mac Lane's *Homology*.

In May, 1965, Robert offered a group theory course using the text by Marshall Hall. This course was taught more informally, but a good deal of group theory was covered. Students were encouraged to raise questions related to the material being presented. There was a sense of discovery involved in participating in the course.

As a continuation of the previous year's course [3], in the fall of 1965 - spring 1966, Robert taught a more advanced graduate course in commutative algebra. The format with the ten problems per week assigned on one Monday, taken up the next Monday, and returned on Wednesday was the same. He covered material from Chapters IV and V of Zariski-Samuel's *Commutative Algebra,* material from Chapter IV of Jacobson's *Lectures in Abstract Algebra vol III* and additional material related to research papers Robert had recently written. For example, in the fall of 1965 several exercises involved the structure of the group of R-automorphisms of the polynomial ring $R[x]$.

Also in the fall of 1965, Joe Mott arrived from the University of Kansas. Joe had been with Robert at Louisiana State where they were both graduate students of Professor Butts. Almost immediately, Joe and Robert organized a Tuesday/Thursday seminar. They started with Chapitre 5 of Bourbaki's *Algèbre Commutative.* Students were encouraged to actively participate. The seminar continued in the fall of 1966 covering material from Chapitre 7 of Bourbaki's *Algèbre Commutative* on divisorial ideals and Krull domains. The usual format for the Tuesday/Thursday seminar was to have students lecture on Tuesdays on the material we were studying, and then Robert and Joe would present problem solutions on Thursdays. A student was assigned to write out the solution to pass out to all participants in the seminar. Several problems covered in the seminar led to further investigations. Joe recalls that a problem

---

[2] Joe Mott tells us that this technique was also used by Professor Hubert Butts at Louisiana State University.

[3] Joe Mott recalls that Robert suggested to the algebra faculty that they follow a procedure with regard to the standard graduate algebra sequence whereby one person would teach the beginning level course one year, continue with the next level course the following year, and so on as long as enrollment was sufficient for successor courses to be offered. Joe remarks that this process has its good points and its bad points. It enabled a faculty member to take a group of students from beginning to end in the basic sequence all the way to the doctorate level. A down side is that students didn't get exposed to other flavors of expertise.

covered on finite character domains led to a joint publication of Joe with Jim Brewer.

Robert taught an algebraic number theory course in May 1966 using the book of Henry Mann. In the summer of 1966, Joe gave a course in algebraic geometry using notes of Artin.

In the fall of 1966, Robert gave an advanced graduate course. The format was the same as for his previous courses. He covered material from Chapters VII and VIII of Zariski-Samuel's *Commutative Algebra* concerning polynomial and power series rings, the Weierstrass Preparation Theorem, dimension theory, local algebra, associated graded rings, completions and Hensel's Lemma.

During this period, Robert began graduating students: Bill Heinzer in December, 1966, Jimmy Arnold, Matt O'Malley and Craig Wood in 1967, and Jim Brewer in March, 1968.

Of course, the above does not fully explain why Robert attracted so many students and collaborators. This seems best explained by personal anecdotal information such as the following:

(J.B.) I was a senior in the fall of 1963 and a student in Robert's linear algebra course. In November, John Kennedy was assassinated and Robert spent hours in his office talking to me about that and other things. With the perspective of someone who has been a professor for almost forty years, I now know what this meant for Robert in terms of time, etc. It made a strong impression on a twenty-one year old. When my first child was born a couple of weeks later, Robert was the first and only non-family member to visit the hospital and he brought my daughter a gift. Such gestures speak to Robert's warmth and humanity. These are but two of the countless examples I have experienced with Robert over the years and his kindness and generosity towards me continues even unto today.

(B.H.) My first course with Robert was the group theory course of May 1965. However, even before this Robert made time to talk with me about exceptional field extensions, and what we at the time called strong integral closure. In reflecting on my time as a graduate student at Florida State, I am impressed by the patience Robert showed in his interactions with me. Robert led by example. The example he set has been an inspiration to me and I believe to all of his students.

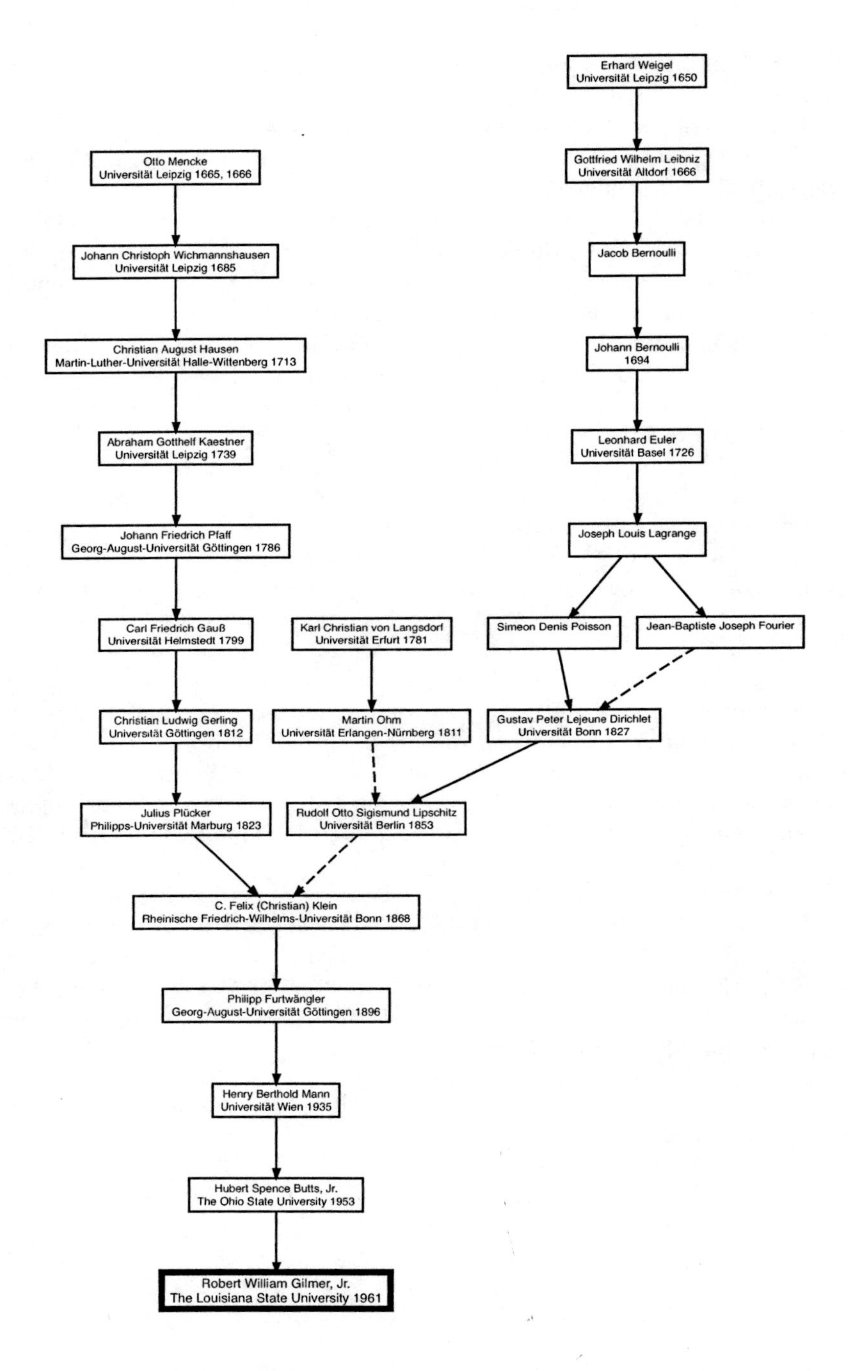

Robert Gilmer's mathematics genealogy

Reprinted with permission of the Mathematics Genealogy Project
http://www.genealogy.math.ndsu.nodak.edu

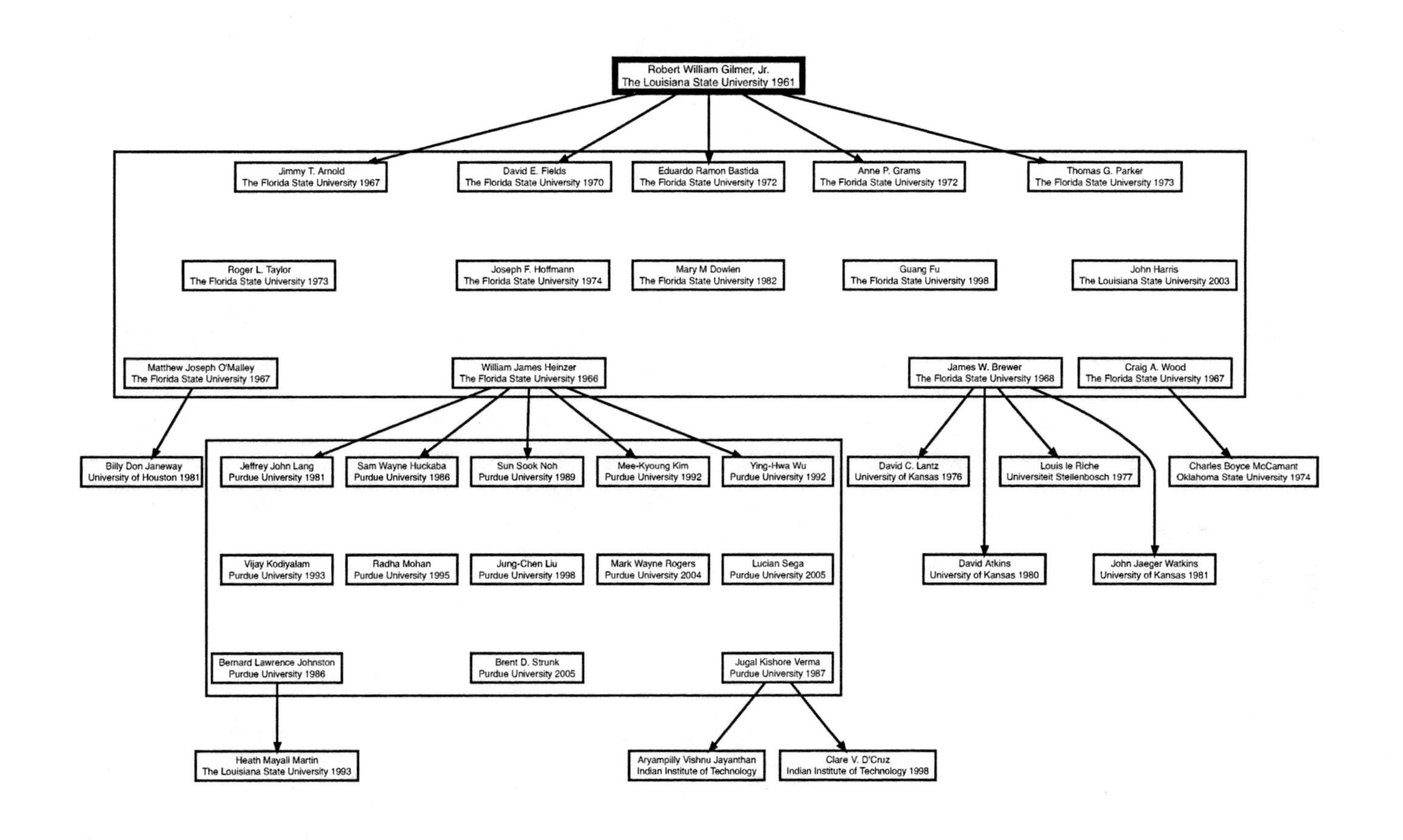
Robert William Gilmer, Jr.
The Louisiana State University 1961
Jimmy T. Arnold
The Florida State University 1967
David E. Fields
The Florida State University 1970
Eduardo Ramon Bastida
The Florida State University 1972
Anne P. Grams
The Florida State University 1972
Thomas G. Parker
The Florida State University 1973
Roger L. Taylor
The Florida State University 1973
Joseph F. Hoffmann
The Florida State University 1974
Mary M Dowlen
The Florida State University 1982
Guang Fu
The Florida State University 1998
John Harris
The Louisiana State University 2003
Matthew Joseph O'Malley
The Florida State University 1967
William James Heinzer
The Florida State University 1966
James W. Brewer
The Florida State University 1968
Craig A. Wood
The Florida State University 1967
Billy Don Janeway
University of Houston 1981
Jeffrey John Lang
Purdue University 1981
Sam Wayne Huckaba
Purdue University 1986
Sun Sook Noh
Purdue University 1989
Mee-Kyoung Kim
Purdue University 1992
Ying-Hwa Wu
Purdue University 1992
David C. Lantz
University of Kansas 1976
Louis le Riche
Universiteit Stellenbosch 1977
Charles Boyce McCamant
Oklahoma State University 1974
Vijay Kodiyalam
Purdue University 1993
Radha Mohan
Purdue University 1995
Jung-Chen Liu
Purdue University 1998
Mark Wayne Rogers
Purdue University 2004
Lucian Sega
Purdue University 2005
David Atkins
University of Kansas 1980
John Jaeger Watkins
University of Kansas 1981
Bernard Lawrence Johnston
Purdue University 1986
Brent D. Strunk
Purdue University 2005
Jugal Kishore Verma
Purdue University 1987
Heath Mayall Martin
The Louisiana State University 1993
Aryampilly Vishnu Jayanthan
Indian Institute of Technology
Clare V. D'Cruz
Indian Institute of Technology 1998

Index

Printed in the USA